Main-Group Elements

					18 VIIIA
					2 **He** 4.003

	13 IIIA	14 IVA	15 VA	16 VIA	17 VIIA	18
	5 **B** 10.81	6 **C** 12.01	7 **N** 14.01	8 **O** 16.00	9 **F** 19.00	10 **Ne** 20.18

10	11 IB	12 IIB	13 **Al** 26.98	14 **Si** 28.09	15 **P** 30.97	16 **S** 32.07	17 **Cl** 35.45	18 **Ar** 39.95
28 **Ni** 58.69	29 **Cu** 63.55	30 **Zn** 65.39	31 **Ga** 69.72	32 **Ge** 72.59	33 **As** 74.92	34 **Se** 78.96	35 **Br** 79.90	36 **Kr** 83.80
46 **Pd** 106.4	47 **Ag** 107.9	48 **Cd** 112.4	49 **In** 114.8	50 **Sn** 118.7	51 **Sb** 121.8	52 **Te** 127.6	53 **I** 126.9	54 **Xe** 131.3
78 **Pt** 195.1	79 **Au** 197.0	80 **Hg** 200.6	81 **Tl** 204.4	82 **Pb** 207.2	83 **Bi** 209.0	84 **Po** (209)	85 **At** (210)	86 **Rn** (222)

63 **Eu** 152.0	64 **Gd** 157.3	65 **Tb** 158.9	66 **Dy** 162.5	67 **Ho** 164.9	68 **Er** 167.3	69 **Tm** 168.9	70 **Yb** 173.0	71 **Lu** 175.0
95 **Am** (243)	96 **Cm** (247)	97 **Bk** (247)	98 **Cf** (251)	99 **Es** (252)	100 **Fm** (257)	101 **Md** (258)	102 **No** (259)	103 **Lr** (260)

ESSENTIAL CONCEPTS

OF CHEMISTRY

SHARON J. SHERMAN
The College of New Jersey

ALAN SHERMAN
Middlesex County College

HOUGHTON MIFFLIN COMPANY   *Boston   New York*

To Robert and Michael and their grandparents

SENIOR SPONSORING EDITOR: Richard Stratton
EDITORIAL ASSISTANT: Katie Shutzer
ASSOCIATE PROJECT EDITOR: Rachel D'Angelo Wimberly
SENIOR PRODUCTION/DESIGN COORDINATOR: Jill Haber
SENIOR MANUFACTURING COORDINATOR: Sally Culler
MARKETING MANAGER: Penelope Hoblyn

Cover design by Walter Kopec, Boston, MA
Cover photography: © Hermann Eisenbeiss, The National Audobon Society Collection/
Photo Researchers, Inc.

Printed in the U.S.A.

Library of Congress Catalog Card Number: 98-72085

ISBN: 0-395-92127-9

456789-HM-04 02 01

CONTENTS OVERVIEW

CONTENTS

x   *Contents*

CHAPTER 10
THE CHEMICAL EQUATION:
Recipe for a Reaction   228

CHAPTER 11
STOICHIOMETRY:
The Quantities in Reactions   257

CUMULATIVE REVIEW / Chapters

CHAPTER 12
HEATS OF REACTION:
Chemistry and Energy   284

CHAPTER 13
THE GASEOUS STATE:
Ideal Behavior   307

PREFACE

Essential Concepts of Chemistry is a new text intended for use in the one-quarter, two-quarter, or one-semester introductory or preparatory chemistry course. The text is a derivative of our well-known text *Basic Concepts of Chemistry,* which has been used by students for more than two decades. That text continues to receive favorable reviews from students and professors for the text's student-oriented approach, which has been its trademark throughout six editions. *Essential Concepts of Chemistry* continues this tradition. However, because it contains sixteen core chapters (instead of the twenty chapters of the longer text) and is published in paperback form, *Essential Concepts of Chemistry* is available to students at half the price of our earlier text.

Like its predecessor, *Essential Concepts of Chemistry* is a text that students will enjoy reading. It is easy to understand and contains a sound problem-solving strategy, numerous examples worked out step-by-step, excellent figures, tables and photos, appendices, and references throughout that demonstrate how chemistry connects to everyday life. We have used a conversational writing style and many pedagogical aids to assist students in mastering the subject matter.

Features of the Book

Applications of Chemistry in Today's World

Essential Concepts of Chemistry contains exciting and interesting examples of chemistry in everyday life to help students connect chemistry and the world around them. This connection is made in three ways—first, through a feature called *Chemical Frontiers.* Each *Chemical Frontiers* feature presents an area of recent research that holds important applications for the future. Topics covered in this feature include using microalgae to produce fuel for vehicles, utilizing coral to repair broken bones, and employing honeybee pollen to detect environmental contaminants. Second, the *Career Sketch* essays describe occupations that draw upon chemistry skills and knowledge of chemistry. Finally, the chapter-opening vignettes introduce the topic of the chapter through the use of a story in which knowledge of basic chemistry plays a role in everyday life. One example is the case of a mother using the scientific method to determine the source of her child's allergy. Other vignettes relate the chapter's topic to a pivotal event in the history of chemistry, like Lavoisier's work to determine that air is a mixture of substances. Each opening vignette is an interesting story in its own right and leads the student into the chapter by demonstrating the relevance of chemistry to daily life.

Polya's Problem-Solving Framework

In addition to a strong program of problem-solving aids, we have also included the general problem-solving framework of the late George Polya of Stanford University. The four-step method is introduced in Chapter 2. We include estimation skills to stress the importance of examining a solution so the student can assess the reasonableness of an answer. In most chapters, at least one problem is solved using Polya's method. We encourage students to apply this method as they solve problems.

Cooperative Learning Opportunities

For several decades, research has shown that students working in small groups improve their academic performance. They can solve problems together and, as part of that process, they discuss the material being studied. Areas that need clarification are identified as students teach one another.

Opportunities for cooperative problem solving are found in the exercises at the end of each chapter. These exercises have been marked with a triangle (◀), indicating that they may be used as conventional problems or for cooperative problem solving. This may be done in class under the direction of the instructor or independently in small study groups. Additional suggestions on implementing cooperative learning are included in the *Instructor's Resource Manual*.

Study Skills Section

A section on study skills appears in the frontmatter of this text. The section includes information about how to study chemistry, how to manage time effectively, and how to set realistic goals. An expanded section on study skills basics is followed by specifics on how to study chemistry. Other topics covered include how to take good notes, and how to take a chemistry test.

Readability

We maintain the readability of the text by using concise and direct sentences, a conversational tone, and paragraphs of manageable length. Students are provided with concrete examples of chemistry applications in everyday life in order to help them connect chemistry to society. In addition, line drawings, diagrams, and photographs are used to emphasize important points.

Problem Solving

In *Essential Concepts of Chemistry*, we address the difficulties students encounter in problem solving. In this text we include a large number of worked-out examples, with accompanying solutions. Each example contains a practice exercise that offers an immediate opportunity to apply the skill just demonstrated. Answers to all practice exercises appear at the end of the book.

End-of-chapter problems are divided into two groups: self-test exercises and extra exercises. More difficult problems are marked with an asterisk to make them easy to identify. Self-test exercises are keyed to the specific learning goals of each chapter and appear in matched pairs. The odd-numbered problems, in most cases, are answered in the back of the book. These problems help reinforce the learning of individual skills. Extra exercises are not keyed to learning goals; these exercises help students review skills in a format resembling a quiz or an exam. The answers to many of these problems appear in the answers section at the end of the book.

Cumulative review problems appear after every two or three chapters. They are designed to test the student's knowledge of the chapters and to reinforce material from earlier in the course. These exercises are also useful in studying for tests, midterms, and final examinations.

Study Aids

Our text was planned with the backgrounds, needs, and learning styles of a wide variety of students in mind. Each chapter begins with a series of learning goals. Throughout each chapter the learning goals appear in the margin next to the discussion of the corresponding material. Key terms appear in boldface and their definitions can be found in the Glossary at the end of the book. A list of key terms appears at the end of each chapter, along with a summary that reviews the most important points discussed in the chapter.

A unit on mathematics (Appendix A) is included at the end of the book. It can be treated as a reference or used at the beginning of the course as a review of skills.

Complete Instructional Package

Essential Concepts of Chemistry is part of a complete instructional package for introductory chemistry. Our *Laboratory Experiments for Basic Chemistry* (now in its sixth edition) contains twenty-five experiments and two laboratory exercises. The laboratory manual also contains important safety and chemical hazards guidelines, which discuss potential hazards and stress safety in the laboratory. Each experiment includes a chemical disposal guide with suggestions for disposing of various chemicals used in the experiment. There is also a section on keeping journals to accompany each experiment. Each journal page has a particular journal prompt, which helps clarify the thinking process. *Laboratory Experiments for Basic Chemistry* also includes the copyrighted game CHEM-DECK, which teaches the student how to write and name chemical compounds while playing card games that are a variation of gin rummy and poker.

The *Instructor's Resource Manual* accompanies both the book and the laboratory manual. It includes a test bank containing more than 700 test questions and answers, including a math pre-test, and a broad selection of final examination questions. Each lab experiment description includes a discussion of the experiment's purpose, answers to questions, and a list of the quantities of chemicals needed to conduct the experiment for a class of twenty-four students.

The *Solutions Manual* contains step-by-step solutions to the more than 1,500 problems covered in the text.

A set of fifty-two, two-color *Transparencies* of figures and tables taken from the text is also available to the instructor upon adoption of the text.

Essential Concepts of Chemistry is also accompanied by a *Study Guide* by James Braun of Clayton State College in Morrow, Georgia, which offers extra explanations and problems for students who need more practice. Each chapter in the *Study Guide* corresponds to a chapter in the textbook, and contains additional worked-out examples and extra exercises for the student to solve. The opportunity for further study is offered in special review exercises that cover groups of chapters.

Multimedia Supplements—Chemistry: Interactive Version 3.0

A dynamic new learning environment: *Chemistry Interactive Version 3.0* CD-ROM is available to supplement this text. This product is designed to deepen students' conceptual understanding and enhance their problem-solving skills. It contains:

- electronically generated problem sets modeled after examples from the text
- tutorials
- animations
- videos
- molecular models

Please see your Houghton Mifflin sales representative or your local bookstore for details.

Acknowledgments

We thank the following people who thoroughly reviewed and commented on portions of the manuscript so that we could continue to improve its quality. Without their time, effort, and helpful suggestions, we could not have forged ahead so easily. They are:

Alan Stevenson, Edgecombe Community College, Tarboro, North Carolina

Alfred Powell, East Arkansas Community College, Forest City, Arkansas

James M. Knight, Alabama Southern Community College, Monroeville, Alabama

Bob Boykin, Bossier Parish Community College

Scott Chaffee, West Virginia State College

Leslie DeVirdi, Colorado State University

Michael Dorneman, Mercer County Community College

John Flowers, City College of New York, Medgar Evers College

John Garland, Washington State University

Ed Groschwitz, Palomar Community College

Marie Hankins, University of Southern Indiana

LarRhee Henderson, Drake University

Kristin Kirschbaum, University of Toledo

John Konitzer, McHenry Community College

Miles Mackey, College of the Redwoods

Mary Taylor, Meridian Community College

Richard Wheet, Texas State Technical College

Once again, we thank those who have supported us, inspired our thoughts, and given us ideas: Professors Charles Oxman, Dominic Macchia, Linda Christopher, and Barbara Drescher of Middlesex County College; Russell Hulse and Christine Ritter of the Princeton University Plasma Physics Laboratory; Wayne Hoy of the Ohio State University; Gregory Camilli of the Rutgers University Graduate School of Education; Suzanne Pasch and Lawrence Marcus of The College of New Jersey; and Booker Alexander, who is now retired.

Finally, we'd like to thank our sons, Robert and Michael, for their continued cooperation and assistance throughout the years. Special thanks go to Robert Scott Sherman for developing all of the *Chemical Frontiers* and *Career Sketches* for the text and for assisting with the preparation of the manuscript.

Key Terms appear in boldface and are accompanied by concise definitions.

Learning Goals appear in the margin next to the appropriate text discussion.

■ LEARNING GOAL 1
Significant figures

To answer this question, we consider the number of significant figures in each measurement. **Significant figures** are *digits that express information that is reasonably reliable. The number of significant figures equals the number of digits written, including the last digit even though its value is uncertain.* For example, the temperature 34.5 degrees given by the first thermometer is known to be correct to three significant figures. These significant figures are the 3, the 4, and the 5. (Expressing the measurement as 34.50 degrees would mean that the temperature is known to four significant figures—the zero on the right is significant.)

The temperature 34.55 degrees given by the second thermometer shows a measurement with confidence to four significant figures. It may not be more accurate than the first measurement, but it is more *precise*. In general, the more significant figures there are, the more precise the measurement is.

Actually, neither of the two temperature measurements may be accurate. (They certainly wouldn't be accurate if the temperature of the liquid were actually, say, 47.51 degrees.) Thus **accuracy** involves *closeness to the actual dimension.* **Precision** is related to the *detail with which a measurement is known (expressed by the number of significant figures.)*

A measuring *instrument* is considered accurate if it consistently provides measurements that are close to the actual amount. The instrument is consistently provides the same measurement
amount. To see the difference, consider
t. If the darts land very close together,
the darts land on the bull's eye, they

ow significant figures play an important
we have to measure a certain piece of
is by measuring the four edges of the
e two rulers, one calibrated in centime-
neters). We set out to measure each edge
ler (the one calibrated in millimeters).
side of the glass, we absent-mindedly
ne calibrated in centimeters). Looking
can see that three edges are measured
urth is measured to one decimal place.
er of the glass? Is it 67.73 cm or 67.7
port only what we know for sure—and
ecimal place. *Our least precise measure-*
ficant figures in our result.
easured quantities, we must report the
measurement. We do this by **rounding**
places. Our result must have no more
quantity.
o one or more digits at the right end of
t the rightmost digit that we keep. In

---

52   **CHAPTER 3**   *Matter and Energy, Atoms and Molecules*

On the day his initial symptoms appeared, Mike was at a friend's house. He and his friend had decided to build a tree house in the backyard, and they worked several hours cutting branches and putting the tree house together. Upon its completion, his friend's Mom bought them pizza to celebrate their fine work, which the boys gobbled down without having washed up.

Joan decided to find out more about the tree house and discovered that the boys built it using juniper branches. Joan called the county health department to find out if junipers could cause allergic reactions. The health officer stated that sticky sap from the pines *could* cause allergic reactions in humans, and he recited a list of symptoms that were identical to Mike's. The puzzle was solved! Mike was allergic to the juniper sap. When Joan later told her son about her findings, Mike recalled getting the sticky sap on his arms, where his rash had formed. He also said that he may have ingested some of the sap when he ate the pizza, which would explain his breathing problems.

Joan called Mike's doctor and told her about the information she had uncovered. Mike was immediately placed on Ventolin to relieve the bronchial spasms. He was also given a nebulizer to help him breathe more easily. Two weeks later Mike's fever was down, the rash had disappeared, and his breathing had returned to normal. Thanks to some good detective work by his mother, Mike was once again healthy.

■■■ **LEARNING GOALS**

After you've studied this chapter, you should be able to:
1. Explain what is meant by the scientific method.
2. Explain the Law of Conservation of Mass and Energy.
3. Explain the difference between physical and chemical properties.
4. Describe the difference between homogeneous and heterogeneous matter, between mixtures and compounds, and between compounds and elements.
5. Describe the difference between an atom and a molecule.
6. Explain the Law of Definite Composition (or Definite Proportions).
7. Explain the terms *atomic mass, formula mass,* and *molecular mass.*
8. Determine the formula or molecular mass of a compound when you are given the formula for the compound.

Chapter-Opening Vignettes stress applications of chemistry.

Learning Goals at the beginning of each chapter outline the objectives being presented.

Self-Test Exercises are keyed to the specific learning goals of each chapter.

SELF-TEST EXERCISES

LEARNING GOAL 1: *Moles from Mass of an Element*

1. Explain the importance of the mole concept in chemistry.
2. (a) Using scientific notation, write the number that represents 1 mole.

(b) Write this number without the use of scientific notation.
3. A sample of gold (Au) has a mass of 19.7 g. How many moles of gold atoms is this?
4. How many moles are there in 5.62 g of silicon?
* 5. The atomic department store is selling uranium atoms at the reduced price of 100,000,000 atoms for 1 cent. What would be the cost in dollars of 1 mg of uranium?

2.4   Significant Figures   **21**

Solution

UNDERSTAND THE PROBLEM

We will add the numbers. Since 13 is the less precise measurement, we must round the sum to a whole number.

DEVISE A PLAN

Use Rule 2, which says if the leftmost dropped digit is 5 or greater, increase the last digit by 1. In this case the leftmost dropped digit is 7, so the 1 in 31 increases to 2, and we have 32.

CARRY OUT THE PLAN

$$\begin{array}{r} 18.7444 \\ +13 \quad\;\; \\ \hline 31.7444 \end{array}$$ Round off to 32 (whole number).

LOOK BACK

The answer makes sense since 18.7444 is close to 19, and $19 + 13 = 32$.

Practice Exercise 2.1   Add 12.4432 and 15. Provide the appropriate number of significant figures in your answer.

EXAMPLE 2.2

Subtract 0.12 from 48.743 and use the appropriate number of significant figures in your answer.

Solution

UNDERSTAND THE PROBLEM

Because we are aware of only two decimal places, 0.12 is the less precise number. Therefore we subtract and round off to two decimal places.

DEVISE A PLAN

Use Rule 1, which says if the leftmost dropped digit is smaller than 5, simply drop the digits. In this case the leftmost dropped digit is the 3 in 48.623.

CARRY OUT THE PLAN

$$\begin{array}{r} 48.743 \\ -0.12\;\; \\ \hline 48.623 \end{array}$$ Round off to 48.62 (two decimal places).

LOOK BACK

Use mental arithmetic, and you will see that the answer makes sense.

Practice Exercise 2.2   Subtract 1.23 from 54.667, giving the appropriate number of significant figures in your answer.

Problem Solving and Worked-Out Examples using Polya's Four-Step Method:

• Understand the Problem
• Devise a Plan
• Carry Out the Plan
• Look Back

Worked-out examples allow students to master the various concepts presented in the chapters.

LEARNING GOAL 7: **Atomic Mass, Formula Mass, and Molecular Mass**

19. Using examples, explain the difference among atomic mass, formula mass, and molecular mass.

20. State which term—atomic mass, formula mass, or molecular mass—is best suited to describe each of the following substances. (a) $C_6H_{12}O_6$ (b) NaCl (c) Fe (d) CO_2 (e) H_2O (f) Al_2O_3 (g) Ca (h) $Ca(OH)_2$

◀ *21. If in the periodic table oxygen were assigned an atomic mass of 1, what would be the atomic mass of sulfur?

*22. If in the periodic table neon were assigned an atomic mass of 1, what would be the atomic mass of bromine?

23. Using the periodic table (inside front cover), look up the atomic masses of the following elements. (For this exercise, round all atomic masses to one decimal place.) (a) Rb (b) Cr (c) U (d) Se (e) As

24. Using the periodic table (inside front cover), look up the atomic masses of the following elements. (For this exercise, round all atomic masses to one decimal place.) (a) S (b) N (c) Li (d)

LEARNING GOAL 8: **Formula or Molecular Mass of a Compound from the Formula**

◀ 25. Determine the molecular or formula mass of the following compounds. (For this exercise, round all atomic masses to one decimal place.) (a) FeO (b) Fe_2O_3 (c) CuI_2 (d) (e) $Mg(OH)_2$ (f) $NiBr_2$ (g) $Hg_3($ (h) $(NH_4)_2CO_3$

26. Determine the molecular or formula mass of the following compounds. (For this exercise, round all atomic masses to one decimal place.) (a) H_2O (b) H_2SO_4 (c) NaCl (d) (e) P_2O_5 (f) $SrSO_4$ (g) C_2H_6O

27. Determine the molecular or formula mass of the following compounds. (For this exercise, round all atomic masses to one decimal place.) (a) SiO_2 (b) H_2SO_3 (c) $Sr(OH)_2$ (e) $Cu(NO_3)_2$ (f) $CoBr_2$ (g) $(NH$ (h) $HC_2H_3O_2$

◀ 28. Determine the molecular or formula mass of the following compounds. (For this exercise, round all atomic masses to decimal place.) (a) LiOH (b) Na_2CO_3

(d) NaBr (e) SO_3 (f) C_2H_6 (g) OF_2 (h) $(NH_4)_2SO_3$

EXTRA EXERCISES

29. State what each of the symbols and subscripts mean in the chemical formulas for (a) H_2O (b) $C_6H_{12}O_6$ (c) $Ca(OH)_2$ (d) H_2

◀ 30. Make a list of heterogeneous mixtures and homogeneous mixtures that you encounter in everyday life. Do the same for elements and compounds.

31. Write the names and symbols for the fourteen elements that have a one-letter symbol.

32. Write the names and symbols of all the metalloids.

◀ 33. How many metals are there in the periodic table? How many nonmetals? How many metalloids?

34. Write the names of the eleven elements whose symbols are not derived from their English names.

35. Name the elements present in each of the following

Indicate whether each of the following statements is true or false.

1. Chemistry is the science that deals with matter and the changes it undergoes.

2. Organic chemicals are those chemicals that have been derived from living or once-living organisms.

3. Sugar and salt are examples of inorganic chemicals.

4. The art of *khemia* flourished until A.D. 1600.

5. Alchemists were able to transmute lead and iron into gold.

6. The idea that every theory must be proved by experiment was advanced by Robert Boyle.

7. In 420 B.C., Greek philosophers were able to prove that atoms existed.

8. Some carbon-containing compounds are classified as inorganic chemicals.

9. The two main branches of chemistry that existed during the 1700s and 1800s still exist today.

10. Nuclear chemistry, biochemistry, and analytical chemistry are three subdivisions into which chemistry can be divided.

Answer the following questions to help sharpen your test-taking skills.

◀ 11. Find the area of a rectangular room that measures 12.5 m by 10.2 m. Use the proper number of significant figures in reporting your answer.

12. Calculate the volume of a cardboard box that measures 15.25 cm long, 12.00 cm wide, and 24.85 cm high.

13. Determine the volume of a cylindrical solid whose radius is 14.50 cm and whose height is 25.05 cm.

14. Convert 0.28 kg to (a) grams, (b) decigrams, (c) milligrams.

15. Convert 6.8 m to (a) decimeters, (b) centimeters, (c) millimeters.

16. Convert 125 mm to (a) centimeters, (b) meters, (c) kilometers.

17. Convert 25,595 mL to (a) liters, (b) deciliters.

*Answers to cumulative-review questions are given in the back of the book.

18. Express 50.0 m in (a) feet, (b) inches.

19. Express 25.5 g in (a) pounds, (b) ounces.

20. Determine the number of cubic centimeters in a cubic inch without using a conversion table.

21. Express 165 g in pounds.

22. Express 12.0 gallons in (a) liters, (b) ounces.

23. Determine the density of a cube that has a mass of 500.0 g and measures 12.0 cm on each side.

24. A spherical balloon with a volume of 113.0 mL is filled with an unknown gas weighing 20.0 g. Will the balloon float on air?

◀ 25. A small metal sphere weighs 90.0 g. The sphere is placed in a graduated cylinder containing 15.0 mL of water. Once the sphere is submerged, the water in the cylinder measures 30.0 mL. What is the density of the sphere?

26. A small metal sphere with a density of 3.50 g/cm³ has a mass of 937.83 g. Calculate the radius of the sphere.

27. Solve the following problems using the proper number of significant figures:
(a) $28.64 + 3.2$
(b) $125.4 \div 13.5$
(c) 6.55×12.1
(d) $98.4 - 0.12$

◀ 28. Express each of the following numbers in scientific notation: (a) 5,000 (b) 0.0005 (c) 602,300,000 (d) 35,000,000

29. How many significant figures are there in each of the following numbers? (a) 5,500.0 (b) 0.5123 (c) 12.000 (d) 3,500

30. Convert each of the following temperatures from °F to °C: (a) 45.0°F (b) −10.0°F (c) 450°F (d) −100.0°F

31. Convert each of the following temperatures from °C to °F: (a) 88.0°C (b) −12.5°C (c) 65.6°C (d) 400°C

◀ 32. Distinguish between an amorphous solid and a crystalline solid.

33. Determine whether each of the following processes involves chemical or physical changes:
(a) A match burns.
(b) Glucose dissolves in water.
(c) Bread becomes moldy.
(d) A piece of wood is sawed.

Extra Exercises help students review skills in a format resembling a quiz or exam.

Cooperative Learning Opportunities are marked with a triangle. These problem-solving exercises can be done in class under the direction of the instructor or independently in small study groups.

Cumulative Review Problems appear after every two or three chapters. These are designed to help students reinforce material from earlier chapters and prepare for exams.

Chemical Frontiers
highlight recent research
that holds applications for
the future.

Self-Test Exercises 91

CHEMICAL FRONTIERS

Warning! Volcano Erupting Soon

Around the planet there are approximately 500 volcanoes that have a record of violent behavior. Currently, however, scientists monitor only about 20 of these volcanoes and only have the resources to occasionally observe one-third of them. As a result, often the first warning signs from a volcano that has been dormant for hundreds of years are the initial blast and the flow of magma down its side. By then, it may be too late to save the lives of those living near the volcano as the avalanche of hot ash, 1,500 degrees Fahrenheit and moving at speeds up to 500 miles per hour, travels down the mountain.

One solution, developed in the early and mid-1990s and still used today, is a satellite-based system that is able to make infrared and radar observations of volcanoes. Such a system monitors changes on a volcano and alerts scientists to a potentially explosive situation. While a great advance in volcano observation, this system does not have the capabilities to warn of impending disaster.

Enter chemistry. Scientists have discovered that before the initial
it inside the volcano, sulfur dioxide
e. Scientists theorize that if they are
fur dioxide coming out of a volcano,
sidents of nearby towns or villages

ary level, satellite-based observations
ellite systems, such as the one just
micals on earth. However, they are
ons of sulfur dioxide that foreshadow
trumentation being developed by a
Alin Krueger, working at the God-
ryland, is proposing a new system,
at every volcano every 15 minutes.
d sound the alarm before the next

respiration, and carbon monoxide, which is
ct of incomplete combustion when gasoline
d. Discuss how Dalton's atomic theory ex-
hese facts.

points in Dalton's atomic theory. Explain
nes had to be corrected after later experi-
vere performed.

e Dalton's atomic theory to the ideas set
Democritus in 420 B.C.

---

118 **CHAPTER 5** *Atomic Theory, Part 2: Energy Levels and the Bohr Atom*

Using this information, the electron configuration, or arrangement of electrons about the nucleus, could be determined. The various energy levels of the atom have been given the letter designations K through Q and the numbers 1 through 7. (Level K or 1 is the lowest energy level.) Associated with each level is a theoretical maximum electron population. Lower energy levels are usually filled before electrons are positioned in higher energy levels. Using the Aufbau principle, electron configurations can be determined and represented using a shorthand notation. The electron configurations endow the elements with many of their properties; elements with similar outer electron configurations exhibit similar chemical behavior.

KEY TERMS

Aufbau principle **(5.8)**
continuous spectrum **(5.1)**
electromagnetic spectrum **(5.1)**
electron configuration **(5.8)**
energy level **(5.3)**
energy sublevel **(5.6)**
energy subshell **(5.6)**
excited state **(5.3)**

ground state **(5.3)**
line spectrum **(5.1)**
orbital **(5.4)**
outermost shell **(5.9)**
quanta **(5.2)**
quantum mechanics **(5.4)**
shell **(5.3, 5.5)**

CAREER SKETCH

Health Careers/Blood Related

Many different careers in the health field require a strong science background. Working with blood is one of those fields.

As a *phlebotomist,* you will be responsible for drawing blood specimens from inpatients and outpatients. Once you gain elementary experience, you will perform more difficult phlebotomy procedures and act as a technical resource for beginning phlebotomists. Important skills for a phlebotomist are good laboratory procedures, strong organizational skills, and adept analytical skills.

A *transfusion services administrator* performs a slightly different role than a phlebotomist. This person serves in a more administrative role, often coordinating busy blood transfusion departments. Important skills for a transfusion services administrator include a knowledge of chemistry, hematology, and microbiology, as well as the organizational and laboratory skills necessary for any professional in the sciences.

Key Terms
(also) appear at the end
of each chapter with
references to the sections
in which they occur.

Career Sketches
highlight occupations that
draw on chemistry skills.

TO THE STUDENT
Tips on Studying Chemistry

Chemistry is an exciting subject. It affects our lives each day in many ways. The foods we eat, the clothes we wear, the materials used to construct our homes, the medications that cure our illnesses, and the energy we depend on are all products of the science of chemistry coupled with technology.

As you learn chemistry, you will gain knowledge that will help you understand more about our technological society. You'll be able to apply your understanding of chemistry to help solve some of the problems you will face in the real world. You'll be able to approach familiar situations with new insight and be able to make decisions based on the knowledge acquired in this course.

Study Skills for Chemistry

How to Use This Book

Your textbook can be an important tool in helping you master the ideas and problem-solving skills you will learn in this course, especially if you know how to use it. Take a minute now to become familiar with the parts of the book. What follows are our suggestions for how you can best use the book to help you succeed in your course.

Begin each chapter by reading the learning goals and the introduction. The learning goals will help you organize the material you'll need to master; they identify topics that are likely to have special emphasis in class.

After you've read the introduction and learning goals, flip through the chapter to get a general sense of the topics it covers. Notice that the learning goals are repeated in the margins near the material to which they apply. The numbered headings in the text will alert you to the major topics of the chapter. Look also at the figures and tables in the chapter.

After your brief flip through the chapter, return to the beginning and start reading. Read each section thoroughly and take notes as you read. At the end of each section, pause to think about what you have just read; if you are not sure you understand everything in the section, go back and read again.

While some of the material in this course requires memorization, a large part of chemistry involves learning how to solve problems. *Essential Concepts of Chemistry* has been written with a major focus on developing problem-solving skills. Throughout the text, you'll find worked-out examples. When you come to them, stop and read through them carefully. Each example presents a problem and then shows you the solution, step by step. Once you have read the worked-out example, try the practice exercise that follows. Check your answers against the answers given at the back of the book. Problem-solving skills are among the most important things you will learn in this course, and working the exercises in the chapter will give you practice and immediate feedback on how well you are mastering them.

At the end of the chapter, you'll find a summary and a list of key terms. Review these, testing yourself to see if you understand the terms. If you need help, the numbers following the terms tell you which section to review. (There is also a glossary at the back of the book giving definitions of the terms.) Finally, work the self-test exercises at the end of the chapter. You will see that they are grouped by learning goal, to help you organize the material. The answers to selected exercises are at the back of the book, to help you test your progress.

How can you be sure, as you study, that you have mastered the material in the chapter? If you are able to work through the examples, practice exercises, and self-test exercises at the end of the chapter, you will have mastered the necessary problem-solving skills. For sections that require you to memorize reactions or descriptive material, self-test exercises are designed to help you master this content.

You'll get the best results if you read the relevant chapter before going to the lecture. Survey the material to be covered and examine the types of problems to be solved. Know what the chapter is about. This preparation will help you get the most from your instructor's lecture.

If you find at the beginning of the course that you need some review of mathematics, turn to Appendix A at the end of this book. It reviews the mathematical skills you'll need in this course.

Taking Good Notes

Note taking is one of the most important skills for a student to develop. In class you will take notes. Review them the next time you study and reread each section of the textbook slowly for understanding. Combine what you learned in class with what you read in the textbook. Where problem solving takes place, go over the step-by-step solutions in the textbook. Then answer the practice exercises to be sure that you understand how to find the solutions yourself. If you're stuck, ask the instructor for help. Do not ignore material that you do not understand.

We have written *Essential Concepts of Chemistry* in a simple, conversational, readable style. You should not have a difficult time understanding the book. Spend some time being sure that you can recall what you have learned. Sometimes it helps to write as you read, recalling what you have learned in your own words on paper. The last step is to reread the entire chapter and go over all of your notes.

Many students find it useful to keep a journal to write about the material being studied. A simple lined notebook can serve as a journal. After each lesson, summarize the point of the lesson. After reading a section in the textbook, write about what you thought was the point of the reading. Read your journal notes to a classmate or, if possible, to a teaching assistant or to the professor. Writing about a topic is an excellent way to find out if you understand the material being studied.

Your notes must be accurate to be useful. You will have to develop your own shorthand in order to take notes quickly. There are many symbols and abbreviations that you can use in devising your own shorthand. Mathematical symbols such as $>$ (greater than) and $<$ (less than), the plus sign $(+)$, and the equals sign $(=)$ are useful. Spelling a word phonetically by leaving out the vowels also speeds things up.

As you work through the exercises in each chapter of the book, clearly indicate each type of problem you are solving. Write a short description of the type of problem and explain why you are solving this particular problem. When there are several variables in a problem, be sure you can solve for each of the variables, depending on the information given.

Combine your lecture notes with your textbook notes. Be sure you understand your notes. To test your knowledge, turn to the end of the chapter and do the self-test exercises. If you can answer them, you're on your way.

The Importance of Attending Class

There is nothing more important than attending class to help you learn chemistry. Regular attendance allows you to have all of the lecture notes you'll need to organize your studies, and the experiments you perform in the laboratory give you hands-on experience in chemistry. Instructors are experts in teaching their subjects. They teach to a variety of learning styles, and you will benefit greatly from the classroom and laboratory experience and the personal interaction with the instructors and teaching assistants. If you need extra help, see your instructor after class and set up an appointment. Some instructors set aside time for weekly review sessions, and you will learn about these in class.

Making the Most of Your Study Group

Becoming part of a study group is one of the best ways to learn the material in the course. When people work together they can share their skills and resources. In this way they can get more done than by working alone. In addition, people often draw strength from groups. Setting a meeting with the study group is like making an appointment to study. It's an appointment that you are more likely to keep, as compared with a solo study session, which can be more easily skipped. With three or four classmates you can practice problem solving and reasoning. Study groups can be organized in many ways. Here are some suggestions for you to try.

The first step in making the most of your study group is to form a group that works for you. Find people who share some of your goals and who face similar challenges. Find people who are as serious about school as you are. This might include those who attend class regularly, take good notes, and keep up in class. If you are a single parent who supports a family, it might be helpful to form a group that contains at least one other member who has similar time constraints and obligations.

At the first meeting of the study group, set a method of operation. It is likely that one member will emerge as the leader. You might decide to have this person remain the leader, or you might share and rotate the leadership role. Decide what material the group will cover during each session. Set an assignment for the next session, so that everyone will be prepared and the time is used efficiently and effectively. Set time limits for discussion of each topic, so that all of the topics that must be covered are addressed. Practice teaching each other the material.

Solving problems as a group is an important task. At the end of each chapter in the self-test exercises, you will find certain problems marked with a triangle. These problems are good for discussion and are ideal for use in the study group. Once each member learns the material, the problems can be solved by the group. The next step is to have each member solve problems individually and then discuss them as a group.

Once the material is discussed and everyone understands the subject matter, you will be ready to test each other. Have each group member choose several questions, make a sample test of your own, and test each other. You might also try to predict what questions or types of questions will be on the test. Don't forget that even with a study group, it is necessary to do a great deal of studying on your own. Working individually is an important part of the learning process.

Taking a Test

Being well prepared and confident is the first step in taking a test. Cramming doesn't work in chemistry. You cannot open the book the night before the exam and learn a few week's worth of material in one sitting. Chemistry must be studied gradually. If you have trouble solving a particular type of problem, you may need to consult the instructor. That's difficult to do if you try to study all at once.

Try to study a little each day. You'll find this method more effective and less stressful. *Essential Concepts of Chemistry* contains an entire package of helpful materials. Besides the textbook, there is a study guide for students, written by James R. Braun, and a solutions manual, containing solutions for all of the problems in the textbook. These ancillary materials can give you the extra practice you might need to achieve success in the course.

Before an exam, it is important to speak with your instructor and find out what material will be covered on the test. Instructors emphasize important concepts in class. Some instructors will help guide your studies by offering information on material that deserves special attention.

Get a good night's sleep before an exam. Be as relaxed as possible so you can think easily. You'll know whether you're prepared or not. If you've studied properly, you should do well.

Once you begin the exam, glance through it to see how long it is and determine how to budget your time. If a problem gives you trouble, skip it and come back to it later. That way you'll have time to get to all of the questions.

Maintaining Good Study Habits

You can successfully complete basic chemistry if you develop good study habits. The sections that follow discuss some common-sense methods of achieving success in college and offer additional advice that can be applied to your day-to-day study of chemistry. As you become a well-prepared and successful student, you will approach examinations with confidence. The successful habits you form will stay with you as you complete your education and will benefit you throughout your life.

Time Management for College Students

You've got an exam on Friday and a paper due on Monday. You know you've got to get to work. Just then the phone rings. A group of friends from your dorm is going out for a snack. You know you shouldn't join them, but you've got to eat anyway, so you go along. When you return home you need a little time to unwind. You turn on the television. Two hours later you're relaxed, but you're also tired. You decide to call it a night. There's always tomorrow!

If this scenario sounds familiar, it should. It happens to all of us now and then. As a college student, however, it's very important to learn to manage your time effectively. The first step in managing your time effectively is to know where you're going. It helps to set goals for yourself. Although we may have vague notions of what we want from life, like being happy, or being a credit to society, or being financially secure, these generalized plans should be made concrete. Goals must be real. They must be examined closely. There are three different types that you should consider: long-range goals, medium-range goals, and short-range goals.

Long-range goals are usually personal wishes. They have to do with your career aims, your educational plans, and your social desires. Think about where you would like to be 5 or 10 years from now. The education you are now receiving in college should be a stepping stone to help you achieve your long-range goals. Besides achieving the benefit of learning, a college education pays off in dollars. College graduates earn about $700,000 more during their lifetimes than their counterparts who have no degrees. Depending on your career plans, the grades you earn in your courses will help determine whether or not you will be able to fulfill your long-range goals. To achieve long-range goals, they need to be broken into smaller parts and examined closely.

Medium-range goals, sometimes called mid-term goals, can be accomplished in one to five years. They help you achieve your long-range goals. They can be set two or three times a year. For example, if you plan to enter medical school after graduation, you will need a considerable number of A's in your courses. A medium-range goal would be to get four or five A's in your courses for four years. Another medium-range goal might be to join a club or improve your skills in your favorite sport. Let's say that your grades last semester weren't the best. A medium-range goal for you might then be to improve your grades. If you're saving money to buy a car, then watching your budget more carefully might be a reasonable goal to set.

Short-range goals, also called short-term goals, can be accomplished in a year or less. These goals involve taking care of your daily tasks and keeping up with your assignments. Reading a chapter in a book, completing an assignment, or writing a paper are examples of short-term goals.

Achieving your goals often requires setting up a plan. Although we'd like to think that we have the willpower to do all that we have to, very often we become distracted. A well-planned, flexible schedule is a useful tool to help us get things done more effectively.

The first step to managing your time is to know yourself. When do you function the best? If you're a morning person, it's best to try and get done as much as you can in the morning, when you are at your best. If you function better late in the day, then you should try to schedule most of your work in the afternoon or evening.

Setting up a monthly plan is the next step in becoming organized. Obtain a calendar and write in, on the appropriate date, the important assignments and events you're responsible for each month of the semester (Figure 1). Term papers, exams, sports meets, and social activities can be handled better if you look at the overall picture. If you know you have two exams scheduled and a paper due close to each other, then you'll have to keep your schedule free to allow for your school work.

After you've completed your monthly plan, devise a weekly plan. Map out a schedule for each day of the week. Think over how much time you need to complete each assignment and to study for your exams, in order to fit all of your responsibilities into your schedule (Figure 2).

■ **FIGURE 1**

Monthly planning calendar

Month _____ Year _____

SUNDAY	MONDAY	TUESDAY	WEDNESDAY	THURSDAY	FRIDAY	SATURDAY

It's usually best to find a quiet place to work. Accomplish your tasks in priority order, breaking the large tasks into smaller ones. Don't try to do too much in one day. It's better to do a little less than to do too much. Complete one task at a time and avoid working on two projects at once.

You may find that after a month or so, you can manage your time effectively without a written schedule. For some students having a written schedule is a necessity. When you are designing your plan, don't forget that you need time for enjoyment and for work. Giving yourself positive reinforcement is a good idea. Reward yourself when you've accomplished something worthwhile. It is important to enjoy what you're doing and to set aside enough time for rest and relaxation.

While some people find it difficult to socialize, there are others who become overly involved in social activities. Attending college can be very exciting, and for some young adults the endless opportunities to work for a cause, socialize with friends, or attend frequent parties are irresistible. It is not uncommon for some college students to use study time for socialization.

Around the time of midterms panic sets in and lasts until final exam time. Don't get caught in this trap. Be sure that you are not becoming involved in too many social activities in order to avoid your studies. If you're serious about receiving a college education, then your study time will be important to you.

■ **FIGURE 2**

Weekly plan

My main goal this week: _____

	Monday	Tuesday	Wednesday	Thursday	Friday	Saturday	Sunday
6 A.M.							
7 A.M.							
8 A.M.							
9 A.M.							
10 A.M.							
11 A.M.							
12 NOON							
1 P.M.							
2 P.M.							
3 P.M.							
4 P.M.							
5 P.M.							
6 P.M.							
7 P.M.							
8 P.M.							
9 P.M.							
10 P.M.							
11 P.M.							
12 MID.							

School-related stress usually results when you allow yourself to become overwhelmed and overloaded. Taking on too much responsibility at one time often results in a high stress level. This can cause a whole range of physical, emotional, and social problems.

When you register for your courses, be sure that you're not taking on too much. Most colleges schedule a full load for you. This means that you sign up for a set number of credits that school personnel believe a full-time student can handle successfully. The full load is calculated so that a full-time student can finish the requirements for an associate's degree in two years, or a bachelor's degree in four years.

Students who work diligently and have good study skills and few other demands on their time can normally handle a full load. Problems often arise when a student signs up for more credits than are recommended for a full load or when there are other important time-consuming responsibilities. When you have too much responsibility to handle at once, stress can readily develop. Instead of completing your course work with ease, you may find yourself playing "catch-up" all semester.

Arranging a course load to fit your needs is one of the most important items to think about as a college student. You can avoid overloading yourself by taking a night course during the summer at a local college or by attending summer school. Spending an extra semester or two in college is a reasonable price to pay in order to achieve success and avoid stress while you're in college.

Your college years are likely to be among the most demanding and enjoyable years of your life. During this time there are activities that will compete for your attention. Studying, developing relationships, and handling your financial affairs are among the most important challenges that will require your energy, creativity, and brain power. Learning to deal successfully with the different facets of your life can be accomplished by developing good coping skills. These are skills that will remain with you throughout your life and can be applied to just about any situation you'll encounter.

Getting Yourself to Study

For many people it's difficult to sit and study. They find countless distractions to avoid getting to the task. If you have a problem getting started, convince yourself that you will be able to get through the material rather quickly if you break it up into small, manageable units. After learning each small unit, stop and take a short break, then continue.

Developing the ability to concentrate is another important study skill. Keep your mind on what you're doing. Don't allow yourself to be distracted. Remove the telephone and the television set and try to eliminate interruptions. Find a quiet place where you won't be disturbed. Sometimes low-level noise like instrumental music or a steady flow of traffic helps stimulate concentration. And don't try to study if you are hungry. That's a sure way to break your concentration.

Some people are procrastinators. They put things off and never get much done. If you tend to procrastinate, you might benefit from examining your behavior. Does procrastination keep you from reaching your goals? If so, being aware of the cost of procrastination might help you break the habit. Perhaps you are one who works best under pressure. If this produces a better work environment for you, perhaps you procrastinate to your benefit.

If you find that procrastination hurts your progress, here are some ways to break the habit. First, look at the benefits of doing the work at hand. Will you feel a sense of accomplishment when the job is done? Will you feel less stressed and more relieved? Next, break the job into smaller parts. Perhaps there is just too much to learn at once. Break the work into 15-minute segments. After 15 minutes, stop and take a rest. Work slowly to build up your capacity to work.

Perhaps you have tried the first two steps and you still can't get started. Contact a friend, roommate, parent, spouse, or child and set a mutual goal.

Work for an hour, then celebrate together. Another strategy is to reward yourself when you are done. If you simply can't break the habit, just sit down and force yourself to complete the task. You will certainly be relieved when it is completed. Finally, it might be necessary to accept the fact that procrastination is a barrier to your success. Be honest with yourself. Then move through the problem and solve it, instead of letting it drain you of your energy.

Final Thoughts

As you begin your study of chemistry, we hope that you do so with an open mind and a positive attitude. You are probably enrolled in a chemistry course because it is part of the path you have chosen in order to reach your career goals. We hope you will enjoy the course and find the textbook a useful tool to help you learn the subject matter. Our goal is to help you get the most out of this course. Chemistry may be new to you, and we hope that some of the tips for studying more effectively will be useful to you. If you have taken a chemistry course before, or if you know of a method of studying that works for you, we invite you to write to us and share your suggestions.

As you enter the fascinating world of chemistry, we hope your experience is a good one and that you come away from the course with an understanding of why we remain so enthusiastic about this exciting subject.

Alan Sherman
Sharon J. Sherman

THE ORIGINS OF CHEMISTRY:
Where It All Began

Imagine being present at the Princeton Plasma Physics Laboratory (PPPL) on the evening of one of the most important scientific experiments of the century—an experiment that if successful would take humans one step closer to harnessing fusion power as a commercial energy source. The scientists at PPPL had been working toward this night for over 40 years. Their goal was to harness the energy source of the stars.

Nuclear fusion is the process of combining nuclei of small atoms to form larger atoms. Our sun and other stars produce their energy by fusing hydrogen atoms into helium atoms. This process creates huge amounts of energy, on the order of tens-of-millions of degrees Celsius. Unlike nuclear fission (splitting atoms), fusion is a very clean process that produces no dangerous long-lasting radioactive isotopes, and the fuel used, hydrogen, is plentiful. If harnessed, fusion power could supply all of humankind's energy needs for thousands and thousands of years, long after fossil fuels have been depleted.

The technical difficulties of creating a device that could reproduce on earth the reactions that take place on the sun and stars appeared at first to be insurmountable. How could such fantastic heat be contained? What vessel or tools could handle it? Moreover, to duplicate the sun's fusion process on earth, the hydrogen atoms would have to form a *plasma* that must reach a temperature of about 100 million degrees Celsius for one second, with a density of about one one-hundred-thousandth that of air at sea level. Under these conditions, the hydrogen nuclei collide and produce helium nuclei and subatomic particles called neutrons. The energy of motion of these neutrons is then converted into electricity. This was the theory, but would it work?

Over the previous four decades, the scientists and engineers at PPPL built several prototype reactors. These reactors utilized magnets to hold the plasma in a "magnetic bottle." Scientists and engineers also developed techniques such as "neutral beam heating" and "radio frequency heating" to heat the plasma to very high temperatures. Each reactor was built to gain knowledge about the behavior of plasmas and to solve the problems associated with plasma heating and confinement. Each reactor helped scientists overcome various technical hurdles, bringing them ever closer to achieving controlled fusion.

In the mid-1970s the PPPL scientists and engineers constructed their largest reactor, the Tokamak Fusion Test Reactor (TFTR). It would take thousands of preliminary experiments over a period of two decades to ready this reactor for its greatest challenge—the production of large amounts of fusion power. The earlier reactors, and the preliminary experiments on TFTR, all used *protium*, a form of hydrogen, as the fuel. To achieve commercial fusion power, however, two different forms of hydrogen atoms, a 50–50 mixture of *deuterium* and *tritium*, would be required. Modifications to the reactor were made, and the researchers at PPPL were at last ready to attempt a fusion breakthrough.

It was a cool evening in Princeton, New Jersey, on December 9, 1993. We arrived at PPPL about 9:00 P.M. There were already about 200 people in the auditorium and main lobby area. There were television monitors everywhere, showing live pictures of TFTR and the project team in the control room. Staff members and their families were on hand to witness this historic event; scores of newspaper reporters and scientists from around the nation were also present. The excitement in the air was tempered with feelings of apprehension. The project team, located in a futuristic-looking control room, began to power up the systems of TFTR. Over the next two hours, these systems would be checked and run through several conditioning experiments.

Finally, at 11:07 P.M. all was ready. TFTR's systems were powered up. A 50–50 mixture of deuterium and tritium was injected into the reactor, and the neutral beam and radio frequency heaters were turned on. A whining sound from the reactor could be heard. A monitor beaming a picture from inside the reactor turned from intense white to intense black. All of the spectators, scientists, and engineers stood silent, as if frozen in time. In a few seconds, which seemed an eternity, it was over. The computer screens came to life, people began to move, while the spectators whispered as the scientists analyzed the information coming from the reactor. Within two minutes, the information was relayed to deputy laboratory director Dr. Dale Meade, who announced to the spectators that three million watts of fusion power had been obtained in a single burst. History was made! Spectators, scientists, and reporters cheered loudly. Dr. Ron Davidson, the director of PPPL said, ''I think this is a historic milestone in fusion energy development, something analogous to the Wright Brothers' first flight.''

The next day, newspaper headlines around the world announced the event. Humankind's dream of finding a clean, safe, and inexhaustible supply of energy took a quantum leap forward. We were at PPPL on the night of December 9 with our son Michael. It was a night that we will never forget.

Sharon and Alan Sherman

LEARNING GOALS

After you've studied this chapter, you should be able to:

1. Discuss the nature of science and scientific thought.
2. Name the substances that Greek philosophers thought were basic to the composition of the earth.
3. Discuss the role of the Egyptians in the development of chemistry.
4. Describe the major goals of the alchemists.
5. Explain how Robert Boyle's book laid the foundations of modern chemistry.
6. Name the two main branches of chemistry.
7. Discuss the uses of chemistry in today's world.

INTRODUCTION

Humans are curious. They ask questions and they try to find answers. The desire to understand their environment has made people search for explanations, principles, and laws, and modern chemistry is one result of that search.

Chemistry is the science that deals with matter and the changes it undergoes. Chemistry also focuses on energy and how it changes when matter is transformed. Chemists seek to understand how all matter behaves and to discover the principles governing this behavior.

Chemists ask fundamental questions and look for solutions in their attempt to unlock the secrets of nature. Many of their questions have been answered, and this information has built the body of knowledge that we call science.

Many of the basic concepts of chemistry have been incorporated into different fields of study. Physicists, biologists, engineers, medical and dental professionals, and nutritionists are just some of the professionals whose fields require chemical knowledge.

In this chapter we first look at the beginnings of chemistry and then briefly touch upon some areas of modern chemical research.

1.1   History

LEARNING GOAL 1
Beginnings of scientific thought

LEARNING GOAL 2
Influence of Greek philosophy on chemistry

Some early alchemical symbols
(The Granger Collection, New York)

LEARNING GOAL 3
Egyptians' role in the development of chemistry

The earliest attempts to explain natural phenomena led to fanciful inventions—to myths and fantasies—but not to understanding. Around 600 B.C., a group of Greek philosophers became dissatisfied with these myths, which explained little. Stimulated by social and cultural changes as well as curiosity, they began to ask questions about the world around them. They answered these questions by constructing lists of logical possibilities. Thus Greek philosophy was an attempt to discover the basic truths of nature by thinking things through, rather than by running laboratory experiments. The Greek philosophers did this so thoroughly and so brilliantly that the years between 600 B.C. and 400 B.C. are called the "golden age of philosophy."

Some of the Greek philosophers believed they could find a single substance from which everything else was made. A philosopher named Thales believed that this substance was water, but another named Anaximenes thought it was air. Some earlier philosophers believed the universe was composed of four elements—earth, air, fire, and water. The Greek philosopher Empedocles took this belief a step further, maintaining that these four elements combine in different proportions to make up all the objects in the universe.

During this period, the Greek philosophers laid the foundation for one of our main ideas about the universe. Leucippus (about 440 B.C.) and Democritus (about 420 B.C.) were trying to determine whether there was such a thing as a smallest particle of matter. In doing so, they established the idea of the atom, a particle so tiny that it could not be seen. At that time there was no way to test whether atoms really existed, and more than 2,000 years passed before scientists proved that they do exist.

While the Greeks were studying philosophy and mathematics, the Egyptians were practicing the art of chemistry. They were mining and purifying the metals gold, silver, and copper. They were making embalming fluids and dyes. They called this art *khemia*, and it flourished until the seventh century A.D., when it was taken over by the Arabs. The Egyptian word *khemia* became the Arabic word *alkhemia* and then the English word *alchemy*. Today our version of the word is used to mean everything that happened in chemistry between A.D. 300 and A.D. 1600.

A major goal of the alchemists was to transmute (convert) "base metals" into gold. That is, they wanted to transform less desirable elements such as

■ LEARNING GOAL 4

Alchemists' major goals

■ LEARNING GOAL 5

Boyle's influence on the development of modern chemistry

■ LEARNING GOAL 6

Branches of chemistry

lead and iron into the element gold. The ancient Arabic emperors employed many alchemists for this purpose, which, of course, was never accomplished.

The alchemists also tried to find the "philosopher's stone" (a supposed cure for all diseases) and the "elixir of life" (which would prolong life indefinitely). Unfortunately they failed in both attempts, but they did have some lucky accidents. They discovered acetic acid, nitric acid, and ethyl alcohol, as well as many other substances used by chemists today.

The modern age of chemistry dawned in 1661 with the publication of the book *The Sceptical Chymist*, written by Robert Boyle, an English chemist, physicist, and theologian. Boyle was "skeptical" because he was not willing to take the word of the ancient Greeks and alchemists as truth, especially about the elements that make up the world. Instead Boyle believed that scientists must start from basic principles, and he realized that every theory had to be proven by experiment. His new and innovative scientific approach was to change the whole course of chemistry.

Table 1.1 chronologically outlines some of the major contributions in chemistry throughout history.

1.2   Chemistry Is a Diverse Field

Robert Boyle
(Science Photo Library/Photo Researchers, Inc.)

During the 1700s and early 1800s, most chemists believed that there were two main branches of chemistry: organic and inorganic. Organic substances were thought to have been derived from living or once-living organisms; inorganic substances were said to have originated from nonliving materials. Sugars, fats, and oils were classified as organic chemicals. Salt and iron were classified as inorganic chemicals.

These two branches of chemistry still exist today, but the rules governing their classification have changed. Chemists now classify organic substances as those containing the element carbon. Inorganic substances are generally those that are composed of elements other than carbon. (Some inorganic substances are exceptions to this rule and do contain carbon.) Sugars, fats, and oils contain the element carbon and are indeed classified as organic substances. Salt and iron do not contain carbon, and both fall into the category of inorganic substances.

Other branches of chemistry include nuclear chemistry, biochemistry, pharmaceutical chemistry, analytical chemistry, environmental chemistry, and physical chemistry.

1.3   How Chemistry Affects Our World

■ LEARNING GOAL 7

Chemistry in today's world

Chemistry and chemical processes are keys to understanding the world in which we live today. Today's chemistry includes areas as diverse as lasers, superconductors, genetic engineering, environmental management, and the synthesis of new pharmaceuticals.

Drawing upon such natural resources as the air, forests, oceans, mines, wells, and farms, the chemical industry produces more than 50,000 different

TABLE 1.1

Time Line of Chemistry*

DATE	PERSON	EVENT
600 B.C.	Thales	Idea that water is the main form of matter
546 B.C.	Anaximenes	Idea that air is the main form of matter
450 B.C.	Empedocles	Idea that the four elements combine in different proportions
420 B.C.	Democritus	Idea of the atom
400 A.D.	Ko Hung	Attempts to find the elixir of life
1000	Avicenna	*Book of the Remedy*
1330	Bonus	*Introduction to the Arts of Alchemy*
1500		*Little Book of Distillation*
1620	Van Helmont	Foundations of chemical physiology
1625	Glauber	Contributions to practical chemistry
1661	Boyle	*The Sceptical Chymist*
1766	Cavendish	Discovery of hydrogen
1775	Lavoisier	Discovery of the composition of air
1787	Lavoisier and Berthollet	A system of naming chemicals
1800	Proust	Law of Definite Composition
1800	Dalton	Proposes an atomic theory
1820	Berzelius	Modern symbols for elements
1829	Döbereiner	Law of Triads
1860	Bunsen and Kirchhoff	Spectroscopic analysis
1869	Mendeleev	Periodic Law
1874	Zeidler	Discovery of DDT
1874	Van't Hoff and Le Bel	Foundations of stereochemistry
1886	Goldstein	Naming of cathode rays
1897	Thomson	Proposes a structure of the atom
1905	Einstein	Matter–energy relationship: $E = mc^2$
1908	Gelmo	Discovers sulfanilamide
1911	Rutherford	Proposes the nuclear atom
1913	Bohr	Proposes energy levels in atoms
1922	Banting, Best, and Macleod	Discover insulin
1928	Fleming	Discovers penicillin
1932	Urey	Discovers deuterium
1942	Fermi	First atomic pile
1945		First atomic bomb; nuclear fission
1950	Pauling	Helical shape of polypeptides
1952		First hydrogen bomb; nuclear fusion
1953	Watson and Crick	Structure of DNA

Time Line of Chemistry (*continued*)

DATE	PERSON	EVENT
1958	Townes and Schawlow	Develop laser beam
1969		Discovery of first complex organic interstellar molecule—formaldehyde
1970	Ghiorso	Synthesis of element number 105
1973	Cohen, Chang, Boyer, and Helling	Initiation of recombinant DNA studies
1974	Seaborg and Ghiorso	Synthesis of element 106
1979	A. Crewe	First color motion pictures of individual atoms
1982	Munzenberg and Armbruster	Synthesis of element 109
1984	Gallico	Development of test-tube skin
1985		Genetically engineered human insulin and human growth hormone become commercially available
1985	Diana	Development of compounds that kill cold viruses
1986		Nearly 800 genes mapped using genetic engineering techniques
1987		New class of high-temperature superconducting oxides with major implications for use in electronics discovered by researchers at the Uni. of Houston and at AT&T Bell Laboratories in Murray Hill, N.J.
1988	Zewail	First snapshots of chemical reactions
1989	Felix	Synthesis of human growth hormone
1990	Filisko	Development of chemical muscles
1991	R. Baughman	Chemists in the United States and several other countries develop techniques for synthesizing large amounts of soccer-ball-shaped C-60 molecules called buckminsterfullerenes ("Buckyballs")
1992	J. Rebek	Synthesis of molecules that mimic some of the essential features of living things
1993		Princeton University's Plasma Physics Laboratory produces 7 megawatts of fusion energy using deuterium and tritium as fuel in their Tokamak Fusion Test Reactor. These experiments represent a major breakthrough for the commercialization of fusion energy. (In 1994 the reactor produced 10.7 megawatts of fusion energy.)
1994		Fermi National Accelerator Laboratory announces that a team of 440 physicists has discovered the Top Quark. This elusive particle, sought after for 17 years, completes the gallery of six quarks. The quarks are some of the most fundamental particles of matter.
1995		Carl Wieman and Eric Cornell of the University of Colorado create a new state of matter, predicted by Albert Einstein over 70 years ago, called the Bose-Einstein Condensate. This type of matter was created at a temperature of 36 billionths of a degree above absolute zero.
1996		NASA scientists report that a meteorite, identified as originating from Mars, may contain fossils of ancient, primitive bacteria from that planet.
1997		The first mammal, a sheep, was cloned from an adult nucleus.

*Taken in part from the *Mallinckrodt Outline of the History of Chemistry*, Mallinckrodt Chemical Works, St. Louis, 1961 chemicals. These chemicals are used to manufacture a variety of products and to provide raw materials for other industries to use in producing their goods. It is by this process that the items required to sustain our basic human needs are made available to us.

Chemistry has played an important role in the processing of foods. Foods are processed so that they remain fresh and free of harmful toxins for a longer period of time. Chemists are hard at work seeking and researching ways to alleviate a world food shortage at a time when human population is approaching 6 billion people. Thousands of drugs to help us treat disease have become available over the last several decades through the application of medical knowledge in the chemical and pharmaceutical industries. Just a few of the chemically based products we use today are plastics, cleansing agents, paper products, textiles, hardware, machinery, building materials, dyes and inks, fertilizers, and paints.

SUMMARY

Scientific thought had its beginnings in the contributions of the Greek philosophers of 600 B.C. to 400 B.C. The earliest chemistry was practiced at about that same time by the Egyptians. Later the art of chemistry—or alchemy—was taken over and extended by the Arabs. "Alchemy" now includes all chemistry from the fourth through the sixteenth centuries A.D. The alchemists made several important discoveries, although they were primarily interested in transmuting base metals into gold. Modern chemistry began with the work of Robert Boyle in the midseventeenth century. Boyle believed that theories must be founded on basic principles and must be supported by experiment.

Modern chemistry and its applications affect us in a variety of ways as we eat or sleep, work or play. Present chemical research is directed toward the solution of problems in many areas, such as physiology, medicine, and the environment.

CAREER SKETCHES*

A strong background in chemistry can help you in almost any career you pursue. Chemists can be found "reaching for the stars" in the aerospace industry or collecting information from the bottom of the sea as an oceanographer. A chemist could have designed the computer

*The career sketch material was adapted from two U.S. Department of Labor publications: Occupational Outlook Handbook and Dictionary of Occupational Titles (1990–1991). In addition, information about careers was obtained from the *Encyclopedia of Careers*, Ninth Edition (Chicago, Ill.: J. G. Ferguson Publishing Co., 1993).

program you're going to use this afternoon or protected the author of that program in court. Careers in chemistry are everywhere!

As a way to expose you to a variety of options for careers in chemistry, we have included career sketches in each chapter. Each "Career Sketch" explores a different path that one might pursue using his or her chemistry knowledge: for example, dietitian, food technologist, or even writer.

Beyond the career sketches, we encourage you to explore other opportunities related to chemistry. You can begin your search in your local or regional newspaper's help wanted section. Or try the Internet. There are dozens of websites dedicated to getting your career search off the ground, such as the Monster Board at http://www.monster.com. Or consult a trade organization for more information. The American Chemical Society, for example, publishes a number of books about careers in chemistry. They can be found in your bookstore or library, or at the ACS web page at http://pubs.acs.org/books/caretoc.htm.

Career opportunities in chemistry are fast paced and exciting! They are on the cutting edge of technology, leading to the creation of tomorrow's greatest innovations. We hope that the 16 sketches that follow pique your interest in the field and encourage you to pursue a career in the wonderful world of chemistry.

CAREER SKETCH

Chemistry Careers

If you are interested in pursuing a career focused on chemistry, consider a career as a *project chemist*. Working for a corporation, you will be responsible for providing technical expertise in troubleshooting, developing analytical methods and sampling techniques, and solving plant and customer quality problems. Additional responsibilities may include customer interface on various product quality issues, communication of technical data by professional reports, and the support of plant laboratory safety goals and quality assurance programs. Most employers will be looking for an extensive background in the sciences, including in-depth knowledge of organic and analytical chemistry, chemistry of production processes, and mathematics.

A *quality assurance chemist* leads organizational redesign initiatives for quality assurance. To achieve this goal, a quality assurance chemist oversees and maintains quality assurance documentation and databases, overviews product releases, works with technical and engineering departments on product-specific projects and issues, and

Chemists work to produce products that improve the quality of life.
(James Holmes/Science Photo Library/Photo Researchers, Inc.)

has responsibility for the statistical analysis of performance and capabilities. This person will also interface with research and development and sales departments on a regular basis. As with the project chemist, an in-depth background in chemistry is necessary for a quality assurance chemist.

CHEMICAL FRONTIERS

What Is a Chemical Frontier?

Scientists discover new uses for an age-old drug that cures common ailments. An organically grown fuel is found that will power your car while being friendly to the environment. Researchers report that a newly synthesized compound exhibits an unusual property. Men and women working in the dynamic field of chemistry have brought a wealth of inventions and innovations to the world. To familiarize you with some of these new findings, we have included Chemical Frontiers in each chapter. By reading about intriguing breakthroughs made in chemistry, you will see the impact chemistry has on our daily lives and gain a glimpse of the exciting future that lies ahead.

CHEMICAL FRONTIERS

*Improving the Way You Drive**

Chemists are working on ways to improve something you use almost every day—the tires on your car. In 1996, worldwide consumption of rubber (the main component of tires) topped 16.5 million metric tons, creating a huge market for elastomers (which are materials that recover their shape after a deforming force is removed).

Recent advances have led to the development of longer-lasting tires. Whereas a decade ago tires may have only lasted 20,000 miles, they can now easily last in excess of 60,000 miles—sometimes up to 80,000 miles. In the early 1990s, a so-called green tire was introduced to consumers. The environmentally friendly tire offers improved fuel economy because it is designed for low rolling resistance.

Materials chemists and scientists are continually working on still more advances in tire design. One proposed advance would add a chemical combination to the sidewalls of tires that would cause the tire to retain the satin, deep black appearance consumers want, even after thousands of miles on the road. Another advance would allow the blending of nitrile with polymers to form an oil-resistant compound (which could reduce the chance of a skid).

Chemists are working to develop new formulations to improve the function of tires. (Gabe Palmer/The Stock Market)

On the horizon are improvements in tire recycling. One New Jersey-based company is experimenting with the addition of a polymer that will allow a tire to be ground into small pieces. The small pieces of rubber would then be molded into parts such as car and truck mud flaps. Also, chemists working for tire and rubber companies are perfecting a run-flat tire—a tire that could run for as many as 50 miles after suffering a puncture or cut and completely losing its air.

Adapted from Chemical & Engineering News, September 8, 1997, pp. 18–21

SELF-TEST EXERCISES

All exercises whose numbers are in underscored, bold-faced italic are answered in the back of the book. Self-test exercises are arranged in matched pairs, one below the other. Difficult problems are marked with an asterisk. Problems marked with a triangle (◄) may be used for cooperative problem solving.

LEARNING GOAL 1: **Nature of Science and Scientific Thought**

◄ 1. Discuss the evolution of science from the golden age of philosophy to the birth of modern chemistry.
 2. What years are called the "golden age of philosophy"? Why are those years considered the beginning of scientific thought?

LEARNING GOAL 2: **Influence of Greek Philosophy on Chemistry**

 3. List the four elements that Empedocles thought made up the world.
 4. What did Thales think the world was composed of?

◄ 5. What contribution did Greek philosopher Democritus make to modern-day society?
 6. Who established the concept of the atom?

 7. If Empedocles had been right and the world were indeed made up of four elements, what two elements would compose paper?
◄ 8. If the world were made up of Empedocles' four elements, what elements would compose vegetation?

LEARNING GOAL 3: **Egyptians' Role in the Development of Chemistry**

 9. List some of the contributions of the Egyptians to the development of chemistry.
 10. The Egyptians mined and purified the metals gold, silver, and antimony. True or false?
 11. While the Greeks were studying philosophy and mathematics, the Egyptians were already practicing the art of chemistry, which they called _____.
 12. Until what time in history did the art of *khemia* flourish?

LEARNING GOAL 4: **Alchemists' Major Goals**

◄13. What was the aim of chemistry from A.D. 300 to A.D. 1600?
 14. Name the metals that the alchemists tried to turn into gold.

 15. The people who tried to turn lead into gold were known as _____.
 16. What is meant by the term *elixir of life*?

LEARNING GOAL 5: **Boyle's Influence on the Development of Modern Chemistry**

 17. What was the message behind Robert Boyle's book *The Sceptical Chymist*?
◄18. How do you think Boyle might have set about proving that water is not an element and that it can be broken down into simpler substances?

LEARNING GOAL 6: *Branches of Chemistry*

19. Name an organic chemical and an inorganic chemical.
◀ 20. Explain by what criterion chemicals are now classified as organic or inorganic.

LEARNING GOAL 7: *Chemistry in Today's World*

◀ 21. Choose one application of chemistry and discuss its impact on society today.
22. Weigh the benefits that chemical research and applications yield to society against the problems caused by the chemical industry. What legislative measures have been taken (or are pending) to solve some of these problems?

EXTRA EXERCISES

23. Make a list of the industries that are based on the science of chemistry.
24. From the "Time Line of Chemistry" (Table 1.1), choose those events in the history of chemistry that you believe have had the greatest social importance. Give reasons for your choices.

25. Explain the benefits of using Biodiesel over gasoline. Discuss the disadvantages of using this material.
◀ 26. If you could extend the "Time Line of Chemistry" 50 years into the future, what would you like to see happen? What advances in research do you believe will take place?
27. Can an educated public, better equipped with scientific knowledge, more easily assess the effects of science on society? Explain your answer.
28. With what is the science of chemistry concerned?
29. According to Empedocles, how would each of the following be classified? (a) copper (b) rain (c) lightning (d) nitrogen gas
◀ 30. Explain in a few sentences how you believe chemistry has affected life in the context of one topic, such as food, clothing, or medicine.
◀ 31. Discuss the contributions of alchemy to the science of chemistry.
32. List four ways in which chemistry has had a positive, healthy effect on life in your home. Explain each.

SYSTEMS OF MEASUREMENT

2

This is the story of Richard Novak, a jeweler in a large metropolitan city, who used something that he had learned in his chemistry class to help him with a business transaction.

One day a customer entered Mr. Novak's store carrying a small burlap bag. The customer approached Mr. Novak and explained that in the bag he had about 200 grams of pure gold nuggets that he wanted to sell. The customer asked Mr. Novak whether he would be interested in purchasing the nuggets. Mr. Novak responded by indicating that he would purchase the nuggets if he could have the material tested to verify that it was pure gold. The customer responded by saying that he didn't want the gold put into any kind of an acid solution. (The usual test for determining if a substance is pure gold is to immerse it in a solution of aqua regia, a mixture of hydrochloric and nitric acids.) Mr. Novak thought for a few minutes and said that he would test the nuggets by placing them in water, instead of acid. The customer consented. Mr. Novak's idea was to determine the density of the nuggets and compare that density to the published value for pure gold. First, Mr. Novak obtained the mass of the nuggets by placing them on a balance. Next, he obtained the volume of the irregularly shaped nuggets by water displacement. To do this he placed the nuggets into a graduated cylinder that was partially filled with water and measured the volume of water displaced. After determining the mass and volume of the nuggets, Mr. Novak calculated the density by dividing the mass of the nuggets by their volume. The nuggets had a density of 18.9 grams/cubic centimeter—exactly the value given for pure gold in the *Handbook of Chemistry and Physics*. Based on the physical appearance of the nuggets and the density information, Mr. Novak purchased the gold nuggets.

■■■ LEARNING GOALS

After you've studied this chapter, you should be able to:

1. Find the number of significant figures in a measurement.
2. Do calculations using the rules for significant figures.
3. Write numbers in scientific notation.
4. Find the area of any square, rectangle, circle, or triangle.
5. Find the volume of any cube, other rectangular solid, cylinder, or sphere.
6. Convert units of mass, length, and volume within the metric system using the factor-unit method.
7. Convert from a metric unit to the corresponding English unit using the factor-unit method.
8. Convert from an English unit to the corresponding metric unit using the factor-unit method.
9. Distinguish between the mass of an object and its weight.
10. Calculate the density, mass, or volume of an object when you are given the other two.
11. Convert temperatures from the Celsius to the Fahrenheit scale and vice versa.

■■■ INTRODUCTION

Chemists, as well as other scientists, use measurements when they do basic research. Measurements represent the dimensions, quantity, or capacity of things. Researchers ask questions and perform experiments. The ability to make accurate measurements of quantities such as mass, volume, temperature, or time enables them to gather the information that leads to the compilation of reliable scientific data. They analyze the data, often with the use of statistics, to look for patterns or regularities in nature. Sometimes the pattern or regularity is basic and can be stated simply to describe some natural phenomenon. The scientist may then devise a hypothesis, or speculative guess, that can explain the phenomenon. The problem-solving process continues as the scientist tests the hypothesis through further experimentation, which leads to the compilation and analysis of more measured data. Sometimes experimental results prove a hypothesis because it successfully explains a regularity or law of nature. The hypothesis then becomes a theory, or detailed explanation, which helps us describe and organize scientific knowledge.

In this chapter we look at systems of measurement so that you will have an understanding of how scientists record and manipulate quantitative data.

We will begin with a look at problem solving so that you will be able to process the data you are given. Familiarity with problem-solving techniques will help you feel more comfortable as you work with units of measurement and tackle problems in all areas of chemistry.

2.1 Problem Solving

Humans have been given the gift of intelligence. We use this intelligence by solving problems. Each day we solve a variety of different problems as we deal with many types of obstacles. When a direct course of action is not clear, we devise ways to reach our goals, regardless of what it takes. Think of a typical day in your life and you will see that you deal with a whole range of problem types. You solve personal problems, social problems, scheduling problems, and financial problems, to name just a few. In your basic chemistry course, you will solve a variety of scientific problems. Like the problems we face in life, the problems you will encounter in basic chemistry require you to use judgment, originality, creativity, and independent thinking.

A great deal of research has gone into determining what makes a successful problem solver. George Polya, who is known as the "father of problem solving," tells us that when we solve problems we must search for some action to attain a goal. In other words, solving a problem does not mean just coming up with an answer. It means coming up with a strategy, or plan, to derive the answer. We will use Polya's four-step model as our general strategy. We will examine each step of Polya's model and show you how to proceed.

In preparation for taking an exam, you may have memorized many different formulas and definitions. Even though you know all the formulas and all the definitions, you may be unsure of where and when to use them. The general problem-solving framework will help you organize your thoughts and allow you to retrieve the relevant background information from your memory. It will help you sort relevant information from irrelevant information and will allow you to apply your knowledge.

One of the most important aspects of the problem-solving process is learning to ask yourself the correct questions. Questions such as "What is the unknown?" and "What information have I been given?" are a good starting point. Another important aspect is your own motivation to become adept at problem solving.

2.2 Polya's Method: The Four-Step General Model

Polya's four-step general model for problem solving will help you organize your thoughts and give you a general framework for solving any type of problem, whether it be a real life situation or a textbook problem.

STEP 1:   **Understand the Problem**

This is a basic step in problem solving. It consists of asking yourself a variety of questions to diagnose the situation. Pinpoint what you are trying to determine. Is there unknown information? What is it? Is any information assumed? Is there any additional information needed? Can you restate the problem in your own words?

STEP 2:   **Devise a Plan**

In this step you come up with a plan to solve the problem. You connect what you know with what is unknown to you. Do you know of a problem that is similar to this one? Is there a similar simpler problem that you can solve? Can you use a formula or write an equation? Can you make a table or draw a diagram to help you solve the problem?

STEP 3:   **Carry Out the Plan**

In this step you actually solve the problem. If you will be using a formula or an equation, you will set it up, plug in the numbers, and solve.

STEP 4:   **Look Back**

In this step you look back at your work and be sure that it is correct. Did you check to be sure that the conditions of the problem were satisfied by the solution? Were all relevant data used? Is your answer reasonable? Compare your solution to your estimate. How do they compare?

When we solve the textbook examples, we will work out at least one solution per chapter using Polya's four-step method. So that you can practice using the process on your own, examine the solutions that do not include Polya's method and try to follow the steps on your own.

2.3 Essential Estimation Skills

Estimation is a close cousin to problem solving. It is the process of producing an answer that is sufficiently close so that decisions can be made. Estimation is an essential skill because it helps make problem solving more meaningful. Instead of coming up with the exact answer to a problem, estimation enables us to determine a ballpark figure for the solution. It encourages us to look at the problem and decide what a sensible answer would be, which gives us a check on our final results. Sometimes when we solve equations we plug numbers into the calculator rather thoughtlessly. We are not always sure when an unreasonable answer has resulted because we don't routinely stop and think of what would be a sensible result and what would be a nonsensical result. Estimation can produce reasonableness about computation. It can encourage us to have a greater appreciation of number size and provide a complement to our routine use of the calculator. Overall, estimation can bring more meaning to the problem-solving process.

The language of estimation includes words such as *about, close to, just about, a little less than,* and *between.* There is no one correct estimate. When you solve the problem you will find the correct answer, which will be exact.

In chemistry, we look for exact answers to problems. To be sure that the answers we come up with are reasonable and sensible, we can use estimation. When the problem is solved and an exact solution is computed, the estimate can be compared with the exact solution. This enables the problem solver to have an additional system of checking the calculation and adds more meaning to the problem-solving process.

2.4   Significant Figures

Suppose that, in an experiment, you are measuring the temperature of a liquid. The thermometer you are using is calibrated, or marked off, only in whole degrees. Imagine that the mercury in the thermometer is halfway between 34 and 35 degrees (Figure 2.1a). You can then estimate that the temperature of the liquid is 34.5 degrees.

Now you pick up another thermometer, which is calibrated in tenths of degrees (Figure 2.1b). When you read this thermometer, it looks as though the temperature of the liquid is 34.55 degrees. It seems that the two thermometers differ in sensitivity, but which measurement is correct?

■ FIGURE 2.1

(a) We can measure the temperature of a liquid using a thermometer calibrated only in degrees. (b) We can also measure the temperature of a liquid using a thermometer calibrated in tenths of degrees.

To answer this question, we consider the number of significant figures in each measurement. **Significant figures** are *digits that express information that is reasonably reliable. The number of significant figures equals the number of digits written, including the last digit even though its value is uncertain.* For example, the temperature 34.5 degrees given by the first thermometer is known to be correct to three significant figures. These significant figures are the 3, the 4, and the 5. (Expressing the measurement as 34.50 degrees would mean that the temperature is known to four significant figures—the zero on the right is significant.)

The temperature 34.55 degrees given by the second thermometer shows a measurement with confidence to four significant figures. It may not be more accurate than the first measurement, but it is more *precise.* In general, the more significant figures there are, the more precise the measurement is.

Actually, neither of the two temperature measurements may be accurate. (They certainly wouldn't be accurate if the temperature of the liquid were actually, say, 47.51 degrees.) Thus **accuracy** involves *closeness to the actual dimension.* **Precision** is related to the *detail with which a measurement is known (expressed by the number of significant figures.)*

A measuring *instrument* is considered accurate if it consistently provides measurements that are close to the actual amount. The instrument is considered to have good precision if it consistently provides the same measurement when it is used to measure the same amount. To see the difference, consider several darts being thrown at a target. If the darts land very close together, the results exhibit precision. If all the darts land on the bull's eye, they exhibit accuracy as well.

Another experiment will show how significant figures play an important role in measurement. Suppose that we have to measure a certain piece of glass to find its perimeter (we do this by measuring the four edges of the glass and adding the lengths). We have two rulers, one calibrated in centimeters and the other calibrated in millimeters. We set out to measure each edge of the glass with the more precise ruler (the one calibrated in millimeters). However, when we get to the fourth side of the glass, we absent-mindedly pick up the less precise ruler (the one calibrated in centimeters). Looking at the glass, shown in Figure 2.2, we can see that three edges are measured to two decimal places and that the fourth is measured to one decimal place. What can we say about the perimeter of the glass? Is it 67.73 cm or 67.7 cm? The answer is 67.7 cm. We can report only what we know for sure—and in this case that is the answer to one decimal place. *Our least precise measurement determines the number of significant figures in our result.*

Whenever we *add or subtract* measured quantities, we must report the results in terms of the least precise measurement. We do this by **rounding off** to the least number of decimal places. Our result must have no more decimal places than our least precise quantity.

To round off a number, we drop one or more digits at the right end of the number and, if necessary, adjust the rightmost digit that we keep. In

■ FIGURE 2.2

Measurement of a piece of glass

some cases, dropped digits must be replaced with zeros. The rules we shall use in rounding numbers are as follows.

RULE 1

If the leftmost dropped digit is smaller than 5, simply drop the digits. Replace dropped digits with zeros as necessary to maintain the magnitude of the rounded number. Thus 7.431 rounded to two significant figures is 7.4. And 7,431 rounded to two significant figures is 7,400.

RULE 2

If the leftmost dropped digit is 5 or greater, increase the last retained digit by 1. Replace dropped digits with zeros as necessary to maintain the magnitude of the rounded number. Thus 93.56 rounded to three significant figures is 93.6. And 9,356 rounded to three significant figures is 9,360.

Here is a rounding rule that is useful in *multiplying or dividing* measured quantities.

STEP 1 Count the number of significant figures in each of the quantities to be multiplied or divided.

STEP 2 Report the result to the least number of significant figures determined in step 1. Round as required.

EXAMPLE 2.1

Add 18.7444 and 13 and report the result to the appropriate number of significant figures.

Solution

UNDERSTAND THE PROBLEM

We will add the numbers. Since 13 is the less precise measurement, we must round the sum to a whole number.

DEVISE A PLAN

Use Rule 2, which says if the leftmost dropped digit is 5 or greater, increase the last digit by 1. In this case the leftmost dropped digit is 7, so the 1 in 31 increases to 2, and we have 32.

CARRY OUT THE PLAN

$$\begin{array}{r} 18.7444 \\ +13 \\ \hline 31.7444 \end{array}$$ Round off to 32 (whole number).

LOOK BACK

The answer makes sense since 18.7444 is close to 19, and 19 + 13 = 32.

Practice Exercise 2.1   Add 12.4432 and 15. Provide the appropriate number of significant figures in your answer.

EXAMPLE 2.2

Subtract 0.12 from 48.743 and use the appropriate number of significant figures in your answer.

Solution

UNDERSTAND THE PROBLEM

Because we are aware of only two decimal places, 0.12 is the less precise number. Therefore we subtract and round off to two decimal places.

DEVISE A PLAN

Use Rule 1, which says if the leftmost dropped digit is smaller than 5, simply drop the digits. In this case the leftmost dropped digit is the 3 in 48.623.

CARRY OUT THE PLAN

$$\begin{array}{r} 48.743 \\ -0.12 \\ \hline 48.623 \end{array}$$ Round off to 48.62 (two decimal places).

LOOK BACK

Use mental arithmetic, and you will see that the answer makes sense.

Practice Exercise 2.2   Subtract 1.23 from 54.667, giving the appropriate number of significant figures in your answer.

EXAMPLE 2.3

What is the area of a square whose side is measured as 1.5 cm?

Solution

The area of a square is equal to the length of its side squared, or

$$A = s \times s = 1.5 \text{ cm} \times 1.5 \text{ cm} = 2.25 \text{ cm}^2$$

Round off to 2.3 cm². (We report to only two significant figures, because the measurement has only two significant figures.)

Practice Exercise 2.3   Calculate the area of a square whose side is measured as 2.5 cm.

EXAMPLE 2.4

Divide 20.8 by 4 and give the result to the proper number of significant figures.

Solution

The result must be rounded to one significant figure because of the one-digit divisor, 4.

$$\frac{20.8}{4} = 5.2 \qquad \text{Round off to 5.}$$

Practice Exercise 2.4   Divide 48.2 by 4, giving the result to the proper number of significant figures.

EXAMPLE 2.5

Multiply 20.8 by 4.1 and report the result to the proper number of significant figures.

Solution

The result must be rounded to two significant figures.

$$20.8 \times 4.1 = 85.28 \qquad \text{Round off to 85.}$$

Practice Exercise 2.5   Multiply 25.5 by 3.2, giving the result to the proper number of significant figures.

Another problem arises when we use significant figures: What do we do about zeros? Are they significant figures or not? Here are some helpful rules. Read them carefully, work through Example 2.6, and then do Practice Exercise 2.6.

RULE 1

Zeros *between* nonzero digits are significant:

4.004 has four significant figures

RULE 2

Zeros to the *left* of nonzero digits are *not* significant, because these zeros show only the position of the decimal point:

0.00254 has three significant figures

0.0146 has three significant figures

0.06 has one significant figure

RULE 3

Zeros that fall at the *end* of a number are not significant unless they are marked as significant. If a zero does indicate the number's precision, we can mark it as significant by *placing a line over it*. Zeros to the right of the decimal place are always significant.

84,000 has two significant figures

84,$\overline{0}$00 has three significant figures

84,000.0 has six significant figures

RULE 4

Exactly defined quantities have an unlimited number of significant figures

4 qt = 1 gal

1 m = 100 cm

EXAMPLE 2.6

In this example, we tell you how many significant figures there are in different numbers. In Practice Exercise 2.6, *you* are asked to tell *us*.

Solution
(a) 0.00087 has two significant figures.
(b) 1.004 has four significant figures.
(c) 873.005 has six significant figures.
(d) 9.00000 has six significant figures.
(e) 320,000 has two significant figures.
(f) 18$\overline{0,0}$00 has four significant figures.
(g) 180,000.0 has seven significant figures.
(h) 2,$\overline{0}$00 has four significant figures.

Practice Exercise 2.6   Find the number of significant figures in each of the following numbers: (a) 0.0023   (b) 5.025   (c) 123.456   (d) 5,000
(e) 5,$\overline{0}$00   (f) 12.000

2.5   Scientific Notation: Powers of 10

■ LEARNING GOAL 3
Scientific notation

In science, we often deal with numbers that are *very* large or *very* small. Numbers like 100,000,000,000 (a hundred billion) or 0.0000008 (eight ten-millionths) arise frequently, and they are troublesome to work with in calculations. There is a shorthand method of writing such numbers, based on powers of 10. The number 100,000,000,000 can be written as 1×10^{11}, and 0.0000008 as 8×10^{-7}. In the first example, we moved the decimal point eleven places to the left:

$$1\,0\,0,\,0\,0\,0,\,0\,0\,0,\,0\,0\,0. = 1 \times 10^{11}$$

Moving the decimal point to the left is compensated for by multiplying by a positive power of 10. In the second example, we moved the decimal point seven places to the right.

$$0.\,0\,0\,0\,0\,0\,0\,8 = 8 \times 10^{-7}$$

Moving the decimal point to the right corresponds to multiplying by a negative power of 10. In both cases, the power that 10 is raised to is called the **exponent** of the **base number** 10. (If this causes you to hesitate, turn to Section A.3 in Appendix A and see that explanation, plus Examples A.1 through A.4.) We have

$$10^{11} \leftarrow \text{Exponent}$$
$$\uparrow\!\rule{1cm}{0.4pt}\ \text{Base number}$$

Here's how to use the shorthand method: To express a number in **scientific notation,** write the number with only one significant figure to the left of the decimal point. Multiply it by 10 raised to the number of places you moved the decimal—positive if it was moved to the left, and negative if it was moved to the right.

$$3,\,8\,0\,0 = 3.8 \times 10^{3}$$

$$0.\,0\,0\,6\,2\,5 = 6.25 \times 10^{-3}$$

$$1\,0\,0,\,0\,0\,0,\,0\,0\,0 = 1 \times 10^{8}$$

$$0.\,0\,0\,0\,0\,0\,1 = 1 \times 10^{-7}$$

The number of significant figures is made clear by using scientific notation. For example, if we know the number 500 to three significant figures, we write 5.00×10^{2}. If we know the number 500 to only one significant figure, we write 5×10^{2}. Writing 5.00×10^{2} is the same as writing $\overline{500}$ (three significant figures). Writing 5×10^{2} is the same as writing 500 (one significant figure).

With a little practice, you'll soon be using scientific notation as easily as you use the more standard notation, but do not deny yourself the practice you need!

EXAMPLE 2.7

Write the following numbers in scientific notation, showing the number of significant figures requested.
(a) 5,$\overline{0}$00 (two significant figures)
(b) 48,$\overline{0}$00 (three significant figures)
(c) 4,090,000 (three significant figures)
(d) 0.000087 (two significant figures)

Solution

(a) 5, $\overline{0}$ 0 0 . = 5.0×10^3 (two significant figures)

(b) 4 8, $\overline{0}$ 0 0 . = 4.80×10^4 (three significant figures)

(c) 4, 0 9 0, 0 0 0 . = 4.09×10^6 (three significant figures)

(d) 0 . 0 0 0 0 8 7 = 8.7×10^{-5} (two significant figures)

Practice Exercise 2.7   Write the following numbers in scientific notation, showing the number of significant figures requested.
(a) 4,200 (two significant figures)
(b) 56,$\overline{0}$00 (three significant figures)
(c) 6,023,000 (four significant figures)
(d) 0.00123 (three significant figures)

2.6 Area and Volume

LEARNING GOAL 4

Areas of geometric figures

Suppose you want to know the area of a tennis court. A flat rectangular surface like a tennis court has just *two* dimensions: length and width, abbreviated *l* and *w*. As Table 2.1 shows, the area of a rectangle is equal to its length *l* times its width *w*. So the area of the tennis court would be *l* × *w*.

LEARNING GOAL 5

Volumes of geometric figures

Now suppose you want to find the volume of a cereal carton. A geometric figure like a carton (a rectangular solid) has three dimensions: length *l*, width *w*, and height *h*. As Table 2.1 and Figure 2.3 show, the volume of your carton is *l* × *w* × *h*.

Area is a *measure of the extent of a surface.* It is a two-dimensional measure that is always stated in squared units such as square feet (ft^2). **Volume** is a *measure of the capacity of an object.* It is a three-dimensional measure that is given in cubed units such as cubic feet (ft^3).

Table 2.1 lists formulas for the areas and volumes of several geometric figures. Example 2.8, which follows, shows you how to use them. Then, in

TABLE 2.1

Areas and Volumes of Various Geometric Figures

Areas (note that areas are always given in squared units: in², ft², and so on.)

1. Square	Area = side × side $A = s \times s$	
2. Rectangle	Area = length × width $A = l \times w$	
3. Circle	Area = π × (radius)² $A = \pi \times r^2$ $\pi = 3.14$	
4. Triangle	Area = $\frac{1}{2}$ × base × height $A = \frac{1}{2} \times b \times h$	

Volumes (note that volumes are always given in cubed units: in³, ft³, and so on.)

5. Cube	$V = s \times s \times s$	
6. Carton (rectangular solid)	$V = l \times w \times h$	
7. Cylinder	$V = \pi \times r^2 \times h$	
8. Sphere	$V = \frac{4}{3}\pi \times r^3$	

■ FIGURE 2.3

Formulas for calculating the volume of a cereal carton and a soup can

Volume = length × width × height
$V = l \times w \times h$

Volume = π × (radius)^2 × height
$V = \pi \times r^2 \times h$
$\pi = 3.14$

Practice Exercise 2.8, you are asked to apply a volume formula on your own. You should take the time to work each practice exercise as you come to it. It will help you learn and apply what you have just read, and it will serve as a quick check of your understanding of each topic. Answers to practice exercises are provided in the back of the book.

EXAMPLE 2.8

The soup can shown in Figure 2.3 has a height of 4.00 inches and a radius of 1.00 inch. What is the volume of the can?

Solution

UNDERSTAND THE PROBLEM

Ask yourself what knowledge is required to arrive at a solution. A soup can has the shape of a cylinder, and therefore you need to know how to calculate the volume of a cylinder.

DEVISE A PLAN

Looking at Table 2.1, you find that the formula for the volume of a cylinder is $V = \pi r^2 h$.

Given: $\pi = 3.14$, $r = 1.00$ inch, $h = 4.00$ inches

CARRY OUT THE PLAN

$$V = \pi r^2 h$$

$$= (3.14)(1.00 \text{ in})^2 (4.00 \text{ in})$$

$$= 12.56 \text{ in}^3 \quad \text{or} \quad 12.6 \text{ in}^3 \quad \text{(Three significant figures)}$$

The can holds 12.6 cubic inches of soup.

LOOK BACK

See if the solution makes sense. Estimation can be used to give an approximate answer, telling us if our calculation is in the ballpark. We can say, "π is roughly 3, so 3 times 1 inch times 1 inch times 4 inches is about 12 cubic inches." Our estimate says that the answer should be about 12 cubic inches. Our exact calculation tells us that the answer is 12.6 cubic inches. We are definitely in the ballpark.

Practice Exercise 2.8   A cylinder has a height of 6.00 inches and a radius of 4.00 inches. What is its volume?

2.7   The English System of Measurement

The English system of units, still used in the United States today, has a built-in problem. It is not an orderly system and is therefore difficult to use. We will look at three units in the English system and then discuss this inherent difficulty.

In the English system, the *foot* is the unit of length and is divided into 12 smaller units called *inches*. The inch can be used to measure short distances, and the foot can be used to measure longer distances. To measure still longer distances there are the *yard* (which equals 3 feet) and the *mile* (which equals 5,280 feet).

The unit of weight in the English system is the *pound*, and the pound is divided into 16 smaller units called *ounces*. To measure larger weights, the *ton* may be used. The ton is equal to 2,000 pounds.

In the English system as used in the United States today, the unit of liquid volume is the *quart*. To measure smaller quantities, the *fluid ounce* is used (32 fluid ounces equal 1 quart). Other units of liquid volume are the *pint* (2 pints equal 1 quart) and the gallon (4 quarts equal 1 gallon).

Table 2.2 lists these units. If you examine that table closely, the problem will be evident: there is no systematic relationship among units used to measure the same property. Consider the length unit. To convert inches to yards, it is necessary to know that 12 inches equal 1 foot and that 3 feet

TABLE 2.2

The English System of Measurement (as Used in the United States)

LENGTH	WEIGHT	VOLUME
12 inches = 1 foot	16 ounces = 1 pound	16 fluid ounces = 1 pint
3 feet = 1 yard	2,000 pounds = 1 ton	2 pints = 1 quart
5,280 feet = 1 mile		4 quarts = 1 gallon

equal 1 yard. You cannot easily move from one unit to the other. The units for weight and volume are just as inconvenient to convert. Such problems led to the development of the metric system of measurement.

2.8   The Metric System of Measurement

In the late 1700s, the French decided to change their own system of measurement. To replace it they developed a logical and orderly system called the **metric system.** The advantages of this system led to its adoption in most countries of the world and in all branches of science. The British held out until 1965 and then began a changeover to the metric system.

In countries using the metric system, almost everything is measured in metric units—distances between cities, the weight of a loaf of bread, the size of a sheet of plywood. Nearly all countries have adopted the metric system. In 1975 the United States Congress passed a bill establishing a policy of voluntary conversion to the metric system and creating the U.S. Metric Board. A program of gradual conversion was the goal of this board, whose funding ended in 1982. The Office of Metric Programs, established by the U.S. Department of Commerce, now has the job of promoting the increased use of metric units of measurement by business and industry.

The metric system consists of (1) a set of standard units of measurement for distance, weight, volume, and so on, and (2) a set of prefixes that are used to express larger or smaller multiples of these units. The prefixes represent multiples of 10. This makes the metric system a decimal system of measurement. In 1960 an international group of scientists modified the metric system by adopting a system of units called the *Système International d'Unités* (International System of Units), or the SI system.

In the SI system there are seven basic units. Table 2.3 lists them. Other units, called *derived units*, are composed of combinations of basic units. (For

TABLE 2.3

Basic SI (Metric) Units

QUANTITY	UNIT	SYMBOL
Length	meter	m
Mass	kilogram	kg
Time	second	s
Electric current	ampere	A
Temperature	kelvin	K
Light intensity	candela	cd
Amount of substance	mole	mol

example, the unit for speed, meters per second, is derived from the basic units *meter* and *second*.) The basic units were all carefully defined. Occasionally one of the definitions is modified to make it more precise or more useful. For example, the meter was first defined as one ten-millionth of the distance between the North Pole and the equator along a meridian of the earth. Later, to ensure that 1 meter meant the same thing everywhere, it was redefined as the distance between two scratches on a platinum bar that is kept at exactly the freezing point of water (zero degrees on the Celsius temperature scale) in a vault outside Paris. Still later, as more exacting measuring instruments were developed, the meter was defined as 1,650,763.73 wavelengths in vacuum of the orange-red line of the spectrum of krypton-86. You will learn more about this in Chapter 5.

LEARNING GOAL 6

Conversion of units within the metric system

Table 2.4 lists the more commonly used metric prefixes. To use one, simply "tack it on" to the front of a metric unit. That gives a related unit that is a multiple of the original unit. For example, the prefix *kilo* has a multiplier of 1,000, so a kilometer is equal to 1,000 meters. The prefix *centi* has a multplier of 0.01, so a centigram is equal to 0.01 gram. Conversely, there are 100 centigrams in a gram.

To make converting from one unit to another simpler, we will use a method called the factor-unit method, or dimensional analysis. With this approach, any problem that requires conversion from one unit to another can be set up and solved in a similar manner. We can say that:

Quantity wanted = quantity given × factor unit

When you multiply the quantity given by the proper factor unit, some of the units cancel to give the desired quantity. Let's see how this works. We'll convert meters to centimeters as we begin to understand this method.

TABLE 2.4

Metric Prefixes

PREFIX	SYMBOL	MULTIPLIER
nano	n	0.000000001
micro	μ (Greek mu)	0.000001
milli	m	0.001
centi	c	0.01
deci	d	0.1
deka	da	10
hecto	h	100
kilo	k	1,000
mega	M	1,000,000

$$\text{Quantity wanted} = \text{quantity given} \times \text{factor unit}$$

$$\text{Centimeters} = \cancel{\text{meters}} \times \frac{\text{centimeters}}{\cancel{\text{meter}}}$$

Note that the factor unit expresses a relationship between the quantity wanted and the quantity given. The factor unit is written in such a way that the given units cancel when you multiply the quantity given times the factor unit. Then you're left with the quantity wanted:

$$\text{Centimeters} = \cancel{\text{meters}} \times \frac{\text{centimeters}}{\cancel{\text{meter}}}$$

$$\text{Centimeters} = \text{centimeters}$$

Of course any factor unit can be expressed in two ways. For example, the relationship 1 meter = 100 centimeters can be expressed as

$$\frac{100 \text{ centimeters}}{1 \text{ meter}} \quad \text{or} \quad \frac{1 \text{ meter}}{100 \text{ centimeters}}$$

You choose the factor unit that will make the proper terms cancel.

We'll now look at a variety of examples in which the factor-unit method is helpful. As you proceed through basic chemistry, you'll find this method useful in many cases. (Appendix A, especially Examples A.5 through A.9, provides many more examples that will help you understand this strategy.)

EXAMPLE 2.9

Change 40 meters to centimeters.

Solution

UNDERSTAND THE PROBLEM

This example requires conversion from one unit to another. We ask "How many centimeters are there in one meter?" Table 2.4 shows that *centi* means 0.01, so there are 100 centimeters in 1 meter.

DEVISE A PLAN

Use the factor-unit method to convert.

CARRY OUT THE PLAN

Remember that there are 100 centimeters in 1 meter. This can be written as

$$\frac{100 \text{ centimeters}}{1 \text{ meter}}$$

which reads "100 centimeters per meter"; the division line reads "per." Next we put in the numbers.

$$? \text{ centimeters} = 40 \cancel{\text{meters}} \times \frac{100 \text{ centimeters}}{\cancel{\text{meter}}} = 4{,}000 \text{ cm}$$

There are 4,000 centimeters in the 40 meters.

LOOK BACK

We want to see if this solution makes sense. We determined that one meter equals 100 centimeters. Does it make sense that 40 meters would be 40 times larger, or 4,000 centimeters? Yes, this does make sense.

Practice Exercise 2.9   Change 35 meters to centimeters.

EXAMPLE 2.10

Change $43\overline{0}$ milligrams to grams.

Solution

UNDERSTAND THE PROBLEM

You will see that it is similar to Example 2.9. Therefore we can use the same strategy to devise a plan.

DEVISE A PLAN

Again we use the factor-unit method. (This time we will use abbreviations.) We ask, "How many milligrams are there in 1 g?" Table 2.4 shows that *milli* means 0.001, so 1 g equals 1,000 mg. This means there is

$$\frac{1 \text{ g}}{1,000 \text{ mg}} \qquad \text{(which reads 1 g per 1,000 mg)}$$

CARRY OUT THE PLAN

We say that

$$g = \cancel{mg} \times \frac{g}{\cancel{mg}}$$

$$? \text{ g} = 43\overline{0} \cancel{mg} \times \frac{1 \text{ g}}{1,000 \cancel{mg}} = 0.430 \text{ g}$$

There is 0.430 g in $43\overline{0}$ mg.

LOOK BACK

We ask if this solution makes sense. If 1,000 mg equals 1 g, how much of a gram would $43\overline{0}$ mg equal? Using estimation we can say that $43\overline{0}$ mg is a little less than half a gram, therefore our answer 0.430 g, which is a little less than half a gram is in the ballpark, and makes sense.

Practice Exercise 2.10   Change $35\overline{0}$ mg to grams.

EXAMPLE 2.11

How many centimeters are there in 8 meters?

Solution

$$cm = \cancel{m} \times \frac{cm}{\cancel{m}}$$

$$? \, cm = 8 \cancel{m} \times \frac{100 \, cm}{1 \cancel{m}} = 800 \, cm$$

There are 800 cm in 8 m.

Practice Exercise 2.11   How many centimeters are there in 18 meters?

EXAMPLE 2.12

Convert 580 millimeters to centimeters.

Solution

$$cm = \cancel{mm} \times \frac{cm}{\cancel{mm}}$$

$$? \, cm = 580 \cancel{mm} \times \frac{1 \, cm}{10 \cancel{mm}} = 58 \, cm$$

There are 58 cm in 580 mm.

If you wonder where the term

$$\frac{1 \, cm}{10 \, mm}$$

came from, look at a metric ruler, and you'll see that there are 10 mm in 1 cm.

Practice Exercise 2.12   Convert 660 mm to centimeters.

EXAMPLE 2.13

Convert 75 millimeters to its corresponding length in   (a) centimeters, (b) meters,   (c) kilometers.

Solution

(a) $? \, cm = 75 \cancel{mm} \times \frac{1 \, cm}{10 \cancel{mm}} = 7.5 \, cm$

(b) $? \, m = 7.5 \cancel{cm} \times \frac{1 \, m}{100 \cancel{cm}} = 0.075 \, m$

(c) $? \, km = 0.075 \cancel{m} \times \frac{1 \, km}{1{,}000 \cancel{m}} = 0.000075 \, km$

Practice Exercise 2.13   Convert 55 millimeters to the corresponding length in   (a) centimeters,   (b) meters,   (c) kilometers.

EXAMPLE 2.14

Convert 2.3 kilograms to the corresponding mass in   (a) grams,   (b) decigrams,   (c) centigrams,   (d) milligrams.

Solution

(a) $? \text{ g} = 2.3 \text{ kg} \times \dfrac{1,000 \text{ g}}{1 \text{ kg}} = 2,300 \text{ g}$

(b) $? \text{ dg} = 2,300 \text{ g} \times \dfrac{10 \text{ dg}}{1 \text{ g}} = 23,000 \text{ dg}$

(c) $? \text{ cg} = 23,000 \text{ dg} \times \dfrac{10 \text{ cg}}{1 \text{ dg}} = 230,000 \text{ cg}$

(d) $? \text{ mg} = 230,000 \text{ cg} \times \dfrac{10 \text{ mg}}{1 \text{ cg}} = 2,300,000 \text{ mg}$

Practice Exercise 2.14   Convert 4.4 kilograms to the corresponding mass in   (a) grams,   (b) decigrams,   (c) centigrams,   (d) milligrams.

You should now realize that conversion between units of the metric system is not difficult. If you want to convert metric units into English units, consult Table 2.5. (More precise conversion factors can be found in Appendix B, Table B.3.)

LEARNING GOAL 7

Conversion of metric to English units

Although it will always be convenient to be able to convert between systems, it is even more important to be able to associate metric measurements with commonly used terms. Start now to "think metric," for only in this way will the metric system become meaningful to you. The following example offers a starting point.

EXAMPLE 2.15

A new car is described in an advertisement as having an overall length of 5.50 meters and an overall width of 1.50 meters. Find the dimensions of the car in feet.

Solution

Consulting Table 2.5, we see that to convert meters to feet it is necessary to multiply by 3.28. In other words, 1 meter = 3.28 feet. Therefore the length and width can be determined as follows:

Length: $? \text{ ft} = 5.50 \text{ m} \times \dfrac{3.28 \text{ ft}}{1 \text{ m}} = 18.0 \text{ ft}$

Width: $? \text{ ft} = 1.50 \text{ m} \times \dfrac{3.28 \text{ ft}}{1 \text{ m}} = 4.92 \text{ ft}$

TABLE 2.5

Conversion of Units

TO CONVERT	INTO	MULTIPLY BY
Length		
inches	centimeters	2.54 cm/in
centimeters	inches	0.39 in/cm
feet	meters	0.30 m/ft
meters	feet	3.28 ft/m
Weight (mass; see Section 2.9)		
ounces	grams	28.35 g/oz
grams	ounces	0.035 oz/g
pounds	grams	454 g/lb
grams	pounds	0.0022 lb/g
Volume		
liters	quarts	1.06 qt/L
quarts	liters	0.946 liter/qt

Practice Exercise 2.15   The dimensions of a new American car are as follows: overall length, 4.20 meters; overall width, 1.25 meters. Is this a compact car? Find its dimensions in feet.

LEARNING GOAL 8
Conversion of English to metric units

Before we work the next example, we need to discuss the SI unit of volume. This unit, a derived unit, has the dimension (length)^3—for example, m^3 or cm^3. However, before the SI system was adopted, the metric unit of volume was the liter. One liter (1 L) is equal in volume to 1,000 cm^3, and 1 mL = 1 cm^3. For some measurements the liter is simpler to use than the SI units, and many people continue to use it in such cases. In this book, we shall use whichever unit best fits a particular situation. In the next example we work with liters, which are used in selling gasoline in all countries using the metric system.

EXAMPLE 2.16

An American family visiting Mexico wants to fill its car with 15 gallons of gasoline. Mexico uses the metric system. How many liters of gasoline should the visitors ask the attendant to put in their tank?

Solution
We know that 15 gallons of gasoline are the same as 6$\overline{0}$ quarts (4 quarts = 1 gallon). Consulting Table 2.5, we see that to convert quarts

to liters we multiply by 0.946 liter/quart. Therefore we may solve the problem as follows:

$$? \text{ liters} = 6\overline{0} \text{ quarts} \times \frac{0.946 \text{ liter}}{1 \text{ quart}} = 57 \text{ liters}$$

Practice Exercise 2.16   To conserve water, the Simon family installed a reducing valve and a water-metering device in their shower. If Mr. Simon uses $2\overline{0}$ gallons of water in his shower, and Mrs. Simon uses 18 gallons in hers, how many liters of water do the Simons use?

Before we leave the topic of measurment and the factor-unit method, we should point out that sometimes we may want to use two (or more) factor units in solving a problem. Let's demonstrate how this works.

EXAMPLE 2.17

A laboratory bench measures exactly 3.00 yards in length. How many inches is this?

Solution

We know that there are exactly 3.00 feet in 1.00 yard. We also know that there are exactly 12.0 inches in 1.00 foot. Therefore, we may use these *two* factor units to solve the problem as follows:

$$? \text{ in} = (3.00 \text{ yd}) \frac{(3.00 \text{ ft})}{1.00 \text{ yd}} \frac{(12.0 \text{ in})}{1 \text{ ft}} = 108 \text{ in}$$

Notice how the various units canceled.

Practice Exercise 2.17   Given that a solar day has exactly 24.0 hours, use the following factor units to compute the number of seconds in a day:

$$1.00 \text{ hour} = 60.0 \text{ minutes}$$

$$1.00 \text{ minute} = 60.0 \text{ seconds}$$

It is valuable to know that a liter is just a little more than a quart. A meter is slightly longer than a yard (it is 1 yard and 3.375 inches). When you are dealing with weights, it is helpful to remember that a nickel weighs about 5 g. If you can keep these three approximations in mind, you'll find it easier to work in the metric system.

2.9   Mass and Weight

LEARNING GOAL 9

Mass and weight

The mass of an object and its weight are often thought of as being the same, so the words *mass* and *weight* are frequently used interchangeably. This is incorrect because, by definition, the two terms have different meanings. **Mass** is *the quantity of matter in an object*, whereas **weight** is *the gravitational force that attracts an object.*

The definition of mass implies that your body's *mass* is constant no matter where you are. Your body's *weight*, on the other hand, varies from planet to planet and even varies slightly at different places on the earth. An astronaut who weighs 180 pounds on earth weighs only 30 pounds on the moon (which has only one-sixth of the gravitational pull of the earth) and has no weight at all in outer space. But *the astronaut's mass is the same in all places.* We can define weight mathematically as

$$\text{Weight} = \text{mass} \times \text{gravity} \qquad \text{or} \qquad W = m \times g$$

The force of gravity varies from planet to planet, so the weight of an object must vary too.

2.10   Density

Which is heavier: glass, iron, or wood from an oak tree? Naturally it depends on the size of each piece. However, what if all three were the same size? In other words, what if we had cubes of glass, iron, and oak wood, each with a volume of 1 cm³? (See Figure 2.4.) Suppose we weighed each cube to determine which was the heaviest. We would find that 1 cm³ of iron weighs 7.9 g, 1 cm³ of glass weighs 2.4 g, and 1 cm³ of oak wood weighs 0.6 g. We would conclude that, *for a particular volume,* iron has the greatest mass.

The concept of density enables us to express this relationship conveniently. **Density** can be defined as *the mass per unit volume of a substance or object* (Table 2.6). The density of a substance can be determined by using the formula

$$D = \frac{m}{V}$$

where D is the substance's density, m is its mass, and V is its volume. If the mass is measured in grams and the volume in cubic centimeters, then the unit of density is

$$\frac{\text{grams}}{\text{cubic centimeters}} \qquad \text{or} \qquad \frac{\text{g}}{\text{cm}^3}$$

(See Appendix A for a review of solving algebraic and word equations.)

■ **FIGURE 2.4**

Three cubes of equal volume

| Iron cube | Glass cube | Oak wood cube |

TABLE 2.6

Densities of Some Common Solids and Liquids

SUBSTANCE	DENSITY AT ROOM TEMPERATURE (g/cm³)
Ethyl alcohol	0.789
Water	1.00
Glycerol (glycerine)	1.261
Sucrose (table sugar)	1.58
Sodium chloride (table salt)	2.17
Aluminum metal	2.70
Iron metal	7.86
Copper metal	8.92
Mercury	13.59
Gold	18.88

EXAMPLE 2.18

A block of iron that is 5.0 cm long, 3.0 cm high, and 4.0 cm wide weighs 474 g. What is the density of iron?

Solution

We first calculate the volume of the block.

$$\text{Volume} = \text{length} \times \text{width} \times \text{height}$$

$$V = 5.0 \text{ cm} \times 4.0 \text{ cm} \times 3.0 \text{ cm} = 6\overline{0} \text{ cm}^3$$

We now know that $V = 6\overline{0} \text{ cm}^3$ and $m = 474$ g, so all we have to do is solve the density formula for D.

$$D = \frac{m}{V} = \frac{474 \text{ g}}{6\overline{0} \text{ cm}^3} = 7.9 \frac{\text{g}}{\text{cm}^3}$$

Practice Exercise 2.18 A block of aluminum is 2.0 cm long, 3.0 cm high, and 5.0 cm wide, and it weighs 81.0 g. What is the density of the aluminum?

EXAMPLE 2.19

Suppose you are told that 400 g of alcohol occupy a volume of 500 mL. What is the density of alcohol?

Solution

In this example we are given the mass and the volume of the substance whose density we are asked to find. Note that the volume of the alcohol is given in milliliters (mL). But *1 milliliter is equal to 1 cubic centi-meter*. That is, 1 mL = 0.001 L = 1 cm^3. So it is acceptable to inter-change the units mL and cm^3. Now let's solve this problem.

$$D = \frac{m}{V} = \frac{400 \text{ g}}{500 \text{ cm}^3} = 0.8 \frac{\text{g}}{\text{cm}^3}$$

Practice Exercise 2.19   Suppose you are told that 880 g of a clear, colorless liquid occupy a volume of 110 mL. What is the density of this liquid?

EXAMPLE 2.20

Determine which is more dense, carbon tetrachloride or chloroform, from the following information:
(a) 16 g of carbon tetrachloride occupy a volume of $1\bar{0}$ mL.
(b) $3\bar{0}$ g of chloroform occupy a volume of $2\bar{0}$ mL.

Solution
We simply calculate the density of each liquid and see which is greater:

$$D \text{ (carbon tetrachloride)} = \frac{m}{V} = \frac{16 \text{ g}}{1\bar{0} \text{ cm}^3} = 1.6 \frac{\text{g}}{\text{cm}^3}$$

$$D \text{ (chloroform)} = \frac{m}{V} = \frac{3\bar{0} \text{ g}}{2\bar{0} \text{ cm}^3} = 1.5 \frac{\text{g}}{\text{cm}^3}$$

The carbon tetrachloride is the more dense liquid.

Practice Exercise 2.20   Determine which is the more dense, liquid A or liquid B, given the following information:
(a) 55.0 g of liquid A occupy a volume of 10.0 mL.
(b) 25.0 g of liquid B occupy a volume of 12.5 mL.

The preceding calculations are straightforward uses of the density for-mula. But what if you were given the density of a material and its mass? Could you calculate its volume? Here's how to do it.

EXAMPLE 2.21

The density of alcohol is 0.8 g/cm^3. Calculate the volume of 1.6 kg of al-cohol.

Solution

First we have to solve the density formula,

$$D = \frac{m}{V}$$

for V. To do so, we multiply both sides of the equation by V. This gives us $D \times V = m$. Now we divide both sides of the equation by D. This gives us the formula we want:

$$V = \frac{m}{D}$$

We are given that $D = 0.8$ g/cm^3 and $m = 1.6$ kg, or 1,600 g. We substitute these numbers into the formula and solve for V.

$$V = \frac{m}{D} = \frac{1,600 \text{ g}}{0.8 \text{ g/cm}^3} = 2,000 \text{ cm}^3 \text{ (or 2,000 mL)}$$

Practice Exercise 2.21   The density of chloroform is 1.5 g/cm^3. Calculate the volume in cm^3 of 2.0 kg of chloroform.

One important fact about density is that most substances *expand when heated*. Therefore, when a certain mass of a substance is hot, it occupies a larger volume than it does when it is cool. This means that the density of the substance decreases as it is warmed. Think about this for a moment, and make sure you understand why it is true. Most densities reported in chemical references, such as the *Handbook of Chemistry and Physics*, are the densities at 20°C, which is about room temperature. Water is an important exception.

2.11 Temperature Scales and Heat

When we heat a substance, we add a quantity of heat to that substance. We can then use a thermometer to measure the **temperature** of the substance. The thermometer measures the *intensity* of the heat; it tells us nothing about the quantity of heat that has entered the substance. (We shall discuss heat quantity in Chapter 12.)

Three different temperature scales are commonly used in measuring heat intensity. Two of these—the Fahrenheit and Celsius scales—are in general use. The third, the Kelvin scale, is used mainly by scientists.

The **Fahrenheit temperature scale** was devised by Gabriel Daniel Fahrenheit, a German scientist, in 1724. On this scale (see Figure 2.5), the freezing point of pure water is at 32 degrees (32°F), and the boiling point of water is at 212 degrees (212°F). There are thus 180 Fahrenheit degrees between the freezing point and the boiling point of water.

The **Celsius temperature scale** was devised in 1742 by Anders Celsius, a Swedish astronomer. His objective was to develop an easier-to-use tempera-

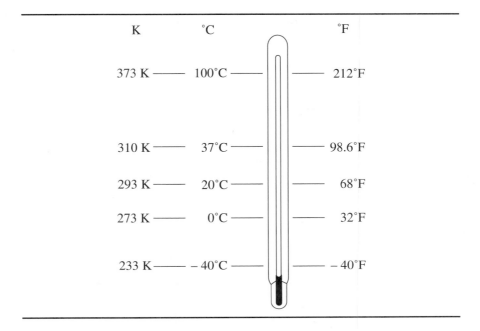

A comparison of Fahrenheit
and Celsius scales

ture scale; he did so by assigning a nice, round 100 Celsius degrees between the freezing and boiling points of pure water. On the Celsius scale (Figure 2.5), the freezing point of water is at zero degrees (0°C), and the boiling point of water is at 100 degrees (100°C). (The Celsius scale is also sometimes referred to as the centigrade scale.)

The **Kelvin temperature scale** is an *absolute* temperature scale. That is, its zero point (0 K) is at absolute zero, the lowest possible temperature theoretically attainable. The divisions of the Kelvin scale are the same size as Celsius degrees, but they are called kelvins (abbreviated K) rather than degrees. Chapter 13 discusses the Kelvin scale in more detail.

2.12   Converting Celsius and Fahrenheit Degrees

LEARNING GOAL 11
**Celsius and Fahrenheit
temperature conversions**

On the Celsius scale, there are 100 divisions between the freezing point and the boiling point of water. On the Fahrenheit scale, there are 180 divisions between these two points. Therefore 100 Celsius degrees cover the same range as 180 Fahrenheit degrees, so that 1 Celsius degree = 1.8 Fahrenheit degrees. Moreover, 0°C is equivalent to 32°F. Formulas for converting from a temperature on one scale to a temperature on the other are based on these facts. The formulas are

$$°F = (1.8 \times °C) + 32 \quad \text{and} \quad °C = \frac{°F - 32}{1.8}$$

Let us use these formulas to convert from Fahrenheit to Celsius degrees and from Celsius to Fahrenheit degrees. When using significant figures remember that exactly defined quantities have an unlimited number of significant figures.

EXAMPLE 2.22

Convert 122°F to degrees Celsius.

Solution

Substitute 122°F into the conversion formula for changing °F to °C.

$$°C = \frac{°F - 32.0}{1.8} = \frac{122 - 32.0}{1.8} = \frac{9\overline{0}}{1.8} = 5\overline{0}$$

$$122°F = 15\overline{0}°C$$

Practice Exercise 2.22   Convert 244°F to degrees Celsius.

EXAMPLE 2.23

Convert 10$\overline{0}$°C to degrees Fahrenheit.

Solution

Substitute 10$\overline{0}$°C into the conversion formula for changing °C to °F.

$$°F = (1.8 \times °C) + 32.0 = (1.8 \times 10\overline{0}) + 32.0 = 18\overline{0} + 32.0 = 212$$

$$10\overline{0}°C = 212°F$$

Practice Exercise 2.23   Convert 15$\overline{0}$°C to degrees Fahrenheit.

SUMMARY

Measurements represent the dimensions, quantity, or capacity of things. Scientists record and analyze measured data as they do research to attain knowledge while engaging in the problem-solving process. Chemistry students, too, engage in problem solving. George Polya's four-step method facilitates this process. The process includes understanding the problem, devising a plan for solution, carrying out the plan, and looking back at the results. Estimation techniques can help students assess the reasonableness of a solution.

In working with measurements, significant figures are used to convey the precision of information. The results of measurements and computations should always be rounded to the appropriate number of significant figures.

Scientific notation, in powers of 10, may be used to indicate significant figures; it also simplifies computations involving very large or very small numbers.

The metric (or SI) system is a system of measurement used in most parts of the world and in all scientific work. It consists of seven basic units and a number of derived units, along with a set of prefixes that indicate multiples of these units. The unit of length is the meter, the unit of mass is the kilogram, and the unit of volume is a derived unit (cubic centimeters or, sometimes, liters). The corresponding units in the English system, as used in the United States, are the foot (length), pound (weight), and quart (volume). The factor-unit method may be used to convert from unit to unit within either system or from system to system. Mass is a measure of the quantity of matter, and weight is a measure of the gravitational force on an object. The density of an object is its mass divided by its volume. Temperatures, which indicate heat intensity, are measured on the Fahrenheit, Celsius, or Kelvin temperature scale and may be converted from one scale to another.

▮▮▮ KEY TERMS

We have listed the major terms that have been defined in this chapter. Be sure that you are familiar with each of these terms. If you need to refresh your memory, refer to the indicated section or use the glossary at the back of the book.

accuracy **(2.4)**	mass **(2.9)**
area **(2.6)**	metric system **(2.8)**
base number **(2.5)**	precision **(2.4)**
Celsius temperature scale **(2.11)**	rounding off **(2.4)**
density **(2.10)**	scientific notation **(2.5)**
exponent **(2.5)**	significant figures **(2.4)**
Fahrenheit temperature scale **(2.11)**	temperature **(2.11)**
Kelvin temperature scale **(2.11)**	volume **(2.6)**
	weight **(2.9)**

CAREER SKETCH

Aerospace Careers

The aerospace industry needs professionals with strong science backgrounds every day. For example, did you ever wonder who designed the material that protects the metallic and nonmetallic surfaces of

Formulation chemists are involved in improving the function of major aircraft components.
(Bruno de Hogues/Tony Stone Images)

aircraft? Or who made sure that the fuel tanks of the airplane were sealed? It was probably the work of a *creative formulation chemist.*

Working for an aerospace company, a creative formulation chemist may be responsible for creating sealants that protect the mating surfaces of aircraft structures to inhibit corrosion, erosion, or damage from ultraviolet rays. These sealants must also maintain cabin pressurization and withstand dynamic stress, weathering, and other environmental conditions. Or the chemist might be asked to design a sealant technology that will be used to seal the fuel tanks of aircraft flying worldwide. Such a substance would have to be chemically engineered to be light in weight, to withstand high temperatures, and to inhibit corrosion.

Another exciting career in the aerospace industry relates to the formulation of coatings on aircraft. A *coatings paint research engineer* is responsible for providing insight and expertise in areas of paint formulation, composition, structure, and property relationships with paint systems. In addition, this person may participate in setting the future development of aerospace coatings and identify and solve aerospace paint problems related to formulation. A coatings paint research engineer may direct projects related to paint application, evaluation of performance, and assessments of new test technologies, which may include conducting scientific experiments.

CHEMICAL FRONTIERS

A Better Glass of Wine*

When one thinks of the wine-making process, grape crushing comes to mind. After all, this has been the basic process for thousands of years. Recently, though, chemists have taken a close look at the wine production process.

Consider the work of chemist Elizabeth J. Waters at the Australian Wine Research Institute. Waters has been working on the problem of hazing—a cloudy suspension that forms in white wine when it is overheated during shipping. The hazing is considered so unsightly that people often refuse to drink the wine. Waters recognized that the problem is caused by grape proteins that precipitate at elevated temperatures. To solve the problem, she used a haze protective factor in the wine.

A more serious problem occurs when certain wines are stored in a warm place. Riesling, for example, forms the odor of kerosene. Peter Winterhalter, a chemist at the University of Würzburg in Germany, discovered that this odor is the result of a compound found in Riesling grapes and leaves. Winterhalter found that he can chemically alter the grapes and leaves, causing them to form a less-noxious fume.

Other studies are underway on how wine grapes differ, what compounds impact their flavor, and when in the vine's life cycle the flavor compounds are produced. Answers to these questions will ultimately lead to a better glass of wine.

*Adapted from *Science News*, Oct 23, 1993, pp. 264–265

SELF-TEST EXERCISES

LEARNING GOAL 1: *Significant Figures*

◄ *1.* How many significant figures are there in each of the following measured quantities?
(a) 5,000.0 cm (b) 300 m (c) 0.204 dL
(d) 8.000 ft (e) 340,000.0 mL (f) 5.0×10^3 g
(g) 8×10^2 yd (h) 6,$\overline{000}$ mm (i) 6,$\overline{000}$ mm

◄ 2. How many significant figures are there in each of the following measured quantities?
(a) 412,$\overline{000}$,000 mm (b) 4.00×10^3 km
(c) 0.8009 mi (d) 0.00036 km
(e) 5.023×10^{-6} kg (f) 703,$\overline{000}$ in (g) 0.009 cL
(h) 5.0000×10^3 cm

3. Determine the number of significant figures in each of the following numbers: (a) 3,$\overline{000}$,000
(b) 3,$\overline{000}$,000 (c) 3×10^6 (d) 3.00×10^6
(e) 3.000000×10^6 (f) 0.0000305 (g) 0.100054
(h) 6.720×10^{-8} (i) 305,075 (j) 35.00

4. Determine the number of significant figures in each of the following numbers: (a) 4,$\overline{000}$,000
(b) 4,$\overline{000}$,000 (c) 2×10^5 (d) 0.00123
(e) 4.02×10^{-5} (f) 23.00 (g) 2×10^{-1}
(h) 750

LEARNING GOAL 2: *Calculations with Significant Figures*

5. Find the perimeter of a four-sided object whose sides have the following dimensions: 4.234 cm, 3.8

cm, 5.67 cm, and 4.00 cm. How many significant figures are there in your answer?

6. Find the perimeter of a triangle whose sides are 3.23 cm, 5.006 cm, and 3 cm. How many significant figures are there in your answer?

7. Find the perimeter of a triangle whose sides are 4.00 cm, 5.000 cm, and 6.0 cm. How many significant figures are in your answer?

8. Find the area of a square whose sides are 4.2 cm. How many significant figures are there in your answer?

◀ **9.** Find the perimeter of a four-sided object whose sides are 7.382 cm, 3.95 cm, 5.4342 cm, and 3.83 cm. (Use the proper number of significant figures.)

10. Find the area of a rectangle whose sides are 10.12 cm and 10.25 cm. (Use the proper number of significant figures.)

11. Find the area of a rectangle whose sides are 20.62 cm and 10.4 cm. How many significant figures are there in your answer?

◀ 12. Find the area of a square whose sides are 2.1 cm each. (Use the proper number of significant figures.)

13. State the number of significant figures in each number given in Exercise 19.

14. Find the perimeter of a square whose sides are 2.1 cm each. (Use the proper number of significant figures.)

LEARNING GOAL 3: **Scientific Notation**

15. Express the following numbers in scientific notation: (a) 600 (b) 6$\overline{0}$0 (c) 600.0 (d) 320,000,000,000 (e) 0.002 (f) 0.0003007 (g) 0.00000015

16. Express the following numbers in scientific notation: (a) 900,000 (b) 45,000 (c) 2,970 (d) 2,546,000 (e) 0.00006 (f) 0.0122 (g) 0.056 (h) 20.0040

17. Express the following numbers in scientific notation: (a) 10,581 (b) 0.00205 (c) 1,000,000 (d) 802

◀ 18. Express the following numbers in scientific notation: (a) 45,000,000 (b) 4$\overline{0}$0 (c) 40$\overline{0}$ (d) 400.0 (e) 425,$\overline{000}$

◀ **19.** Express the following numbers in scientific notation: (a) 850,000,000 (b) 0.00000607 (c) 6,308,000 (d) 0.06005 (e) 500 (f) 50$\overline{0}$

(g) 5$\overline{00}$ (h) 500.0 (i) 23,$\overline{000}$,000 (j) 0.0000000930

20. Express the following numbers in scientific notation: (a) 0.123 (b) 0.006 (c) 0.00601

LEARNING GOAL 4: **Areas of Geometric Figures**

◀ **21.** Find the area of a rectangular room that measures 8.00 m by 3.00 m. Report your answer to the proper number of significant figures. If the height of the room is 4.00 m, what is the volume of the room?

22. Find the area of a rectangular room that measures 5.0 m by 4.0 m. Use the correct number of significant figures in reporting your answer.

23. Determine the area of this page in (a) square centimeters, (b) square inches.

24. Find the area of this page in (a) square meters, (b) square feet.

25. Find the area occupied by a rectangular house that measures 3$\overline{0}$ m by 2$\overline{0}$ m.

26. Find the area of a table that measures 1.5 meters by 2.0 meters.

LEARNING GOAL 5: **Volumes of Geometric Figures**

27. Find the volume of a milk carton that measures 50.0 mm long, 6.00 cm wide, and 1.50 dm high. Report your answer in cm^3 to the proper number of significant figures.

28. Find the volume of a box that measures 5.0 m long, 4.0 m wide, and 2.0 m high.

29. Determine the volume of a cylinder-shaped can that measures 60.0 cm high and has a radius of 15.0 cm.

30. Determine the volume of a metal cylinder that is 25 cm high and has a radius of 3.0 cm.

◀ **31.** A residential swimming pool in the shape of a rectangle measures 16.0 feet by 32.0 feet. The average height of such a pool is 5.00 feet. What is the volume of this pool in ft^3? If 1.00 foot = 30.5 cm, what is the volume of the pool in cm^3? Report your answers to the proper number of significant figures.

32. There are 100 cm in 1 m. How many cubic centimeters (cm^3) are there in 2 cubic meters (2 m^3)?

33. If the pool in Exercise 31 is filled with water, how many liters of water can it hold? Knowing that

1.00 L = 1.06 quarts and that 4.00 quarts = 1.00 gallons, how many gallons of water can the pool hold? Report all answers to the proper number of significant figures.

◄ 34. Find the volume of a cylinder-shaped jar of applesauce that measures $1\overline{0}$ cm high and has a radius of 5.0 cm.

35. There are 12 inches in 1 foot. From this information, calculate the number of cubic inches in 1 cubic foot.

36. A cookie jar has the dimensions $4\overline{0}$ cm by 25 cm by 15 cm. Calculate its volume.

LEARNING GOAL 6: *Conversion of Units Within the Metric System*

37. Convert 0.25 kg to　(a) grams,　(b) decigrams,　(c) milligrams,　(d) micrograms.

38. Convert 4.17 kg to　(a) grams,　(b) decigrams,　(c) milligrams,　(d) micrograms.

39. Convert 3.1 m to　(a) decimeters,　(b) centimeters,　(c) millimeters.

40. Convert 35,000 mm to　(a) centimeters,　(b) decimeters,　(c) meters,　(d) kilometers.

41. Convert 149 mm to　(a) centimeters,　(b) meters,　(c) kilometers.

42. Convert 5.5 m to　(a) decimeters,　(b) centimeters,　(c) millimeters.

◄ **43.** Convert 7.850 m to　(a) millimeters,　(b) centimeters,　(c) decimeters,　(d) kilometers.

44. Convert 125 mm to　(a) centimeters,　(b) meters,　(c) kilometers.

45. Convert 0.34 km to　(a) meters,　(b) decimeters,　(c) centimeters,　(d) millimeters.

46. Convert 1.234 m to　(a) millimeters,　(b) centimeters,　(c) decimeters,　(d) kilometers.

47. Convert 2,185 mg to　(a) centigrams,　(b) decigrams　(c) grams,　(d) kilograms.

48. Convert 7.5 km to　(a) meters,　(b) decimeters,　(c) centimeters.

49. Convert $3,5\overline{0}0$ mL to　(a) liters,　(b) deciliters.
50. Convert $1,5\overline{0}0$ mL to　(a) liters,　(b) deciliters.

51. There are $1\overline{0}0$ cm in 1 meter. From this information, calculate the number of cubic centimeters (cm^3) in 1 cubic meter (m^3).

52. Convert $25,55\overline{0}$ mg to　(a) centigrams,　(b) decigrams,　(c) grams,　(d) kilograms.

LEARNING GOAL 7: *Conversion of Metric to English Units*

53. Express 25.0 m in　(a) feet,　(b) inches.
54. Express 100.0 m in　(a) feet,　(b) inches.

55. Express 1.2 L in　(a) gallons,　(b) quarts.
56. Express 10.0 L in　(a) gallons,　(b) quarts.

57. Express 5.00 g in　(a) pounds,　(b) ounces.
58. Express 100.0 kg in　(a) pounds,　(b) tons.

◄ **59.** Without looking at a conversion table, determine the number of cubic feet in a cubic meter.

◄ 60. Express $25\overline{0}$ m in　(a) yards,　(b) feet.

61. Express $1\overline{0}0$ m in　(a) yards,　(b) feet.

62. Express 2.50 g in pounds.

63. There are 3.28 ft in one meter (1.00 m). From this information, calculate the number of square feet (ft^2) in one square meter (m^2).

64. Express 50.0 L in gallons.

LEARNING GOAL 8: *Conversion of English to Metric Units*

65. Convert your height from feet and inches to meters.

66. Express $2\overline{0}$ gal in　(a) liters,　(b) milliliters.

67. Convert your weight from pounds to kilograms.

68. Express 100.0 lb in　(a) kilograms,　(b) grams.

◄ **69.** Express $1\overline{0}$ ft in　(a) meters,　(b) centimeters,　(c) millimeters.

70. Express 5.0 ft in　(a) meters,　(b) centimeters.

71. Express 6.00 ft in　(a) meters,　(b) centimeters.
72. Express 25 ft in　(a) meters,　(b) centimeters.

73. Express $1\overline{0}0$ yd in　(a) meters,　(b) centimeters.
74. Express 15 gal in　(a) liters,　(b) milliliters.

75. Express 5.00 gal in　(a) liters,　(b) milliliters.
◄ 76. Express $22\overline{0}$ lb in　(a) kilograms,　(b) grams.

LEARNING GOALS 9 AND 10: *Calculation of Density, Mass, and Volume*

77. Determine the density of this textbook. You'll have to weigh it and determine its volume. Report your answer in grams per cubic centimeter.

78. A solid cube is 6.00 cm on each side and has a mass of 0.583 kg. What is its density in g/cm³?

79. A cube is 5.00 cm on each side and has a mass of 600.0 g. What is its density?

80. A block of aluminum with a density of 2.70 g/cm³ weighs 274.5 g. What is the volume of the block?

81. A block of aluminum with a density of 2.7 g/cm³ and a mass of 549 g. What is the volume of the block?

◀ 82. The element barium has a density of 3.5 g/cm³. What would be the mass of a rectangular block of barium with the dimensions 2.0 cm × 3.0 cm × 4.0 cm?

◀ **83.** The element barium, which is a soft, silvery-white metal, has a density of 3.5 g/cm³. What would be the mass of a rectangular block of barium with the dimensions 1.0 cm × 3.0 cm × 5.0 cm?

84. Calculate the mass of a cube that has a density of 2.50 g/cm³ and is 1.5 cm on each side. Report your result to the proper number of significant figures.

***85.** A spherical balloon is filled with helium, a gaseous element. Helium has a density of 0.177 g/L. The balloon has a radius of 3.0 cm. What is the mass of the helium in the balloon? (*Hint:* Volume of a sphere = $\frac{4}{3}\pi r^3$.)

86. A cube has a density of 0.250 g/cm³ and a mass of 15.0 g. What is the volume of this cube?

87. A 1.00-L container holds a block of cesium that is 1.00 cm × 2.00 cm × 3.00 cm and whose density is 1.90 g/cm³; 14.0 g of iron, whose density is 7.86 g/cm³; and 0.500 L of mercury, whose density is 13.6 g/cm³. The rest of the container is filled with air, whose density is 1.18×10^{-3} g/cm³. Calculate the *average* density of the contents of the container. *Hint:* To calculate average density, use the formula

$$\text{Average density} = \frac{m_a + m_b + \cdots}{V_a + V_b + \cdots}$$

88. A small metal sphere has a mass of 75.0 g. The sphere is placed in a graduated cylinder containing water. The water level in the cylinder changes from 10.0 cm³ to 20.0 cm³ when the sphere is submerged. What is the density of the sphere?

***89.** A fish tank whose dimensions are 1.50 ft × 1.00 ft × 0.500 ft contains six fish. Fish one and two weigh 2.00 g each, fish three weighs 2.50 g,

and the combined weight of fish four, five, and six is 9.80 g. The bottom of the fish tank is covered with gravel, which occupies one-eighth of the volume of the tank. The density of this gravel is 3.00 g/cm³. The rest of the fish tank is filled with water (density 1.00 g/cm³). Assuming that the volume occupied by the fish is negligible, calculate the average density of the contents of the fish tank.

◀ 90. Determine the density of a rectangular solid that has the dimensions 8.0 cm × 2.0 cm × 3.0 cm and a mass of 192.0 g.

◀ **91.** Calculate the mass of a cube that has a density of 1.61 g/cm³ and is 4.1 cm on each side. Express your answer to the proper number of significant figures.

92. A balloon contains a gaseous compound. The radius of the balloon is 1.06 cm, and the mass of the gaseous compound in the balloon is 9.80 g. What is the density of the gaseous compound?

***93.** Suppose you are drinking a chocolate milk shake. Its average density is 2.00 g/cm³. The volume of the container that holds it is $25\overline{0}$ cm³. The milk shake consists of 90.0% milk products and 10.0% chocolate powder by volume. Assuming that the density of the milk products is 1.50 g/cm³, calculate the amount of chocolate powder that was used to make the drink.

94. Determine the length of a metallic cube that has a density of 10.5 g/cm³ and a mass of 672 g.

95. A cube measures 3.50 cm on each side and has a mass of $40\overline{0}$ g. What is its density?

96. The element copper has a density of 8.92 g/cm³. What is the mass of a cylinder-shaped piece of copper that has a radius of 2.00 cm and a height of 10.00 cm?

97. Gold has a density of 18.9 g/cm³. A certain block of gold is a rectangular solid that measures 30.0 cm by 5.00 cm by 10.0 cm. What is its mass?

98. An empty graduated cylinder has a mass of 80.00 g. The empty cylinder is filled with exactly 30.0 cm³ of a liquid and weighed again. The cylinder with the liquid in it has a mass of 488 g. What is the density of the liquid? If you are told that this liquid is an element, can you determine what element it is?

99. The density of lead is 11.3 g/cm³. What is the volume of a chunk of lead that has a mass of $25\overline{0}$ g?

100. A small stone that appears to be an irregularly shaped diamond is weighed and found to have a mass of 21.06 g (about 105 carats if it is indeed a diamond). The stone is placed in a graduated cylinder that initially has 10.0 cm³ of water in it. When the stone is immersed in the water, the water level rises to 16.0 cm³. What is the density of the stone? Could the stone be a diamond? (*Hint:* Check the *Handbook of Chemistry and Physics.*)

101. A small rock has a mass of 55.0 g. The rock is placed in a graduated cylinder containing water. The water level in the cylinder changes from 25.0 cm³ to 40.0 cm³ when the rock is submerged. What is the density of the rock?

102. A piece of copper that has a mass of 120.0 g is melted and poured into some liquid silver that has a mass of 200.0 g. The two metals are mixed and allowed to solidify. What is the average density of this metal alloy? [*Hint:* D(copper) = 8.92 g/cm³ and D(silver) = 10.5 g/cm³.]

*103. An object in the shape of a sphere—for example, a hollow globe—is filled with ethyl alcohol. The sphere has a radius of 18.0 cm, and ethyl alcohol has a density of 0.800 g/cm³. What is the mass of ethyl alcohol in the sphere? (*Hint:* Volume of a sphere = $\frac{4}{3}\pi r^3$.)

104. Determine the mass in grams of one gallon of gasoline if the density of gasoline at 25°C is 0.56 g/cm³.

105. A cube has a density of 0.500 g/cm³ and a mass of 25.0 g. What is the volume of this cube?

106. What mass of lead [D(lead) = 11.3 g/cm³] occupies the same volume as 100 g of aluminum [D(aluminum) = 2.70 g/cm³]?

LEARNING GOAL 11: *Celsius and Fahrenheit Temperature Conversions*

107. Convert 100.0°F to degrees Celsius and to Kelvins.
108. Convert 2,200°F to degrees Celsius and to Kelvins.

109. Convert −40°F to degrees Celsius. What is unusual about this temperature?
◄110. Convert −20.0°F to degrees Celsius.

111. An individual who is ill has a temperature of 40.0°C. Normal body temperature is 37.0°C. This represents a 3.0°C rise in temperature. What type of increase above normal body temperature does

this represent in °F? What is the individual's body temperature in °F?

112. Convert 23.0°C to degrees Fahrenheit.

◄***113.*** Change each of the following temperatures from degrees Fahrenheit to degrees Celsius. (a) 50°F (b) −94°F (c) 419°F (d) −130°F

114. Convert −20.0°C to degrees Fahrenheit.

115. Change each of the following temperatures from degrees Celsius to degrees Fahrenheit. (a) 95.0°C (b) −80.0°C (c) 80.0°C (d) 210°C

116. Convert 15.0°C to degrees Fahrenheit.

EXTRA EXERCISES

117. Would a carefully weighed object weigh the same in Death Valley as it would on the top of Mount McKinley?

◄118. In filling out an application form for a company that uses the metric system, a job applicant recorded his height as 4.00 m and his weight as 200.0 kg. Did he fill the application out correctly?

119. An art dealer needs to know the volume of a piece of irregularly shaped sculpture. A chemist friend tells the dealer to submerge the object in water and measure the amount of water displaced by it. The dealer finds that 48 g of water are displaced. If the density of the water is 1.0 g/cm³, what is the volume of the art object?

120. A chemistry student decides to cook a pizza. The instructions call for cooking the pizza for 10 minutes at 425°F. The oven dial, however, is set in degrees Celsius. At what temperature should the dial be set so that the pizza will be done in 10 minutes?

121. Choose a room where you live and measure its dimensions in English units; then convert each dimension to metric units.

122. Express the following relationships in scientific notation:
(a) _____ g = 1 kg
(b) _____ kg = 1 g
(c) _____ μg = 1 g
(d) _____ mg = 1 g

◀ **123.** Assuming that the following numbers are from experimental measurements, calculate the answer and state it to the proper number of significant figures.

$$\frac{(3.12 \times 10^6)(8.123 \times 10^{-4})}{3.1}$$

124. A temperature is measured as 16.0°F. What is this temperature in degrees Celsius?

125. Change the following temperatures to °F:
(a) 16°C   (b) 2$\overline{00}$°C

126. Change the following temperatures to °C:
(a) 3$\overline{00}$°F   (b) −15$\overline{0}$°F

127. The 1976 world's record for the 100-meter dash was 9.90 seconds. What would this speed be in miles per hour?

MATTER AND ENERGY, ATOMS AND MOLECULES

3

We all learn the scientific method when we take a chemistry course, yet few of us think that we'll ever have to use it in our personal lives. This is the story of Joan Penn, who used the training and knowledge she received in a chemistry course to help her 11-year-old son Mike over a serious illness.

Our story begins on a Sunday afternoon in late September. Mike returned home from playing soccer complaining that he had a headache. By morning he felt better, so Mike went off to school. When he came home, he was coughing and running a low-grade fever. Two days later his mother took him to the doctor who diagnosed a viral infection. The doctor recommended that Mike rest until the virus passed.

After two days, his conditioned worsened. He was coughing and running a high fever. In addition, a rash appeared on his arms. A strep test proved positive and Mike was placed on an antibiotic. After two days on the medication, his condition continued to worsen, and he was now coughing severely and having bronchial spasms.

Joan was quite concerned. After all, the antibiotic Mike was on should have certainly wiped out any strep infection, yet Mike was still getting worse. She thought that perhaps her son was actually not suffering from a virus, but rather that he was having an allergic reaction to something environmental—for instance, something in the park, where he played every day. She decided to retrace Mike's schedule over the past two weeks to look for something—anything—that stood out from his regular schedule.

Joan checked Mike's school for any unusual incidents. Nothing. She checked the parks where he had played; she called the township to see if pesticides were being sprayed in the parks. All results proved negative. Joan decided to ask Mike if there was anything he had done over the past two weeks that he usually didn't do. One event stood out as slightly unusual.

On the day his initial symptoms appeared, Mike was at a friend's house. He and his friend had decided to build a tree house in the backyard, and they worked several hours cutting branches and putting the tree house together. Upon its completion, his friend's Mom bought them pizza to celebrate their fine work, which the boys gobbled down without having washed up.

Joan decided to find out more about the tree house and discovered that the boys built it using juniper branches. Joan called the county health department to find out if junipers could cause allergic reactions. The health officer stated that sticky sap from the pines *could* cause allergic reactions in humans, and he recited a list of symptoms that were identical to Mike's. The puzzle was solved! Mike was allergic to the juniper sap. When Joan later told her son about her findings, Mike recalled getting the sticky sap on his arms, where his rash had formed. He also said that he may have ingested some of the sap when he ate the pizza, which would explain his breathing problems.

Joan called Mike's doctor and told her about the information she had uncovered. Mike was immediately placed on Ventolin to relieve the bronchial spasms. He was also given a nebulizer to help him breathe more easily. Two weeks later Mike's fever was down, the rash had disappeared, and his breathing had returned to normal. Thanks to some good detective work by his mother, Mike was once again healthy.

LEARNING GOALS

After you've studied this chapter, you should be able to:
1. Explain what is meant by the scientific method.
2. Explain the Law of Conservation of Mass and Energy.
3. Explain the difference between physical and chemical properties.
4. Describe the difference between homogeneous and heterogeneous matter, between mixtures and compounds, and between compounds and elements.
5. Describe the difference between an atom and a molecule.
6. Explain the Law of Definite Composition (or Definite Proportions).
7. Explain the terms *atomic mass, formula mass,* and *molecular mass.*
8. Determine the formula or molecular mass of a compound when you are given the formula for the compound.

This chapter is really the beginning of your study of chemistry. We start with discussions of the most elementary concepts, those of matter and energy. Then, in this chapter and later chapters, we build on and extend these concepts. As we do this, we shall be discussing the results of centuries of scientific research—the theories and laws of modern chemistry. These theories and laws are sometimes presented to students as though each resulted from a quick flash of insight on the part of some scientist. Actually they are the fruit of years—and sometimes decades or centuries—of hard work by many people.

3.1   The Scientific Method

Chemistry is an experimental science that is concerned with the behavior of matter. Much of the body of chemical knowledge consists of abstract concepts and ideas. Without application, these concepts and ideas would have little impact on society. Chemical principles are applied for the benefit of society through technology. Useful products are developed by the union of basic science and applied technology.

LEARNING GOAL 1

The scientific method

Over the past 200 years, science and technology have moved forward at a rapid pace. Ideas and applications of these ideas are developed through carefully planned experimentation, in which researchers adhere to what is called the **scientific method.** The scientific method is composed of a series of logical steps that allow researchers to approach a problem and try to come up with solutions in the most effective way possible. It is generally thought of as having four parts:

1. *Observation and classification.* Scientists begin their research by carefully observing natural phenomena. They carry out experiments, which are observations of natural events in a controlled setting. This allows results to be duplicated and rational conclusions to be reached. The data the scientists collect are analyzed, and the facts that emerge are classified.
2. *Generalization.* Once observations are made and experiments carried out, the researcher seeks regularities or patterns in the results that can lead to a generalization. If this generalization is basic and can be communicated in a concise statement or a mathematical equation, the statement or equation is called a *law.*
3. *Hypothesis.* Researchers try to find reasons and explanations for the generalizations, patterns, and regularities they discover. A hypothesis expresses a tentative explanation of a generalization that has been stated. Further experiments then test the validity of the hypothesis.
4. *Theory.* The new experiments are carried out to test the hypothesis. If they support it without exception, the hypothesis becomes a theory. A

theory is a tested model that explains some basic phenomenon of nature. It cannot be proven to be absolutely correct. As further research is performed to test the theory, it may be modified or a better theory may be developed.

The scientific method represents a systematic means of doing research. There are times when discoveries are made by accident, but most knowledge has been gained via careful, planned experimentation. In your study of chemistry you will examine the knowledge and understanding that researchers using the scientific method have uncovered.

3.2   Matter and Energy

We begin with the two things that describe the entire universe: *matter* and *energy*. **Matter** is *anything that occupies space and has mass*. That includes trees, clothing, water, air, people, minerals, and many other things. Matter shows up in a wide variety of forms.

Energy is the *ability to perform work*. Like matter, energy is found in a number of forms. Heat is one form of energy, and light is another. There are also chemical, electrical, and mechanical forms of energy. And energy can change from one form to another. In fact, matter can also change form or change into energy, and energy can change into matter, but not easily.

3.3   Law of Conservation of Mass and Energy

LEARNING GOAL 2
Law of Conservation of Mass and Energy

The **Law of Conservation of Mass** tells us that when a chemical change takes place, no detectable difference in the mass of the substances is observed. In other words, mass is neither created nor destroyed in an ordinary chemical reaction. This law has been tested by extensive experimentation in the laboratory, and the work of the brilliant French chemist-physicist Antoine Lavoisier provides evidence for this conclusion. Lavoisier performed many experiments involving matter. In one instance he heated a measured amount of tin and found that part of it changed to a powder. He also found that the *product* (powder plus tin) weighed *more* than the original piece of tin. To find out more about the added weight, he heated metals in sealed jars, which, of course, contained air. He measured the mass of his starting materials (*reactants*), and when the reaction concluded and the metal no longer changed to powder, he measured the mass of the products. In every such reaction, the mass of the reactants (oxygen from the air in the jar plus the original metal) equaled the mass of the products (the remaining metal plus the powder). Today we know that the reaction actually stopped when all of

■ **FIGURE 3.1**

An experiment like Lavoisier's. An experimenter puts a test tube containing a lead nitrate solution into a flask containing a potassium chromate solution. The experimenter weighs the flask and contents, then turns the flask upside down to mix the two solutions. A chemical reaction takes place, producing a yellow solid. The experimenter weighs the flask and contents again and finds no change in mass.

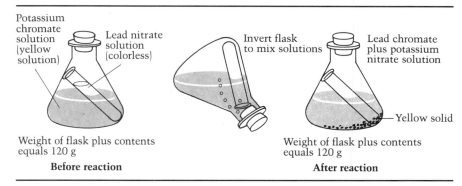

Potassium chromate solution (yellow solution)

Lead nitrate solution (colorless)

Invert flask to mix solutions

Lead chromate plus potassium nitrate solution

Yellow solid

Weight of flask plus contents equals 120 g

Before reaction

Weight of flask plus contents equals 120 g

After reaction

the oxygen in the sealed jar combined with the metal to form the powder. Lavoisier concluded that when a chemical change occurs, *matter is neither created nor destroyed, it just changes from one form to another* (Figure 3.1), which is a statement of the Law of Conservation of Mass.

Whenever a chemical change occurs, it is accompanied by an energy transformation. In the 1840s, more than half a century after Lavoisier, three scientists—the Englishman James Joule and the Germans Julius von Mayer and Hermann von Helmholtz—performed a number of experiments in which energy transformations were studied. They provided experimental evidence that led to the discovery of the **Law of Conservation of Energy.** The law tells us that *in any chemical or physical change, energy is neither created nor destroyed, it is simply converted from one form to another.*

An auto engine provides a good example of how one form of energy is converted to a different form. *Electrical* energy from the battery generates a spark that contains *heat* energy. The heat ignites the gasoline–air mixture, which explodes, transforming chemical energy into heat and *mechanical* energy. The mechanical energy causes the pistons to rise and fall, rotating the engine crankshaft and moving the car.

At the same time, in the same engine, matter is changing from one form to another. When the gasoline explodes and burns, it combines with oxygen in the cylinders to form carbon dioxide and water vapor. (Unfortunately, carbon monoxide and other dangerous gases may also be formed. This is one of the major causes of air pollution.

To appreciate the significance of these facts, think of the universe as a giant chemical reactor or system. At any given time there are certain amounts of matter and energy present, and the matter has a certain mass. Matter is always changing from one form to another, and so is energy. Besides that, matter is changing to energy and energy to matter. But *the sum of all the matter (or mass) and energy in the universe always remains the same.* This repeated observation is called the **Law of Conservation of Mass and Energy.**

3.4 Potential Energy and Kinetic Energy

Which do you think has more energy, a metal cylinder held 1 foot above the ground or an identical cylinder held 5 feet above the ground? If you dropped them on your foot, you would know immediately that the cylinder with more energy was the one that was 5 feet above the ground. But where does this energy come from?

Work had to be done to raise the two cylinders to their respective heights—to draw them up against the pull of gravity. And energy was needed to do that work. The energy used to lift each cylinder was "stored" in each cylinder. The higher the cylinder was lifted, the more energy was stored in it—due to its position. *Energy that is stored in an object by virtue of its position* is called **potential energy.**

If we drop the cylinders, they fall toward the ground. As they do so, they lose potential energy because they lose height. But now they are moving; their potential energy is converted to "energy of motion." The more potential energy they lose, the more energy of motion they acquire. *The energy that an object possesses by virtue of its motion* is called **kinetic energy.** The conversion of potential energy to kinetic energy is a very common phenomenon. It is observed in a wide variety of processes, from downhill skiing to the generation of hydroelectric power.

3.5 The States of Matter

Matter may exist in any of the three physical states: solid, liquid, and gas.

A **solid** has a definite shape and volume that it tends to maintain under normal conditions. The particles composing a solid stick rigidly to one another. Solids most commonly occur in the **crystalline** form, which means they have a fixed, regularly repeating, symmetrical internal structure. Diamonds, salt, and quartz are examples of crystalline solids. A few solids, such as glass and paraffin, do not have a well-defined crystalline structure, although they do have a definite shape and volume. Such solids are called **amorphous solids,** which means they have no definite internal structure or form.

A **liquid** has a definite volume but does not have its own shape since it takes the shape of the container in which it is placed. Its particles cohere firmly, but not rigidly, so the particles of a liquid have a great deal of mobility while maintaining close contact with one another.

A **gas** has no fixed shape or volume and eventually spreads out to fill its container. As the gas particles move about they collide with the walls of their container causing *pressure,* which is a force exerted over an area. Gas particles move independently of one another. Compared with those of a liquid or solid, gas particles are quite far apart. Unlike solids and liquids, which cannot be compressed very much at all, gases can be both compressed and expanded.

Often referred to as the fourth state of matter, **plasma** is *a form of matter composed of electrically charged atomic particles.* Many objects found in the earth's outer atmosphere, as well as many celestial bodies found in space (such as the sun and stars), consist of plasma. A plasma can be created by heating a gas to extremely high temperatures or by passing a current through it. A plasma responds to a magnetic field and conducts electricity well.

3.6 ## Physical and Chemical Properties

■ LEARNING GOAL 3
Physical and chemical properties

Matter—whether it is solid, liquid, or gas—possesses two kinds of properties: physical and chemical. These unique properties separate one substance from another and ensure that no two substances are alike in every way. The **physical properties** are those that can be observed or measured without changing the chemical composition of the substance. These properties include state, color, odor, taste, hardness, boiling point, and melting point. A **physical change** is one that alters at least one of the physical properties of the substance without changing its chemical composition. Some examples of physical change are (1) altering the physical state of matter, such as what occurs when an ice cube is melted; (2) dissolving or mixing substances together, such as what happens when we make coffee or hot cocoa; and (3) altering the size or shape of matter, such as what happens when we grind or chop something.

Chemical properties stem from the ability of a substance to react or change to a new substance that has different properties. This often occurs in the presence of another substance. For example, iron reacts with oxygen to produce iron(III) oxide (rust). This is an example of a **chemical change.** The chemical properties can be observed or measured when a substance undergoes chemical change. The rusting of iron is an example of a chemical property of iron. When we pass an electric current through water, it decomposes to form hydrogen gas and oxygen gas. This reaction is an example of a chemical property of water.

Sometimes it is difficult to differentiate a chemical change from a physical change. In fact, physical changes almost always accompany chemical changes. Some of the signs of physical change that tell us that a chemical change has occurred include the presence of a large amount of heat or light, the presence of a flame, the formation of gas bubbles, a change in color or odor, or the formation of a solid material that settles out of a solution.

3.7 ## Mixtures and Pure Substances

Since matter consists of all the material things that compose the universe, many distinctly different types of matter are known. Matter that has a definite and *fixed composition* is called a **pure substance,** which is a substance that cannot be separated into any other form of matter by physical change. Some of the pure substances that you are familiar with are helium, oxygen, table salt, water, gold, and silver. Two or more pure substances can be combined to form a **mixture,** whose *composition can be varied.* The substances in a mixture can be separated by physical means; we can separate the substances without chemical change.

Matter can also be classified as heterogeneous or homogeneous (Figure 3.2). **Homogeneous matter** has the same parts with the same properties throughout, and **heterogeneous matter** is made up of different parts with different properties. A combination of salt and pepper is an example of heterogeneous matter, whereas a teaspoonful of sugar is an example of homogeneous matter. Another example of homogeneous matter is a teaspoonful of salt dissolved in a glass of water. We call this a **homogeneous mixture** because it is a uniform blend of two or more substances, and its proportion can be varied. In a homogeneous mixture every part is exactly like every other part. The salt can be separated from the water by physical means. Seawater and air are also examples of homogeneous mixtures.

We know that there are two types of homogeneous matter: pure substances and homogeneous mixtures. According to this classification scheme, matter can be broken down even further. Let's look at both homogeneous mixtures and heterogeneous mixtures. In a **heterogeneous mixture** different parts have different properties. A salt-sand mixture is heterogeneous because it is composed of two substances, *each of which retains its own unique properties.* It does not have the same composition or properties throughout, and its composition can be varied. A salt-sand mixture can be separated by physical means. If the mixture is placed in water, the salt will dissolve. The sand can be filtered out, and the salt can be recovered by heating the saltwater until the water evaporates (Figure 3.3).

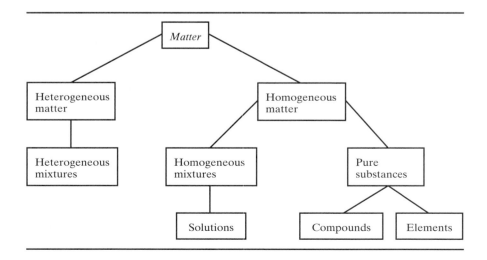

■ **FIGURE 3.2**

Classification of matter

■ **FIGURE 3.3**

Separating sand from saltwater

We call any part of a system with uniform composition and properties a **phase.** A **system** is the body of matter being studied. A heterogeneous mixture is composed of two or more phases separated by physical boundaries. Additional examples of heterogeneous mixtures are oil and water (two liquids) and a tossed salad (several solids). It is important to note that although a pure substance is always homogeneous in composition, it may actually exist in more than one phase in a heterogeneous system. Think of a glass of ice water. This is a two-phase system composed of water in the solid phase and water in the liquid phase. In each phase the water is homogeneous in composition, but since two phases are present, the system is heterogeneous.

3.8   Solutions

Solutions are homogeneous mixtures. That is, they are uniform in composition. Every part of a solution is exactly like every other part. The salt-and-water mixture described in Section 3.7 is a solution. Even if we add more water to the solution, it will still be homogeneous, because the salt particles will continue to be distributed evenly throughout the solution. We would get the same result if we added more salt to the solution. However, we couldn't do this indefinitely. Homogeneity would end when the solution reached *saturation* (the point at which no more salt could dissolve in the limited amount of water). We will discuss this further in Chapter 15.

TABLE 3.1

Names and Symbols of Some Common Elements

ELEMENT	SYMBOL	ELEMENT	SYMBOL
Aluminum	Al	Iodine	I
Bromine	Br	Magnesium	Mg
Calcium	Ca	Nickel	Ni
Carbon	C	Nitrogen	N
Chlorine	Cl	Oxygen	O
Chromium	Cr	Phosphorus	P
Fluorine	F	Silicon	Si
Helium	He	Sulfur	S
Hydrogen	H	Zinc	Zn

3.9   Elements

An **element** is *a pure substance that cannot be broken down into simpler substances, with different properties, by physical or chemical means.* The elements are the basic building blocks of all matter. There are now 109 known elements.* Each has its own unique set of physical and chemical properties. (The elements are tabulated on the inside front cover of this book, along with their *chemical symbols*—a shorthand notation for their names. Some common elements are listed in Table 3.1.) The elements can be classified into three types: **metals, nonmetals,** and **metalloids.**

Examples of metallic elements are sodium (which has the symbol Na), calcium (Ca), iron (Fe), cobalt (Co), and silver (Ag). These elements are all classified as metals because they have certain properties in common. They have luster (in other words, they are shiny), they conduct electricity well, they conduct heat well, and they are malleable (can be pounded into sheets) and ductile (can be drawn into wires).

Some examples of nonmetals are chlorine, which has the symbol Cl (note that the second letter of this symbol is a lower-case "el" and not the numeral *one*), oxygen (O), carbon (C), and sulfur (S). These elements are classified as nonmetals because they have certain properties in common. They don't shine, they don't conduct electricity well, they don't conduct heat well, and they are neither malleable nor ductile.

The metalloids have some properties like those of metals and other properties like those of nonmetals. Some examples are arsenic (As), germanium (Ge), and silicon (Si). These particular metalloids are used in manufacturing transistors and other semiconductor devices (Table 3.1).

*In early 1995 a team of German and Russian scientists reported the synthesis of elements 110 and 111.

3.10   Atoms

■ LEARNING GOAL 5
Difference between atom and molecule

Suppose we had a chunk of some element, say gold, and were able to divide it again and again, into smaller and smaller chunks. Eventually we could get a particle that could not be divided any further without losing its identity. This particle would be an atom of gold. An **atom** is *the smallest particle of an element that enters into chemical reactions.*

The atom is the ultimate particle that makes up the elements. Gold is composed of gold atoms, iron of iron atoms, and neon of neon atoms. These atoms are so small that billions of them are needed to make a speck large enough to be seen with a microscope. In 1970, Albert Crewe and his staff at the University of Chicago's Enrico Fermi Institute took the first black-and-white pictures of single atoms, using a special type of electron microscope. In late 1978, Crewe and his staff took the first time-lapse moving pictures of individual uranium atoms.

3.11   Compounds

■ LEARNING GOAL 6
Law of Definite Composition

A **compound** is *a pure substance that is made up of two or more elements chemically combined in a definite proportion by mass.* Unlike mixtures, compounds have a definite composition. Water, for instance, is made up of hydrogen and oxygen in the ratio of 11.1% hydrogen to 88.9% oxygen by mass. No matter what the source of the water, it is always composed of hydrogen and oxygen in this ratio. This idea, that *every compound is composed of elements in a certain fixed proportion,* is called the **Law of Definite Composition** (or the Law of Definite Proportions). It was first proposed by the French chemist Joseph Proust in about 1800.

The properties of a compound need not be similar to the properties of the elements that compose it. For example, water is a liquid, whereas hydrogen and oxygen are both gases. When two or more elements form a compound, they truly form a new substance.

Compounds can be broken apart into elements only by chemical means—unlike mixtures, which can be separated by physical means. More than thirteen million compounds have been reported to date, and millions more may be discovered. Some compounds we are all familiar with are sodium chloride (table salt), which is composed of the elements sodium and chlorine, and sucrose (cane sugar), which is composed of the elements carbon, hydrogen, and oxygen.

Uranium single atoms and microcrystals obtained from a solution of uranyl acetate (Courtesy Albert V. Crewe)

3.12   Molecules

We have discussed what happens when a chunk of an element is continually divided: We eventually get down to a single atom. What happens when we keep dividing a chunk of a compound? Suppose we do so with the compound

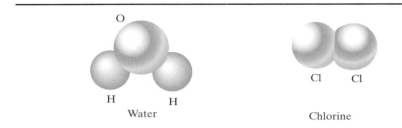

sugar. As we continue to divide a sugar grain, we eventually reach a small
particle that can't be divided any further without losing the physical and
chemical properties of sugar. This ultimate particle of a compound, *the
smallest particle that retains the properties of the compound,* is called a **molecule.** Like atoms, molecules are extremely small, but with the aid of an
electron microscope we can observe some of the very large and more complex
molecules. Molecules are uncharged particles. That is, they carry neither a
positive nor a negative electrical charge.

Molecules, as you might have guessed, are made up of two or more
atoms. They may be composed of different kinds of atoms (for instance,
water contains hydrogen and oxygen) or the same kind of atoms (for instance,
a molecule of chlorine gas contains two atoms of chlorine). (See Figure 3.4.)
*The Law of Definite Composition states that the atoms in a compound are
combined in definite proportions by mass.* We can see in Figure 3.4 that they
are also combined in definite proportions by number. For example, water
molecules always contain two hydrogen atoms for every oxygen atom.

3.13 Molecular Versus Ionic Compounds

As we just learned, a molecule is the smallest *uncharged* part of a compound
formed by the chemical combination of two or more atoms. Such compounds
are known as *molecular compounds,* and we usually say that such compounds
are composed of molecules. Water is a molecular compound composed of
water molecules. Many molecular compounds are composed of atoms of
nonmetallic elements that are chemically combined.

However, there are many compounds composed of oppositely charged
ions. An ion is *a positively or negatively charged atom or group of atoms.*
Compounds composed of ions are known as *ionic compounds.* These compounds are held together by attractive forces between the positive and negative ions that compose the compound. Ordinary table salt, sodium chloride,
is such a compound. For these compounds, it is more proper to talk about
a **formula unit** of the compound, rather than a molecule, as being *the smallest
part of an ionic compound that retains the properties of the compound.* Ionic
compounds are composed of metallic and nonmetallic elements. (We'll have

much more to say about ions and ionic compounds when we discuss chemical bonding in Chapter 7.)

3.14   Symbols and Formulas of Elements and Compounds

We have already mentioned the chemical symbols—the shorthand for the names of the elements. In some cases, as in the symbol O (capital "oh") for oxygen, the chemical symbol is the first letter of the element's name, capitalized. Often, though, the symbol for an element contains two letters. In these cases only the first letter is capitalized; the second letter is never capitalized. For instance, the symbol for neon is Ne and the symbol for cobalt is Co. (Be careful of this: CO does not represent cobalt, but a combination of the elements carbon and oxygen, which is a compound.) Some symbols come from the Latin names of the elements: iron is Fe, from the Latin *ferrum,* and lead is Pb, from the Latin *plumbum.* (See Table 3.2.)

Because compounds are composed of elements, we can use the chemical symbols as a shorthand for compounds too. We use the symbols to write the **chemical formula,** which shows the elements that compose the compound. For example, sodium chloride (table salt), an ionic compound, contains one atom of sodium (Na) and one atom of chlorine (Cl). The formula unit for sodium chloride is NaCl. Water, a molecular compound, is another example. The water molecule contains two atoms of hydrogen (H) and one atom of oxygen, so we write it as H_2O. The number 2 in the formula indicates that there are two atoms of hydrogen in the molecule; note that it is written as a *subscript.* A molecule of ethyl alcohol contains two atoms of carbon, six

TABLE 3.2

Elements with Symbols Derived from Latin Names

ELEMENT	SYMBOL	LATIN NAME
Copper	Cu	Cuprum
Gold	Au	Aurum
Iron	Fe	Ferrum
Lead	Pb	Plumbum
Mercury	Hg	Hydrargentum
Potassium	K	Kalium
Silver	Ag	Argentum
Sodium	Na	Natrium
Tin	Sn	Stannum

of hydrogen, and one of oxygen. It is written as follows:

EXAMPLE 3.1

State the number of atoms of each element in a molecule or formula unit of the following compounds: (a) $C_6H_{12}O_6$ (b) $Ca(OH)_2$ (c) $C_3H_6O_2$ (d) $Al_2(SO_4)_3$

Solution

UNDERSTAND THE PROBLEM

We ask, "What does the subscript mean?" Subscripts tell us the number of atoms in a molecule or formula unit of the substance.

DEVISE A PLAN

Our plan will be to look at the subscripts noted in each case. We can then determine the number of atoms of each element in a molecule or formula unit of the compound.

CARRY OUT THE PLAN

(a) There are 6 atoms of C, 12 atoms of H, and 6 atoms of O.
(b) There is 1 atom of Ca, 2 atoms of O, and 2 atoms of H. (In this case, the subscript 2 means multiply everything inside the parentheses by 2.)
(c) There are 3 atoms of C, 6 atoms of H, and 2 atoms of O.
(d) There are 2 atoms of Al, 3 atoms of S, and 12 atoms of O. (In this case, the subscript 3 means multiply everything inside the parentheses by 3.)

LOOK BACK

Recheck to be sure that you have followed the steps correctly.

Practice Exercise 3.1   State the number of atoms of each element in a molecule or formula unit of the following compounds: (a) C_2H_7N (b) $(NH_4)_2SO_4$

3.15 Atomic Mass

Suppose we want to find the relative masses of the atoms of the various elements. Suppose also that we have a double-pan balance that can weigh a single atom. We begin by assigning an arbitrary mass to one of the elements. Let's say we decide to assign a mass of 1 unit to hydrogen. (This is what chemists did originally, because hydrogen was known to be the lightest element even before the relative atomic masses were determined.)

Then, to find the relative mass of, say, carbon, we place an atom of carbon on one pan. We place hydrogen atoms on the other pan, one by one, until the pans are exactly balanced. We find that it takes 12 hydrogen atoms to balance 1 carbon atom, so we assign a relative atomic mass of 12 to carbon. In the same way for an oxygen atom, we find that it takes 16 hydrogen atoms to balance 1 oxygen atom. So we assign a relative mass of 16 to oxygen. By doing this experiment for all the other elements, we can determine all the masses relative to hydrogen (which is assigned mass 1).

Unfortunately this kind of balance has never existed, and more elaborate means had to be developed to find the relative atomic masses of the elements. But the logic we used here is the same as that used by the many scientists who determined the relative masses. (The atomic mass scale has been revised. The present scale is based on a particular type of carbon atom, called carbon-12. This carbon is assigned the value of 12 atomic mass units, amu.) The periodic table (inside the front cover) lists the relative atomic masses of the elements below their symbols. From now on, instead of referring to "relative atomic mass," we will use the simpler term **atomic mass.**

EXAMPLE 3.2

Using the periodic table (inside the front cover), look up the atomic masses of the following elements. (For this exercise, round all atomic masses to one decimal place.)   (a) I   (b) Ba   (c) As   (d) S

Solution

Look up the atomic mass of each element in the periodic table. Remember, the atomic mass is the number below the symbol of the element.

(a) I is 126.9

(b) Ba is 137.3

(c) As is 74.9

(d) S is 32.1

Practice Exercise 3.2   Using the periodic table (inside the front cover), look up the atomic masses of the following elements. (For this exercise, round all atomic masses to one decimal place.)   (a) La   (b) Fe   (c) Ar   (d) Sn

---

3.16   Formula Mass and Molecular Mass

LEARNING GOAL 8

Formula or molecular mass of a compound from the formula

The **molecular mass** of a compound is *the sum of the atomic masses of all the atoms that make up a molecule of the compound.* The term *molecular mass* is applied to compounds that exist as molecules. For example, the molecule P_2O_5 has two phosphorus atoms and five oxygen atoms. The atomic mass of each P (to one decimal place) is 31.0, and the atomic mass of each O (to one decimal place) is 16.0. Therefore the molecular mass of the

The masses of individual atoms are determined with a mass spectrometer. Electrons are removed from atoms (or molecules), and the resultant ions are accelerated through a magnetic field. The amount of bending in the path of the ions is related to the mass and the charge of the ions. (Photo: Courtesy of Hewlett-Packard)

compound is

$$(2 \times 31.0) + (5 \times 16.0) = 62.0 + 80.0 = 142.0$$

(Check this calculation and see whether you get the same answer. If not, you may have forgotten that in an equation like this, multiplication and division are done *before* addition and subtraction.)

The **formula mass** of a compound is *the sum of the atomic masses of all the ions that make up a formula unit of the compound.* The term *formula mass* is applied to compounds that are written as formula units and exist mostly as ions (charged atoms or groups of atoms). For example, the formula unit of aluminum oxide is Al_2O_3. This means that a formula unit of aluminum oxide has two aluminum atoms (actually aluminum ions) and three oxygen atoms (actually oxide ions). The atomic mass of aluminum (to one decimal place) is 27.0, and the atomic mass of oxygen (to one decimal place) is 16.0. Therefore the formula mass of Al_2O_3 is

$$(2 \times 27.0) + (3 \times 16.0) = 54.0 + 48.0 = 102.0$$

Now let's try to determine the formula and molecular masses of some additional compounds.

EXAMPLE 3.3

Find the molecular or formula masses (to one decimal place) of the following compounds: (a) H_2O (b) NaCl   (c) $Ca(OH)_2$ (d) $Zn_3(PO_4)_2$
Note: In a chemical formula, parentheses followed by a subscript mean that everything inside the parentheses is multiplied by the subscript. For example, in one formula unit of $Ca(OH)_2$, there is one Ca atom plus two O atoms and two H atoms.

Solution

We must find the atomic mass of each element in the periodic table and then add the masses of all the atoms in each compound.

(a) The atomic mass of H is 1.0, and the atomic mass of O is 16.0.

Molecular mass of $H_2O = (2 \times 1.0) + (1 \times 16.0)$

$$= 2.0 + 16.0 = 18.0$$

(b) The atomic mass of Na is 23.0, and the atomic mass of Cl is 35.5.

Formula mass of $NaCl = (1 \times 23.0) + (1 \times 35.5)$

$$= 23.0 + 35.5 = 58.5$$

(c) The atomic mass of Ca is 40.1, the atomic mass of O is 16.0, and the atomic mass of H is 1.0.

Formula mass of $Ca(OH)_2 = (1 \times 40.1) + (2 \times 16.0) + (2 \times 1.0)$

$$= 40.1 + 32.0 + 2.0 = 74.1$$

(d) The atomic mass of Zn is 65.4, the atomic mass of P is 31.0, and the atomic mass of O is 16.0.

Formula mass of $Zn_3(PO_4)_2 = (3 \times 65.4) + (2 \times 31.0) + (8 \times 16.0)$

$$= 196.2 + 62.0 + 128.0 = 386.2$$

Practice Exercise 3.3   Find the molecular or formula masses (to one decimal place) of the following compounds: (a) Na_2CO_3 (b) $CoCl_2$ (c) Cl_2O (d) N_2O_4

■■■ SUMMARY

The body of knowledge called science has been developed through the scientific method: observation and classification of data, generalization of observations, and testing of generalizations. Of importance in all the sciences is the idea that the universe is made up of only matter and energy. Matter and energy can be neither created nor destroyed, but they can be changed to other forms. And matter can be transformed into energy, and vice versa. Potential energy is energy that is stored in a body because of its position. Kinetic energy is energy that is due to motion.

Matter may exist in any of three states—solid, liquid, or gas—and may be either heterogeneous (nonuniform) or homogeneous (uniform). Mixtures are combinations of two or more kinds of matter, each retaining its own chemical and physical properties. Mixtures too may be either homogeneous or heterogeneous, and solutions are homogeneous mixtures. Elements are the basic building blocks of matter. The 109 known elements are, in turn, made up of atoms. A compound is a substance that is made up of two or more elements chemically combined in definite proportions by mass. Molecules are the smallest particles that retain the properties of a compound, and atoms are the smallest particles that enter into chemical reactions.

Each element has its own symbol, and every compound has its own chemical formula. Each element has a unique atomic mass. The atomic mass of an element is found in the periodic table. The formula mass or molecular mass of a compound is the sum of the atomic masses in a formula unit or molecule of the compound.

KEY TERMS

amorphous solid **(3.5)**
atom **(3.10)**
atomic mass **(3.15)**
chemical change **(3.6)**
chemical formula **(3.14)**
chemical property **(3.6)**
chemical symbols **(3.9)**
compound **(3.11)**
crystalline **(3.5)**
element **(3.9)**
energy **(3.2)**
formula mass **(3.16)**
formula unit **(3.13)**
gas **(3.5)**
heterogeneous matter **(3.7)**
heterogeneous mixture **(3.7)**
homogeneous matter **(3.7)**
homogeneous mixture **(3.7)**
kinetic energy **(3.4)**
Law of Conservation of Energy **(3.3)**
Law of Conservation of Mass **(3.3)**

Law of Conservation of Mass and Energy **(3.3)**
Law of Definite Composition **(3.11)**
liquid **(3.5)**
matter **(3.2)**
metal **(3.9)**
metalloid **(3.9)**
mixture **(3.7)**
molecular mass **(3.16)**
molecule **(3.12)**
nonmetal **(3.9)**
phase **(3.7)**
physical change **(3.6)**
physical property **(3.6)**
plasma **(3.5)**
potential energy **(3.4)**
pure substance **(3.7)**
scientific method **(3.1)**
solid **(3.5)**
solution **(3.8)**
system **(3.7)**

CAREER SKETCH

Writing Careers

If you possess strong writing skills, there is a plethora of science-related positions you might consider (other than writing college-level chemistry textbooks, of course).

You might choose to work as a *freelance writer,* producing articles for a newspaper, a magazine, or a journal. If you are interested in instructional writing, you might consider a job writing and developing multimedia instructional programs. In this type of job, you could write about a variety of diseases and therapies for clinicians, physicians, and patients. Topics might include cardiology, imaging, ophthalmology, oncology, and neurology, to name a few. Or you could be responsible for writing and editing user training documentation on various science-related products and services for an entire company.

A different type of writing involves working for a medical provider or other type of medical company. These organizations often seek writers with science backgrounds to prepare clinical trial study reports for submission to regulatory agencies. Here you will write and edit clinical documents, such as clinical reports, regulatory summaries, and scientific articles for publication.

For any job related to science writing, excellent writing skills and a degree in a field of science are almost always required. Also essential are your abilities to analyze, organize, and interpret data—skills you are currently sharpening with your study of chemistry.

CHEMICAL FRONTIERS

New Coatings Technology Leads to Plastic Car Windows*

Scientists are busy working on new technology that will virtually eliminate all glass from automobiles. If they are successful, polycarbonate windows may soon replace the glass that has surrounded drivers and passengers since the invention of the automobile.

Glass has been the material of choice for automobile makers because it is strong, does not scratch easily, and is relatively inexpensive. However, researchers in Michigan are working on improved coatings technology that could make polycarbonate an even better option for manufacturers. Two ideas are currently being worked on: one method would create an inorganic plasma coating to cover the polycarbonate windows; the second would embed microscopic ceramic particles in an elastomeric siloxane coating.

Companies that manufacture polycarbonate glazing hope that the price of polycarbonate windows will be roughly equivalent to that of glass. Further, they point out that the new windows will have several advantages over glass. Polycarbonate will allow auto designers to integrate large window areas into a car design, since the material weighs 40% less than glass. Also, because the material is nearly impossible to break, the new windows will deter theft and protect occupants from being thrown from cars in side-impact and rollover accidents.

*Adapted from *Chemical & Engineering News*, June 23, 1997, p. 11

SELF-TEST EXERCISES

LEARNING GOAL 1: *The Scientific Method*

1. (a) Explain the difference between a theory and a hypothesis.
   (b) Explain the difference between a theory and a scientific law.

2. Suppose that you are a researcher and you believe that you have found a vaccine that can prevent AIDS. How would you use the scientific method to determine whether the vaccine is effective?

LEARNING GOAL 2: *Law of Conservation of Mass and Energy*

◄ 3. According to the Law of Conservation of Energy, in any chemical or physical change, energy is neither created nor destroyed. Then why are we always worried about running out of energy in the future?

4. What is the significance of the Law of Conservation of Mass and Energy in terms of your study of chemistry?

LEARNING GOALS 3 & 4: *Physical and Chemical Changes/Types of Matter*

<u>5.</u> Match each word on the left with its definition on the right.

   (a) Homogeneous   1. The basic building block of matter

   (b) Heterogeneous   2. The word used to describe matter that is uniform throughout

   (c) Mixture   3. A type of matter in which each part retains its own properties

   (d) Compound   4. A chemical combination of two or more elements

   (e) Element   5. The word used to describe matter that is not uniform throughout

◄ 6. (a) Distinguish among an element, a compound, and a mixture.
   (b) What type of matter is uniform throughout?
   (c) What type of matter is not uniform throughout?

◄ <u>7.</u> State whether each of the following processes involves physical or chemical changes: (a) Shredding paper (b) Burning paper (c) Cooking an egg (d) Mixing egg whites with egg yolk (e) Digesting food (f) Toasting bread

8. Determine whether each of the following processes involves chemical or physical processes:
   (a) Ice melts.
   (b) Sugar dissolves in water.
   (c) Milk sours.
   (d) Eggs become rotten.
   (e) Water boils.
   (f) An egg is hard-cooked.

LEARNING GOAL 5: *Difference Between Atom and Molecule*

9. Describe the difference between an atom and a molecule.

◄ 10. (a) What is the smallest particle of matter that can enter into a chemical combination?
   (b) What is the smallest uncharged individual unit of a compound that is composed of two or more atoms?

<u>11.</u> Classify each of the following elements as metal, metalloid, or nonmetal: (a) Ba (b) Si (c) O (d) Hg (e) Ge (f) In (g) U

12. Classify each of the following elements as metal, metalloid, or nonmetal: (a) Mn (b) Nd (c) Al (d) At (e) Pt (f) Cl (g) Ra

<u>13.</u> Explain the difference between Co and CO.
<u>14.</u> Explain the difference between Si and SI.

LEARNING GOAL 6: *Law of Definite Composition*

◄ <u>15.</u> How does the following information obtained from several experiments confirm the Law of Definite Composition?
   Experiment 1: $\overline{100}$ g of water are decomposed by electrolysis into its elements: 88.9 g of oxygen gas and 11.1 g of hydrogen gas are obtained.
   Experiment 2: 25.0 g of water are decomposed by electrolysis into its elements: 22.2 g of oxygen gas and 2.8 g of hydrogen are obtained.
   Experiment 3: $\overline{500}$ g of water are decomposed by electrolysis into its elements: 444.5 g of oxygen gas and 55.5 g of hydrogen are obtained.

16. Give an example of the Law of Definite Composition (Proportions).

<u>17.</u> State the number of atoms of each element in a molecule or formula unit of the following compounds: (a) $C_{12}H_{22}O_{11}$ (b) K_2CrO_4 (c) $H_8N_2O_3S_2$ (d) $Zn(NO_3)_2$

◄ 18. State the number of atoms of each element in a molecule or formula unit of the following compounds: (a) H_2SeO_4 (b) $C_{21}H_{27}FO_6$ (c) $(NH_4)_3PO_4$ (d) $Fe_3(AsO_4)_2$

LEARNING GOAL 7: *Atomic Mass, Formula Mass, and Molecular Mass*

19. Using examples, explain the difference among atomic mass, formula mass, and molecular mass.
20. State which term—atomic mass, formula mass, or molecular mass—is best suited to describe each of the following substances. (a) $C_6H_{12}O_6$
    (b) $NaCl$ (c) Fe (d) CO_2 (e) H_2O
    (f) Al_2O_3 (g) Ca (h) $Ca(OH)_2$

◄ *21. If in the periodic table oxygen were assigned an atomic mass of 1, what would be the atomic mass of sulfur?
    *22. If in the periodic table neon were assigned an atomic mass of 1, what would be the atomic mass of bromine?

23. Using the periodic table (inside front cover), look up the atomic masses of the following elements. (For this exercise, round all atomic masses to one decimal place.) (a) Rb (b) Cr (c) U
    (d) Se (e) As
24. Using the periodic table (inside front cover), look up the atomic masses of the following elements. (For this exercise, round all atomic masses to one decimal place.) (a) S (b) N (c) Li (d) Cs (e) Au

LEARNING GOAL 8: *Formula or Molecular Mass of a Compound from the Formula*

◄ 25. Determine the molecular or formula mass of each of the following compounds. (For this exercise, round all atomic masses to one decimal place.)
    (a) FeO (b) Fe_2O_3 (c) CuI_2 (d) Na_3PO_4
    (e) $Mg(OH)_2$ (f) $NiBr_2$ (g) $Hg_3(PO_4)_2$
    (h) $(NH_4)_2CO_3$
26. Determine the molecular or formula mass of each of the following compounds. (For this exercise, round all atomic masses to one decimal place.)
    (a) H_2O (b) H_2SO_4 (c) $NaCl$ (d) $Ca_3(PO_4)_2$
    (e) P_2O_5 (f) $SrSO_4$ (g) C_2H_6O (h) SO_2
27. Determine the molecular or formula mass of each of the following compounds. (For this exercise, round all atomic masses to one decimal place.)
    (a) SiO_2 (b) H_2SO_3 (c) $Sr(OH)_2$ (d) RbF
    (e) $Cu(NO_3)_2$ (f) $CoBr_2$ (g) $(NH_4)_3PO_4$
    (h) $HC_2H_3O_2$
◄ 28. Determine the molecular or formula mass of each of the following compounds. (For this exercise, round all atomic masses to one decimal place.) (a) $LiOH$ (b) Na_2CO_3 (c) $CoCl_2$

(d) $NaBr$ (e) SO_3 (f) C_2H_6 (g) OF_2
(h) $(NH_4)_2SO_3$

EXTRA EXERCISES

29. State what each of the symbols and subscripts mean in the chemical formulas for (a) H_2O
    (b) $C_6H_{12}O_6$ (c) $Ca(OH)_2$ (d) H_2
◄ 30. Make a list of heterogeneous mixtures and homogeneous mixtures that you encounter in everyday life. Do the same for elements and compounds.
31. Write the names and symbols for the fourteen elements that have a one-letter symbol.
32. Write the names and symbols of all the metalloids.
◄ 33. How many metals are there in the periodic table? How many nonmetals? How many metalloids?
34. Write the names of the eleven elements whose symbols are not derived from their English names.
35. Name the elements present in each of the following compounds: (a) $MgCl_2$ (b) N_2O
    (c) $(NH_4)_2SO_4$ (d) H_3PO_4
36. Write the chemical formula of each of the following, given the number of atoms in a molecule or formula unit of the compound: (a) one nitrogen atom, two oxygen atoms (nitrogen dioxide) (b) two sodium atoms, one sulfur atom (sodium sulfide) (c) three potassium atoms, one arsenic atom, four oxygen atoms (potassium arsenate) (d) two phosphorus atoms, five oxygen atoms (diphosphorus pentoxide)

37. Classify each of the following as an element, a compound, or a mixture:
    (a) gold
    (b) air
    (c) carbon dioxide
    (d) wine
    (e) table salt
38. State whether each of the following involves a physical or chemical change:
    (a) toasting bread
    (b) water freezing
    (c) tearing paper
    (d) burning wood
39. Determine the molecular or formula mass of each of the following compounds. (For this exercise, round all atomic masses to one decimal place.)
    (a) OsO_4 (b) HNO_3 (c) $Fe(OH)_2$
    (d) $Ba_3(PO_4)_2$
40. Explain the difference between Hf and HF.

Indicate whether each of the following statements is true or false.

1. Chemistry is the science that deals with matter and the changes it undergoes.

2. Organic chemicals are those chemicals that have been derived from living or once-living organisms.

3. Sugar and salt are examples of inorganic chemicals.

4. The art of *khemia* flourished until A.D. 1600.

5. Alchemists were able to transmute lead and iron into gold.

6. The idea that every theory must be proved by experiment was advanced by Robert Boyle.

7. In 420 B.C., Greek philosophers were able to prove that atoms existed.

8. Some carbon-containing compounds are classified as inorganic chemicals.

9. The two main branches of chemistry that existed during the 1700s and 1800s still exist today.

10. Nuclear chemistry, biochemistry, and analytical chemistry are three subdivisions into which chemistry can be divided.

Answer the following questions to help sharpen your test-taking skills.

◀ 11. Find the area of a rectangular room that measures 12.5 m by 10.2 m. Use the proper number of significant figures in reporting your answer.

12. Calculate the volume of a cardboard box that measures 15.25 cm long, 12.00 cm wide, and 24.85 cm high.

13. Determine the volume of a cylindrical solid whose radius is 14.50 cm and whose height is 25.05 cm.

14. Convert 0.28 kg to   (a) grams,   (b) decigrams,   (c) milligrams.

15. Convert 6.8 m to   (a) decimeters,   (b) centimeters,   (c) millimeters.

16. Convert 125 mm to   (a) centimeters,   (b) meters,   (c) kilometers.

17. Convert 25,595 mL to   (a) liters,   (b) deciliters.

*Answers to cumulative-review questions are given in the back of the book.

18. Express 50.0 m in   (a) feet,   (b) inches.

19. Express 25.5 g in   (a) pounds,   (b) ounces.

20. Determine the number of cubic centimeters in a cubic inch without using a conversion table.

21. Express 165 g in pounds.

22. Express 12.0 gallons in   (a) liters,   (b) ounces.

23. Determine the density of a cube that has a mass of 500.0 g and measures 12.0 cm on each side.

24. A spherical balloon with a volume of 113.0 mL is filled with an unknown gas weighing 20.0 g. Will the balloon float on air?

◀ 25. A small metal sphere weighs 90.0 g. The sphere is placed in a graduated cylinder containing 15.0 mL of water. Once the sphere is submerged, the water in the cylinder measures 30.0 mL. What is the density of the sphere?

26. A small metal sphere with a density of 3.50 g/cm^3 has a mass of 937.83 g. Calculate the radius of the sphere.

27. Solve the following problems using the proper number of significant figures:
   (a) 28.64 + 3.2
   (b) 125.4 ÷ 13.5
   (c) 6.55 × 12.1
   (d) 98.4 − 0.12

◀ 28. Express each of the following numbers in scientific notation: (a) 5,000   (b) 0.0005   (c) 602,300,000   (d) 35,000,000

29. How many significant figures are there in each of the following numbers?   (a) 5,500.0   (b) 0.5123   (c) 12.000   (d) 3,500

30. Convert each of the following temperatures from °F to °C: (a) 45.0°F   (b) −10.0°F   (c) 450°F   (d) −100.0°F

31. Convert each of the following temperatures from °C to °F: (a) 88.0°C   (b) −12.5°C   (c) 65.6°C   (d) 400°C

◀ 32. Distinguish between an amorphous solid and a crystalline solid.

33. Determine whether each of the following processes involves chemical or physical changes:
   (a) A match burns.
   (b) Glucose dissolves in water.
   (c) Bread becomes moldy.
   (d) A piece of wood is sawed.

34. Distinguish between heterogeneous and homogeneous matter.

35. State the number of atoms of each element in a molecule or formula unit of the following compounds: (a) $Zn(C_2H_3O_2)_2$ (b) $(NH_4)_2CrO_4$

36. Why do we use the term *molecular mass* for some compounds and *formula mass* for other compounds?

37. If in the periodic table calcium were assigned an atomic mass of 1, what would be the atomic mass of mercury?

38. Using the periodic table (inside front cover), look up the atomic masses of the following elements. (For this exercise, round all atomic masses to one decimal place.) (a) Yb (b) At (c) P (d) Ag

39. Determine the molecular or formula mass of each of the following compounds. (For this exercise, round all atomic masses to one decimal place.) (a) $Fe(C_2H_3O_2)_3$ (b) $(NH_4)_2SO_4$ (c) $C_9H_8O_4$ (d) $Co(SO_4)_2$

40. Explain in detail what the chemical formula $Al_2(SO_4)_3$ means.

41. Determine whether each of the following is an element, a compound, or a mixture:
(a) air
(b) arsenic
(c) carbon dioxide
(d) water
(e) gold
(f) root beer soda
(g) gasoline

42. Determine whether each of the following is an example of a heterogeneous or homogeneous mixture:
(a) beach sand
(b) ethyl alcohol and water
(c) tossed salad
(d) soda water (club soda)

43. Using the periodic table (inside front cover), write the names and symbols for all elements whose symbol begins with the letter A. (*Hint:* There are eight.)

44. Write the names and symbols of all the nonmetallic elements.

45. Name the elements present in each of the following compounds: (a) K_2S (b) Ag_2CrO_4 (c) $KMnO_4$ (d) $Hg_3(PO_4)_2$

46. Write the chemical formula of each of the following, given the number of atoms in a molecule or formula unit of the compound:
(a) two nitrogen atoms, one oxygen atom (dinitrogen monoxide)
(b) two potassium atoms, one chromium atom, four oxygen atoms (potassium chromate)
(c) one nitrogen atom, three hydrogen atoms (ammonia)
(d) one strontium atom, one sulfur atom, four oxygen atoms (strontium sulfate)

47. State whether each of the following involves a physical or chemical change:
(a) a candle burning
(b) table sugar dissolving in water
(c) tooth decaying
(d) snow melting

48. Explain the difference between No and NO.

49. A piece of paper is weighed and is then burned. The resulting ash weighs less than the paper. Is this a violation of the Law of Conservation of Mass? Explain.

50. You are given a mixture of sodium chloride (table salt) and sand. Explain how you would separate this mixture.

Joe and Susan Townes, married for six years, had spent the initial four years of their marriage building their careers and saving for a home in which they could eventually raise their family. Joe was a successful engineer and Susan a well-established and respected real estate agent.

Then, two years ago, Susan, no longer content only with her career, told Joe she wanted to start a family. Suffering from extreme anxiety at work and unusually irritable when she lost a real estate deal, Susan was becoming depressed and tired and had already suffered a 7-kg (15.4-lb) weight loss. She told Joe that this "burn-out" could be cured by setting about to start the family they had always wanted.

Joe, sympathetic to Susan's problems and surprised Susan hadn't raised the issue before, nevertheless stated that he had no intention of starting a family at that time, much to Susan's dismay. Joe said that Susan's anxiety and irritability had affected their own previously solid marital relationship. Her erratic behavior, unpredictable mood swings, and the general breakdown in communication between them convinced Joe that starting a family should be postponed until Susan felt better and their relationship had returned to its former happy state.

Three days later, Susan, who had become even more depressed, lost a major real estate deal. She became extremely upset, with shaking hands, pounding in her chest, profuse sweating, and difficulty with breathing. Realizing that she was ill, Susan requested that a colleague take her to the nearest hospital emergency room.

At approximately 2:00 P.M. Dr. Philip Doherty met Susan in the Simon Memorial Hospital Emergency Room. He observed a fragile, pale young woman who was jittery and irritable. A complete history and physical revealed that Susan had a significant weight loss over the past three months accompanied by weakness, fatigue, emotional instability, and finally episodes of heart palpitations and breathlessness.

Based on the physical exam, Dr. Doherty immediately suspected that Susan was experiencing an endocrine disorder known as hyperthyroidism (Graves' Disease). His first course of action was to perform several blood tests to determine if Susan was, in fact, suffering from this condition. The test results came back positive; Susan had hyperthyroidism.

Treatment for Susan was directed toward reducing output of her thyroid hormone. The chosen method of treatment was radioactive iodine therapy. One of the most effective ways to administer this therapy is through the use of iodine-131 (I-131), a radioactive isotope of iodine. The isotope was given to Susan by mouth in the form of a single, one-time radioactive "cocktail." The I-131 atoms act chemically like the more common, nonradioactive iodine-126 (I-126) atoms. I-126 is an important nutritional component of our diets, because the thyroid gland requires iodine to produce the hormone thyroxine. The I-131 atoms, like the I-126 atoms, are quickly absorbed in the stomach and become concentrated in the thyroid. Unlike the I-126 atoms, however, radiation from the I-131 atoms destroys some of the cells of the thyroid, thereby decreasing the thyroid's activity.

Just three weeks after Susan was given the radioactive thyroid treatment her symptoms began to subside. Within two months, tests revealed that Susan's thyroid function was normal, and her behavioral and physical problems disappeared.

Last June, Susan and Joe Townes became the proud parents of a 4-kg (8.8-lb) baby boy.

LEARNING GOALS

After you've studied this chapter, you should be able to:

1. Describe Dalton's atomic theory, Thomson's model of the atom, and Rutherford's model of the atom.

2. Give the charge and mass of the electron, proton, and neutron.

3. Find the number of protons, electrons, and neutrons in an atom of an element.

4. Explain what isotopes are and give examples.

5. Use the standard isotopic notation for mass number and atomic number.

6. Calculate the average atomic mass of an element when you know the relative abundances of its isotopes.

7. Calculate the percentage abundance of two isotopes when you know the atomic mass of each isotope and the average atomic mass.

8. Use the table of relative abundances of isotopes to calculate the number of atoms or grams of a particular isotope present in a given sample.

■■ INTRODUCTION

Around 420 B.C., Democritus developed the idea that the atom—indestructible and indivisible—was the smallest particle of matter. Twenty-two hundred years later, John Dalton reinforced this idea in his atomic theory. Yet less than a century after Dalton presented his theory, scientists had adopted a different model, in which the atom was made up of several even smaller particles. The development of that model is the subject of this chapter.

4.1   What Is a Model?

■ LEARNING GOAL 1

Atomic theory and models

Before we try to discuss the atom, we should recall that until very recently no one had ever seen one. How, then, did we know what an atom was like? We really didn't, exactly. But scientists have put together observations, experimental results, and a good deal of reasoning and have come up with a *model*. This model of the atom is just a representation of what scientists believe the atom is like. Models can have different degrees of accuracy. Take a model of a car, for instance. It could be carved from a solid block of wood or plastic and represent just the shape of the car. Or a more detailed model could show the interior of the car as well. A still more detailed model might have a gasoline-powered motor and move just like a real car. This model would represent the car most accurately. The car model would be a physical model, whereas models of the atom are *intellectual* or *thought* models; however, they too have become more accurate as scientists have gathered more information about atoms.

4.2   Dalton's Atomic Theory

John Dalton (1766–1844) was a British schoolteacher who studied physics and chemistry. On the basis of facts and experimental evidence, he proposed an atomic theory. This theory was an attempt to explain all the different forms of matter. It has turned out to be one of the greatest contributions to chemistry since the time of the ancient Greeks. Dalton theorized that:

1. All elements are composed of tiny, indivisible particles called atoms. This idea was similar to Democritus' idea. Both men believed there was an ultimate particle—that all matter was composed of tiny indestructible and indivisible spheres.

John Dalton
(Reprinted with permission from
Torchbearers of Chemistry by
Henry Monmouth Smith,
© Academic Press, Inc., 1949)

2. All matter is composed of combinations of these atoms. In Chapter 3 we discussed how the atoms of different elements combine to produce formula units that make up compounds. For example, two atoms of hydrogen and one atom of oxygen combine to form a molecule of water.
3. Atoms of different elements are different. Dalton believed gold was different from silver because somehow the atoms of gold were different from the atoms of silver. For example, he thought they differed in mass.
4. Atoms of the same element have the same size, mass, and form. Dalton believed all gold atoms, for example, were the same as all other gold atoms in every respect.

Dalton's atomic theory laid the foundation for what is believed today. But his basic idea about the indivisibility of the atom was eventually shown to be incorrect, as was his idea that all atoms of the same element have the same size, mass, and form. Revisions of his theory have led to our current model of the atom. Let us review how our present-day model of the atom was developed.

4.3   The Discovery of the Electron

In the mid-1800s, scientists began to question whether the atom was really indivisible. They also wanted to know why atoms of different elements had different properties. Some of the answers came with the invention of the Crookes tube, or *cathode-ray tube* (Figure 4.1).

The scientists of the time knew that some substances conduct electric current (that is, they are conductors), whereas other substances do not. And, with enough electrical power, a current can be driven through any substance—solid, liquid, or gas. Electric current results from the accumula-

■ **FIGURE 4.1**

A Crookes or cathode-ray tube

tion of electric charge. Electric charge may be transferred from one object to another. We find that electric charge is either positive or negative. Experiments tell us that unlike charges attract, and like charges repel. In the cathode-ray tube, a high-voltage electric current is driven through a nearly empty tube, a near **vacuum**. The tube contains two pieces of metal called *electrodes*. Each electrode is attached by a wire to the source of an electric current. The source has two *terminals,* positive and negative. *The electrode attached to the positive electric terminal* is called the **anode;** *the electrode attached to the negative terminal* is called the **cathode.** Crookes showed that when the current was turned on, a beam moved from the cathode to the anode; in other words, the beam moved from the negative to the positive terminal. Because the beam originated at the *negative* terminal, and because it was attracted to the *positive* terminal, the beam had to be negative in nature.

LEARNING GOAL 2

Mass and charge of electrons, protons, and neutrons

What was this beam? Was it made of particles or waves? Did it come from electricity or from the metal electrodes? Physicists in Crookes's time were not sure about the answers to these questions, but they did make guesses. Whatever this beam was, it traveled in straight lines (they knew that because it cast sharp shadows). For lack of a better name, the German physicist Eugen Goldstein called the beams *cathode rays,* because they came from the cathode.

The German physicists in Crookes's time favored the *wave theory* of cathode rays because the beam traveled in straight lines, like ocean waves. But the English physicists favored the *particle theory.* They said that the beam was composed of tiny particles that moved very quickly—so quickly that they were hardly influenced by gravity. That was why the particles moved in a straight path. (Note how a single experimental observation led to two different theories.)

Crookes proposed a method to solve the dilemma. If the beam was composed of negative particles, a magnet would deflect them (Figure 4.2). But if the beam was a wave, a magnet would cause almost no deflection. Particles would also be more easily deflected by an electric field. In 1897 the English physicist J. J. Thomson used both these techniques—magnetic and electric—to show that the rays were composed of particles (Figure 4.3). Today we call these particles **electrons.** Moreover, we understand that an electric current is actually a flow of electrons.

Were the electrons in the beam coming from the metal electrodes or from the current source? Unless it was the metal that gave off these electrons, the atom could still be the smallest particle of matter. Proof that metals do give off electrons came from the laboratories of Phillipp Lenard, a German physicist. In 1902 he showed that ultraviolet light directed onto a metal makes the metal send out, or emit, electrons. This effect, known as the *photoelectric effect,* indicated that the metal atoms—and the atoms of other elements—contain electrons.

In 1911 a young American physicist named Robert Millikan calculated the mass of the electron: 9.11×10^{-28} grams. (To get an idea of how very small this is, consider the fact that it would take about

■ **FIGURE 4.2**

A cathode-ray tube in operation. Compare these pictures with Figure 4.3. (Fundamental Photographs)

■ **FIGURE 4.3**

The Thomson experiment: (1) A beam of electrons (dashed line) moves from cathode toward anodes. (2) Some electrons pass between anodes. (3) Electric field causes beam of electrons to bend. This is visible on the screen. (4) But by adding a magnet to counteract deflection caused by electric field, one can make the beam follow a straight path. (5) One can then measure the strengths of the electric and magnetic fields and calculate the charge-to-mass ratio (e/m) of the electron.

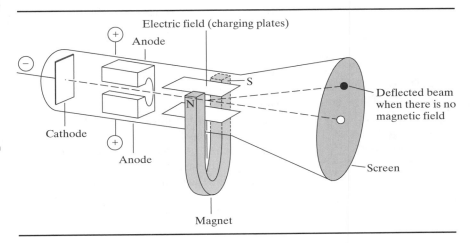

16,000,000,000,000,000,000,000,000,000, or 1.6×10^{28}, electrons to equal the weight of a standard half-ounce chocolate bar.)

4.4   The Proton

The discovery of the electron as part of the atom and as a *negative* particle of electricity was very important, but it raised many questions. Because electrons are part of all atoms, and because atoms in their normal state are electrically *neutral,* scientists reasoned, there must be something else in the atom that balances the negative electron. Otherwise, if matter had a negative charge, you would get a shock every time you touched anything.

This "something else," it seemed, would have to be some other sort of particle that has the same amount of charge as the electron but is positive in nature. Such reasoning led scientists to suppose that there were positive particles, which they called **protons,** in the atom. But then why could atoms give off only negative particles (electrons) and not positive ones (protons)? The British physicist J. J. Thomson suggested an answer. He thought of the atom as a sphere made up of positive electricity in which electrons were embedded (Figure 4.4). The English called this the *plum-pudding theory,* likening the electrons to the raisins in plum pudding. As Americans, we can find a more familiar analogy in a scoop of chocolate-chip ice cream. The chocolate chips represent the electrons. The ice cream represents a sea of positive electricity (composed of protons). Each element is different from the others because each has a different number of electrons and protons arranged in a different way—like different scoops of ice cream with different numbers of chocolate chips. For example, hydrogen atoms have one electron and one proton. Helium atoms have two electrons and two protons. Thomson seemed to have given scientists an idea they could agree on, but his theory still needed experimental proof.

Lord Rutherford, a former student and colleague of Thomson, attempted to test Thomson's theory. He had been working with positive particles called **alpha particles** (α particles), which are actually helium atoms with their electrons removed. He reasoned that if he shot a stream of these positive particles through a gold foil, the neutral and symmetrical "plum-pudding" gold atoms would not greatly affect the path of the alpha particles. Most of the particles would move straight through the foil (Figure 4.5). Rutherford used gold foil in his experiment because it could be hammered into very thin sheets, making the thickness of the foil resemble a single layer of atoms. However, when Rutherford actually performed the experiment, he found that about one alpha particle in 20,000 ricocheted, or bounced back out of the foil! As Rutherford said, "It was almost as incredible as if you fired a fifteen-inch shell at a piece of tissue paper and it came back and hit you."

To explain these peculiar results, Rutherford devised a new model for the atom, called the nuclear model. We subscribe to this model today. Rutherford said that an atom must have a *center of positive charge,* where all the protons must be located. He called this center of positive charge the **nucleus**

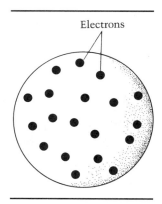

Electrons

■ **FIGURE 4.4**

The Thomson model of the atom

Lord Rutherford
(UPI/Corbis-Bettmann)

of the atom. Rutherford felt that the nucleus of the atom must be very small compared to the overall size of the atom. He also said that the nucleus must be where the mass of the atom is concentrated. Rutherford's reasoning was that some alpha particles were tremendously deflected (Figure 4.6) because they were repelled by a high concentration of positive charge with an immovable mass. He also decided that the electrons were located around the nucleus instead of in it and that the mass of the electrons was very small compared to the mass of the protons. His reasoning here was that for so many of the particles to go through the foil undeflected, the electrons had to be located in a relatively large space outside the nucleus but still within the atom.

Later experiments similar to Rutherford's showed that the diameter of the nucleus of an atom is approximately 10^{-13} cm, whereas the diameter of the whole atom is up to 100,000 times as great. Suppose we could expand an atom so that the nucleus was 1 millimeter in diameter (about the size of this dot: ·). Then the nearest electron could be as far as 50 meters (half the length of a football field) away from the nucleus! And between the nucleus and the electrons is only vast empty space.

Rutherford's model of the atom, based as it was on new experimental evidence, made the Thomson model of the atom obsolete.

■ **FIGURE 4.5**

According to the Thomson model of the atom, positive particles ought to pass through a metal foil without being deflected.

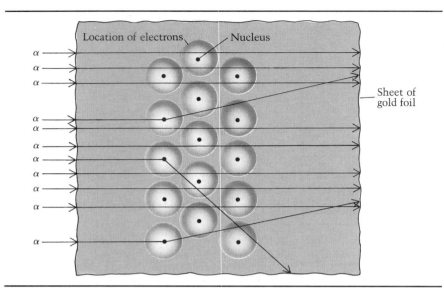

Location of electrons   Nucleus

Sheet of gold foil

■ FIGURE 4.6

The actual way that positive particles pass through metal foil, as shown by Rutherford

4.5 The Neutron

Thus, before 1920, scientists had experimentally confirmed the existence of two basic *subatomic* particles (particles that are fundamental constituents of an atom): electrons, which are negatively charged and have a very small mass, and protons, which are positively charged and contain most of the atom's mass. But a problem came up. Scientists had defined a unit of mass called the **atomic mass unit** (abbreviated amu). They knew that the proton had a mass of 1.0072766 amu and that the electron had a much smaller mass of 0.0005486 amu. These numbers all seemed fine at first. The mass of hydrogen—which is made up of one proton and one electron—was very close to the sum of the masses of a proton and an electron. (Remember that protons account for just about all of the atom's mass; when we are calculating the mass of an atom we can usually ignore the mass of the electrons.) The problem arose when scientists considered helium. The helium atom has two electrons and two protons, so the scientists thought it should have a mass of about 2 amu. But its mass is 4 amu. And when they looked at other elements, they were even more puzzled. Carbon has six electrons and six protons but a mass of 12 amu. Where was the extra mass?

To account for the extra mass in the atom, scientists proposed another particle, one that had no charge but had a mass equal to the proton's. In 1932, the British physicist James Chadwick (1891–1974) was able to prove the existence of a new subatomic particle called the **neutron,** which has a mass of 1.008665 amu. If this particle were also present in the atom, it would add to the mass but would not affect the charge (see Table 4.1). Scientists

TABLE 4.1

The Particles in the Atom

PARTICLE	SYMBOL	CHARGE	MASS (amu)	MASS (grams)
Proton	p	+1	1.0073	1.672×10^{-24}
Electron	e	−1	0.000549	9.107×10^{-28}
Neutron	n	0	1.0087	1.675×10^{-24}

could now picture helium as being composed of two electrons, two protons, and two neutrons, so they could understand helium's mass of 4 amu.

4.6   Atomic Number

LEARNING GOAL 3

Number of protons, electrons, and neutrons

11   ← Atomic number
Na
22.990 ← Atomic mass

■ **FIGURE 4.7**

In the periodic table used in this text, an element's atomic number appears above its symbol.

The factor that makes one element different from another element is that atoms of different elements have different numbers of electrons and protons. *The number of protons in an atom of an element* is called the **atomic number** of that element and is symbolized by Z.

The atomic number of an element is also equal to the positive charge on the nucleus and to the number of electrons in the neutral atom of that element. On any periodic table there is a whole number identified with each element. This is the atomic number Z. In the periodic table inside the front cover, the atomic number is given above the symbol for each element. An atom of sodium (Na), which has the atomic number 11, has 11 electrons and 11 protons (Figure 4.7). An atom of magnesium (Mg), which has the atomic number 12, has 12 electrons and 12 protons; and an atom of uranium (U), which has the atomic number 92, has 92 electrons and 92 protons. Note that the elements are listed in the periodic table in the order of their atomic numbers.

4.7   Isotopes

LEARNING GOAL 4

Definition of an isotope

If you look at the periodic table, you will see that the atomic masses of most elements are not whole numbers. For example, sodium (Na) has a mass of 22.990 amu, titanium (Ti) a mass of 47.90 amu, and rubidium (Rb) a mass of 85.468 amu. Why is this so? If each proton and each neutron gives a mass of almost exactly 1 amu to the atom, then the atomic masses of the elements certainly ought to be very close to whole numbers.

Investigation of this question led to the discovery that *each element can exist in more than one form.* When John Dalton said that the atoms of an element must be identical in size, mass, and shape, he did not know that *isotopes,* or different forms of an element, could exist. **Isotopes** are *forms*

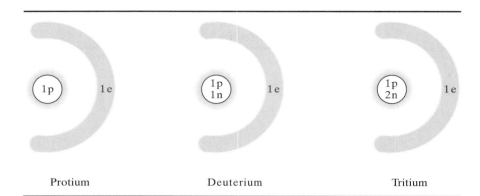

Protium   Deuterium   Tritium

of an element that have the same number of electrons and protons but different numbers of neutrons.

Although the numbers of negative and positive charges of isotopes are identical, isotopes differ in mass. Hydrogen, for example, exists in three isotopic forms (Figure 4.8). The common form of hydrogen (called protium) contains one electron and one proton. Another form of hydrogen (called deuterium) contains one electron, one proton, and one neutron. A third form of hydrogen (called tritium) contains one electron, one proton, and two neutrons.

Note that all isotopes of an element have the same number of protons and thus have the same atomic number. However, it is possible to distinguish among isotopes by means of mass numbers. The **mass number** of an isotope is simply *the sum of the number of protons and the number of neutrons.*

A standard notation has been developed for giving the mass number and atomic number of an isotope. It consists of writing the symbol of the element with its mass number as a *superscript* (above the line) and its atomic number as a *subscript* (below the line). Both numbers are usually written to the left of the symbol for the element.

LEARNING GOAL 5

Standard notation for mass number and atomic number

$$\text{Mass number} \rightarrow \quad {}^{16}_{8}\text{O} \quad {}^{12}_{6}\text{C} \quad {}^{35}_{17}\text{Cl}$$
$$\text{Atomic number} \rightarrow$$

With this notation we can immediately tell the composition of the isotope in terms of protons, electrons, and neutrons. The number of protons (or, in the neutral atom, the number of electrons) is simply the atomic number of the element. The number of neutrons is the difference between the mass number and the atomic number. This notation allows us to distinguish among the isotopes of an element.

EXAMPLE 4.1

Give the number of electrons, protons, and neutrons in each of the following neutral isotopes: (a) ${}^{238}_{92}\text{U}$ (b) ${}^{235}_{92}\text{U}$ (c) ${}^{84}_{36}\text{Kr}$

Solution

The atomic number (subscript) gives the number of electrons in the neutral atom as well as the number of protons. Subtracting the atomic number from the mass number (superscript) gives the number of neutrons.

(a) Because the atomic number is 92, there are 92 electrons and 92 protons. And subtracting the atomic number from the mass number gives the number of neutrons: $238 - 92 = 146$ neutrons. So there are 92 electrons, 92 protons, and 146 neutrons in an atom of $^{238}_{92}U$.

(b) Again there are 92 electrons and 92 protons. And $235 - 92 = 143$ neutrons. So there are 92 electrons, 92 protons, and 143 neutrons in an atom of $^{235}_{92}U$. Note that $^{238}_{92}U$ and $^{235}_{92}U$ are an *isotope pair*.

(c) Here we have 36 electrons and 36 protons. And $84 - 36 = 48$ neutrons. So there are 36 electrons, 36 protons, and 48 neutrons in an atom of $^{84}_{36}Kr$.

Practice Exercise 4.1   Give the number of electrons, protons, and neutrons in each of the following isotopes: (a) $^{28}_{14}Si$ (b) $^{31}_{15}P$ (c) $^{6}_{3}Li$

The following notations can be used to distinguish among the three isotopes of hydrogen:

$$^{1}_{1}H \qquad\qquad ^{2}_{1}H \qquad\qquad ^{3}_{1}H$$

Protium   Deuterium   Tritium

LEARNING GOAL 6

Average atomic mass of an element

The deuterium and tritium forms of hydrogen are not as common as the protium form. If a random sample of all hydrogen atoms found on the earth were taken, it would be found to consist of 99.985% protium atoms, 0.015% deuterium atoms, and a vanishingly small amount of tritium atoms. The atomic mass given in the periodic table is the *weighted average of the atomic masses* of all three isotopes. *When calculating the average atomic mass for any element, we must take into account the relative abundance of each isotope of that element and its exact isotopic mass.* For example, the exact isotopic mass of $^{1}_{1}H$ is 1.0078 amu, and the exact isotopic mass of $^{2}_{1}H$ is 2.0141. It is for this reason that the atomic mass of hydrogen shown in the periodic table is 1.008 amu. This is nearly the same as the mass of protium, the most common isotope of hydrogen. Table 4.2 gives a breakdown of the isotopes of the first fifteen elements and their isotopic masses. This averaging process also explains why the atomic masses of the elements in the periodic table are not whole numbers.

LEARNING GOAL 7

Percentage abundance of isotopes

LEARNING GOAL 8

Table of relative abundances of isotopes

If we know the percentage abundance of each isotope of an element, we can calculate the average atomic mass of the element. The following example shows how to do this. In the example we use the exact isotopic mass.

EXAMPLE 4.2

Using Table 4.2, calculate the atomic mass of boron.

TABLE 4.2

Naturally Occurring Isotopes of the First Fifteen Elements

NAME	SYMBOL	ATOMIC NUMBER	MASS NUMBER	ISOTOPIC MASS	PERCENTAGE NATURAL ABUNDANCE
Hydrogen-1	$_1^1H$	1	1	1.007825	99.985
Hydrogen-2	$_1^2H$	1	2	2.01410	0.015
Hydrogen-3	$_1^3H$	1	3	3.01605	Negligible
Helium-3	$_2^3He$	2	3	3.016	0.00013
Helium-4	$_2^4He$	2	4	4.003	99.99987
Lithium-6	$_3^6Li$	3	6	6.015	7.42
Lithium-7	$_3^7Li$	3	7	7.016	92.58
Beryllium-9	$_4^9Be$	4	9	9.012	100
Boron-10	$_5^{10}B$	5	10	10.013	19.6
Boron-11	$_5^{11}B$	5	11	11.009	80.4
Carbon-12	$_6^{12}C$	6	12	12.0000	98.89
Carbon-13	$_6^{13}C$	6	13	13.003	1.11
Nitrogen-14	$_7^{14}N$	7	14	14.003	99.63
Nitrogen-15	$_7^{15}N$	7	15	15.000	0.37
Oxygen-16	$_8^{16}O$	8	16	15.995	99.759
Oxygen-17	$_8^{17}O$	8	17	16.999	0.037
Oxygen-18	$_8^{18}O$	8	18	17.999	0.204
Fluorine-19	$_9^{19}F$	9	19	18.998	100
Neon-20	$_{10}^{20}Ne$	10	20	19.992	90.92
Neon-21	$_{10}^{21}Ne$	10	21	20.994	0.257
Neon-22	$_{10}^{22}Ne$	10	22	21.991	8.82
Sodium-23	$_{11}^{23}Na$	11	23	22.9898	100
Magnesium-24	$_{12}^{24}Mg$	12	24	23.9850	78.70
Magnesium-25	$_{12}^{25}Mg$	12	25	24.9858	10.13
Magnesium-26	$_{12}^{26}Mg$	12	26	25.9826	11.17
Aluminum-27	$_{13}^{27}Al$	13	27	26.9815	100
Silicon-28	$_{14}^{28}Si$	14	28	27.9769	92.21
Silicon-29	$_{14}^{29}Si$	14	29	28.9765	4.70
Silicon-30	$_{14}^{30}Si$	14	30	29.9738	3.09
Phosphorus-31	$_{15}^{31}P$	15	31	30.9738	100

Solution

From Table 4.2, the percentage abundance of ^{10}B is 19.6%, and its exact isotopic mass is 10.013 amu. That of ^{11}B is 80.4%, and its exact isotopic mass is 11.009 amu. Therefore the atomic mass of

$$B = (10.013 \text{ amu})(0.196) + (11.009 \text{ amu})(0.804)$$
$$= 1.96 \text{ amu} + 8.85 \text{ amu} = 10.81 \text{ amu}$$

For a review of changing percentages to decimal equivalents, see Example A.19 in Appendix A.

Practice Exercise 4.2   Using Table 4.2, calculate the atomic mass of lithium.

The concept of isotopes with different abundances allows us to solve two other kinds of quantitative problems. These are illustrated by the next two examples.

EXAMPLE 4.3

A hypothetical element Q exists in two isotopic forms, ^{270}Q and ^{280}Q. The atomic mass of Q is 276 amu. What is the percentage abundance of each isotope? (Assume that the exact isotopic mass is the same as the mass number for each isotope.)

Solution

UNDERSTAND THE PROBLEM

If we know the percentage abundance of each isotope of an element, we can calculate the average atomic mass of the element. This problem requires that we determine the percentage abundance of each element when given the atomic mass.

DEVISE A PLAN

We will use algebra and the ability to solve an algebraic equation. We will let x equal the fraction of ^{270}Q and $(1 - x)$ equal the fraction of ^{280}Q. Then we have

$$270x + 280(1 - x) = 276$$

CARRY OUT THE PLAN

$$x = 0.40$$
$$1 - x = 0.60$$

In other words, $\overline{40}$% of the element is ^{270}Q and $6\overline{0}$% of the element is ^{280}Q.

LOOK BACK

We ask if the solution makes sense. Using estimation we can reason that if 50% of the element were ^{270}Q and 50% were ^{280}Q, then the atomic mass of the element should be 275. Since it is more than 275, there should be a slightly greater amount of ^{280}Q present, and that is the case with this solution.

Practice Exercise 4.3   A hypothetical element Y exists in two isotopic forms, ^{100}Y and ^{110}Y. The atomic mass of Y is 108. What is the percentage abundance of each isotope? (Assume that the exact isotopic mass is the same as the mass number for each isotope.)

EXAMPLE 4.4

Using Table 4.2, find how many ^{17}O atoms there are in a sample containing 1,000,000 oxygen atoms.

Solution

From Table 4.2, we find that the natural abundance of ^{17}O is 0.037%. Therefore the number of ^{17}O atoms in 1,000,000 O atoms equals

$$(0.00037)(1,000,000) = 370 \text{ atoms of } {}^{17}O$$

Practice Exercise 4.4   Using Table 4.2, find how many ^{20}Ne atoms there are in a sample containing 1,000,000 neon atoms.

4.8   What Next for the Atom?

We have traced the development of a model for the atom from the time of Dalton, who believed that the atom was an indivisible unit of matter, to the time of Rutherford, who suggested that the atom has a nucleus composed of protons and neutrons and an outer part for the electrons. But many questions were still left unanswered. How were the electrons arranged in an atom? Where were the electrons? How were they moving? How fast were they going? Could they account for the formation of compounds? Why didn't these negative electrons "fall into" the positive nucleus of the atom? After all, opposite charges attract! We will answer these questions in Chapter 5.

▮▮▮ SUMMARY

John Dalton's atomic theory was an attempt to explain all the different forms of matter. He theorized that (1) the atom is indivisible; (2) all atoms of the same element have the same mass, size, and form; (3) all matter is composed of combinations of atoms; and (4) atoms of different elements are different. Although the first two of these ideas were later shown to be incorrect, most of Dalton's theory is the basis for what is still believed today. Dalton's model of the atom was modified as various subatomic particles—protons, electrons, and neutrons—were discovered. Thomson's theory of the atom followed his "discovery" of electrons. This British physicist theorized that the atom is a sphere made up of positive electricity in which negatively charged electrons are embedded. Lord Rutherford, a colleague of Thomson, tested Thomson's theory with an experiment in which a stream of positive (alpha) particles was supposed to pass straight through the atoms in a gold foil with only a few particles being slightly deflected. Instead, some of the alpha particles bounced back. To explain this result, Rutherford theorized that an atom has a relatively very small center where all the positively charged protons and most of the atomic mass are located. Electrons, he postulated, are located outside the nucleus, perhaps relatively far from it.

One difficulty arises from the two-particle atomic model. For most atoms, the masses of all the protons and electrons did not account for the total mass of the atom. In 1932 James Chadwick was able to prove the existence of a new subatomic particle called the neutron. The neutron has about the same mass as a proton but carries no charge. With neutrons, scientists were able to account for the total mass of atoms.

The number of protons in the atom of an element is called the atomic number of that element. The sum of the number of protons and the number of neutrons is called the mass number of the element. All the atoms of any element have the same atomic number, but they may have different mass numbers. Two or more atoms with the same atomic number but different mass numbers are called isotopes.

KEY TERMS

alpha particle (4.4)

anode (4.3)

atomic mass unit (4.5)

atomic number (4.6)

cathode (4.3)

electron (4.3)

isotope (4.7)

mass number (4.7)

neutron (4.5)

nucleus (4.4)

proton (4.4)

vacuum (4.3)

CAREER SKETCH

Health Careers/Radiation Related

The medical profession offers a number of different opportunities for people with backgrounds in science, particularly for those who are

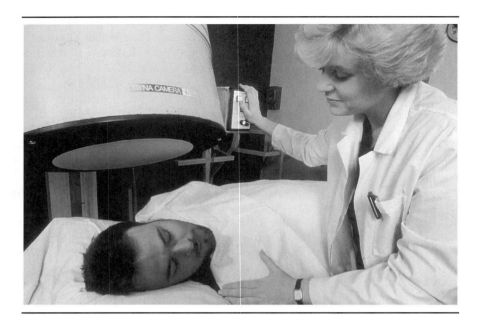

This therapeutic radiation physicist uses state-of-the-art equipment for diagnostic purposes.
(Larry Mulvehill/Photo Researchers, Inc.)

interested in the field but do not want to become doctors. Careers dealing with radiation lend themselves to those with a science education.

Consider the post of *therapeutic radiation physicist*. This person works closely with physicians, therapists, and others. Responsibilities may include scheduling computerized radiation treatments or making physics measurements on equipment used to treat patients.

Another health career related to radiation is that of *radiation safety officer*. This person's responsibilities include ensuring the safety of the treatment machines, generating tables to be used in treatment calculations, and planning patient treatments using the latest technologies.

Ultimately, the goal of almost any radiation scientist is to increase survival from cancer. To reach this goal, a radiation scientist may be responsible for such complex tasks as effective integration of local–regional therapy with ionizing radiation and/or resection and systemic therapy with cytotoxic drugs and hormones. The prerequisite to completing these tasks is a strong background in the sciences.

For more information on careers as a radiation scientist, point your web browser at the Radiation Therapy Oncology Group, http://www.rtog.org. For general information about careers in medicine, try http://www.medsearch.com.

Warning! Volcano Erupting Soon

Around the planet there are approximately 500 volcanoes that have a record of violent behavior. Currently, however, scientists monitor only about 20 of these volcanoes and only have the resources to occasionally observe one-third of them. As a result, often the first warning signs from a volcano that has been dormant for hundreds of years are the initial blast and the flow of magma down its side. By then, it may be too late to save the lives of those living near the volcano as the avalanche of hot ash, 1,500 degrees Fahrenheit and moving at speeds up to 500 miles per hour, travels down the mountain.

One solution, developed in the early and mid-1990s and still used today, is a satellite-based system that is able to make infrared and radar observations of volcanoes. Such a system monitors changes on a volcano and alerts scientists to a potentially explosive situation. While a great advance in volcano observation, this system does not have the capabilities to warn of impending disaster.

Enter chemistry. Scientists have discovered that before the initial blast, as the gases rise up a conduit inside the volcano, sulfur dioxide is emitted from the volcano's dome. Scientists theorize that if they are able to measure the amount of sulfur dioxide coming out of a volcano, they will be able to forewarn residents of nearby towns or villages about impending blast.

To accomplish this on a planetary level, satellite-based observations are being considered. Current satellite systems, such as the one just described, are able to detect chemicals on earth. However, they are not able to detect the small emissions of sulfur dioxide that foreshadow a volcanic eruption. But new instrumentation being developed by a NASA scientist could change this. Alin Krueger, working at the Goddard Space Flight Center in Maryland, is proposing a new system, which would check SO_2 emissions at every volcano every 15 minutes. If successful, such a system could sound the alarm before the next volcanic eruption on Planet Earth.

SELF-TEST EXERCISES

LEARNING GOAL 1: *Atomic Theory and Models*

1. Explain the idea of a model. Give an example of a physical model and an intellectual model.

◀ 2. The elements carbon and oxygen form two different compounds, carbon dioxide, which is a product of animal respiration, and carbon monoxide, which is a product of incomplete combustion when gasoline is burned. Discuss how Dalton's atomic theory explains these facts.

3. List the points in Dalton's atomic theory. Explain which ones had to be corrected after later experiments were performed.

◀ 4. Compare Dalton's atomic theory to the ideas set forth by Democritus in 420 B.C.

5. Contrast the Thomson model and the Rutherford model of the atom.

6. Give a general outline of the construction of a Crookes tube.

◄ 7. Explain how a television set is similar to a Crookes tube. (You may want to check an outside reference source for help in answering this question.)

8. Discuss the experiment that was conducted to test the wave and particle theories of cathode rays.

LEARNING GOAL 2: *Mass and Charge of Electrons, Protons, and Neutrons*

9. How many electrons would it take to equal the mass of one proton? Use Table 4.1 to obtain the masses of these subatomic particles.

10. Which of the three subatomic particles has the greastest mass? Which has the least mass?

11. Name the three major subatomic particles. Give their charges and approximate atomic masses (in amu).

◄ 12. Compare the charges of the three subatomic particles and briefly explain how each charge was determined.

LEARNING GOAL 3: *Number of Protons, Electrons, and Neutrons*

13. Find the number of protons (p), electrons (e), and neutrons (n) in neutral atoms of the following: (a) $^{244}_{94}$Pu (b) $^{48}_{22}$Ti (c) $^{262}_{103}$Lr

14. Find the number of protons (p), electrons (e), and neutrons (n) in neutral atoms of the following: (a) $^{75}_{33}$As (b) $^{266}_{109}$Mt (c) $^{222}_{86}$Rn

15. Find the number of protons, electrons, and neutrons in neutral atoms of the following: (a) $^{16}_{8}$O (b) $^{17}_{8}$O (c) $^{18}_{8}$O (d) $^{20}_{10}$Ne (e) $^{21}_{10}$Ne (f) $^{22}_{10}$Ne

16. Find the number of protons, electrons, and neutrons in neutral atoms of the following: (a) $^{31}_{15}$P (b) $^{22}_{10}$Ne (c) $^{24}_{12}$Mg

17. From the information that follows, write the isotopic notation for each atom, then decide which one is not an isotope of the other two atoms. (a) 8p, 8e, 8n (b) 7p, 7e, 8n (c) 8p, 8e, 9n

◄ 18. Which of the following describes an atom that is not an isotope of the other two? (a) 12p, 12e, 14n (b) 12p, 12e, 12n (c) 13p, 13e, 12n

19. Find the number of protons, electrons, and neutrons in neutral atoms of the following: (a) $^{57}_{25}$Mn (b) $^{60}_{27}$Co (c) $^{80}_{36}$Kr (d) $^{128}_{52}$Te

20. Determine the number of protons, electrons, and neutrons in neutral atoms of the following hypothetical isotopes: (a) $^{35}_{16}$X (b) $^{300}_{90}$Y

◄ **21.** The alchemists wanted to turn lead into gold. Suppose that you had the isotope of lead $^{208}_{82}$Pb, and you wanted to turn it into the isotope of gold $^{197}_{79}$Au. How many protons, electrons, and neutrons would you have to remove from the lead atom to turn it into this gold atom?

22. Which of the following describes an atom that is not an isotope of the other two? (a) 14p, 14e, 15n (b) 15p, 15e, 14n (c) 15p, 15e, 15n

◄ **23.** Fill in the blanks in the following table. (Use the periodic table when necessary.)

SYMBOL	PROTONS	ELECTRONS	NEUTRONS	MASS NUMBER	ATOMIC NUMBER
$^{174}_{70}$Yb					
	59			141	
			60		44
			24	45	
		22	28		
$^{2}_{12}$Mg			13		

24. State the number of protons, electrons, and neutrons in $^{42}_{20}$Ca.

LEARNING GOAL 4: *Definition of an Isotope*

25. (a) Can two different isotopes of a particular element have the same mass number? Explain.
    (b) Can atoms of two different elements have the same mass number? Explain.

26. Atoms of an element that have the same number of electrons and protons but different numbers of neutrons are called _____.

27. Isotopes are atoms with the same number of _____ and _____ but different numbers of _____. (Place the words *protons, electrons,* and *neutrons* in the proper places.

28. (a) The number of protons in an atom is called the _____ _____ of that element.
    (b) The sum of the numbers of protons and neutrons in an isotope is called the _____ _____ of that isotope.

LEARNING GOAL 5: *Standard Notation for Mass Number and Atomic Number*

29. Write the standard isotopic notation for the following atoms: (a) 80p, 80e, 120n (b) 7p, 7e, 8n (c) 13p, 13e, 14n (d) 107p, 107e, 155n

30. Write the standard isotopic notation for the following atoms: (a) 3p, 3e, 3n (b) 9p, 9e, 10n (c) 14p, 14e, 16n

LEARNING GOAL 6: *Average Atomic Mass of an Element*

31. The element gallium (Ga) exists in two isotopic forms with the following abundances: 60.16% ^{69}Ga and 39.84% ^{71}Ga. Calculate the atomic mass of gallium. (The exact isotopic mass of ^{69}Ga is 68.9257. The exact isotopic mass of ^{71}Ga is 70.9249.)

32. The element carbon (C) exists in two isotopic forms with the following abundances: 98.89% ^{12}C and 1.11% ^{13}C. Calculate the atomic mass of carbon.

33. The element chlorine (Cl) exists naturally in two isotopic forms with the following abundances: 75.53% ^{35}Cl and 24.47% ^{37}Cl. Calculate the atomic mass of chlorine. (The exact isotopic mass of ^{35}Cl is 34.9689. The exact isotopic mass of ^{37}Cl is 36.9659.)

34. The element magnesium (Mg) exists in three isotopic forms with the following abundances: 78.70% ^{24}Mg, 10.13% ^{25}Mg, and 11.17% ^{26}Mg. Calculate the atomic mass of magnesium. Use Table 4.2.

35. The element chromium (Cr) exists naturally in four isotopic forms with the following abundances: 4.31% ^{50}Cr, 83.76% ^{52}Cr, 9.55% ^{53}Cr, and 2.38% ^{54}Cr. Calculate the atomic mass of chromium. (The exact isotopic mass of ^{50}Cr is 49.9461. The exact isotopic mass of ^{52}Cr is 51.9405. The exact isotopic mass of ^{53}Cr is 52.9407. The exact isotopic mass of ^{54}Cr is 53.9389.)

36. The element copper exists in two isotopic forms with the following abundances: 69.09% ^{63}Cu and 30.91% ^{65}Cu. Calculate the atomic mass of copper. (The exact isotopic mass of ^{63}Cu is 62.9296. The exact isotopic mass of ^{65}Cu is 64.9278.)

LEARNING GOAL 7: *Percentage Abundance of Isotopes*

***37.** Bromine exists in two isotopic forms, ^{79}Br and ^{81}Br. The atomic mass of bromine is 79.90 amu. Calculate the percentage abundance of each isotope. (The exact isotopic mass of ^{79}Br is 78.9183. The exact isotopic mass of ^{81}Br is 80.9163.)

38. Boron (B) exists in two isotopic forms, ^{11}B and ^{10}B. The atomic mass of boron is 10.81 amu. Calculate the percentage abundance of each isotope. Use Table 4.2.

***39.** The element antimony occurs naturally in two isotopic forms, ^{121}Sb and ^{123}Sb. The atomic mass of antimony is 121.75 amu. Calculate the percentage abundance of each isotope. (The exact isotopic mass of ^{121}Sb is 120.9038. The exact isotopic mass of ^{123}Sb is 122.9041.)

40. Chlorine exists in two isotopic forms, ^{35}Cl and ^{37}Cl. The atomic mass of chlorine is 35.453 amu. Calculate the percentage abundance of each isotope. (The exact isotopic mass of ^{35}Cl is 34.9689 and that of ^{37}Cl is 36.9659.)

LEARNING GOAL 8: *Table of Relative Abundances of Isotopes*

41. Using Table 4.2, determine how many ^{22}Ne atoms you will find in 1.00×10^9 neon atoms.

42. Using Table 4.2, determine how many ^{29}Si atoms you will find in 5.00×10^6 silicon atoms.

_43._ Using Table 4.2, determine how many atoms of oxygen you would need to get 7,400 atoms of ^{17}O.

44. Using Table 4.2, determine how many atoms of oxygen you would need to get 102 atoms of the isotope ^{18}O.

_45._ Using Table 4.2, determine how many ^{6}Li atoms you would find in 100,000,000 lithium atoms.

46. Using Table 4.2, determine how many ^{7}Li atoms you would find in 1,000 lithium atoms.

◄**_47._** Using Table 4.2, determine how many ^{26}Mg atoms you would find in 5.00×10^{6} Mg atoms.

48. Using Table 4.2, determine how many ^{14}N atoms you would find in 1.000×10^{14} nitrogen atoms.

_49._ Using Table 4.2, determine how many ^{13}C atoms you would find in 2.00×10^{4} C atoms.

◄50. Using Table 4.2, determine how many atoms of ^{22}Ne you would find in a sample of neon that contains 1×10^{6} neon atoms.

_51._ Using Table 4.2, determine how many ^{15}N atoms you would find in 1.0×10^{15} N atoms.

52. Using Table 4.2, determine how many ^{6}Li atoms you would find in 1,000 atoms of Li.

_53._ Using Table 4.2, determine how many atoms of hydrogen you would need to get 150,000 atoms of ^{2}H.

54. Using Table 4.2, determine how many ^{14}N atoms you would find in exactly 100,000,000 atoms of N.

EXTRA EXERCISES

55. The existence of isotopes became known to scientists around the year 1900, about 100 years after Dalton first proposed his atomic theory. The proof that isotopes existed were associated more with physics experiments than with chemistry experiments. What does this tell you about the chemical properties of isotopes? Explain.

◄**_56._** Consider the following unknown atoms.

$$^{200}_{84}A \qquad ^{208}_{82}B \qquad ^{222}_{86}C \qquad ^{184}_{74}D$$

(a) Which atom has the most electrons in the neutral state?
(b) Which atom has the fewest neutrons?
(c) What is the mass number of element C?
(d) Which atom has the fewest protons?
(e) Identify element A, using a periodic table.

_57._ What is the relationship between the atomic number and the number of electrons in a neutral atom?

58. A student produces a beam of rays and can't decide whether these rays are cathode rays or alpha particles. Explain a simple experiment that the student might perform to determine what these rays are.

_59._ What experiment gave the proof that atoms of all elements contain electrons?

_60._ Indicate the number of protons, electrons, and neutrons that each of the following atoms contains: (a) $^{13}_{6}C$ (b) $^{10}_{5}B$ (c) $^{14}_{7}N$

_61._ What was the problem with Dalton's statement that all the atoms of a particular element are identical in every respect?

_62._ How many neon-21 atoms are there in 1 million neon atoms?

_63._ How many electrons are needed to equal the mass of one proton?

◄64. Can the atomic number of an element exceed its mass number? Explain.

65. Make a list of some of our modern conveniences that would not function if atoms were indeed indivisible.

66. Referring to Table 4.2, determine all the possible molecular masses for H_2O, using the various isotopes of hydrogen and oxygen.

ATOMIC THEORY, PART 2:
Energy Levels and the Bohr Atom

Let's journey back to the beginning of the nineteenth century. It is the year 1814 and the German optician Joseph von Fraunhofer is testing a prism of excellent quality he has just manufactured. Fraunhofer allows light to pass through a slit and then through his triangular prism. When the light is projected onto a white screen, it forms a series of colored lines that are separated by dark lines. Fraunhofer counts almost 600 lines and notes the position of each relative to the others.

It is 30 years later: Robert Kirchhoff and Robert Bunsen use Fraunhofer's technique to show that when specific elements are heated in a flame they produce a unique series of lines—*a line spectrum*. The device they've developed is called a *spectroscope*, used by Bunsen and Kirchhoff for "fingerprinting" elements by the light patterns they produce. Of even more importance, Bunsen and Kirchhoff are able to work backward and deduce the elements in an unknown substance from its line spectrum.

The spectroscope is also used to identify new elements. In 1860 Bunsen and Kirchhoff test a mineral that produces spectral lines different from all known elements. Additional chemical testing shows that this new element is similar to sodium and potassium. They name the element *cesium*, from the Latin *sky blue*, which is the most prominent line in its spectrum.

Other chemists also use this technique and discover new elements. In 1875, the French chemist Emile Lecoq de Boisbaudran discovers a new element contained in zinc ore. He names the element gallium (see Chapter 6). Chemists at the time are not sure why each element has its own unique line spectrum, but many chemists feel that it has something to do with the internal structure of the atom. The nineteenth century, which began with Dalton's atomic theory of an indivisible atom with no internal structure, is about to undergo a major refinement.

After you've studied this chapter, you should be able to:

1. Discuss the difference between a line spectrum and a continuous spectrum.

2. Describe the Bohr model of the atom.

3. Explain the significance of shells and energy levels.

4. Differentiate between the ground state and an excited state of an atom.

5. Describe the quantum mechanical model of the atom.

6. Explain the significance of energy sublevels.

7. Define and discuss electron orbitals.

8. Write the electron configurations for various elements using the filling-order diagram.

9. Explain the importance of electron configuration.

■■■ **INTRODUCTION**

By the early 1900s, scientists had developed a model of the atom with protons and neutrons in the center, or nucleus, and electrons outside the nucleus. But where, exactly, were the electrons? And why weren't the negative electrons attracted into the nucleus by the positive protons? In 1913 the Danish physicist Niels Bohr proposed a model to answer these and other questions. His model has since been modified in the light of later observations and research, but it still provides an excellent basis for understanding atomic structure. In this chapter, we discuss the observations that led Bohr to propose his model, as well as the model itself.

5.1 Spectra

■ LEARNING GOAL 1

Line spectra and continuous spectra

In the mid-1600s the English scientist, astronomer, and mathematician Sir Isaac Newton allowed a beam of sunlight to pass through a prism. He found that the prism separated the light into a series of different colors, which we call the *visible spectrum* (Figure 5.1). His results were published in *Opticks* in 1704. Later experiments showed that a specific amount of energy is associated with each color of light, because light is a form of energy.

In the 1850s the German physicist Gustav Robert Kirchhoff and the German chemist Robert Bunsen developed an instrument called the spectroscope. Its main parts were a heat source (the Bunsen burner), a prism, and a telescope (Figure 5.2). Kirchhoff and Bunsen heated samples of various elements. In each case, as an element became hotter and hotter and began to glow, it produced light of its own characteristic color. When the light from

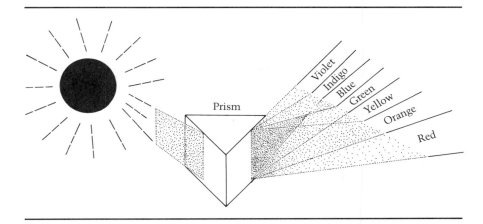

■ FIGURE 5.1

Sunlight passing through a prism

the heated element was allowed to pass through the prism, it separated into a series of bright, distinct lines of various colors.

The spectrum produced by sunlight is called a **continuous spectrum**, because one color merges into the next without any gaps or missing colors. The spectrum that Bunsen and Kirchhoff found is called a **line spectrum** (Figure 5.3), because it is a series of bright lines separated by dark bands. Each of the elements produces its own unique line spectrum.

Scientists find these line spectra very useful for identifying elements in unknown samples. Analysis of line spectra has enabled scientists to identify the elements that compose the sun, the stars, and other extraterrestrial bodies. This technique is also used in chemical laboratories to identify the elements present in unknown samples.

The visible spectrum produced by sunlight is only a small part of what is called the **electromagnetic spectrum** (Figure 5.4). The existence of electromagnetic waves was predicted by the British physicist James Clerk Maxwell in 1864. He assumed that electric and magnetic fields acting together produced radiant energy. He believed that radiant energy took the form of electromagnetic waves and that visible light was only one form of electromagnetic wave (Figure 5.5). We distinguish one form of electromagnetic radiation from another in terms of *wavelength*. The wavelength is the distance from one point on a wave to the corresponding point on the next wave. For electromagnetic waves, the shorter the wavelength, the higher the energy. Included in the electromagnetic spectrum were both visible waves and invisible waves, each associated with a different amount of energy. For example, x rays are invisible waves with enough energy to pass through skin but not bones. Microwaves are also invisible waves, but they have enough energy to heat food. Electromagnetic waves can travel through the vacuum of space, as well as through air and water.

Robert Bunsen
(Archive Photos)

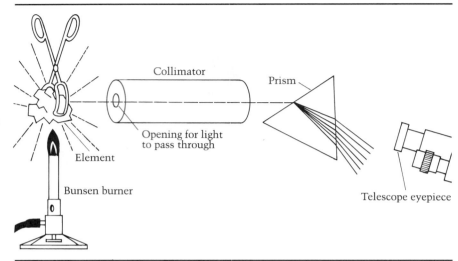

■ **FIGURE 5.2**

A spectroscope works as follows: An element is heated in the flame of a Bunsen burner. Some light from the heated element passes through a collimator, which is simply a brass tube with a narrow slit at both ends. The collimator directs a fine beam of light onto the prism. When light passes through the prism, it is broken into its different parts—a spectrum—and observers can see the spectrum through a telescope.

Maxwell's predictions were verified by the German physicist Heinrich R. Hertz in 1887. By varying an electric charge, he produced electromagnetic waves that were longer than visible light waves. This discovery eventually led to the development of radio and television.

The full impact of these experiments with light was not felt until 1913, when it was discovered that the existence of line spectra held the key to the structure of the atom.

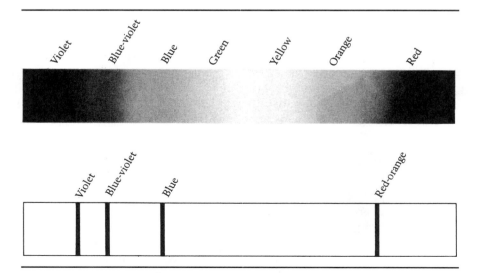

■ **FIGURE 5.3**

(Above) A continuous spectrum produced by sunlight. (Below) A line spectrum produced by hydrogen.

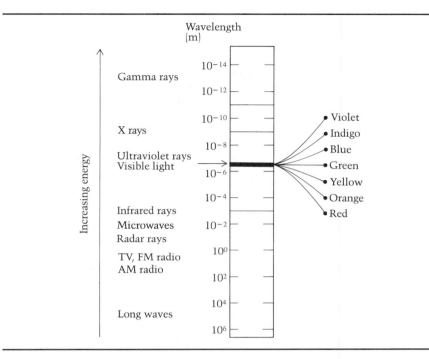

■ FIGURE 5.4

The electromagnetic spectrum

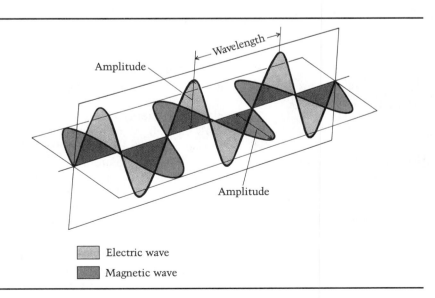

■ FIGURE 5.5

Light waves are one form of electromagnetic wave. They consist of an electric field and a magnetic field vibrating at right angles to each other.

5.2 Light As Energy

We have noted that light is a form of energy and that energy can be converted from one form to another, For example, a photoelectric cell converts light energy to electrical energy (in the form of electric current). Energy may also be absorbed and then re-emitted. For example, the luminous dial on a watch absorbs energy when it is exposed to a strong light. When it is removed from the light source, the dial re-emits this energy in the form of light: It glows.

In the experiments of Kirchhoff and Bunsen, each element was made to absorb energy in the form of heat. When the element got hot enough, it emitted this energy as light. Scientists still wondered why the elements emitted line spectra rather than continuous spectra, and they wondered what subatomic particle caused the line spectra.

In 1900, the German physicist Max Planck suggested that although light seems continuous, it is not. He said that electromagnetic radiation can be thought of as a stream of minute packets of energy called **quanta** (singular, *quantum*). The word quanta comes from the Latin, *quantus*, meaning ''how much.'' According to Planck's quantum hypothesis, these bundles of energy (also referred to as *photons*) are the smallest packets of energy associated with a particular form of electromagnetic radiation, and each contains a specific, fixed amount of energy and has its own wavelength. Niels Bohr used the concept of quanta to explain line spectra of elements.

5.3 The Bohr Atom

LEARNING GOAL 2
Bohr model of the atom

LEARNING GOAL 3
Shells and energy levels

LEARNING GOAL 4
Difference between ground state and excited state of an atom

In 1913 Bohr suggested that electrons are responsible for line spectra. He proposed the idea that electrons travel around the nucleus of the atom in **shells**, which were described as imaginary spherical surfaces roughly concentric with the nucleus. Bohr also proposed that each shell is associated with a particular **energy level** and that shells farther away from the nucleus are associated with higher energy levels. According to the Bohr model, electrons whirl around the nucleus in specific shells, just as planets travel in specific orbits around the sun. But electrons, unlike planets, are able to jump from one shell (at one energy level) to another (at a higher energy level) when they absorb enough energy from an outside source to do so (Figure 5.6). They can also fall back to their original shells by emitting this energy.

Look at the hydrogen atom in Figure 5.7. The single electron is moving around the nucleus in the shell closest to the nucleus—the shell associated with the lowest energy. In this situation, the electron is said to be in its *lowest energy level*, and the atom is in its **ground state**. As the electron moves, it neither gains nor loses energy. Its energy of motion exactly counterbalances the attraction of the nucleus. For this reason, it is not pulled into the nucleus.

Now suppose we heat our hydrogen atom. According to Bohr, the electron will jump to a higher-energy shell, farther from the nucleus, when it has absorbed a certain amount of heat energy. It must absorb the entire amount before it can leave the shell it occupies. It *cannot* absorb some

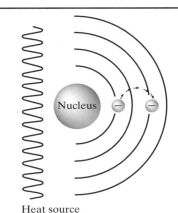

Heat source

■ FIGURE 5.6

In the Bohr model of the atom, electrons "jump" from one shell to another when an outside source gives them enough energy to do so.

energy, move out toward the next shell, absorb more energy and move closer to that shell, and so on until it reaches the next shell. It must absorb all the energy it needs and then instantaneously jump to the next shell. It may then absorb enough energy at that level to jump to another level (Figure 5.8). Or it may absorb enough energy to jump two or three levels at a time. But in each case it will absorb the energy and jump instantaneously. When an electron is at an energy level above its lowest level, the atom is said to be in an **excited state**. (An electron can also absorb so much energy that it escapes completely from the atom. This process, called *ionization*, is discussed in Chapter 6.)

Eventually, the electron emits some or all of the energy that was absorbed, and it falls back to a lower energy level. The energy is emitted in quanta, the difference in energy between the higher-energy shell where the electron was and the lower-energy shell where it lands. If the electron spiraled down without a level-to-level requirement, it would emit a continuous stream

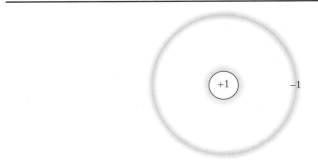

■ FIGURE 5.7

A hydrogen atom in its ground state

An excited hydrogen atom. Note that the electron is in energy level 3, and thus the atom is in an excited state.

of energy. But it moves only from one level to another and emits its energy in specific bursts that may be seen as light energy. It is this light, being emitted by electrons falling back to lower energy levels, that produces the unique line spectrum of hydrogen (or of any other element).

It is difficult to visualize electrons moving instantaneously from one energy level to another. Picture a marble rolling down a flight of stairs. The marble can rest on each step (shell), but it can never stop part of the way between steps. And, as the marble falls from step to step, it also loses (emits) energy in bursts.

5.4 Quantum Mechanical Model

LEARNING GOAL 5

Quantum mechanical model of the atom

The Bohr model of the atom contributed a great deal to our knowledge of atomic structure. The concept of energy levels for electrons, as well as the relationship between the observed spectral lines and the shifting of electrons from one energy level to another for the hydrogen atom were important to the development of atomic theory. Unfortunately, Bohr's calculations did not hold true for heavier atoms. Scientists continued to search for information to understand atomic structure.

In the decades that followed, revolutionary discoveries were made in physics. These discoveries had an impact on chemical knowledge. Experimental evidence showed that the laws of physics that applied to large objects did not apply to very small objects. In 1924, the French physicist Louis de Broglie explained that electrons in motion had properties of waves and also had mass. Two years later the Austrian physicist Erwin Schrödinger introduced a complicated mathematical equation that described electrons as having dual characteristics. He showed that in some ways the properties of electrons are best described in terms of waves, and in other ways their properties are most similar to those of particles having mass. Physicists formalized laws that could be used to characterize the motion of electrons in what we call **quantum mechanics.**

Quantum mechanical concepts have replaced the ideas Bohr presented in his model, where electrons are thought to revolve about the nucleus in planetary orbits. In this model of the atom, we think of an electron *not* as being in a carefully defined orbit, but instead as occupying a volume of space. This volume of space is called an **orbital**. The orbital represents a region of space in which the electron can be found with a 95% probability.

5.5   Energy Levels of Electrons

As we learned from both the Bohr model and the quantum mechanical model, we are most likely to find an electron at a certain specified distance from the nucleus, called an **energy level** or electron **shell**. The electrons in each shell have *approximately* the same amount of energy. We will use the terms *shell* and *energy level* interchangeably.

There are two ways to identify energy levels. One method uses the letters of the alphabet beginning with the letter K and continuing sequentially (Figure 5.9). The farther away a level is from the nucleus, the higher the energy of an electron at that level. The K energy level is associated with electrons at the lowest level, whereas the Q energy level holds the electrons of the highest level known. The newer method of identifying electron shells uses the *principal quantum number*, n, to designate the energy levels allowed for electrons. The shell having the lowest energy level is designated $n = 1$, then the next higher level is $n = 2$, and so on, until we reach $n = 7$. We know of no atom in its ground state having electrons in an energy level higher than the seventh. Notice that the K energy level and the $n = 1$ energy level are exactly the same, as are the L energy level and the $n = 2$ energy level, and so on.

Each energy level can hold only a certain number of electrons at any one time. Table 5.1 gives the *theoretical maximum number* of electrons for each level. The lowest energy level ($n = 1$) holds a maximum of 2 electrons, while the second, third, fourth, and fifth energy levels hold 8, 18, 32, and 50

■ **FIGURE 5.9**

Names of the energy levels in atoms

TABLE 5.1

Theoretical Maximum Number of Electrons for Each Energy Level

ENERGY LEVEL		MAXIMUM NUMBER
Letter designation	*Number*	OF ELECTRONS
K	1	2
L	2	8
M	3	18
N	4	32
O	5	50
P	6	72
Q	7	98

electrons, respectively. The maximum number of electrons in each level can also be determined using a simple mathematical calculation.

Maximum number of electrons for an energy level $= 2n^2$

EXAMPLE 5.1

Calculate the maximum number of electrons permitted in the following: (a) *M* shell ($n = 3$)   (b) *P* shell ($n = 6$)

Solution

UNDERSTAND THE PROBLEM

Using the number and letter designations, calculate the electron capacity for each shell.

DEVISE A PLAN

Use the mathematical equation $2n^2$.

CARRY OUT THE PLAN

(a) For $n = 3$, $2(3)^2 = 2(9) = 18$
(b) For $n = 6$, $2(6)^2 = 2(36) = 72$

LOOK BACK

Check the mathematics and determine if these solutions make sense.

Practice Exercise 5.1   Calculate the maximum number of electrons permitted in each of the following: (a) *N* shell ($n = 4$) (b) *Q* shell ($n = 7$)

5.6 Electron Subshells

Within each energy level we find electrons whose energies are very close in magnitude, but not exactly the same. The range in energies can be explained by the existence of **energy subshells** or **energy sublevels**. Electrons within each sublevel have exactly the same amount of energy.

To visualize the concept of energy sublevels, think of a tall office building with the unusual architectural feature depicted in Figure 5.10. The "building" is upside down. Each floor of the building represents an energy level [K ($n = 1$), L ($n = 2$), and so on]. Offices on each floor represent the energy sublevels. Each floor is a different size, so each has a different number of offices. In the same way, each energy level has a different number of sublevels. The sublevels are represented by the letters s, p, d, f, g, h, and i. The K energy level ($n = 1$), which is on the ground floor, has only one sublevel: the s sublevel; the L energy level ($n = 2$) has two: the s and p sublevels; and so on.

Each office of our "energy level building" is a different size and can hold only a certain number of desks. That is, each sublevel is a different size and can hold only a certain number of electrons. The s sublevel can hold a maximum of two electrons, and the p sublevel a maximum of six. Figure 5.10 shows the maximum number of electrons that each sublevel can hold. We discussed the maximum number of electrons for each energy level (2, 8, 18, 32, 50, 72, 98), and now we see that the total number of electrons that a main energy level can hold is simply the sum of the numbers of electrons that all of its *sublevels* can hold.

■ **FIGURE 5.10**

Energy levels and sublevels, showing maximum number of electrons in each

$Q(n = 7)$ can have 98	s 2	p 6	d 10	f 14	g 18	h 22	i 26
$P(n = 6)$ can have 72		s 2	p 6	d 10	f 14	g 18	h 22
$O(n = 5)$ can have 50			s 2	p 6	d 10	f 14	g 18
$N(n = 4)$ can have 32				s 2	p 6	d 10	f 14
$M(n = 3)$ can have 18					s 2	p 6	d 10
$L(n = 2)$ can have 8						s 2	p 6
$K(n = 1)$ can have 2							s 2

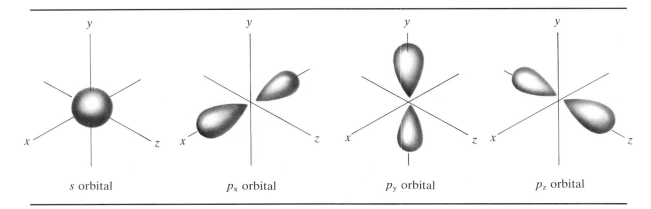

s orbital	p_x orbital	p_y orbital	p_z orbital

■ FIGURE 5.11

The shapes of the *s*, p_x, p_y, and p_z orbitals. The shape of the orbital defines the region of space in which an electron may be found with a probability of 95%. Darker areas are regions of higher probability.

5.7 Electron Orbitals

■ LEARNING GOAL 7

Electron orbitals

We have already looked at energy levels and sublevels. We will now look at the final concept that will help us understand how electrons are arranged about the nucleus of an atom. Electron *orbitals* are the regions of space around the nucleus of an atom where we are most likely to find an electron with a specific amount of energy. Each type of orbital has a specific shape and can hold a maximum of *two* electrons (Figure 5.11). For example, all *s* orbitals are spherical. However, the different levels of *s* orbitals occupy

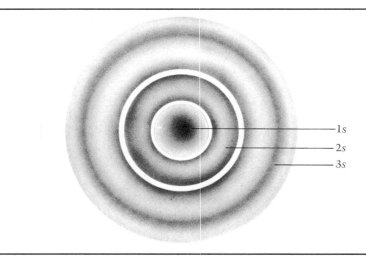

1*s*
2*s*
3*s*

■ FIGURE 5.12

The relationship of the 1*s*, 2*s*, and 3*s* orbitals to each other

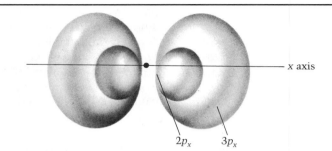

The spatial relationship of the $2p_x$ and $3p_x$ orbitals to each other

different regions of space around the nucleus. Figure 5.12 shows how the $1s$, $2s$, and $3s$ orbitals are related to each other.

The p orbitals are dumbbell shaped, as Figure 5.11 shows. There are three p orbitals in each major energy level except the first. They are called p_x, p_y, and p_z, and they correspond to three axes in space. Each of the p orbitals can hold a maximum of two electrons, so the three p orbitals combined can hold a maximum of six electrons. (This total corresponds to the maximum number of electrons that a p sublevel can hold.) The p orbitals that are associated with the different energy levels occupy different regions of space around the nucleus. Figure 5.13 shows the relationship between the $2p_x$ and $3p_x$ orbitals, and Figure 5.14 shows the relationship among the $1s$, $2s$, and $2p_x$ orbitals.

Orbitals of the d and f type can hold a maximum of 10 and 14 electrons, respectively. There are five orbitals for a d sublevel and seven orbitals for an f sublevel.

To clarify the idea of an electron being found in a specific orbital with a probability of 95%, consider the following analogy. George, a bank teller, works at the first National Bank between 10:00 A.M. and 3:00 P.M. The probability of finding George anywhere in the bank between these hours on a given day is extremely high, let us say 95%. (It is not 100% because there is always the chance that he may be ill or on vacation on a given day.)

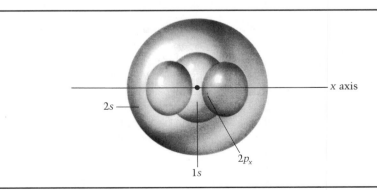

The relationship among the $1s$, $2s$, and $2p_x$ orbitals

Now suppose we make a map of the inside of the bank. We mark George's position every 10 minutes during the course of the workday. For most of the day we would find him at window three. This is where we would find most of our marks on the map. However, there would also be other marks on the map where he might have walked.

This map would be very much like the orbital pictures shown in Figures 5.11 to 5.14. Where there are many marks or darker shading, we have a better chance of finding George and the electron—for example, at window three or at the edges of the orbitals in Figure 5.13. Where there are fewer marks or lighter shading, the probability of finding George or an electron is low. And the probability of finding an electron any place within the orbital (or George any place within the bank) is 95%. Just as George can take a day off, an electron can move out of its orbital and then—just as easily—move back in.

EXAMPLE 5.2

For the third energy level determine (a) the maximum number of electrons the level could hold, (b) the number of sublevels, (c) the maximum number of electrons each sublevel could hold, and (d) the number of orbitals within each sublevel.

Solution
(a) For the third energy level, $n = 3$, therefore $2(n)^2 = 2(3)^2 = 2(9) = 18$.
(b) For the third level, the sublevels are $3s$, $3p$, and $3d$.
(c) The $3s$ sublevel can hold a maximum of 2 electrons, $3p$ can hold a maximum of 6 electrons, and $3d$ can hold a maximum of 10 electrons.
(d) For $3s$ there is one orbital, three orbitals for $3p$, and five orbitals for $3d$.

Practice Exercise 5.2   Determine all of the above for the fourth energy level.

5.8 Writing Electron Configurations

LEARNING GOAL 8

Writing electron configurations

The term **electron configuration** refers to the way the electrons fill the various energy levels of the atom and is what determines how an atom behaves chemically. As shown in Figure 5.7, unexcited hydrogen has its one electron in the first energy level. This seems reasonable, since the first level is associated with the lowest energy. Both of helium's electrons are also in the first energy level (Figure 5.15). Lithium, however, has three electrons. Two of them go into the first energy level, but one must go in the second level, the next-highest level. This is consistent with the fact that the first level can hold only two electrons. Atoms with 4 to 10 electrons have 2 electrons in the first level and the remainder in the second level. Neon, with 10 electrons, fills both the first and second levels (Table 5.2). Sodium, with 11 electrons, has full first and second levels and 1 electron in the third level.

FIGURE 5.15

Some ground-state electron configurations

TABLE 5.2

Ground-State Electron Configurations of the First 21 Elements

ELEMENT	ATOMIC NUMBER	ELECTRON CONFIGURATION			
		K (1)	L (2)	M (3)	N (4)
Hydrogen	1	1			
Helium	2	2			
Lithium	3	2	1		
Beryllium	4	2	2		
Boron	5	2	3		
Carbon	6	2	4		
Nitrogen	7	2	5		
Oxygen	8	2	6		
Fluorine	9	2	7		
Neon	10	2	8		
Sodium	11	2	8	1	
Magnesium	12	2	8	2	
Aluminum	13	2	8	3	
Silicon	14	2	8	4	
Phosphorus	15	2	8	5	
Sulfur	16	2	8	6	
Chlorine	17	2	8	7	
Argon	18	2	8	8	
Potassium	19	2	8	8	1
Calcium	20	2	8	8	2
Scandium	21	2	8	9	2

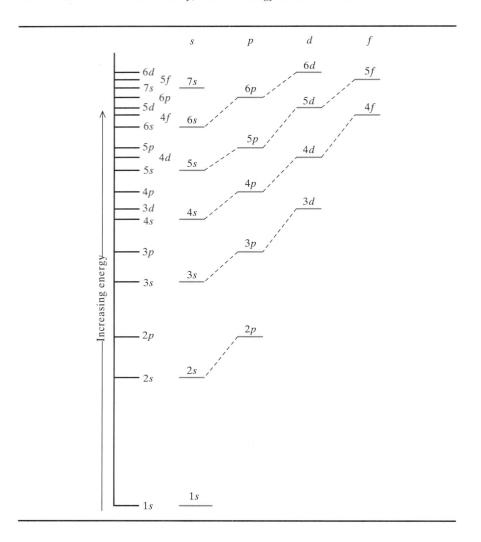

Table 5.2 shows that (up to element 18) the electrons fill the energy levels in the order that we might predict. The lowest-numbered level is filled before any electrons are positioned in the next-higher-numbered level. But when we write the electron configuration for $_{19}$K the pattern changes. We would expect the ground-state configuration for potassium to include two electrons in the first level, eight electrons in the second level, and nine electrons in the third level. Instead we find that potassium includes two electrons in the first level, eight electrons in the second level, eight electrons in the third level, and one electron in the fourth energy level. Why should this last electron go into the fourth level when the third level hasn't been filled?

To understand electron configuration more fully we must look further into energy sublevels. These sublevels group electrons according to energy, and the electron configuration shows how many electrons an atom has at the various energies. Since electrons do not occupy orbitals in a random fashion, we must know in exactly what order they do occupy the orbitals. Up to element number 18, we can write electron configurations by filling the energy levels closest to the nucleus first. For elements above number 18, this method does not produce the correct configuration. We can then use the **Aufbau principle** to guide us in determining electron configuration. Aufbau comes from the German word *aufbauen*, which means to build. We must consider the overlapping of energy levels, shown in Figure 5.16. For example, the 4*s* sublevel has a lower energy than the 3*d* sublevel. Therefore the 4*s* sublevel fills with electrons before the 3*d* sublevel does, even though the 4*s* sublevel is farther away from the nucleus than the 3*d* sublevel. And the 5*s* sublevel fills with electrons before the 4*d* and 4*f* sublevels.

Now you can write the electron configurations for any element with any atomic number—by simply filling the lowest energy *sublevel* first. The Aufbau diagram in Figure 5.17 will help you do this. All you have to do is follow the diagonal arrows, starting at the upper left. That is, the 1*s* sublevel is filled first, then the 2*s* sublevel, followed by 2*p*, 3*s*, 3*p*, 4*s*, 3*d*, and 4*p* sublevels, and so on. Unfortunately, even using the Aufbau diagram you will not be able to predict the electron configurations for *all* elements. There are a few exceptions.

When writing electron configurations we do not use words, we use a system involving shorthand. We represent the *K* energy level by the number 1, the *L* energy level by the number 2, the *M* energy level by the number

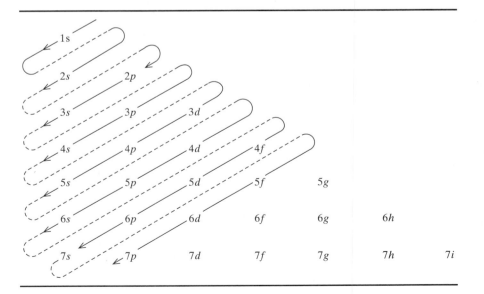

■ FIGURE 5.17

The order in which electrons fill energy sublevels

3, and so on. Then, for example, a $2p$ electron is an electron in the p sublevel of level 2.

Here are two examples of the notation we will use to write electron configurations:

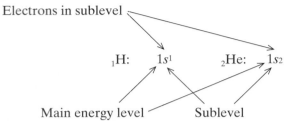

Electrons in sublevel

$_1$H: $1s^1$ $_2$He: $1s^2$

Main energy level   Sublevel

Note that the superscript to the right of the sublevel letter tells us the number of electrons in that sublevel.

Here are some more examples:

$_3$Li: $1s^2 2s^1$ $_{18}$Ar: $1s^2 2s^2 2p^6 3s^2 3p^6$

$_4$Be: $1s^2 2s^2$ $_{19}$K: $1s^2 2s^2 2p^6 3s^2 3p^6 4s^1$

$_5$B: $1s^2 2s^2 2p^1$ $_{20}$Ca: $1s^2 2s^2 2p^6 3s^2 3p^6 4s^2$

$_6$C: $1s^2 2s^2 2p^2$ $_{21}$Sc: $1s^2 2s^2 2p^6 3s^2 3p^6 4s^2 3d^1$

Because the $4s$ sublevel is lower in energy than the $3d$ sublevel, the last electron in potassium (K) is in the $4s$ sublevel and not the $3d$ sublevel (check Figure 5.17 again). The same is true of calcium (Ca). But beginning with scandium (Sc), the $3d$ sublevel starts to fill. This is because the $4s$ sublevel is now completely filled, and the $3d$ sublevel is now the lowest unfilled energy sublevel. Sublevels continue to fill in this manner. Table 5.3 gives the resulting electron configurations for all elements, and Figure 5.18 shows the pattern of filling in the periodic table.

EXAMPLE 5.3

Using Figure 5.17 write the electron configuration for each of the following elements: (a) $_8$O   (b) $_{30}$Zn

Solution

Follow the arrows in Figure 5.17. Remember not to place more than two electrons in an s sublevel, six in a p sublevel, ten in a d, and so on.
(a) $_8$O: $1s^2 2s^2 2p^4$
(b) $_{30}$Zn: $1s^2 2s^2 2p^6 3s^2 3p^6 4s^2 3d^{10}$

Practice Exercise 5.3   Using Figure 5.17 write the electron configuration for each of the following elements: (a) $_{12}$Mg   (b) $_{34}$Se

5.9 The Importance of Electron Configuration

■ LEARNING GOAL 9

Importance of electron configuration

When we know the electron configurations of elements, we can predict how elements will react with each other. The number of electrons in the **outermost shell** (the shell farthest away from the nucleus, by position) controls the chemical properties of an element. Elements with similar outer electron configurations behave in very similar ways. For instance, the elements Ne, Ar, Kr, Xe, and Rn are unreactive; in other words, they are almost chemically inert. Their atoms all have *eight* electrons in their outermost energy levels. On the other hand, the elements Li, Na, K, Rb, Cs, and Fr all react violently with water; their atoms all have *one* electron in their outermost energy levels.

The electron configurations of the elements also help us predict how the atoms of some elements will bond to each other. We will discuss that in Chapter 7.

■ **FIGURE 5.18**

The general pattern for filling energy levels of the atoms of each element. (*Note:* There are some exceptions to this general pattern for certain transition and inner-transition elements, as shown in Table 5.3.)

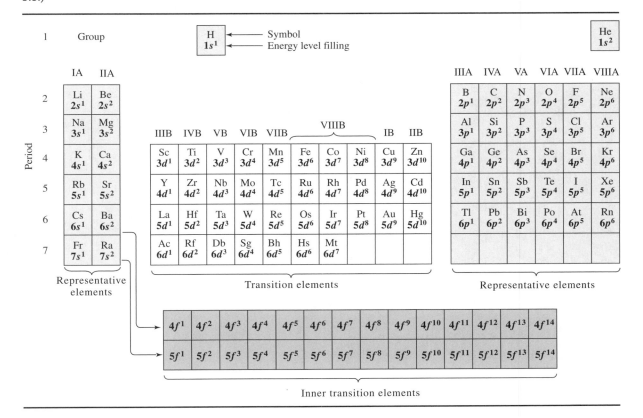

TABLE 5.3

Electron Configuration of the Elements

ATOMIC NUMBER	ELEMENT	1	2		3			4				5				6				7
		s	s	p	s	p	d	s	p	d	f	s	p	d	f	s	p	d	f	s
1	H	1																		
2	He	2																		
3	Li	2	1																	
4	Be	2	2																	
5	B	2	2	1																
6	C	2	2	2																
7	N	2	2	3																
8	O	2	2	4																
9	F	2	2	5																
10	Ne	2	2	6																
11	Na	2	2	6	1															
12	Mg	2	2	6	2															
13	Al	2	2	6	2	1														
14	Si	2	2	6	2	2														
15	P	2	2	6	2	3														
16	S	2	2	6	2	4														
17	Cl	2	2	6	2	5														
18	Ar	2	2	6	2	6														
19	K	2	2	6	2	6		1												
20	Ca	2	2	6	2	6		2												
21	Sc	2	2	6	2	6	1	2												
22	Ti	2	2	6	2	6	2	2												
23	V	2	2	6	2	6	3	2												
24	Cr	2	2	6	2	6	5	1												
25	Mn	2	2	6	2	6	5	2												
26	Fe	2	2	6	2	6	6	2												
27	Co	2	2	6	2	6	7	2												
28	Ni	2	2	6	2	6	8	2												
29	Cu	2	2	6	2	6	10	1												
30	Zn	2	2	6	2	6	10	2												
31	Ga	2	2	6	2	6	10	2	1											
32	Ge	2	2	6	2	6	10	2	2											
33	As	2	2	6	2	6	10	2	3											

TABLE 5.3

Electron Configuration of the Elements *(continued)*

ATOMIC NUMBER	ELEMENT	1	2		3			4				5				6				7
		s	*s*	*p*	*s*	*p*	*d*	*s*	*p*	*d*	*f*	*s*	*p*	*d*	*f*	*s*	*p*	*d*	*f*	*s*
34	Se	2	2	6	2	6	10	2	4											
35	Br	2	2	6	2	6	10	2	5											
36	Kr	2	2	6	2	6	10	2	6											
37	Rb	2	2	6	2	6	10	2	6			1								
38	Sr	2	2	6	2	6	10	2	6			2								
39	Y	2	2	6	2	6	10	2	6	1		2								
40	Zr	2	2	6	2	6	10	2	6	2		2								
41	Nb	2	2	6	2	6	10	2	6	4		1								
42	Mo	2	2	6	2	6	10	2	6	5		1								
43	Tc	2	2	6	2	6	10	2	6	6		1								
44	Ru	2	2	6	2	6	10	2	6	7		1								
45	Rh	2	2	6	2	6	10	2	6	8		1								
46	Pd	2	2	6	2	6	10	2	6	10										
47	Ag	2	2	6	2	6	10	2	6	10		1								
48	Cd	2	2	6	2	6	10	2	6	10		2								
49	In	2	2	6	2	6	10	2	6	10		2	1							
50	Sn	2	2	6	2	6	10	2	6	10		2	2							
51	Sb	2	2	6	2	6	10	2	6	10		2	3							
52	Te	2	2	6	2	6	10	2	6	10		2	4							
53	I	2	2	6	2	6	10	2	6	10		2	5							
54	Xe	2	2	6	2	6	10	2	6	10		2	6							
55	Cs	2	2	6	2	6	10	2	6	10		2	6			1				
56	Ba	2	2	6	2	6	10	2	6	10		2	6			2				
57	La	2	2	6	2	6	10	2	6	10		2	6	1		2				
58	Ce	2	2	6	2	6	10	2	6	10	2	2	6			2				
59	Pr	2	2	6	2	6	10	2	6	10	3	2	6			2				
60	Nd	2	2	6	2	6	10	2	6	10	4	2	6			2				
61	Pm	2	2	6	2	6	10	2	6	10	5	2	6			2				
62	Sm	2	2	6	2	6	10	2	6	10	6	2	6			2				
63	Eu	2	2	6	2	6	10	2	6	10	7	2	6			2				
64	Gd	2	2	6	2	6	10	2	6	10	7	2	6	1		2				
65	Tb	2	2	6	2	6	10	2	6	10	9	2	6			2				
66	Dy	2	2	6	2	6	10	2	6	10	10	2	6			2				

TABLE 5.3

Electron Configuration of the Elements *(continued)*

ATOMIC NUMBER	ELEMENT	1	2		3			4				5				6				7
		s	*s*	*p*	*s*	*p*	*d*	*s*	*p*	*d*	*f*	*s*	*p*	*d*	*f*	*s*	*p*	*d*	*f*	*s*
67	Ho	2	2	6	2	6	10	2	6	10	11	2	6			2				
68	Er	2	2	6	2	6	10	2	6	10	12	2	6			2				
69	Tm	2	2	6	2	6	10	2	6	10	13	2	6			2				
70	Yb	2	2	6	2	6	10	2	6	10	14	2	6			2				
71	Lu	2	2	6	2	6	10	2	6	10	14	2	6	1		2				
72	Hf	2	2	6	2	6	10	2	6	10	14	2	6	2		2				
73	Ta	2	2	6	2	6	10	2	6	10	14	2	6	3		2				
74	W	2	2	6	2	6	10	2	6	10	14	2	6	4		2				
75	Re	2	2	6	2	6	10	2	6	10	14	2	6	5		2				
76	Os	2	2	6	2	6	10	2	6	10	14	2	6	6		2				
77	Ir	2	2	6	2	6	10	2	6	10	14	2	6	7		2				
78	Pt	2	2	6	2	6	10	2	6	10	14	2	6	9		1				
79	Au	2	2	6	2	6	10	2	6	10	14	2	6	10		1				
80	Hg	2	2	6	2	6	10	2	6	10	14	2	6	10		2				
81	Tl	2	2	6	2	6	10	2	6	10	14	2	6	10		2	1			
82	Pb	2	2	6	2	6	10	2	6	10	14	2	6	10		2	2			
83	Bi	2	2	6	2	6	10	2	6	10	14	2	6	10		2	3			
84	Po	2	2	6	2	6	10	2	6	10	14	2	6	10		2	4			
85	At	2	2	6	2	6	10	2	6	10	14	2	6	10		2	5			
86	Rn	2	2	6	2	6	10	2	6	10	14	2	6	10		2	6			
87	Fr	2	2	6	2	6	10	2	6	10	14	2	6	10		2	6			1
88	Ra	2	2	6	2	6	10	2	6	10	14	2	6	10		2	6			2
89	Ac	2	2	6	2	6	10	2	6	10	14	2	6	10		2	6	1		2
90	Th	2	2	6	2	6	10	2	6	10	14	2	6	10		2	6	2		2
91	Pa	2	2	6	2	6	10	2	6	10	14	2	6	10	2	2	6	1		2
92	U	2	2	6	2	6	10	2	6	10	14	2	6	10	3	2	6	1		2
93	Np	2	2	6	2	6	10	2	6	10	14	2	6	10	4	2	6	1		2
94	Pu	2	2	6	2	6	10	2	6	10	14	2	6	10	6	2	6			2
95	Am	2	2	6	2	6	10	2	6	10	14	2	6	10	7	2	6			2
96	Cm	2	2	6	2	6	10	2	6	10	14	2	6	10	7	2	6	1		2
97	Bk	2	2	6	2	6	10	2	6	10	14	2	6	10	8	2	6	1		2
98	Cf	2	2	6	2	6	10	2	6	10	14	2	6	10	10	2	6			2
99	Es	2	2	6	2	6	10	2	6	10	14	2	6	10	11	2	6			2

TABLE 5.3

Electron Configuration of the Elements *(continued)*

ATOMIC NUMBER	ELEMENT	1	2		3			4				5				6				7
		s	*s*	*p*	*s*	*p*	*d*	*s*	*p*	*d*	*f*	*s*	*p*	*d*	*f*	*s*	*p*	*d*	*f*	*s*
100	Fm	2	2	6	2	6	10	2	6	10	14	2	6	10	12	2	6			2
101	Md	2	2	6	2	6	10	2	6	10	14	2	6	10	13	2	6			2
102	No	2	2	6	2	6	10	2	6	10	14	2	6	10	14	2	6			2
103	Lr	2	2	6	2	6	10	2	6	10	14	2	6	10	14	2	6	1		2
104	Db	2	2	6	2	6	10	2	6	10	14	2	6	10	14	2	6	2		2
105	Jl	2	2	6	2	6	10	2	6	10	14	2	6	10	14	2	6	3		2
106	Rf	2	2	6	2	6	10	2	6	10	14	2	6	10	14	2	6	4		2
107	Bh	2	2	6	2	6	10	2	6	10	14	2	6	10	14	2	6	5		2
108	Hn	2	2	6	2	6	10	2	6	10	14	2	6	10	14	2	6	6		2
109	Mt	2	2	6	2	6	10	2	6	10	14	2	6	10	14	2	6	7		2

■■■ SUMMARY

When a beam of sunlight passes through a glass prism, the sunlight separates into a rainbow of colors called the visible spectrum. Each color of light is associated with a specific amount of energy. The visible spectrum is part of the electromagnetic spectrum, and both are continuous spectra—their various parts merge into each other. In the 1850s, Kirchhoff and Bunsen showed that when a sample of an element is heated until it glows, it produces a line spectrum of visible light. Each element produces its own characteristic line spectrum.

In 1913 Niels Bohr proposed a model of the atom to explain these line spectra. The model consisted of shells, or imaginary spherical surfaces, roughly concentric with the nucleus. Each shell was associated with a particular energy level. Bohr postulated that electrons travel around the nucleus in their orbits. However, an electron can jump from one shell to another at a higher energy level when it absorbs enough energy. It can fall back to the original shell by emitting this energy in bursts. These separate bursts show up as line spectra.

Revolutionary discoveries occurred in physics in the decades that followed. These discoveries gave rise to the quantum mechanical model of the atom, which is based on a series of mathematical equations. The electrons in an atom were now thought of as occupying energy levels, with each energy level being composed of sublevels. The probability of finding an electron in an orbital, or region of space about the nucleus, was postulated.

Using this information, the electron configuration, or arrangement of electrons about the nucleus, could be determined. The various energy levels of the atom have been given the letter designations *K* through *Q* and the numbers 1 through 7. (Level *K* or 1 is the lowest energy level.) Associated with each level is a theoretical maximum electron population. Lower energy levels are usually filled before electrons are positioned in higher energy levels. Using the Aufbau principle, electron configurations can be determined and represented using a shorthand notation. The electron configurations endow the elements with many of their properties; elements with similar outer electron configurations exhibit similar chemical behavior.

KEY TERMS

Aufbau principle **(5.8)**

continuous spectrum **(5.1)**

electromagnetic spectrum **(5.1)**

electron configuration **(5.8)**

energy level **(5.3)**

energy sublevel **(5.6)**

energy subshell **(5.6)**

excited state **(5.3)**

ground state **(5.3)**

line spectrum **(5.1)**

orbital **(5.4)**

outermost shell **(5.9)**

quanta **(5.2)**

quantum mechanics **(5.4)**

shell **(5.3, 5.5)**

CAREER SKETCH

Health Careers/Blood Related

Many different careers in the health field require a strong science background. Working with blood is one of those fields.

As a *phlebotomist,* you will be responsible for drawing blood specimens from inpatients and outpatients. Once you gain elementary experience, you will perform more difficult phlebotomy procedures and act as a technical resource for beginning phlebotomists. Important skills for a phlebotomist are good laboratory procedures, strong organizational skills, and adept analytical skills.

A *transfusion services administrator* performs a slightly different role than a phlebotomist. This person serves in a more administrative role, often coordinating busy blood transfusion departments. Important skills for a transfusion services administrator include a knowledge of chemistry, hematology, and microbiology, as well as the organizational and laboratory skills necessary for any professional in the sciences.

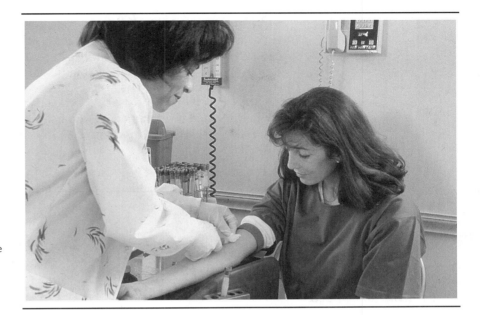

A phlebotomist is responsible for drawing blood samples from patients.
(Blair Seitz/Photo Researchers, Inc.)

*"Fifteen Day Fireworks Finale"**

Researchers at the University of California, Riverside, had much to celebrate on a recent Independence Day. What they were celebrating (and testing) on that July 4 was a revolutionary new method they had developed to monitor airborne particles. The method, called "aerosol time-of-flight mass spectrometry," rapidly provides information on both particle size and chemical composition of impurities in the air.

To test the new method, researcher Kimberly A. Prather and her colleagues monitored the air outside their lab from July 3 to July 7, searching for airborne particles. The group believed that the distinctive chemical signature of fireworks, as well as the well-defined start time and location of a nearby display, would provide an ideal testing ground for their new method.

Most of the particles the researchers detected arrived at the lab, located just three miles from the show, on the morning of July 5. The researchers were able to detect what they considered the peak number of airborne particles from the fireworks display 12 hours after the start of the Independence Day celebration! Perhaps most surprising to the researchers, though, was that they were able to detect particles from the fireworks on July 19—a full 15 days after the show.

The University of California researchers' findings, which appear in the May 15, 1997, issue of *Analytical Chemistry,* can lead to important new safety developments, primarily as a result of the speed with which results can be obtained. While traditional air quality measurements can take hours or days to furnish results, because the filters that collect particles must be sent away for expensive and time-consuming analysis, the new method gives immediate results. This can be a life-saving time difference when one is trying to detect chemical or biological toxins in the air.

*Adapted from *Science News,* July 5, 1997, p. 15

SELF-TEST EXERCISES

LEARNING GOAL 1: *Line Spectra and Continuous Spectra*

1. How does a line spectrum differ from a continuous spectrum?

◀ 2. What is a spectrum? What is the difference between the electromagnetic spectrum and the visible spectrum?

◀ <u>3.</u> How could a spectroscope help scientists analyze the sun and other extraterrestrial bodies?

4. Which end of the visible spectrum has the highest energy, red or violet? Which end has the longest wavelength?

5. Can you discover the origin of the word *helium?* (*Hint:* See the *Handbook of Chemistry and Physics.*)

6. Do you think line spectra could be used to identify compounds? Explain.

LEARNING GOAL 2: *Bohr Model of the Atom*

7. Explain the major differences among the Thomson, Rutherford, and Bohr models of the atom.

8. Explain how the *ground state* of an atom differs from its *excited state.*

<u>9.</u> What is Bohr's most important contribution to our ideas about the structure of the atom?

10. According to the Bohr model of the atom, do electrons "spiral" or "jump" from one energy level to another? What type of spectrum would be produced by each of these types of movements by electrons?

LEARNING GOAL 3: *Shells and Energy Levels*

<u>11.</u> What subatomic particles produce the line spectrum of an element?

12. Complete the following sentence with the word *higher* or *lower:* The farther away an energy level is from the nucleus, the _____ the energy of an electron in that level.

◀ <u>13.</u> What is meant by shells and energy levels? How many shells are known to exist in atoms? State their letter and number designations.

14. Using the $2n^2$ rule, calculate the maximum number of electrons that can populate each of the first three energy levels.

LEARNING GOAL 4: *Difference Between Ground State and Excited State of an Atom*

15. What do we mean by the ground state of an atom? What do we mean by the excited state of an atom?

16. Describe the ground-state configuration of a lithium atom. What changes could take place if the lithium atom were in an excited state?

17. If a neutral hydrogen atom has an electron in the second energy level, is the atom in its ground state or an excited state?

◀ 18. A particular neutral magnesium atom has two electrons in the first shell, eight electrons in the second shell, and two electrons in the third shell. Would you characterize this atom as being in the ground state or in an excited state?

LEARNING GOAL 5: *Quantum Mechanical Model of the Atom*

19. Why was the Bohr model of the atom inadequate to explain the behavior of the atom?
20. What new knowledge regarding the laws of physics was discovered in the early 1900s?
21. Discuss Louis de Broglie's contributions to the field of atomic theory.
22. Discuss Erwin Schrödinger's contributions to the knowledge of the behavior of electrons.

LEARNING GOAL 6: *Energy Sublevels*

◀ **23.** How many energy sublevels does each of the following energy levels contain?
   (a) energy level 1
   (b) energy level 2
   (c) energy level 3
   (d) energy level 4
24. How many energy sublevels does energy level 7 contain? Name them.
25. How is the total number of electrons an energy level contains related to the number of electrons a sublevel can hold?
26. The third energy level can hold a maximum of 18 electrons. How many sublevels does this energy level contain? How many electrons does each sublevel contain?
◀ **27.** Determine the following for the sixth energy level of an atom:
   (a) number of subshells it contains
   (b) maximum number of electrons this energy level can hold
28. Determine the following for the fourth energy level of an atom:
   (a) number of subshells it contains
   (b) maximum number of electrons this energy level can hold

LEARNING GOAL 7: *Electron Orbitals*

◀ 29. Using an analogy similar to the one in the text, explain what is meant by 95% probability and describe how it relates to the quantum mechanical model of an atom.
30. What is meant by the "region of highest probability of finding an electron"?
31. For the fifth energy level, determine the number of orbitals in the first two sublevels.
32. For the sixth energy level, determine the number of orbitals in the first three sublevels.
33. Describe the shape of the $2s$, $3s$, and $4s$ electron orbitals.
34. Describe the shape of the $4p$ orbitals.
35. Which of the following does not exist? (a) $2s$ (b) $2p$ (c) $2d$ (d) $2f$
36. Which of the following does not exist? (a) $3s$ (b) $3p$ (c) $3d$ (d) $3f$

LEARNING GOAL 8: *Writing Electron Configurations*

37. Using Figure 5.17 write the electron configurations for the following elements: (a) $_7$N (b) $_{23}$V (c) $_{54}$Xe (d) $_{51}$Sb
38. Using Figure 5.17 write the electron configurations for the following elements: (a) $_{50}$Sn (b) $_{27}$Co (c) $_{86}$Rn (d) $_{35}$Br
39. Name the neutral elements with the following electron configurations. (Use Table 5.3.)
   (a) $1s^22s^22p^63s^23p^63d^{10}4s^24p^64d^{10}5s^25p^66s^2$
   (b) $1s^22s^22p^63s^23p^64s^2$
   (c) $1s^22s^22p^63s^23p^63d^{10}4s^24p^65s^2$
   (d) $1s^22s^2$
40. Name the neutral elements with the following electron configurations. (Use Table 5.3.)
   (a) $1s^2$
   (b) $1s^22s^22p^3$
   (c) $1s^22s^22p^63s^23p^63d^{10}4s^1$
   (d) $1s^22s^22p^63s^23p^63d^{10}4s^24p^4$
◀ **41.** A certain neutral element has 2 electrons in the first level, 8 electrons in the second level, 18 electrons in the third level, and 3 electrons in the fourth level. List the following information for this element:
   (a) atomic number
   (b) total number of s electrons

(c) total number of *p* electrons
(d) total number of *d* electrons
(e) number of protons

42. A certain neutral element has 2 electrons in the first level, 8 electrons in the second level, 18 electrons in the third level, and 2 electrons in the fourth level. List the following information for this element:
    (a) atomic number
    (b) total number of *s* electrons
    (c) total number of *p* electrons
    (d) total number of *d* electrons
    (e) number of protons

43. Write electron configurations to determine which of the following elements belong to the same groups: elements with atomic numbers 6, 19, 17, 14, 3, and 35.

44. Write electron configurations to determine which of the following elements belong to the same groups: elements with atomic numbers 5, 18, 16, 36, 10, and 13.

45. Use the $2n^2$ rule to calculate the maximum number of electrons that the seventh energy level can hold.

46. Use the $2n^2$ rule to calculate the maximum number of electrons that the fourth energy level can hold.

47. Using Figure 5.17, write the electron configurations for each of the following elements: (a) $_{31}$Ga (b) $_{51}$Sb (c) $_{82}$Pb (d) $_{88}$Ra

48. Using Figure 5.17 write the electron configurations for the following elements: (a) $_3$Li (b) $_{12}$Mg (c) $_{56}$Ba

49. Name the neutral elements with the following electron configurations. (Use Table 5.3.)
    (a) $1s^2 2s^2 2p^6 3s^2$
    (b) $1s^2 2s^2 2p^6 3s^2 3p^6 3d^{10} 4s^2 4p^6 5s^2$
    (c) $1s^2 2s^2 2p^6 3s^2 3p^6 3d^{10} 4s^2 4p^6 4d^{10} 4f^{14} 5s^2 5p^6 6s^2$
    (d) $1s^2 2s^2 2p^6 3s^2 3p^6 3d^{10} 4s^2 4p^6 4d^{10} 4f^{14} 5s^2 5p^6 5d^{10} 6s^2 6p^1$

50. Name the neutral elements with the following electron configurations. (Use Table 5.3.)
    (a) $1s^2 2s^2 2p^6 3s^1$
    (b) $1s^2 2s^2 2p^6 3s^2 3p^6 3d^{10} 4s^2 4p^6 4d^{10} 4f^{12} 5s^2 5p^6 6s^2$
    (c) $1s^2 2s^2 2p^6 3s^2 3p^6 3d^{10} 4s^2 4p^6 4d^{10} 4f^{14} 5s^2 5p^6 6s^2$
    (d) $1s^2 2s^2 2p^6 3s^2 3p^6 3d^{10} 4s^2 4p^6 4d^{10} 4f^{14} 5s^2 5p^6 5d^{10} 6s^2 6p^6$

51. A given neutral element has 2 electrons in the first level, 8 electrons in the second level, 18 electrons in the third level, 22 electrons in the fourth level, 8 electrons in the fifth level, and 2 electrons in the sixth level. List the following information for this element:
    (a) atomic number
    (b) total number of *s* electrons
    (c) total number of *p* electrons
    (d) total number of *d* electrons
    (e) total number of *f* electrons
    (f) number of protons

52. A given neutral element has 2 electrons in the first level, 8 electrons in the second level, 18 electrons in the third level, 32 electrons in the fourth level, 9 electrons in the fifth level, and 2 electrons in the sixth level. List the following information for this element:
    (a) atomic number
    (b) total number of *s* electrons
    (c) total number of *p* electrons
    (d) total number of *d* electrons
    (e) total number of *f* electrons
    (f) number of protons

LEARNING GOAL 9: *Importance of Electron Configuration*

53. Explain the importance of electron configuration with regard to the chemical behavior of atoms.

54. Complete the following sentence: The number of electrons in the _____ shell determines the chemical properties of an element.

55. The elements in Group VIIIA of the periodic table (this group is the vertical column of elements farthest to the right in the periodic table) have eight electrons in their outermost energy level. (The exception to this is helium, which has only two electrons in its outermost energy level.) What is chemically unique about the Group VIIIA elements?

56. Separate the following into groups (vertical columns) of elements that exhibit similar chemical behavior: (a) Li (b) Ne (c) K (d) Rb (e) Kr (f) Na (g) Cs (h) Rn (i) Xe

EXTRA EXERCISES

57. Write electron configurations for all elements with atomic numbers between 1 and 20.

58. Write electron configurations for all elements with atomic numbers between 1 and 20 that have the same number of electrons in the outermost energy level as a magnesium atom.

◄ 59. What was the reasoning behind replacement of the Rutherford model of the atom with Bohr's model?

60. Explain why the line spectrum of an element is sometimes referred to as its fingerprint.

61. Explain briefly how electrons produce a line spectrum.

62. In Chapter 4 you learned that not all atoms of an element need be identical and that these different atoms are called isotopes. Do isotopes of the same element produce the same line spectrum? Explain.

63. In this chapter we stated that a marble rolling down a flight of stairs is similar to an electron jumping from one energy level to the next. Can you think of other analogous phenomena?

64. What can happen to an electron in an atom when it absorbs too much energy while being excited?

65. What would be the atomic number of an element whose energy levels 1, 2, 3, 4, 5, 6, and 7 were filled with the maximum number of electrons?

THE PERIODIC TABLE:
Keeping Track of the Elements

This is the story of a physician who tracked down the cause of death of one of his patients in order to try to save other people who might also be at risk. Our story begins in 1955. A 49-year-old woman, Mary Brown, visited her doctor complaining of pain in her jaw and a severe lack of energy. Routine blood tests showed that she was anemic. Anemia may be caused by bleeding, poor nutrition, drugs, or inadequate red blood cell production caused by tumors in the bone marrow.

After elimination of the first three possibilities, her physician, Dr. Ingram, suspected a problem with the bone marrow. A bone marrow biopsy was performed and Mary was diagnosed as having a malignant tumor of the bone marrow. The pain in her jaw was caused by inflammation of the bone marrow of the jaw.

Malignant tumors invade surrounding tissues and send out cells that form new tumors at distant sites. The formation of new tumors at distant sites is called *metastasis.* One year after diagnosis, Mary Brown, then 50 years old, died due to widespread metastasis of her cancer.

Dr. Ingram, upset and puzzled about how his patient of ten years, who seemed to have lived a healthy lifestyle, developed bone marrow cancer, decided to investigate. Dr. Ingram's investigation began with Mary's surviving family members, who gave the doctor a more complete history of Mary's life. The mystery was soon solved. Dr. Ingram determined that the cause of Mary's illness and subsequent death was her previous occupation. In the early 1920s, Mary was employed by a company that applied a radium-based luminous paint on watch and instrument dials. Workers who performed this task would keep a fine point on their paint brushes by turning the bristles on their tongues. When the dials were painted with this material, they glowed in the dark. Such items became very popular with consumers.

Although radium-based paint was very popular with consumers, it proved to be toxic to people like Mary who worked with it. Dr. Ingram knew that radium produces ionizing radiation that damages the body and can cause cancer. Unfortunately, working with radium for five years and ingesting small amounts of it took its toll on Mary and many of her coworkers.

Dr. Ingram recalled that radium has chemical properties that resemble those of calcium because the arrangements of electrons (which determine the properties of elements) are very similar in radium and calcium atoms. These two elements are so close in electron arrangement that the body is unable to distinguish them from one another. So radium can be incorporated into the bones and teeth just as easily as calcium is. Once inside the body, however, radium produces damaging ionizing radiation.

Working with radium is an occupational hazard. Unfortunately, classifying occupational carcinogens is not always easy because there is often a long period between exposure to the carcinogen and development of cancer. This was the case with radium. Many workers were exposed to radium-based paint while manufacturing watches and instrument dials, and only years later were the dangers of radium recognized. Dr. Ingram hoped that his research would help those people who were exposed to the radium so that they could be told of the risks that they face and could seek proper medical attention.

LEARNING GOALS

After you've studied this chapter, you should be able to:
1. Discuss the historical basis and significance of the periodic table.
2. Distinguish between a period and a group in the periodic table.
3. State the difference between an A-group (representative) element and a B-group element (transition metal).
4. Predict trends for properties such as atomic radius, ionization potential, and electron affinity.

INTRODUCTION

In preceding chapters, we referred to the periodic table for information about the elements, such as atomic number and mass. The periodic table can be used to predict many properties that can be useful to us. In this chapter, we trace the development of the periodic table and see what kinds of additional information it holds.

6.1 History

From the time of the ancient Greeks up to 1800, only 14 elements were known. Then, in a short span of 10 years, from 1800 to 1810, 14 more elements were discovered. By 1830, 45 elements were known. As new elements were discovered, chemists probably began to feel insecure. All these elements had different properties, and there didn't seem to be any relationship among them. And chemists wondered how many elements actually existed. To answer this question they had to find some relationship among the known elements.

One of the first chemists to notice some order among the elements was the German scientist Johann Döbereiner. He published an account of his observations in 1829. It occurred to Döbereiner that bromine had chemical and physical properties somewhere between those of chlorine and iodine and that bromine's atomic mass was almost midway between those of chlorine and iodine (Figure 6.1). Could this be nothing more than coincidence?

Döbereiner searched and found two more groups of similar elements. The first was made up of calcium (Ca), strontium (Sr), and barium (Ba); the

Element	Color	State	M.P.	B.P.	Atomic mass	Chemical symbol
Chlorine	Green-yellow	Gas	−101°C	−34.6°C	35.5	Cl
Bromine	Brown	Liquid	−7.2°C	58.8°C	79.9	Br
Iodine	Purple	Solid	113.5°C	184.4°C	126.9	I

■ **FIGURE 6.1**

Some similar elements

Element	Color	State	M.P. (°C)	B.P. (°C)	Atomic mass (amu)
Ca	Silver-white	Solid	842.8	1487	40.08
Sr	Silver white to pale yellow	Solid	769	1384	87.62
Ba	Yellow-silver	Solid	725	1140	137.34
S	Yellow	Solid	113	444	32.06
Se	Blue-gray	Solid	217	685	98.96
Te	Silver-white	Solid	452	1390	127.6

■ **FIGURE 6.2**

More similar elements

second included sulfur (S), selenium (Se), and tellurium (Te) (Figure 6.2). He called these groups *triads* (sets of three).

In 1864 the English chemist John Newlands arranged the known elements in order of increasing atomic mass. He came upon the idea of arranging them in vertical columns. Because he noticed that the eighth element had chemical and physical properties that were similar to those of the first, he let the eighth element start a new column (Figure 6.3). Newlands called his arrangement the *Law of Octaves*. But there were many places in his arrangement where dissimilar elements were next to each other.

In 1869 the German chemist Julius Lothar Meyer devised an incomplete periodic table consisting of 56 elements. In the same year the Russian chemist Dmitri Mendeleev (pronounced "Menduh-LAY-eff") arranged the

■ **FIGURE 6.3**

Newlands's Law of Octaves

Li	←— Similar —→	Na	←— Similar —→	K
Be	←— Similar —→	Mg	←— Similar —→	Ca
B	←— Similar —→	Al	←— Not Similar —→	Cr
C	←— Similar —→	Si	←— Not Similar —→	Ti
N	←— Similar —→	P	←— Not Similar —→	Mn
O	←— Similar —→	S	←— Not Similar —→	Fe
F	←— Similar —→	Cl	←— Not Similar —→	Co

↑— Increasing atomic mass —↓

Dmitri Mendeleev (1834–1907). In 1869 he predicted the chemical and physical properties of the element gallium, which he called eka-aluminum. (Science Photo Library/Photo Researchers, Inc.)

elements in order of increasing atomic mass. In cases where discrepancies arose, *the chemical properties were used to place the elements.* Mendeleev established horizontal rows, or *periods.* Hydrogen by itself made up the first period. Each of the next two periods contained seven elements. (The noble gases had not yet been discovered.) The periods after that contained more than seven elements. This is the point at which Mendeleev's table differed from Newlands's table.

In arranging the table, Mendeleev occasionally put heavier elements before lighter ones to keep elements with the same chemical properties in the same vertical column, or *group* (Figure 6.4). Sometimes he left open spaces in the table, where he reasoned that unknown elements should go. This was the case with gallium. In 1869, when he made up his table, the

| | | | | | Group | | | | | |
Period	I	II	III	IV	V	VI	VII	VIII		
1	1 H							2		
2	3 Li	4 Be	5 B	6 C	7 N	8 O	9 F	10		
3	11 Na	12 Mg	13 Al	14 Si	15 P	16 S	17 Cl	18		
4	19 K	20 Ca	21	22 Ti	23 V	24 Cr	25 Mn	26 Fe	27 Co	28 Ni
	29 Cu	30 Zn	31	32	33 As	34 Se	35 Br	36		
5	37 Rb	38 Sr	39 Y	40 Zr	41 Nb	42 Mo	43	44 Ru	45 Rh	46 Pd
	47 Ag	48 Cd	49 In	50 Sn	51 Sb	52 Te	53 I	54		
6	55 Cs	56 Ba	57 La	72	73 Ta	74 W	75	76 Os	77 Ir	78 Pt
	79 Au	80 Hg	81 Tl	82 Pb	83 Bi	84	85	86		
7	87	88	89	104	105					

58 Ce	59	60	61	62	63	64	65 Tb	66	67	68 Er	69	70	71
90 Th	91	92 U	93	94	95	96	97	98	99	100	101	102	103

■ **FIGURE 6.4**

Mendeleev's original periodic table arranged in modern form

	Eka-aluminum	Gallium
Atomic mass →	Predicted: 68	Observed: 69.9
Oxide formula →	Predicted: E_2O_3	Observed: Ga_2O_3
Oxide in acid →	Predicted: should dissolve in acid	Observed: dissolves in acid
Salt formation →	Predicted: should form salts as EX_3	Observed: forms salts as GaX_3

■ **FIGURE 6.5**

Predicted and actual properties of eka-aluminum, which was named gallium

Paul Emile Lecoq de Boisbaudran (1838–1912). In 1875 he discovered and named the element gallium. (Reprinted with permission from *Torchbearers of Chemistry* by Henry Monmouth Smith, © Academic Press, Inc., 1949)

element gallium was unknown; however, Mendeleev *predicted* its existence. He based his prediction on the properties of aluminum (which appeared directly above gallium in the table). Mendeleev even went so far as to predict the melting point, boiling point, and atomic mass of the then-unknown gallium, which he called eka-aluminum. Six years later, while analyzing zinc ore, the French chemist Lecoq de Boisbaudran discovered the element gallium. Its properties were almost identical to those Mendeleev had predicted (Figure 6.5).

The periodic tables proposed by both Mendeleev and Meyer were based on increasing atomic masses. Mendeleev used similarities among chemical properties to account for discrepancies among elements. Certain irregularities in these tables were corrected by the work of Henry G. J. Moseley, a British physicist. Moseley was able to determine the nuclear charge on the atoms of known elements. He concluded that elements should be arranged by increasing atomic number rather than by atomic mass. This research led to the development of the *periodic law,* which states that the chemical properties of the elements are periodic functions of their atomic numbers. Elements with similar chemical properties recur at regular intervals, and they are placed accordingly on the periodic table.

6.2   The Modern Periodic Table

■ LEARNING GOAL 2

Periods and groups in the periodic table

The periodic table of today is similar to Mendeleev's but has many more elements—those that have been discovered since 1869. Today's table consists of seven horizontal rows called **periods** and a number of vertical columns called **groups** (or **families**) (Table 6.1).

The groups are numbered with Roman numerals. All the elements in each group have the same number of electrons in their outermost shells, so they all behave similarly. For example, the Group IA elements react violently when they come into contact with water. And all elements in Group IA have one electron in their outermost shell.

Some of the groups in the periodic table are labeled with a Roman numeral followed by A, others with a Roman numeral followed by B.

TABLE 6.1

TABLE 6.1

Periodic Table of the Elements

Group IA																	VIIIA
1 H 1.00794	IIA											IIIA	IVA	VA	VIA	VIIA	2 He 4.002602
3 Li 6.941	4 Be 9.01218			Transition Elements								5 B 10.811	6 C 12.011	7 N 14.0067	8 O 15.9994	9 F 18.998403	10 Ne 20.179
11 Na 22.98977	12 Mg 24.305	IIIB	IVB	VB	VIB	VIIB	VIIIB			IB	IIB	13 Al 26.98154	14 Si 28.0855	15 P 30.97376	16 S 32.066	17 Cl 35.453	18 Ar 39.948
19 K 39.0983	20 Ca 40.078	21 Sc 44.9559	22 Ti 47.88	23 V 50.9415	24 Cr 51.9961	25 Mn 54.9380	26 Fe 55.847	27 Co 58.9332	28 Ni 58.69	29 Cu 63.546	30 Zn 65.38	31 Ga 69.723	32 Ge 72.59	33 As 74.9216	34 Se 78.96	35 Br 79.904	36 Kr 83.80
37 Rb 85.4678	38 Sr 87.62	39 Y 88.9059	40 Zr 91.22	41 Nb 92.9064	42 Mo 95.94	43 Tc (98)	44 Ru 101.07	45 Rh 102.9055	46 Pd 106.42	47 Ag 107.8682	48 Cd 112.41	49 In 114.82	50 Sn 118.710	51 Sb 121.75	52 Te 127.60	53 I 126.9045	54 Xe 131.29
55 Cs 132.9054	56 Ba 137.33	57–71 La– Lu	72 Hf 178.49	73 Ta 180.9479	74 W 183.85	75 Re 186.207	76 Os 190.2	77 Ir 192.22	78 Pt 195.08	79 Au 196.9665	80 Hg 200.59	81 Tl 204.383	82 Pb 207.2	83 Bi 208.9804	84 Po (209)	85 At (210)	86 Rn (222)
87 Fr (223)	88 Ra (226)	89–103 Ac– Lr	104 Rf (261)	105 Db (262)	106 Sg (263)	107 Bh (262)	108 Hs (265)	109 Mt (267)									

Period (1–7 along left side)

Lanthanides

57 La 138.9055	58 Ce 140.12	59 Pr 140.9077	60 Nd 144.24	61 Pm (145)	62 Sm 150.36	63 Eu 151.96	64 Gd 157.25	65 Tb 158.9254	66 Dy 162.50	67 Ho 164.9304	68 Er 167.26	69 Tm 168.9342	70 Yb 173.04	71 Lu 174.967

Actinides

89 Ac (227)	90 Th 232.0381	91 Pa (231)	92 U 238.0289	93 Np (237)	94 Pu (244)	95 Am (243)	96 Cm (247)	97 Bk (247)	98 Cf (251)	99 Es (252)	100 Fm (257)	101 Md (258)	102 No (259)	103 Lr (260)

```
  1   ←———— Atomic number
  H   ←———— Symbol
1.00794 ←———— Atomic mass
```

Metal	Metalloid	Nonmetal

LEARNING GOAL 3

A-group and B-group elements

The A groups are called the **representative elements.** The B groups are called the **transition metals.** You can also see that as we move from the top to the bottom of the periodic table (in other words, from period 1 to period 7), the periods get larger—they have more elements in them. In fact, periods 6 and 7 are so large that to fit the table on one page, we have to write part of each period *below* the rest of the table (the 15 lanthanides and the 15 actinides).

The most significant changes to the periodic table in the last few decades have been the names of the newest elements. Traditionally, the creator or discoverer of an element has the privilege of proposing its name. And choosing a name is no trivial matter! It may be influenced by issues ranging from national pride to professional rivalry. But skepticism about which labs were actually the first to create specific elements led to worldwide disagreement about the names of elements 104 through 109. The American Chemical Society proposed one set of names, but the International Union of Pure and Applied Chemistry (IUPAC) proposed a different set.

After nearly two decades of disagreement, however, international agreement was reached in late 1997. IUPAC developed what it considered a "fair compromise between the various claims and suggestions" of the American, German, and Russian teams that discovered the elements. Finally, publishers of chemistry textbooks and scientific papers could avoid the confusion caused by five different versions of the periodic table.

The final names for the new elements (which are listed in the newest version of the periodic table, on the inside front cover of this book) are the following: Rutherfordium (Rf, Element 104) named for New Zealand–born physicist Ernest Rutherford; Dubnium (Db, Element 105) named for the Joint Institute for Nuclear Research at Dubna, Russia, where some of the new elements were created; Seaborgium (Sg, Element 106) named for Glenn T. Seaborg, an American chemist and Nobel laureate whose team created 10 new elements during and after World War II; Bohrium (Bh, Element 107) named for Niels Bohr, founder of quantum physics; Hassium (Hs, Element 108) named for the German province Hess, where elements 109 and 110 were created; and Meitnerium (Mt, Element 109) named for Austrian-born physicist Lise Meitner. Work continues on naming the most recently discovered elements, atomic numbers 110, 111, and 112.

6.3   Periodic Trends

LEARNING GOAL 4
Periodic trends

When we examine the periodic table, we can see certain trends in the properties of the elements. For example, the elements in Group IA (at the left) are all metals. As we move across to the right, we continue to encounter metals through more than half the width of the table. Then we reach elements that have properties intermediate between metals and nonmetals; these are the metalloids. As we move still further to the right, we finally reach the nonmetals. (See Section 3.9 for a review of these terms.)

Other trends in properties such as atomic radius, the tendency to lose or gain electrons, and some physical properties are related to the elements' position in the periodic table because of their electron configuration. In the remainder of this chapter, we discuss these trends as well as the properties themselves and their structural basis.

6.4   Periodicity and Electron Configuration

All the elements within each A group have the same number of electrons in their outermost shell. Moreover, for the A-group elements, *the group number tells us how many outer electrons there are.* For example, all the elements in Group VIA have six electrons in their outermost energy level. All the elements in Group IA have one electron in their outermost energy level.

For the B-group elements, there is no clear connection between group number and the number of electrons in the outermost energy level. For example, Group IVB elements *do not* have four electrons in their outermost energy level.

6.5   Similarities Among Elements in a Group and Period

The configuration of the outermost energy level determines the chemical behavior of an element. Thus all elements in the same group should behave (react) similarly, because they all have the same outer-shell electron configuration. In other words, the properties that these elements exhibit as a result of their interaction with other substances are similar. But what about elements in the same period? In general, the elements in any period have different outer-shell electron configurations, so these elements have different chemical properties.

For example, in period 3, Al (with two $3s$ electrons and one $3p$ electron) is a metal, whereas Cl (with two $3s$ electrons and five $3p$ electrons) is a nonmetal. However, this is not so for the transition metals in periods 4 through 7. Consider the elements Mn, Fe, Co, and Ni in period 4. Their electron configurations differ because they have different numbers of electrons in their $3d$ sublevels. However, level 3 is *not* their outermost energy level. The outermost level for these transition metals is the $4s$ sublevel. We would therefore expect these elements to exhibit somewhat similar behavior, and they do. Moreover, the same is true of the transition elements in periods 5 through 7.

To summarize:

1. Elements in any one group behave in similar ways, because they have (with few exceptions) *identical electron configurations in the outermost shell.*
2. Elements in any one period generally do not behave similarly.
3. However, because of the manner in which electrons fill their shells (filling some inner shells after outer shells), the transition metals—that is, the B groups—show somewhat similar chemical behavior.

6.6   Atomic Radius

The arrangement of the periodic table is helpful in predicting the relative sizes of atoms within any period or group. We shall consider the **atomic radius** (radius of an atom) to be the distance from the center of the nucleus to the outermost electron. Atomic radii are measured with a unit called the *angstrom,* named after a Swedish physicist. One angstrom (1 Å) is equal to 10^{-10} m, or one ten-billionth of a meter.

In any group in the periodic table, each element has one more energy level than the element above it (Figure 6.6). Therefore, as we move down through a particular group, the atomic radii of the elements should increase.

■ **FIGURE 6.6**

Relative sizes of some Group IIA elements

Beryllium, Be   Magnesium, Mg   Calcium, Ca

But what happens to the atomic radii of the elements as we move from left to right across a period? You might assume that because the number of electrons in the atom is increasing, the size of the atom should also increase. This is not the case. Remember that as we move across a period, the number of protons increases along with the number of electrons. Therefore there is a greater pull on the electrons of the inner energy levels, which usually causes a shrinking of the atomic radius (Figure 6.7).

Using the elements in period 2, we can explain what happens. In lithium (atomic number 3), three protons attract two K electrons and one L electron. In beryllium (atomic number 4), four protons attract two K electrons and two L electrons. In boron, five protons attract two K electrons and three L electrons. This increased positive charge acting on the K electrons makes the K energy level shrink. The L energy level also shrinks, and this results in a decreased atomic radius (Figure 6.8).

Also keep in mind that for the transition metals, atomic radii change very little because electrons are filling the d and f subshells. Table 6.2 shows the atomic radii of all the elements.

■ **FIGURE 6.7**

The atomic radii of the elements decrease as we move from left to right across a period. The increased nuclear charge results in a greater pull on the electrons of the inner energy levels, so the atom shrinks.

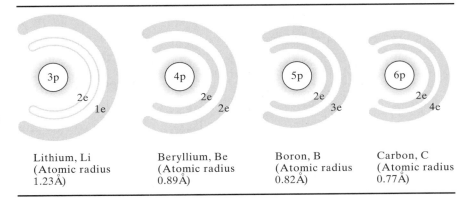

Lithium, Li
(Atomic radius
1.23Å)

Beryllium, Be
(Atomic radius
0.89Å)

Boron, B
(Atomic radius
0.82Å)

Carbon, C
(Atomic radius
0.77Å)

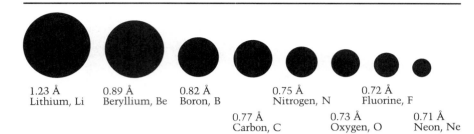

1.23 Å	
Lithium, Li	
0.89 Å	
Beryllium, Be	
0.82 Å	
Boron, B	
0.75 Å	
Nitrogen, N	
0.72 Å	
Fluorine, F	

0.77 Å
Carbon, C

0.73 Å
Oxygen, O

0.71 Å
Neon, Ne

■ **FIGURE 6.8**

The decrease of atomic radii within a period

TABLE 6.2

Periodic Table of Atomic Radii (in Angstroms)

Period	Group IA																	VIIIA
1	1 H 0.32	IIA											IIIA	IVA	VA	VIA	VIIA	2 He 0.31
2	3 Li 1.23	4 Be 0.89			Transition Elements								5 B 0.82	6 C 0.77	7 N 0.75	8 O 0.73	9 F 0.72	10 Ne 0.71
3	11 Na 1.54	12 Mg 1.36	IIIB	IVB	VB	VIB	VIIB	VIIIB			IB	IIB	13 Al 1.18	14 Si 1.11	15 P 1.06	16 S 1.02	17 Cl 0.99	18 Ar 0.98
4	19 K 2.03	20 Ca 1.74	21 Sc 1.44	22 Ti 1.32	23 V 1.22	24 Cr 1.18	25 Mn 1.17	26 Fe 1.17	27 Co 1.16	28 Ni 1.15	29 Cu 1.17	30 Zn 1.25	31 Ga 1.26	32 Ge 1.22	33 As 1.20	34 Se 1.17	35 Br 1.14	36 Kr 1.12
5	37 Rb 2.16	38 Sr 1.91	39 Y 1.62	40 Zr 1.45	41 Nb 1.34	42 Mo 1.30	43 Tc 1.27	44 Ru 1.25	45 Rh 1.25	46 Pd 1.28	47 Ag 1.34	48 Cd 1.48	49 In 1.44	50 Sn 1.40	51 Sb 1.40	52 Te 1.36	53 I 1.33	54 Xe 1.31
6	55 Cs 2.35	56 Ba 1.98	57–71 La– Lu	72 Hf 1.44	73 Ta 1.34	74 W 1.30	75 Re 1.28	76 Os 1.26	77 Ir 1.27	78 Pt 1.30	79 Au 1.34	80 Hg 1.49	81 Tl 1.48	82 Pb 1.47	83 Bi 1.46	84 Po 1.46	85 At 1.45	86 Rn
7	87 Fr	88 Ra 2.20	89–103 Ac– Lr	104 Rf	105 Db	106 Sg	107 Bh	108 Hs	109 Mt									

57 La 1.69	58 Ce 1.65	59 Pr 1.64	60 Nd 1.64	61 Pm 1.63	62 Sm 1.62	63 Eu 1.85	64 Gd 1.62	65 Tb 1.61	66 Dy 1.60	67 Ho 1.58	68 Er 1.58	69 Tm 1.58	70 Yb 1.70	71 Lu 1.56
89 Ac 2.0	90 Th 1.65	91 Pa	92 U 1.42	93 Np	94 Pu	95 Am	96 Cm	97 Bk	98 Cf	99 Es	100 Fm	101 Md	102 No	103 Lr

6 C 0.77

← Atomic number

← Atomic radius

EXAMPLE 6.1

List the following elements in order of *increasing* atomic radius: Mg, K, and Ca. Use only the periodic table (Table 6.1); don't look up the atomic radii.

Solution

UNDERSTAND THE PROBLEM

We must order the elements in terms of atomic radius without knowing the specific atomic radius of each element. In other words, we must look at periodic trends.

DEVISE A PLAN

Locate the position of each element in the periodic table, remembering that the atomic radius increases from the top to the bottom of a group of elements and decreases from left to right across a period of elements.

CARRY OUT THE PLAN

Remember that magnesium (Mg) is in period 3, whereas potassium (K) and calcium (Ca) are in period 4. Therefore, an atom of Mg has the smallest atomic radius of the three atoms in question. Next we look at K and Ca, the two elements that are in the same period. Since K is in Group IA and Ca is in Group IIA, a K atom has a larger atomic radius than a Ca atom. Therefore, of the three atoms, Mg is the smallest, Ca is larger, and K is the largest.

LOOK BACK

We glance at the periodic table to confirm our reasoning.

Practice Exercise 6.1   List the following elements in order of *decreasing* atomic radius: S, O, N, and B. Use only the periodic table (Table 6.1); don't look up the atomic radii.

6.7 Ionization Potential

When an electron is pulled completely away from a neutral atom, what remains behind is a positively charged particle called an *ion*. A **positive ion** (also called a **cation**) is *an atom (or group of atoms) that has lost one or more electrons*. A **negative ion** (also called an **anion**) is *an atom (or group of atoms) that has acquired one or more electrons*.

The process of removing an electron from a neutral atom is called *ionization*. Here is what happens when a neutral atom of potassium becomes ionized:

$$K \rightarrow K^{1+} + 1e^{1-}$$

■ FIGURE 6.9

The decrease of ionization potential within a group

Hydrogen, H	Lithium, Li	Sodium, Na
314 kcal/mole	124 kcal/mole	119 kcal/mole

The equation shows the neutral potassium atom (on the left) losing one electron to become a potassium ion (on the right). The superscript 1+ on the potassium shows that it has become a *positively charged particle.* That is, it has lost an electron ($1e^{1-}$), which is negatively charged. Thus the potassium is left with one more positive charge (proton) than negative charge (electron). It is no longer electrically neutral.

The **ionization potential** (also called *ionization energy*) of an element is *the energy needed to pull an electron away from an isolated ground-state atom of that element.* The farther away from the nucleus an electron is, the less it is attracted by the positive charge of the nucleus and the less energy is needed to pull that electron away from the atom. For this reason, the electron is always pulled away from the outermost energy level. In any group of the periodic table, the ionization potential decreases as we move down the group (Figure 6.9).

Ionization energy is usually measured in units called calories. A calorie is the amount of heat needed to raise the temperature of one gram of water one degree Celsius. A discussion of the calorie appears in Chapter 12.

Within each period, ionization potential generally increases from left to right (Table 6.3). This results from the decrease in atomic radius (and thus the stronger attraction between nucleus and outer electrons) as we move from left to right in the periodic table. The ionization potentials of the transition metals in each period do not vary much, because their atomic radii do not vary much (see Section 6.6).

EXAMPLE 6.2

List the following elements in order of *increasing* ionization potential: Ca, Mg, and K. Use only the periodic table (Table 6.1); don't look up the atomic radii.

Solution

Locate the position of each element in the periodic table. Remember that the ionization potential decreases from the top to the bottom of a

TABLE 6.3

Periodic Table of Ionization Potentials (in Kilocalories per Mole)

Period	Group IA																	VIIIA
1	1 H 314	IIA											IIIA	IVA	VA	VIA	VIIA	2 He 566
2	3 Li 124	4 Be 215			Transition Elements								5 B 191	6 C 260	7 N 335	8 O 312	9 F 402	10 Ne 498
3	11 Na 119	12 Mg 176	IIIB	IVB	VB	VIB	VIIB	VIIIB			IB	IIB	13 Al 138	14 Si 188	15 P 254	16 S 239	17 Cl 300	18 Ar 363
4	19 K 100	20 Ca 141	21 Sc 151	22 Ti 158	23 V 155	24 Cr 156	25 Mn 171	26 Fe 182	27 Co 181	28 Ni 176	29 Cu 178	30 Zn 216	31 Ga 138	32 Ge 187	33 As 242	34 Se 225	35 Br 273	36 Kr 323
5	37 Rb 96	38 Sr 131	39 Y 152	40 Zr 160	41 Nb 156	42 Mo 166	43 Tc 172	44 Ru 173	45 Rh 178	46 Pd 192	47 Ag 174	48 Cd 207	49 In 133	50 Sn 169	51 Sb 199	52 Te 208	53 I 241	54 Xe 280
6	55 Cs 90	56 Ba 120	57–71 La– Lu	72 Hf 127	73 Ta 140	74 W 184	75 Re 181	76 Os 201	77 Ir 212	78 Pt 208	79 Au 212	80 Hg 241	81 Tl 141	82 Pb 171	83 Bi 184	84 Po 196	85 At	86 Rn 248
7	87 Fr	88 Ra 122	89–103 Ac– Lr	104 Rf	105 Db	106 Sg	107 Bh	108 Hs	109 Mt									

57 La 129	58 Ce 159	59 Pr 133	60 Nd 145	61 Pm	62 Sm 129	63 Eu 131	64 Gd 142	65 Tb 155	66 Dy 157	67 Ho	68 Er	69 Tm	70 Yb 143	71 Lu 115
89 Ac 162	90 Th	91 Pa	92 U 92	93 Np	94 Pu	95 Am	96 Cm	97 Bk	98 Cf	99 Es	100 Fm	101 Md	102 No	103 Lr

6
C
260

⟵ ——— Atomic number

⟵ ——— Ionization potential

group of elements and increases from left to right across a period of elements. Magnesium (Mg) is in period 3, whereas potassium (K) and calcium (Ca) are in period 4. Therefore, an atom of Mg has the highest ionization potential of the three atoms in question.

Now consider K and Ca, the two elements that are in the same period. Because K is in Group IA and Ca is in Group IIA, a Ca atom has a higher ionization potential than a K atom.

Therefore, of the three atoms, K has the lowest ionization potential, Ca has a higher ionization potential, and Mg has the highest ionization potential. (*Note:* Look back at the solution of Example 6.1. Notice how the ionization potential varies inversely with the atomic radius. In other words, as the atomic radius of an atom gets larger, the ionization potential gets smaller, and vice versa.)

Practice Exercise 6.2   List the following elements in order of decreasing ionization potential: S, O, N, and B. Use only the periodic table (Table 6.1); don't look up the atomic radii.

6.8 Electron Affinity

An atom's ability to acquire an additional electron is called its **electron affinity** and is another important factor that determines its chemical properties. The process in which a neutral atom acquires an electron can be represented in this way:

$$Cl + e^{1-} \rightarrow Cl^{1-}$$

The equation shows that when a neutral chlorine atom picks up an electron, it becomes a negatively charged chloride ion. When the chlorine atom picks up an electron to form a chloride ion, it gives up something in return: *energy*. The amount of energy the chlorine atom releases is a measure of its electron affinity. Electron afffinity depends on the attraction between an electron and the nucleus of an atom.

As we move down a group, the electron affinity of the elements generally decreases. If we look at some Group VIIA elements, we can see a decrease in electron affinity as we move from chlorine to iodine (Table 6.4). This is because in iodine, the additional electron is added to the fifth energy level, whereas in chlorine the additional electron is added to the third energy level. (Remember that the closer the electron is to the nucleus, the more it is attracted to the nucleus, and electron affinity depends on how strongly an electron is attracted to the nucleus.) Fluorine's unexpectedly low electron affinity is caused by the special nature of this element. (Some scientists feel that the electron affinity value for fluorine is reasonable but that the electron affinity value for chlorine is too high.)

TABLE 6.4

Electron Affinities of Group VIIA Elements

ELEMENT	ELECTRON AFFINITY (IN ELECTRON VOLTS)	OUTER ELECTRON CONFIGURATION
F	3.6	$2s^2\,2p^5$
Cl	3.75	$3s^2\,3p^5$
Br	3.53	$4s^2\,4p^5$
I	3.2	$5s^2\,5p^5$

TABLE 6.5

Summary of Trends in the Periodic Table

TREND OF	TOP TO BOTTOM IN A GROUP	LEFT TO RIGHT IN A PERIOD
Atomic radius	Increases	Decreases
Ionization potential	Decreases	Increases
Electron affinity	Decreases	Increases

If we examine the trend of electron affinity within a period, we find that it *increases* as we go from left to right in a period, and the attraction between the nucleus and the electron becomes greater.

Table 6.5 summarizes what we have discussed about trends in the periodic table. As your study of chemistry progresses, you will find that you rely more and more on information conveyed by the periodic table.

SUMMARY

As more and more elements were discovered during the eighteenth and nineteenth centuries, scientists attempted to find relationships among the properties of the various known elements. Finally, in 1869, Dmitri Mendeleev arranged the elements in a grid or table form, with horizontal rows called periods and vertical columns called groups. Elements with similar chemical properties were positioned in the same group. The table ordered the known elements and their properties so well that Mendeleev was able to use it to predict the properties of yet undiscovered elements. The periodic table of today is similar to Mendeleev's. It consists of seven periods and sixteen groups. For elements in the A groups (representative elements), the group number is the same as the number of electrons in the outermost shell. For elements in the B groups (transition metals), there is not a simple connection between group number and number of electrons in the outermost energy level. However, because the transition metals fill inner shells after outer shells, they show quite similar chemical behavior.

The atomic radius of an atom is the distance from the center of the nucleus to the outermost electron. As we move from the top to the bottom of any group in the periodic table, atomic radius generally increases. And as we move from left to right across a period, atomic radius generally decreases. The ionization potential for an element is the energy needed to pull an outershell electron away from an isolated atom of that element. As we move from the top to the bottom of a group, the ionization potential decreases. And as we move from left to right across a period, the ionization potential generally increases. Electron affinity, a measure of an atom's ability to acquire additional electrons, is related to the energy released when an additional electron is added to a neutral atom. As we move down a group in the periodic table, the electron affinity of the elements generally decreases, whereas as we move from left to right across a period, the electron affinity increases.

KEY TERMS

anion **(6.7)**

atomic radius **(6.6)**

cation **(6.7)**

electron affinity **(6.8)**

group (family) **(6.2)**

ionization potential **(6.7)**

negative ion **(6.7)**

period **(6.2)**

positive ion **(6.7)**

representative element **(6.2)**

transition metal **(6.2)**

CAREER SKETCH

Oceanographer

Studying the oceans of Planet Earth can help scientists understand many phenomena that occur, not only at sea, but also on land. Scientists who study the ocean, called *oceanographers,* use the principles of chemistry, physics, mathematics, and engineering.

Oceanographers study both the oceans and their plant and animal life. They use surface ships, aircraft, or underwater craft, using specialized instruments to measure and record findings. As an oceanographer, your research may help develop methods to forecast weather, expand fisheries, mine ocean resources, or improve national defense capabilities.

Many oceanographers perform their work at sea. Others work almost entirely in laboratories on land.

A strong background in the physical or biological sciences is advantageous if you want to become an oceanographer. The Woods Hole Oceanographic Institute has more information at its website at http://www.whoi.edu.

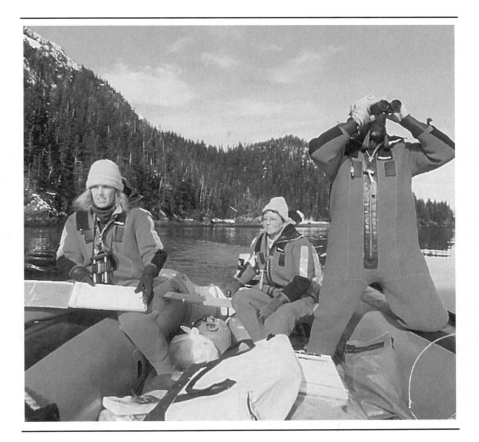

Oceanographers use principles from chemistry, physics, mathematics, and engineering to study the ocean.
(Randy Brandon/Peter Arnold, Inc.)

*"Searching for Chemicals with a Buzz"**

Before the advent of sophisticated sensors to detect methane gas, canaries accompanied workers into coal mines. If a canary died in its cage, workers knew to be cautious of the dangerous methane gas, which could ignite and explode. Today, researchers at the University of Montana in Missoula are using a similar method to test for pollution.

The researchers, led by Jerry J. Bromenshenk, believe that they can use honeybees to test for environmental contaminants. The theory behind the honeybee experiments is rather simple. On journeys from the hive, bees pick up water, nectar, pollen, and air. When they return to the hive, they bring these materials with them. There, scientists can check for contaminants. As very efficient and undiscriminating collectors, bees seem to be able to provide information about environmental contaminants in broad geographic areas.

Two initial tests seem to support Bromenshenk's hypothesis. In one experiment, he measured the amounts of arsenic, cadmium, lead, zinc, copper, and fluoride the bees brought back with them to the hive. He then compared his results with maps that local agencies had drawn to analyze soil samples. The two corresponded. A second bee test at the Army's Aberdeen Proving Ground in Maryland also corresponded with known contaminants.

In the future, the researchers hope to be able to use artificial neural networks in their analysis. This will allow them to analyze the enormous amount of data from the beehives more efficiently. They also hope that their bee research will have two practical uses. First, researchers hope to be able to analyze the behavior of bees to identify the existence of a chemical problem. Second, they hope to be able to use the bees to perform ongoing chemical sampling, replacing expensive sampling programs in effect today.

*Adapted from *Science News*, May 24, 1997, pp. 324–325

SELF-TEST EXERCISES

LEARNING GOAL 1: *Historical Basis and Significance of Periodic Table*

◀ 1. Julius Lothar Meyer published a periodic table at about the same time that Dmitri Mendeleev published his table. Why do you think that Mendeleev received the major portion of the credit for discovering the periodic law?

2. How did knowledge of the periodic law help future scientists discover new elements?

◀ 3. Locate at least two instances in the periodic table (Table 6.1) in which a heavier element appears before a lighter one.

4. What was Mendeleev's reason for placing some heavier elements before lighter ones in his table? Why didn't this upset the "logic" of his table?

◀ 5. Why aren't elements 57 to 71 and 89 to 103 shown with the other elements in the body of Table 6.1?

6. Discuss the contributions to the development of the periodic table made by (a) Döbereiner, (b) Newlands, (c) Mendeleev.

LEARNING GOAL 2: *Periods and Groups in the Periodic Table*

7. Give the name and symbol of the element from the information that follows.
(a) The element is in period 4 and group IIIA.
(b) The element is in period 3 and group VIA.
(c) The element is in period 6 and group IIB.
(d) The element is in period 7 and group VIIB.

8. In the periodic table, a horizontal row is called a _____. A vertical column is called a _____.

9. What similarity is there among the outermost energy levels of elements within the same group?

10. "Elements in the B groups (the transition metals) show somewhat similar chemical behavior." Explain why this statement is true.

LEARNING GOAL 3: *A-Group and B-Group Elements*

11. What does the group number tell us about the A-group elements?

◀ *12.* How many elements are in the following periods?
   (a) period 2
   (b) period 3
   (c) period 4
   (d) period 5
   (e) period 6
   (f) period 7

13. The *representative* elements are also known as the _____ (A/B) group elements. The *transition metals* are also known as the _____ (A/B) group elements.

14. "Elements within any one group have the same number of electrons in their outermost energy levels." Is this statement true for the A-group or the B-group elements?

LEARNING GOAL 4: *Periodic Trends*

◀ *15.* List the following elements in order of increasing ionization potential: Br, Sr, I, Te, and Ba. Use only the periodic table (Table 6.1); don't look up the ionization potentials.

16. What is an angstrom, and how big is it?

17. List the following elements in order of increasing atomic radius: In, P, I, Sb, and As. Once again, use only the periodic table.

18. Does atomic radius increase or decrease as we move from the top to the bottom of a group in the periodic table? Why?

19. Define *electron affinity*.

20. As we move from left to right across a period in the periodic table, does atomic radius increase or decrease? Why?

◀ 21. After you've removed one electron from a neutral atom, would you need more or less energy to remove a second electron? Why?

22. List the following elements in order of increasing atomic radius: In, Tl, H, F, and S. (Use the periodic table.)

23. Where in the periodic table do we find the following?
   (a) elements with the highest electron affinity
   (b) elements with the lowest ionization potential
   (c) the most unreactive elements

24. What name is given to the energy needed to pull an electron away from an isolated atom?

25. Complete the following statements:
   (a) The ionization potential _____ (increases/decreases) as one moves down a *group* of elements.
   (b) The ionization potential _____ (increases/decreases) as one moves from left to right across a *period* of elements.
   (c) The atomic radius _____ (increases/decreases) as one moves from left to right across a *period* of elements.

◀ 26. Explain why it is more difficult to remove an electron from an atom that is located at the top of a group in the periodic table than to remove an electron from an atom that is located at the bottom of a group.

◀ *27.* List the following elements in order of increasing atomic radius: Cl, Ca, Sr, Rb, and S. Use only the periodic table (Table 6.1); don't look up the atomic radii.

◀ 28. The elements sodium and chlorine are both in period 3 of the periodic table. Sodium has an ionization potential of 119 kcal/mole, and chlorine has an ionization potential of $30\overline{0}$ kcal/mole. Why is the ionization potential of chlorine so much greater?

29. List the following elements in order of increasing ionization potential: P, I, Sb, and As. Once again, use only the periodic table.

30. List the following elements in order of increasing ionization potential: Br, Ge, Se, Cs, and F. Use only the periodic table.

31. A positive ion is an atom (or group of atoms) that has (a) gained a proton, (b) lost a proton, (c) gained an electron, (d) lost an electron.

32. The quantity that depends on the attraction between an electron and the nucleus of an atom is called _____.

33. A Cl^{1-} ion is a Cl atom that has (a) gained a proton, (b) lost a proton, (c) gained an electron, (d) lost an electron.

34. Electron affinity _____ (increases/decreases) as one moves down a *group* of elements.

35. An Mg^{2+} ion is an Mg atom that has (a) gained two protons, (b) lost two protons, (c) gained two electrons, (d) lost two electrons.

36. Electron affinity _____ (increases/decreases) as one moves from left to right across a *period* of elements.

EXTRA EXERCISES

37. (a) Explain why a bromine atom is smaller than a potassium atom, even though a bromine atom contains more subatomic particles.
    (b) Explain why the ionization potential for a sodium atom is greater than the ionization potential for a rubidium atom.

38. What relationship is there with regard to chemical properties among
    (a) all elements in a particular period?
    (b) all elements in a particular group?

39. Without using the periodic table, find the electron configurations of the following elements. State which ones are chemically similar. (a) $_{11}Na$ (b) $_{13}Al$ (c) $_8O$ (d) $_{16}S$ (e) $_5B$ (f) $_3Li$

40. Without using the periodic table, find the electron configurations of the following elements. State which ones are chemically similar. (a) $_1H$ (b) $_3Li$ (c) $_4Be$ (d) $_{12}Mg$ (e) $_{20}Ca$

41. What would you expect the atomic number to be for the noble gas that follows radon?

42. Based on periodic trends, which group of atoms is most likely to form (a) 1+ ions? (b) 1− ions?

43. Explain the difference between a chlorine atom and a chloride ion.

44. Which member of each of the following pairs should have the larger radius? Why? (a) K, Br (b) K, Na (c) K, K^{1+}

45. Determine how many electrons are in the outermost energy level of a neutral atom of the following elements: (a) sodium (b) carbon (c) nitrogen

46. Which has the higher ionization potential, calcium or magnesium? Why?

47. How many elements are in the (a) first period? (b) second period? (c) third period? (d) fourth period? (e) fifth period? (f) sixth period?

48. Which element of each pair has the smaller ionization potential? (a) K, Ca (b) O, Se

49. Which element of each pair has the larger atomic radius? (a) K, Ca (b) O, Se

50. What would be the group number for element 114?

51. Why are the atomic masses of elements 94 through 109 listed in parentheses?

52. If you had the responsibility of naming the next discovered element, what would you name it?

53. We know that the number of electrons in the outermost energy level of the Group IIIB elements is 3 because the group number reflects the number of electrons in the outermost energy level. True or false? Why?

54. We know that the number of electrons in the outermost energy level of the Group IIIA elements is 3 because the group number reflects the number of electrons in the outermost energy level. True or false? Why?

55. By writing electron configurations, determine which of the following elements belong to the same groups: elements with atomic number 15, 38, 50, 56, 14, and 33.

56. By writing electron configurations, determine which of the following elements belong to the same groups: elements with atomic number 16, 36, 39, 8, and 34.

1. How are protons, neutrons, and electrons distributed in the atom?

2. How did Eugen Goldstein choose the term *cathode rays?*

3. What experiment did Sir William Crookes perform to determine whether cathode rays were composed of particles or waves?

4. What is the photoelectric effect?

5. What reasoning led physicist J. J. Thomson to develop the "plum-pudding" theory of the atom?

6. When Rutherford tested Thomson's theory of the atom with his own experiments, what results led him to devise a new model of the atom?

7. Why did Rutherford's model of the atom render Thomson's atomic model obsolete?

8. What type of particle adds to the mass of the atom without affecting its charge?

9. John Dalton said that atoms of an element must be identical in size, mass, and shape. How does the existence of isotopes disprove this theory?

◀ 10. Give the number of protons that are found in each of the following isotopes: (a) $^{59}_{27}$Co (b) $^{90}_{40}$Zr (c) $^{74}_{34}$Se

11. Using Table 4.2, calculate the atomic mass of neon.

12. Using Table 4.2, determine how many ^{15}N atoms there are in a sample containing 100,000 nitrogen atoms.

◀ 13. Write the standard isotopic notation for the following elements. (Use the periodic table to get the symbols of the elements.) (a) 17p, 17e, 18n (b) 14p, 14e, 14n

14. The hypothetical element Z exists in two isotopic forms, ^{50}Z and ^{52}Z. The atomic mass of Z is 50.5. What is the percentage abundance of each isotope?

15. State the number of protons, electrons, and neutrons in $^{73}_{32}$Ge.

16. How many electrons are there in a neutral $^{60}_{26}$Fe atom?

17. Of the three isotopes of oxygen, which is the most abundant: oxygen-16, oxygen-17, or oxygen-18?

18. Using Table 4.2, determine how many ^{7}Li atoms there are in a sample containing 2.0×10^3 lithium atoms.

◀ 19. Using Table 4.2, determine how many atoms of ^{7}Li you would find in a sample of lithium that has 1.000×10^6 atoms of Li.

20. Using Table 4.2, determine how many atoms of Si you would need to get exactly 470,000 atoms of ^{29}Si.

21. Do electrons, protons, or neutrons produce the line spectrum of an element?

22. As an electron travels to an energy level farther from the nucleus, the energy of that electron _____.
(a) increases   (b) decreases   (c) remains the same

23. How many electrons are in the outermost energy level of an oxygen atom?

24. Without using the periodic table, determine the electron configurations of the following elements:
(a) $_1$H   (b) $_3$Li   (c) $_{11}$Na   (d) $_{19}$K

25. A neutral atom of magnesium has _____ electrons in its outermost energy level. (You may consult the periodic table.)

26. Name the element that has the following ground-state configuration. You may consult the periodic table.

    K, 2 electrons; L, 8 electrons; M, 6 electrons

◀ 27. The chemical properties of an element are determined by the number of electrons in the _____.
(a) energy level closest to the nucleus
(b) outermost energy level
(c) second energy level
(d) none of these

28. Why is phosphorus in Group VA?

◀ 29. Write electron configurations for all elements with atomic numbers between 1 and 20 that have the same number of electrons in the outermost energy level as a lithium atom.

30. What is the atomic number of an element whose energy levels K, L, M, and N are filled to capacity?

31. When a rainbow is produced naturally, what acts as the prism?

32. How are line spectra used in chemical laboratories today?

33. In a detective story it was discovered that the victim died of arsenic poisoning. This information was made

available by analysis of samples of the victim's hair. How could spectral analysis have made this discovery possible?

34. What is the major source of radiant energy?

35. Waves on a lake travel through water, and sound waves travel through air. Through what medium do electromagnetic waves travel?

36. What types of waves are included in the electromagnetic spectrum? (a) visible (b) invisible (c) both of these (d) neither of these

37. How did Heinrich R. Hertz verify the theory of electromagnetic radiation proposed by James Clerk Maxwell?

38. Give an example of light energy being absorbed and re-emitted.

39. Give an example of heat energy being absorbed and re-emitted as light energy.

◀ **40.** Bundles of energy absorbed or emitted by electrons are called _____.

41. State the periodic law and explain how it is the basis of the modern-day periodic table.

◀ **42.** What was the contribution of Henry G. J. Moseley in terms of the periodic table?

43. The element with atomic number 107 is an artificially produced radioactive element. Both Soviet and German scientists claim to have discovered it, but neither claim has been officially accepted. To produce an isotope of the element with atomic number 107, a sample of bismuth-209 was bombarded with chromium-54. Write the electron configuration of the element with atomic number 107. What is its group number?

44. In 1974 a team of U.S. scientists working under Albert Ghiorso at the Lawrence Berkeley Laboratory in Berkeley, California, bombarded the element californium-249 with a beam of oxygen-18 to produce an isotope of the element with atomic number 106.

Write the electron configuration for the element with atomic number 106. What is its group number?

45. Our modern periodic table contains cases where elements are not in proper order according to atomic mass. Where do these cases occur?

46. Ionization potential increases as one moves down a group of elements. True or false?

47. Atomic radius decreases as one moves from left to right across a period of elements. True or false?

◀ **48.** A Ca^{2+} ion is a calcium atom that has gained two protons. True or false?

49. The element with atomic number 108 should behave like what other elements in the periodic table?

50. Name the neutral element with the electron configuration $1s^2 2s^2 2p^6 3s^1$.

51. The elements in the periodic table show properties that are related to their atomic numbers. True or false?

◀ **52.** The element with the electron configuration $1s^2 2s^2 2p^6$ belongs in group VIA. True or false?

53. Lithium and fluorine are both in period 2 of the periodic table. Which element has the higher ionization potential? Why?

54. Write the electron configuration for $_{14}Si$.

55. The volume of space occupied by an electron is called an orbital. True or false?

◀ **56.** On the basis of periodic trends, which group of atoms is most likely to form (a) 2+ ions? (b) 2− ions?

57. Which member of each of the following pairs should have the larger radius? (a) Li, Na (b) Ca, Ca^{2+}

58. Which element has the lower ionization potential, Mg or Cl?

59. Which element has the larger atomic radius, Li or Cs?

60. Which element has the larger atomic radius, Li or F?

CHEMICAL BONDING:
How Atoms Combine

The time is about five billion years ago. You are suspended in the void of space. But look, our sun is not there. Neither is the Earth or any of the other planets in our present solar system. In the distance we see what looks like a giant star. Its light flickers, and it is about to explode.

There is no noise in space, but we see a tremendous explosion that seems to envelop us. The black void is filled with blinding light. It is our solar system forming before our eyes, out of the remains of an exploding star! All of the basic elements that we are familiar with in the periodic table are present. The most abundant element is hydrogen. A tremendous amount of hydrogen forms our sun. Other elements that are present begin to coalesce and form planets. Many of these elements combine to form compounds.

On one particular planet the conditions are just right so that, over the course of the next five billion years, the simple compounds that are formed react with each other to form more complex compounds. Eventually these compounds form living organisms of a very complex nature. This planet is the third planet from the sun, and we are just some of those complex organisms found living on it.

LEARNING GOALS

After you've studied this chapter, you should be able to:

1. Write Lewis (electron-dot) structures for the A-group elements.
2. Define and give an example of a covalent bond and a coordinate covalent bond.
3. Name the diatomic elements from memory.
4. Write electron-dot structures for various covalent compounds.
5. Define and give an example of an ionic bond.

147

6. Use the concept of electronegativity to find the ionic percentage and covalent percentage of a chemical bond.
7. Distinguish between a polar and a nonpolar bond.
8. Distinguish between a polar and a nonpolar molecule.

INTRODUCTION

Matter is held together by forces acting between atoms. When chemical reactions occur, atoms can combine to form compounds. In doing so, they join together in such a way as to attain more stable configurations that have lower levels of chemical potential energy.

Atoms have the ability to gain, lose, or share electrons. This enables them to form compounds in two different ways. When electrons are gained and lost, the compounds formed are held together by forces that bind oppositely charged ions. When electrons are shared, the compounds formed are held together by forces that bind atoms together to form molecules.

Whether the binding forces are due to electron loss and gain or to sharing, when atoms combine they are said to be joined together by a **chemical bond.** They have attained a more stable arrangement—usually eight electrons in their outermost shell. In this chapter, we examine the types of bonds that tie atoms together in molecules and unite elements in compounds.

7.1 Lewis (Electron-Dot) Structures

LEARNING GOAL 1
Lewis structures

Because it is the electrons in the outermost energy levels that form the bonds, we will be concerned only with these electrons in the elements we study. A very useful notation for showing the outermost electrons of an atom (and the bonds they form) was devised in 1916 by G. N. Lewis, an American physical chemist. In this *electron-dot* notation, the outer electrons are shown as dots on the sides of the symbol for the element. (Figure 7.1 provides several examples.) Each dot corresponds to an outer-shell electron. Although dot placement isn't critical, one method is to place the dots one to a side, beginning at the top and moving clockwise. A second dot is then placed on a side only after there is one dot on each of the four sides of the symbol. If you move from left to right in Figure 7.1, you will see how each additional dot (as necessary) is placed on the diagram. Electron-dot

■ FIGURE 7.1

Electron-dot diagrams of some A-group elements

IA	IIA	IIIA	IVA	VA	VIA	VIIA
Ḣ						
L̇i	Be·	Ḃ·	·Ċ·	·N̈·	·Ö:	·F̈:

structures make it easier to see the role outer-shell electrons play in chemical bonding.

Recall now that all the elements in the same A group have the same number of electrons in their outermost energy level. This means that all the elements in any A group have the same electron-dot diagram. Figure 7.2 shows the diagrams for fluorine and bromine, which are both in Group VIIA. Both have seven dots. And Figure 7.3 shows the diagrams for nitrogen and arsenic; both are in Group VA, so both have five dots.

Now that you know how to write the electron-dot notation for an atom of an element, we'll learn how to use this notation to draw the *electron-dot structure* of a compound. The **electron-dot structure** is *a schematic representation of the bonding in a molecule of a compound*. It does not represent the physical position of the electrons in the molecule. However, the electron-dot notation is useful for understanding how bonding occurs.

7.2   The Covalent Bond: The Octet Rule

■ LEARNING GOAL 2
Covalent bonds

It has long been known that the Group VIIIA elements tend to be very unreactive and stable. Because of their tendency to keep to themselves, these elements are called the *noble gases.* Researchers believe that this stability is connected with the fact that each of these Group VIIIA elements has eight electrons in its outermost energy level. (The element helium is an exception because its atom has only two electrons, but these two electrons fill the outermost energy level of the helium atom.) The atoms of other elements can achieve the same type of stability by obtaining eight electrons in their outermost energy levels. We state this relationship formally as the **octet rule,** which tells us that when forming compounds, atoms of elements gain, lose, or share electrons such that a noble gas configuration is achieved for each atom. When atoms share electrons with other atoms, a *covalent bond* is formed. A **covalent bond** is *a chemical bond in which two atoms share a pair of electrons.*

To see how electrons are shared, imagine a sample of fluorine atoms in a closed container. A fluorine atom has seven electrons in its outermost energy level. It needs eight electrons in that energy level to complete its octet. One way to obtain another electron is to share an outer electron belonging to another fluorine atom (Figure 7.4). (Examine Figure 7.4, and think of the arrow as indicating that the items on the left combine to form what is on the right. We will discuss such chemical equations in more detail in Chapter 10.) When a fluorine atom shares one of its electrons with another fluorine atom, a fluorine molecule (F_2) is formed. But now each atom has a stable octet. Evidence of this stability is the fact that fluorine is always found as F_2 molecules.

The type of covalent bond formed between the two fluorine atoms is called a **single covalent bond** because each atom has shared a *single* electron with the other to form one bond. The single bond is the most common type of covalent bond.

■ FIGURE 7.4

Bonding between fluorine
atoms

The Diatomic Elements

LEARNING GOAL 3
Diatomic elements

Diatomic elements are elements that are found naturally in molecules with
two atoms each. Fluorine is thus a *diatomic element*. In fact, all the Group
VIIA elements are diatomic. They all form single covalent bonds (Figure
7.5). Hydrogen, oxygen, and nitrogen also exist in diatomic forms (Figure
7.6) because they share each other's electrons.

An easy way to remember the diatomic elements is with the memory
aid HONClBrIF (pronounced "HON kel brif"). Remember that these atoms
exist as diatomic molecules because doing so yields a stable octet of electrons
or, in the case of hydrogen, a stable duet (two electrons, like helium).

Other Molecules Having Single Covalent Bonds

Now let's see how some elements obtain eight electrons (or, in the case
of hydrogen, two electrons) in their outermost energy levels by forming
compounds. We will use the electron-dot notation to show the bonding.

Water, which has the formula H_2O, bonds in this way:

$$\dot{H} + \dot{H} + {}_x^x\overset{x}{\underset{x}{O}} \longrightarrow H{}^x_x\overset{xx}{\underset{x}{O}}$$
$$\dot{H}$$

In this diagram an X is used to represent the oxygen's electrons. The two
hydrogens share their single electrons with the oxygen. So the oxygen obtains
eight electrons in its outermost energy level, and each hydrogen now has its
two electrons.

In ammonia (NH_3), methane (CH_4), and hydrogen chloride (HCl), the
bonding looks like this:

$$3\dot{H} + {}_x^x\overset{x}{N}{}^x \longrightarrow H{}^x_x\overset{xx}{N}{}^x_xH$$
$$\overset{\cdot x}{H}$$

$$4\dot{H} + {}_x\overset{x}{C}{}^x \longrightarrow H{}^x_x\overset{x}{C}{}^x_xH$$
$$\overset{x}{\underset{\cdot}{H}}$$

$$\dot{H} + {}_{xx}^{xx}\overset{x}{Cl}{}^x \longrightarrow H{}^x_x\overset{xx}{Cl}{}^x_{xx}$$

The N, C, and Cl atoms now have eight electrons in their outermost energy
level. The H atoms complete their outer shells with two electrons. The
molecules are more stable than the atoms from which they are formed.

■ FIGURE 7.5

Bonding diagrams of Group
VIIA diatomic elements

$$:\ddot{F}\cdot + \cdot\ddot{F}: \longrightarrow :\ddot{F}{}^x_x\overset{xx}{F}{}^x_{xx}$$

$$:\ddot{C}l\cdot + \cdot\ddot{C}l: \longrightarrow :\ddot{C}l{}^x_x\overset{xx}{C}l{}^x_{xx}$$

$$:\ddot{B}r\cdot + \cdot\ddot{B}r: \longrightarrow :\ddot{B}r{}^x_x\overset{xx}{B}r{}^x_{xx}$$

$$H\cdot + {}^xH \longrightarrow H{}^xH$$
Hydrogen (single bond)

$$:\ddot{O}\cdot + {}_x^x\overset{x}{O}{}^x \longrightarrow :\ddot{O}{}^x_x\overset{xx}{O}{}^x$$
Oxygen (double bond)

$$\cdot\dot{N}\cdot + {}_x^x\overset{x}{N}{}^x \longrightarrow \dot{N}{}^x_x\overset{xx}{N}$$
Nitrogen (triple bond)

■ FIGURE 7.6

Bonding diagrams for other
diatomic elements. Hydrogen
needs only two electrons to
complete its outer energy
level. In O_2, two electrons
from each atom contribute to
the bond. Three electrons
from each nitrogen atom
contribute to the bond.

The Use of Dashes

LEARNING GOAL 4

**Electron-dot structures
of covalent compounds**

Instead of using electron-dot notation to show the bonding in compounds, we sometimes use dashes. The dash (—) has the same meaning as a double dot (:), but it simplifies the diagram. The dash, then, represents a single bond, or a shared pair of electrons. The following examples will show you how the dash is used in combination with dot notation.

$$:\ddot{C}l{}^x_x\ddot{C}l{}^x_x \qquad \text{or} \qquad :\ddot{C}l - \ddot{C}l{}^x_x$$

$$H{}^x\ddot{O}{}^x_x \qquad \text{or} \qquad H - \ddot{O}{}^x_x$$
$$\overset{..}{\underset{H}{}} \qquad\qquad\qquad\qquad\qquad | $$
$$\qquad\qquad\qquad\qquad\qquad\qquad H$$

$$H{}^x_x N{}^x_x H \qquad \text{or} \qquad H - \overset{xx}{N} - H$$
$$\overset{xx}{H} \qquad\qquad\qquad\qquad\qquad\qquad |$$
$$\qquad\qquad\qquad\qquad\qquad\qquad H$$

$$\qquad\qquad H \qquad\qquad\qquad\qquad\qquad H$$
$$\qquad\qquad{}^{x.} \qquad\qquad\qquad\qquad\qquad |$$
$$H{}^x_x C{}^x_x H \qquad \text{or} \qquad H - C - H$$
$$\qquad\qquad{}^{x}_{.} \qquad\qquad\qquad\qquad\qquad |$$
$$\qquad\qquad H \qquad\qquad\qquad\qquad\qquad H$$

$$H{}^x_x \ddot{C}l{}^x_x \qquad \text{or} \qquad H - \ddot{C}l{}^x_x$$

This notation is especially useful when you want to represent the structures of compounds that have double and triple bonds.

The Double Bond

A **double covalent bond** is *one in which two pairs of electrons are shared between two atoms.* For example, in oxygen (O_2) the bonding is

$$\cdot\ddot{O}: + {}^x_x\ddot{O}{}^x_x \longrightarrow :\ddot{O}:{}^x_x\ddot{O}{}^x_x \quad \text{or} \quad :\ddot{O} = \ddot{O}{}^x_x$$

By sharing two electrons, each oxygen atom obtains eight electrons in its outermost energy level. (Actually, this bonding structure for O_2 is simplified and not entirely correct, but the simplified structure communicates the general idea.)

The bonding in the compound ethylene (C_2H_4), which is used to make polyethylene plastic, can be written as

$$\begin{matrix} H & H \\ {}^x_x & {}^{xx} \\ C & :{}^x_x C \\ {}^x_x & {}^{xx} \\ H & H \end{matrix} \quad \text{or} \quad \begin{matrix} H & & H \\ | & & | \\ C & = & C \\ | & & | \\ H & & H \end{matrix}$$

(In this diagram, an x is used to represent the electron in each hydrogen.) Here the hydrogen atoms are satisfied with two electrons, and each of the carbon atoms has the eight electrons it needs.

The Triple Bond

A **triple covalent bond** is *one in which three pairs of electrons are shared between two atoms*. For example, in nitrogen (N_2) the bonding can be diagrammed in this way:

$$\cdot \ddot{N} \cdot + {}^{x}_{x}\ddot{N}^{x} \longrightarrow \dot{N} \overset{xx}{\underset{xx}{:}} N \quad \text{or} \quad \dot{N} \equiv \overset{xx}{N}$$

With the triple bond, each nitrogen atom obtains eight electrons in its outermost energy level. The original N atom has only five electrons, but when two N atoms get together by triple bonding, each has a share in eight electrons.

In the compound acetylene (C_2H_2), the bonding is

$$H{:}C{:}{}^{x}_{x}C{:}H \quad \text{or} \quad H - C \equiv C - H$$

EXAMPLE 7.1

Write the electron-dot structures for each of the following covalent compounds.
(a) CO_2 (Each oxygen is bonded to the carbon, but the oxygens aren't bonded to each other.)
(b) CH_3I (The iodine and all the hydrogens are bonded to the carbon.)
(c) $COCl_2$ (This is phosgene. The oxygen and both chlorines are bonded to the carbon.)
(d) $AsCl_3$ (Each chlorine is bonded to the arsenic.)
(e) H_2S (Both hydrogens are bonded to the sulfur.)
(f) CH_2O (This is formaldehyde. The oxygen and both hydrogens are bonded to the carbon.)
(g) C_2Cl_2 (One chlorine is bonded to each carbon.)

Solution

(a) $:\ddot{O} = C = \overset{xx}{\underset{x}{O}}{}^{x}$

(b) $H - \overset{\displaystyle H}{\underset{\displaystyle H}{\overset{|}{\underset{|}{C}}}} - \overset{xx}{\underset{xx}{I}}{}^{x}$

(c) ${}^{x}_{x}\overset{xx}{Cl} - \overset{|}{\underset{|}{C}} = \overset{xx}{O}{}^{x}$
$\quad\quad :\underset{\cdot\cdot}{Cl}:$

(d) $:\ddot{Cl} - \overset{|}{\underset{|}{As}} - \overset{xx}{\underset{xx}{Cl}}{}^{x}$
$\quad\quad :\underset{\cdot\cdot}{Cl}:$

(e) $H - \overset{\cdot\cdot}{\underset{\cdot\cdot}{S}} - H$

(f) $H - \overset{|}{\underset{|}{C}} = \overset{xx}{O}{}^{x}$
$\quad\quad\quad H$

(g) $:\ddot{Cl} - C \equiv C - \overset{xx}{\underset{xx}{Cl}}{}^{x}$

Practice Exercise 7.1   Write the electron-dot structures for the following covalent compounds: (a) NH_3 (b) HCl (c) SO_2

7.3 The Coordinate Covalent Bond

In the examples of covalent bonds that we have discussed so far, both atoms furnished electrons to form the bond. This can be thought of as a "Dutch-treat" bond, in which each atom pays its share of the bill. Another type of covalent bond exists. It is called a **coordinate covalent bond,** *a covalent bond in which only one atom donates the electrons to form the bond.*

A coordinate covalent bond can be thought of as a "you-treat" bond: one atom pays the entire bill. An example of this type of bond is found in the compound phosphoric acid (H_3PO_4). A coordinate covalent bond forms between the phosphorus atom and the single oxygen atom, to which it gives two electrons.

Ordinary (not coordinate) covalent bonds form between the phosphorus atom and the other three oxygen atoms (those with hydrogens attached to them). Once formed, the coordinate covalent bond is indistinguishable from an ordinary covalent bond. However, in drawing bonding diagrams of compounds that have coordinate covalent bonds, we sometimes use an arrow to designate the coordinate covalent bond. For example, H_3PO_4 may be diagrammed as follows:

$$
\begin{array}{c}
:\ddot{O}: \\
\uparrow \\
H - \ddot{O} - P - \ddot{O} - H \\
| \\
:O: \\
| \\
H
\end{array}
$$

Note that the arrow does not mean that the bond is in any way different.

EXAMPLE 7.2

Draw bonding diagrams for the following covalent compounds that have coordinate covalent bonds included in their structures.
(a) SO_3 (Each oxygen is bonded to the sulfur.)
(b) H_2SO_4 (Each oxygen is bonded to the sulfur, and each of the hydrogens is bonded to a different oxygen.)

Solution

(a)
$$\ddot{O}:$$
$$\|$$
$$:\ddot{O} \leftarrow S \rightarrow \ddot{O}:$$

There are two coordinate covalent bonds in this compound.

(b)
$$:\ddot{O}:$$
$$\uparrow$$
$$H-\ddot{O}-S-\ddot{O}-H$$
$$\downarrow$$
$$:\ddot{O}:$$

There are two coordinate covalent bonds in this compound.

Practice Exercise 7.2   Draw a bonding diagram for H_2SO_3, sulfurous acid, which has a coordinate covalent bond included in its structure. (*Hint:* Each oxygen is bonded to the sulfur, and each of the hydrogens is bonded to a different oxygen.)

7.4   Ionic Bonding

LEARNING GOAL 5
Ionic bonds

One atom can actually *give* electrons to another atom so that they both obtain an octet of electrons in their outermost energy level. This process is called *ionic bonding*. An example of ionic bonding is the formation of sodium fluoride (Figure 7.7). A bond formed by sharing electrons (a covalent bond) would help the fluorine *but not the sodium*. If a sodium atom shared a pair of electrons, it would have only two electrons in its outermost energy level:

$$Na \overset{xx}{\underset{xx}{x}} \overset{}{F} \overset{}{x}$$

This does not satisfy the octet rule.

However, note what happens if the sodium *gives* or *transfers* its outermost electron to the fluorine. First, the fluorine obtains eight electrons in its outermost energy level. And, in the sodium atom, the second energy level

■ **FIGURE 7.7**

Ionic bonding in sodium fluoride

FIGURE 7.8

Ionic bonding in calcium chloride

$$\dot{\text{Ca}}\cdot + \overset{xx}{\underset{xx}{^{x}\text{Cl}^{x}_{x}}} + \overset{xx}{\underset{xx}{^{x}\text{Cl}^{x}_{x}}} \longrightarrow \text{Ca}^{2+} + \overset{xx}{\underset{xx}{\dot{x}\text{Cl}\dot{x}^{1-}_{x}}} + \overset{xx}{\underset{xx}{\dot{x}\text{Cl}\dot{x}^{1-}_{x}}}$$

(which contains eight electrons) now becomes the outermost energy level; so the sodium too has a completed octet.

But what holds the atoms together? Remember that the fluorine now has an extra electron, which used to belong to the sodium. This means that the fluorine now has a total of ten electrons and still has only nine protons. Therefore the fluorine atom *has a net electric charge* of 1− (because it has one extra electron). The sodium atom has given up an electron, so it now has eleven protons and only ten electrons. Therefore the sodium atom *has a net electric charge* of 1+ (because it has one less electron). The two atoms have become *ions*. An **ion** is *an atom or group of atoms that has gained or lost electrons and therefore has a net negative or positive charge*. A **monatomic ion** is *an ion composed of a single atom*. The positively charged sodium atom is called a *sodium ion,* written Na^{1+}. The negatively charged fluorine atom is called a fluor*ide* ion, written F^{1-}. Opposite charges attract, so the Na^{1+} and F^{1-} ions are attracted to each other. This attraction makes them form the neutral salt sodium fluoride (NaF). The positive-negative attraction is what holds these ions together. An **ionic bond** is *a chemical bond in which one atom transfers one or more electrons to another atom and the resulting ions are held together by the attraction of opposite charges*.

Here's another example: Calcium and chlorine combine by forming an ionic bond to make calcium chloride (CaCl_2), but it takes two chlorines to do the job (Figure 7.8). In this compound, the calcium atom gives up its two outermost electrons, transferring one electron to each chlorine atom. The calcium atom becomes the positively charged Ca^{2+} ion, because it still has 20 protons and now only 18 electrons. Each chlorine atom now has an extra electron, for a total of 18, but it still has only 17 protons. As a result, each chlorine atom becomes a negatively charged chloride ion. The calcium ion and the chloride ions are attracted to each other, forming CaCl_2.

FIGURE 7.9

Covalent bonding in boron trifluoride

FIGURE 7.10

Covalent bonding in arsenic pentachloride

7.5 Exceptions to the Octet Rule

There are exceptions to every rule, and not all elements obey the octet rule. For example, in the compound boron trifluoride (BF_3), the boron atom has only six electrons in its outermost energy level (Figure 7.9).

Arsenic pentachloride is another exception to the rule. (This name means that one arsenic is bonded to five chlorines; *penta* is from the Greek word meaning "five.") In this compound, arsenic has *ten* outer electrons (Figure 7.10). In fact, many of the transition metals fail to follow the octet rule. Still, in the vast majority of cases, the octet rule will give you the correct bonding structure.

7.6 Covalent or Ionic Bonds: The Concept of Electronegativity

At this point you may be asking, "How can I tell whether two elements bond ionically or covalently?" To answer this question, we have to introduce the concept of electronegativity. **Electronegativity** is *the attraction that an atom has for the electrons it is sharing with another atom.* The relative electronegativities for atoms of the different elements are listed in Table 7.1. (Note that no electronegativity values are listed for the Group VIIIA elements, because they tend not to form chemical compounds.)

The electronegativity scale was devised by Linus Pauling, winner of the 1954 Nobel Prize in chemistry and the 1962 Nobel Peace Prize. The scale is

TABLE 7.1

Periodic Table of Electronegativities

Period	Group IA	IIA	IIIB	IVB	VB	VIB	VIIB	VIIIB			IB	IIB	IIIA	IVA	VA	VIA	VIIA	VIIIA
1	1 H 2.1																	2 He
2	3 Li 1.0	4 Be 1.5				Transition Elements							5 B 2.0	6 C 2.5	7 N 3.0	8 O 3.5	9 F 4.0	10 Ne
3	11 Na 0.9	12 Mg 1.2											13 Al 1.5	14 Si 1.8	15 P 2.1	16 S 2.5	17 Cl 3.0	18 Ar
4	19 K 0.8	20 Ca 1.0	21 Sc 1.3	22 Ti 1.5	23 V 1.6	24 Cr 1.6	25 Mn 1.5	26 Fe 1.8	27 Co 1.8	28 Ni 1.8	29 Cu 1.9	30 Zn 1.6	31 Ga 1.6	32 Ge 1.8	33 As 2.0	34 Se 2.4	35 Br 2.8	36 Kr
5	37 Rb 0.8	38 Sr 1.0	39 Y 1.2	40 Zr 1.4	41 Nb 1.6	42 Mo 1.8	43 Tc 1.9	44 Ru 2.2	45 Rh 2.2	46 Pd 2.2	47 Ag 1.9	48 Cd 1.7	49 In 1.7	50 Sn 1.8	51 Sb 1.9	52 Te 2.1	53 I 2.5	54 Xe
6	55 Cs 0.7	56 Ba 0.9	57–71 La– Lu	72 Hf 1.3	73 Ta 1.5	74 W 1.7	75 Re 1.9	76 Os 2.2	77 Ir 2.2	78 Pt 2.2	79 Au 2.4	80 Hg 1.9	81 Tl 1.8	82 Pb 1.8	83 Bi 1.9	84 Po 2.0	85 At 2.2	86 Rn
7	87 Fr 0.7	88 Ra 0.9	89–103 Ac– Lr	104 Rf	105 Db	106 Sg	107 Bh	108 Hs	109 Mt									

57 La 1.1	58 Ce 1.1	59 Pr 1.1	60 Nd 1.1	61 Pm 1.1	62 Sm 1.1	63 Eu 1.1	64 Gd 1.1	65 Tb 1.1	66 Dy 1.1	67 Ho 1.1	68 Er 1.1	69 Tm 1.1	70 Yb 1.1	71 Lu 1.2
89 Ac 1.1	90 Th 1.3	91 Pa 1.5	92 U 1.7	93 Np 1.3	94 Pu 1.3	95 Am 1.3	96 Cm 1.3	97 Bk 1.3	98 Cf 1.3	99 Es 1.3	100 Fm 1.3	101 Md 1.3	102 No 1.3	103 Lr

6
C
2.5

← Atomic number

← Electronegativity

based on fluorine having an assigned value of 4.0. Fluorine is the most electronegative element. It tends to attract electrons more strongly than any other element does.

The metal elements generally have low electronegativities, whereas the nonmetals have high electronegativities. This means that the atoms of metals tend to lose electrons more readily than the atoms of nonmetals. In fact, the atoms of nonmetals have a strong tendency to pick up electrons. By looking at Table 7.1, you can see the trend for electronegativity, which generally increases as one moves from left to right across a given period of elements and decreases as one moves from the top to the bottom of a group.

To decide whether a bond is ionic or covalent, we find the difference between the electronegativities of the elements forming the bond. Chemists generally agree that if the difference between the electronegativities of the two elements is 2.0 or greater, the bond between the two elements is primarily ionic. If the difference is less than 2.0, the bond is primarily covalent. (Note that 2.0 is simply a convenient number to use as the cutoff point between ionic and covalent bonding. There is nothing special about this number. In fact, we will discuss a more sophisticated way of looking at ionic and covalent bonds shortly.)

EXAMPLE 7.3

Determine whether the bonds in the following compounds are ionic or covalent (use Table 7.1):   (a) H_2S (b) KCl   (c) MgO   (d) H_2

Solution

To solve these problems, we use the specific strategy just described.

(a) There are two H—S bonds in hydrogen sulfide (H_2S). We want to determine whether these are ionic or covalent. From Table 7.1, we find that the electronegativity of H is 2.1, and that of S is 2.5. The difference in electronegativities is

$$2.5 - 2.1 = 0.4$$

Because the difference is only 0.4, the bond between hydrogen and sulfur is covalent. (If the difference is less than 2.0, the bond is considered covalent.)

(b) From Table 7.1, we find that the electronegativity of K is 0.8, and that of Cl is 3.0. The difference in electronegativities is

$$3.0 - 0.8 = 2.2$$

Because the difference is 2.2, the bond between potassium and chlorine is ionic.

(c) From Table 7.1, we find that the electronegativity of Mg is 1.2, and that of O is 3.5. The difference in electronegativities is

$$3.5 - 1.2 = 2.3$$

Because the difference is 2.3, the bond between magnesium and oxygen is ionic.

(d) From Table 7.1, we find that the electronegativity of H is 2.1. The difference between the electronegativities of the two H atoms is

$$2.1 - 2.1 = 0.0$$

The bond between the two H atoms is very definitely a covalent bond. (It's as covalent as it can get!)

Practice Exercise 7.3   Determine whether the bonds in each of the following compounds are ionic or covalent (use Table 7.1):   (a) H_2O (b) KF   (c) MgS   (d) Cl_2

7.7   Ionic Percentage and Covalent Percentage of a Bond

So far we have thought of compounds as either ionically bonded or covalently bonded. We decided which *type* of bond by calculating the difference in electronegativities of the elements forming the bond. A difference of less than 2.0 between the electronegativities of the two elements meant that the bond was covalent, whereas a difference of 2.0 or more meant that the bond was ionic. But this is an oversimplified approach.

Chemists find it better to express chemical bonds as a certain percentage ionic and a certain percentage covalent. Table 7.2 relates difference in electronegativities to the ionic and covalent percentages of any bond. With this table, we can state that the hydrogen-oxygen bond in water (with an electronegativity difference of 1.4) is 39% ionic and 61% covalent. (The percentages listed in Table 7.2 come directly from Pauling's work.)

EXAMPLE 7.4

Determine the ionic and covalent percentages of the bonds in each of the following compounds: (a) NH_3 (b) SO_2 (c) HCl   (d) NaF

Solution

(a) Look up the electronegativities of nitrogen and hydrogen, determine the difference, and use Table 7.2 to find the ionic and covalent percentages of the bond. The electronegativity of N is 3.0, and that of H is 2.1. The difference between these electronegativities is

$$3.0 - 2.1 = 0.9$$

This corresponds to a bond that is 19% ionic and 81% covalent.

(b) The electronegativity of S is 2.5, and that of O is 3.5. The difference between these electronegativities is

$$3.5 - 2.5 = 1.0$$

This corresponds to a bond that is 22% ionic and 78% covalent.

TABLE 7.2

The Relationship Between Electronegativity Difference and the Ionic Percentage and Covalent Percentage of a Chemical Bond

DIFFERENCE IN ELECTRONEGATIVITY	IONIC PERCENTAGE	COVALENT PERCENTAGE
0.0	0.0	100
0.1	0.5	99.5
0.2	1.0	99.0
0.3	2.0	98.0
0.4	4.0	96.0
0.5	6.0	94.0
0.6	9.0	91.0
0.7	12.0	88.0
0.8	15.0	85.0
0.9	19.0	81.0
1.0	22.0	78.0
1.1	26.0	74.0
1.2	30.0	70.0
1.3	34.0	66.0
1.4	39.0	61.0
1.5	43.0	57.0
1.6	47.0	53.0
1.7	51.0	49.0
1.8	55.0	45.0
1.9	59.0	41.0
2.0	63.0	37.0
2.1	67.0	33.0
2.2	70.0	30.0
2.3	74.0	26.0
2.4	76.0	24.0
2.5	79.0	21.0
2.6	82.0	18.0
2.7	84.0	16.0
2.8	86.0	14.0
2.9	88.0	12.0
3.0	89.0	11.0
3.1	91.0	9.0
3.2	92.0	8.0

(c) The electronegativity of H is 2.1, and that of Cl is 3.0. The difference between these electronegativities is

$$3.0 - 2.1 = 0.9$$

This corresponds to a bond that is 19% ionic and 81% covalent.

(d) The electronegativity of Na is 0.9, and that of F is 4.0. The difference between these electronegativities is

$$4.0 - 0.9 = 3.1$$

This corresponds to a bond that is 91% ionic and 9% covalent.

Practice Exercise 7.4 Determine the ionic and covalent percentages of the bonds in each of the following compounds: (a) K_2O (b) $CaCl_2$ (c) Mg_3N_2 (d) HF

EXAMPLE 7.5

List the following compounds in order of decreasing covalence of their bonds: H_2S, H_2O, CO_2, and N_2.

Solution

Using Tables 7.1 and 7.2, determine the percentage covalence of each compound. In the compound H_2S, the H—S bond has an electronegativity difference of 0.4, which means it is 96% covalent.

In the compound H_2O, the O—H bond has an electronegativity difference of 1.4, which means it is 61% covalent.

In the compound CO_2, the C=O bond has an electronegativity difference of 1.0, which means it is 78% covalent.

In the compound N_2, the N≡N bond has an electronegativity difference of 0.0, which means it is 100% covalent.

Therefore the order of covalence is

N_2	H_2S	CO_2	H_2O
Most covalent			Least covalent

Practice Exercise 7.5 List the following compounds in order of decreasing covalence of their bonds: CO_2, $AsCl_3$, CS_2, and H_2S.

7.8 Shapes and Polarities of Molecules

Up to now, we have used only two-dimensional diagrams to show how atoms are bonded together in molecules. But our world is *three*-dimensional, and atoms and molecules, like other objects, have three dimensions. In fact, the

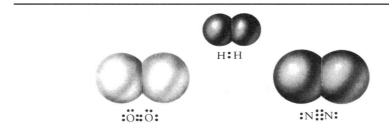

H:H

■ FIGURE 7.11

Three-dimensional models of some diatomic elements

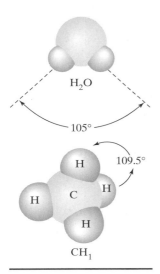

HCl

H₂O

105°

109.5°

CH₁

■ FIGURE 7.12

Three-dimensional models of hydrogen chloride (HCl), water (H₂O), and methane (CH₄) molecules

three-dimensional shapes of some molecules lead to additional properties of covalently bonded compounds.

Three-Dimensional Characteristics of Molecules

If we could enlarge a single molecule of any diatomic element (for instance, H_2, O_2, or N_2), it would appear to be linear (in a line) (Figure 7.11). This would also be true for covalently bonded compounds consisting of two *un*like atoms, such as HCl (Figure 7.12).

But if we could enlarge a single molecule of water, we would see a bent molecule (Figure 7.12). Experiments have shown that the angle between the hydrogen atoms is about 105 degrees.

If we looked at a molecule of methane (CH_4), we would see a pyramid-shaped symmetrical molecule (Figure 7.12). The angle between neighboring hydrogen atoms in methane has been shown to be 109.5 degrees.

Molecules of other compounds exhibit many other shapes. The important thing about these shapes is that they can affect bonding and the properties of the molecules.

The Not-So-Covalent Bond

Recall that when two atoms form a covalent bond, they share electrons. But sometimes they don't share the electrons equally. For example, in the compound hydrogen bromide (HBr), the bromine atom has a greater attraction for the electrons than the hydrogen atom does. In other words, bromine has a greater electronegativity than hydrogen. Because of the bromine's greater attraction for electrons, the two electrons being shared spend more time with the bromine than with the hydrogen. This unequal sharing makes the hydrogen seem to have a partial positive charge, and the bromine a partial negative charge (Figure 7.13). This kind of covalent bond, in which there is unequal sharing of electrons, is called a **polar covalent bond.** This is what we meant when we said a bond was a certain percentage ionic.

Most covalent bonds are polar. However, bonds formed by the diatomic elements (H—H, O=O, and so on) are nonpolar. This is because there is no difference in electronegativity between the two atoms forming the bond. In a **nonpolar covalent bond,** the electrons are shared equally by the atoms.

EXAMPLE 7.6

Determine whether each of the following covalent bonds is polar or nonpolar. (*Hint:* Use Table 7.1 to determine the electronegativity of each element.) (a) N—O bond   (b) S—O bond   (c) Cl—Cl bond   (d) P—O bond   (e) C—S bond

Solution

We look up the electronegativity value of each element in Table 7.1. If a difference in electronegativity exists between the two atoms forming the bond, then the bond is polar. If no difference in electronegativity exists, then the bond is nonpolar.

(a) N————O   An electronegativity difference exists, so the
     3.0   3.5   bond is *polar*.

     0.5 difference

(b) S————O   An electronegativity difference exists, so the
     2.5   3.5   bond is *polar*.

     1.0 difference

(c) Cl————Cl   No electronegativity difference exists, so the
     3.0   3.0   bond is *nonpolar*.

     0.0 difference

(d) P————O   An electronegativity difference exists, so the
     2.1   3.5   bond is *polar*.

     1.4 difference

(e) C————S   No electronegativity difference exists, so the
     2.5   2.5   bond is *nonpolar*.

     0.0 difference

Practice Exercise 7.6   Determine whether each of the following covalent bonds is polar or nonpolar. (*Hint:* Use Table 7.1.)   (a) F—F bond   (b) C—O bond   (c) B—O bond

---

δ^+ δ^-

H Br

■ **FIGURE 7.13**

A three-dimensional drawing of HBr. The δ^+ indicates that the hydrogen side of the molecule has a partial positive charge. The δ^- shows that the bromine side of the molecule has a partial negative charge. (δ is the lower-case Greek letter delta.)

Polar and Nonpolar Molecules

LEARNING GOAL 8

Polar and nonpolar molecules

Just as bonds can have polarity, molecules can also have polarity. Even if a molecule contains polar bonds, the molecule itself must not necessarily be a polar molecule. To determine whether or not a molecule is polar we must look at the geometry of the molecule and the polarity of its bonds.

As we mentioned earlier, molecular geometry tells us how the atoms of a molecule are arranged in space in relation to one another. You are familiar

Three-dimensional models of water and carbon tetrachloride molecules. The water molecule is polar; the carbon tetrachloride is nonpolar.

with the linear arrangement of the diatomic elements, the bent shape characteristic of a water molecule, and the pyramidal shape of the methane molecule. Such information is necessary when determining molecular polarity. We will look at the polarities of three molecules—HBr, H_2O, and CCl_4—to understand the combined effects of molecular geometry and bond polarity.

The linear HBr molecule (Figure 7.13) contains a polar covalent bond. There is a shift in electronic charge such that the hydrogen side of the molecule is positive and the bromine side of the molecule is negative. A **polar molecule** (or **dipole**) is a molecule that is positive on one side and negative on the other side. Thus HBr is a polar molecule.

And then there is water, a compound with two polar bonds. A water molecule is also polar, because there is a center of positive charge between the two hydrogen atoms and a center of negative charge on the oxygen atom (Figure 7.14).

A molecule of carbon tetrachloride (CCl_4) has four carbon-chlorine bonds. Each bond is a *polar* covalent bond. But the molecule itself is *nonpolar* (Figure 7.14). Here's why: The carbon tetrachloride molecule is shaped like a pyramid. The positive charge is on the carbon atom, but so is the center of the negative charges of the four Cl atoms. The two charges coincide and cancel each other, so the molecule is not polar.

EXAMPLE 7.7

Determine whether the bonds in each of the following molecules are polar or nonpolar. Then determine whether the molecule itself is polar or nonpolar.

(a) F_2 (linear in shape)

(b) $CHCl_3$, chloroform (shaped like a pyramid)

(c) *cis*-dichloroethene (The *cis* means that the two chlorines are on the same side of the double bond.)

$$\begin{array}{ccc} H & & H \\ \diagdown & & \diagup \\ & C = C & \\ \diagup & & \diagdown \\ Cl & & Cl \end{array}$$

(d) *trans*-dichloroethene (The *trans* means that the two chlorines are on opposite sides of the double bond.)

Solution

(a) Fluorine is a diatomic element; therefore the F—F bond is nonpolar and the F_2 molecule is nonpolar.

(b) The bonding in chloroform is similar to the bonding in carbon tetrachloride. Each bond in chloroform is a polar covalent bond. But the molecule of chloroform, unlike that of carbon tetrachloride, is polar. This is because the center of positive charge and the center of negative charge do not coincide. The molecule is not symmetrical:

(c) In *cis*-dichloroethene, each C—H and C—Cl bond is polar. The C=C bond is nonpolar. The molecule is polar because it has a center of positive charge between the two hydrogens and a center of negative charge between the two chlorines. Because the centers of positive charge and negative charge do not coincide, the molecule is polar.

(d) In *trans*-dichloroethene, the C—H and C—Cl bonds are polar and the C=C bond is nonpolar. The molecule is also nonpolar because it is shaped in such a way that the center of positive charge coincides with the center of negative charge. The charges coincide as shown in the double bond between the two carbon atoms.

Practice Exercise 7.7 Determine whether the bonds in each of the following molecules are polar or nonpolar. Then determine whether the molecule itself is polar or nonpolar.

Molecule	Shape

BF_3

F   F
     B
     |
     F

Cl_2O

     O
Cl   Cl

CO_2

$O = C = O$

SUMMARY

In a Lewis (electron-dot) diagram of an element, the symbol for an atom of the element is surrounded by dots. The dots correspond to the electrons in the outermost energy level of an atom of the element. Electron-dot diagrams are used to show how atoms combine to form molecules.

It has long been known that Group VIIIA elements tend to be very stable. Scientists believe this stability is due to the fact that each element in this group (except helium) has eight electrons in its outermost energy level. The atoms of other elements can achieve this same sort of stability by obtaining eight electrons in their outermost energy level. To do so, two atoms may share electrons in a covalent bond. A single covalent bond is formed when each atom donates one electron to the bond, a double covalent bond is formed when each atom donates two electrons, and a triple covalent bond is formed when each atom donates three electrons.

In a coordinate covalent bond, only one atom donates the two electrons required to form the bond. However, both atoms obtain an octet of electrons in their outermost shell.

Elements that are found in molecules with two atoms each (such as H_2) are called diatomic elements; these elements are hydrogen, oxygen, nitrogen, and the Group VIIA elements. The bonds between their atoms are covalent bonds.

An ion is an atom or group of atoms that has gained or lost electrons and therefore has a negative or positive electric charge. In ionic bonding, atoms of one element donate electrons to the atoms of a second element. Both become ions, but they are ions of opposite charges. These opposite charges attract, and this attraction holds the atoms together in a molecule.

Electronegativity is the attraction that an atom has for the electrons it is sharing with another atom. If the difference between the electronegativities of two elements is 2.0 or greater, the bond between the two elements is said to be ionic. If the difference is less than 2.0, the bond is covalent.

The concept of electronegativity is also used in determining the ionic and covalent percentages of bonds.

Because of their three-dimensional character, some molecules exhibit unequal sharing of electrons, resulting in polar covalent bonds. Such molecules may have a positive side and a negative side; if so, they are called polar molecules, or dipoles.

KEY TERMS

chemical bond **(Introduction)**	ionic bond **(7.4)**
coordinate covalent bond **(7.3)**	monatomic ion **(7.4)**
covalent bond **(7.2)**	nonpolar covalent bond **(7.8)**
diatomic element **(7.2)**	octet rule **(7.2)**
double covalent bond **(7.2)**	polar covalent bond **(7.8)**
electron-dot structure **(7.1)**	polar molecule (dipole) **(7.8)**
electronegativity **(7.6)**	single covalent bond **(7.2)**
ion **(7.4)**	triple covalent bond **(7.2)**

CAREER SKETCH

Food Technology Careers

If you're interested in a career in which your lunch hour lasts all day, consider a career in the food sciences. A variety of positions is available, ranging from food development to food regulation.

A *food technologist* is responsible for product development. Most often, this person will work for a product supplier, manufacturing, or processing company. Responsibilities can include research and development in exciting areas such as fruit preps, variegates, chocolate, toppings, sweet sauce, condiments using gum, starch, pectin, gelatin, hydrocolloids, etc. A bachelor's degree or a master's degree is usually necessary.

A somewhat related career involves troubleshooting for food manufacturing companies. If you are good at problem solving and have a strong science background, a position as a *technical service representative* may be a career to consider. This person will travel to the food customer's plant, acting as a liaison between research and development and sales, conducting plant trials, and assisting with troubleshooting.

Expect to work with dairy, fruit applications, fruit preps, chocolate sauce, toppings, cultured products, yogurt, ice cream, and frozen desserts as a technical service representative.

If you're interested in a job that is a bit less hands-on, consider the regulatory field. A person in this position must be a good writer and have strong communication skills. This person will have to deal with and manage regulatory issues, often representing his or her company before external agencies such as the Food and Drug Administration. Issues this person is likely to encounter include environmental quality, environmental and human safety, drug and cosmetics regulations, and product and labeling standards.

CHEMICAL FRONTIERS

*From Soccer Balls to Superconductors**

Three scientists won the 1996 Nobel Prize in Chemistry for their discovery that 60 carbon atoms can roll themselves into a pattern similar to that of a soccer ball. The molecule had an alternating pattern of hexagons and pentagons held together by stable "5–6 bonds." Because the C_{60} molecule resembled the domes designed by architect R. Buckminster Fuller, scientists named it "buckminsterfullerene."

Investigations have expanded to include other hollow, cagelike molecules (fullerenes), often called buckyballs. Scientists recognized that there were many uses for this stable, hollow molecule, but first they had to break the strong molecular bonds and get inside.

Recently, scientists have used lasers to break these bonds. When the laser is pointed at the 5–6 bonds, they open to allow one or more atoms inside.

The possible uses for this new technology are numerous. Buckyballs can superconduct, lubricate, and absorb light, making them useful in perfecting conductors. Placing drugs inside the buckyballs may lead to cures for certain diseases. Or doctors may use filled molecules to obtain better internal pictures without x-rays. All of this may be possible once chemists fully understand fullerene molecules.

*Adapted from *Science News,* February 12, 1994, p. 101, and *Science News,* October 19, 1996, p. 247

SELF-TEST EXERCISES

LEARNING GOAL 1: *Lewis Structures*

◀ *1.* Write the electron-dot structure for each of the following atoms: (a) As (b) Cl (c) Ra (d) Cs (e) Rn

◀ 2. Write the electron-dot structure for each of the following elements: (a) F (b) Ca (c) K (d) C (e) B

3. Write the electron-dot structure for (a) an oxygen atom, (b) an oxide ion.

4. Write the electron-dot structure for (a) a bromine atom, (b) a fluoride ion.

5. Write the electron-dot structure for (a) a magnesium atom, (b) a magnesium ion.

6. Write the electron-dot structure for (a) a barium atom, (b) a barium ion.

◀ 7. Atoms are said to be *isoelectronic* when they have the same electron configuration. Write the electron-dot structures of two ions that are isoelectronic with an argon atom.

8. Write the electron-dot structure for each of the following elements: (a) Si (b) Te (c) At

9. Write the electron-dot structure for each of the following ions: (a) magnesium ion (b) sodium ion (c) aluminum ion

10. Write the electron-dot structure for each of the following ions: (a) calcium ion (b) cesium ion (c) boron ion

11. Write the electron-dot structures of two ions that are isoelectronic (have the same electron configuration) as a calcium ion.

12. Write the electron-dot structure for each of the following ions: (a) sulfide ion (b) telluride ion (c) nitride ion

LEARNING GOAL 2: *Covalent Bonds*

◀ 13. Explain why covalent bonds are called electron-sharing bonds.

14. Define the term *coordinate covalent bond*.

15. When each atom donates *two* electrons to form a covalent bond, the bond formed is a _____ (single, double, quadruple) covalent bond.

16. When each atom donates *three* electrons to form a covalent bond, the bond formed is a _____ (single, triple, sextuple) covalent bond.

LEARNING GOAL 3: *Diatomic Elements*

17. Name the diatomic elements from memory.

◀ 18. Which of the diatomic elements contain single covalent bonds? Which contain double bonds? Which contain triple bonds?

LEARNING GOAL 4: *Electron-Dot Structures of Covalent Compounds*

◀ **19.** Write the electron-dot structure for each of the following covalent compounds: (a) $CHBr_3$ (b) C_2H_2 (c) PH_3 (d) HCN

20. Write the electron-dot structure for each of the following covalent compounds: (a) HCl (b) CBr_4 (c) SO_3 (d) $SbCl_3$

21. Write the electron-dot structure for each of the following covalent compounds: (a) CF_4 (b) C_2H_4 (c) AsH_3 (d) H_2S

◀ 22. Write the electron-dot structure for each of the following covalent compounds: (a) CS_2 (b) CO_2 (c) Cl_2O (d) C_3H_8

◀ **23.** Not all compounds follow the octet rule. Draw the bonding diagram for the covalent compound PCl_5. How many electrons surround the P atom in this compound?

24. Not all compounds follow the octet rule. Draw the bonding diagram for the compound BF_3. How many electrons surround the B atom in this compound?

◀ **25.** Draw the bonding diagram for each of the following covalent compounds, which have coordinate covalent bonds: (a) HNO_3 (b) H_2SO_3

26. Draw the bonding diagram for the compound chloric acid, $HClO_3$, which has a coordinate covalent bond. (*Hint:* The chlorine is bonded to three oxygen atoms, and the hydrogen is bonded to one of the oxygen atoms.)

LEARNING GOAL 5: *Ionic Bonds*

◀ 27. Define the term *ionic bond*.

28. How is an ionic bond different from a covalent bond?

LEARNING GOAL 6: *Ionic and Covalent Bonds*

29. Determine whether each of the following compounds is bonded ionically or covalently: (a) $CsCl$ (b) CH_4 (c) HCl (d) SO_3

30. Determine whether each of the following compounds is bonded ionically or covalently: (a) MgO (b) $BaCl_2$ (c) BaO (d) CaF_2 (e) SO_2

31. Find the ionic and covalent percentages of the bonds in each of the following compounds: (a) CH_4 (b) N_2 (c) FeO (d) Al_2O_3

32. Find the ionic and covalent percentages of the bonds in each of the following compounds: (a) HCl (b) NH_3 (c) KI (d) CS_2

◀ **33.** List the following compounds in order of *decreasing* covalence: HI, HCl, HF, and HBr.

34. List the following compounds in order of *decreasing* covalence: Br_2, SO_2, NH_3, HCl.

35. Find the ionic and covalent percentages of the bonds in each of the following compounds: (a) CF_4 (b) AsH_3 (c) $CsCl$ (d) CuO

36. Find the ionic and covalent percentages of the bonds in each of the following compounds: (a) CO_2 (b) F_2 (c) $MgCl_2$ (d) BaF_2

37. List the following compounds in order of *increasing* covalence: H_2Se, H_2O, H_2Te, and H_2S.

◀ 38. List the following compounds in order of *increasing* covalence: HF, NH_3, BH_3, CH_4, and H_2O.

39. Which of the following compounds is the *most* ionic? (a) $LiCl$ (b) $NaCl$ (c) KCl (d) $RbCl$ (e) $CsCl$

40. Which of the following compounds is the *most* ionic? HF HCl HBr HI

41. Determine whether the bonds in each of the following compounds are ionic or covalent (use Table 7.1): (a) SO_3 (b) $BiCl_3$ (c) LiF (d) NO

42. Determine whether the bonds in each of the following compounds are ionic or covalent (use Table 7.1): (a) LiH (b) OF_2 (c) CaI_2 (d) MgH_2

LEARNING GOALS 7 AND 8: *Polar and Nonpolar Bonds and Molecules*

◀ 43. Explain why some molecules are polar and others are nonpolar.

44. Determine whether the bond in each of the following molecules is polar or nonpolar: (a) H_2 (b) NO (c) HCl

◀ **45.** Determine whether the *bonds* in each of the following molecules are polar or nonpolar. Then determine whether the *molecule* itself is polar or nonpolar.

Molecule	Shape	Molecule	Shape
(a) PBr_3	Br – P (Br, Br)	(b) Cl_2O	Cl – O – Cl
(c) CH_3Cl	H, H – C – H, Cl	(d) O_2	$O = O$

Molecule	Shape
(a) H_2	H — H
(b) CO_2	O = C = O
(c) C_6H_6	
(d) CH_2O	H — C = O, H

46. Determine whether the *bonds* in each of the following molecules are polar or nonpolar. Then determine whether the *molecule* is polar or nonpolar.

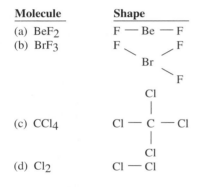

Molecule	Shape
(a) BeF_2	F — Be — F
(b) BrF_3	F, F, Br, F
(c) CCl_4	Cl — C — Cl (Cl, Cl)
(d) Cl_2	Cl — Cl

47. Determine whether the *bonds* in each of the following molecules are polar or nonpolar. Then determine whether the *molecule* itself is polar or nonpolar.

Molecule	Shape	Molecule	Shape
(a) C_2H_2	H — C ≡ C — H	(b) NH_3	H, N, H, H
(c) CH_4	H — C — H (H, H)	(d) N_2	N ≡ N

◀ 48. Determine whether the *bonds* in each of the following molecules are polar or nonpolar. Then determine whether the *molecule* is polar or nonpolar.

EXTRA EXERCISES

49. "In both ionic and covalent bonds, *electron deficiencies are satisfied.*" Explain what is meant by this statement.

◀ **50.** Draw the electron-dot diagram for each of the following: (a) $(PH_4)^{1+}$ (b) CO_2 (c) OF_2

51. Explain why the formula for sodium chloride, NaCl, involves one sodium ion and one chloride ion but the formula for calcium chloride, $CaCl_2$, has two chloride ions with every one calcium ion.

52. Determine the order of increasing bond polarity among the following groups: O=O, N—O, C—O

53. Which of the following bonds is the most covalent? C—H, C—O, N—O, C—S

◀ **54.** Draw the electron-dot diagram for each of the following: (a) $(NH_4)^{1+}$ (b) ClBr

55. Predict the compound that is *most likely* to be formed from each of the following pairs of elements: (a) Mg and O   (b) Al and O   (c) Cs and S

56. Use the octet rule to explain the following:
(a) The formula unit of aluminum chloride is $AlCl_3$.
(b) The formula unit of sodium oxide is Na_2O.
(c) The formula unit of magnesium nitride is Mg_3N_2.

57. Explain how a molecule that has polar covalent bonds can be a nonpolar molecule.

◀ 58. Use the octet rule to predict the formula unit of each of the following compounds: (a) K and S (b) Ca and Br (c) Li and O (d) Al and Cl

<u>**59.**</u> Explain why a covalently bonded molecule with the formula H_3 should not exist.

60. Using the octet rule, predict the formula and draw the electron-dot diagram for each of the covalently bonded molecules that form between the following atoms: (a) C and F (b) As and Cl (c) I and Cl (d) Te and H

CHEMICAL NOMENCLATURE:
The Names and Formulas of Chemical Compounds

Sometimes, the names and chemical formulas of substances can be very useful to individuals in many fields outside of the chemical industry. As an example, take this story, which involved a family in the pest-control business and several local physicians.

Tom Daniels, a four-year-old boy, wandered into the family garage to play "explorer." He had been warned by his parents not to play in the garage because his dad, Ed Daniels, was in the pest-control business and kept many chemicals at home.

Tom was in the garage for only a few minutes when he came running into the house crying to his mother. "Mom," he said, "my mouth is burning." His mother detected a garlicky odor and she also noticed a white powder on his hands and face. Just then Tom vomited some material that had a bloody dark color. Working swiftly, Mrs. Daniels brushed some white powder from Tom's body and placed it in a plastic bag. She washed Tom's hands and face and had him rinse his mouth, making sure that he did not swallow anything. Then she took Tom into the garage and had him point to the container that he was playing with. Lifting the cover, she discovered a white powder with the same odor that was emanating from Tom—it was rat poison. Mrs. Daniels asked Tom if he had swallowed any of the powder and he said that he may have swallowed some.

Mrs. Daniels called Tom's pediatrician, Dr. Williams, and told him what had happened. Dr. Williams told Mrs. Daniels to take Tom and the samples of powder to the emergency room of their local hospital. Within 10 minutes Tom and his mother were at the hospital, where Dr. Bellen, the emergency room physician, had been alerted by a call from Dr. Williams, and was waiting for Tom to arrive. She took a history from Mrs. Williams and began to examine Tom. She also read the label on the container of rat poison and had a sample sent to the laboratory. (The powder that was brushed from Tom's body and placed in a plastic bag was also sent to the laboratory.) The label indicated that the active ingredient in the rat poison was white phosphorus.

Dr. Williams, Tom's pediatrician, soon arrived at the hospital. Both he and Dr. Bellen consulted with the poison control center, which advised that the best treatment for phosphorus poisoning was to give the patient frequent dosages of 0.2% copper(II) sulfate solution. This solution, if given in time, would prevent systemic poisoning by phosphorus. (This prevention occurs because the copper bonds with the phosphorus and envelops the phosphorus particles with an insoluble coating of copper(II) phosphide.) They were also advised by the poison control center that after treatment with the copper(II) sulfate, the patient should ingest a laxative solution of sodium sulfate, so that the enveloped phosphorus particles are quickly eliminated from the body. Liquid petrolatum is also administered to the patient to coat the stomach. By the time the physicians finished talking to the poison control center, the laboratory had confirmed that the powder was a phosphorus-based poison, and the physicians immediately gave Tom the prescribed treatment.

Mrs. Daniels was told that Tom would have to stay in the hospital for a few days for observation, because in some cases of phosphorus poisoning there is a quiet period from one to three days, followed by symptoms such as nausea, vomiting, an enlarged liver, skin eruptions, jaundice, and reduced blood pressure. The physicians commended Mrs. Daniels for her quick action at home that may have prevented the poison from getting past Tom's stomach. She was told that, based on his symptoms and vital signs, Tom may not have swallowed too much of the poison, and that the copper(II) sulfate treatment should handle the small amount that might have entered his stomach or intestines.

During Tom's stay in the hospital his diet was free of all edible oils and fats, or substances containing them, such as milk, because these substances promote the absorption of phosphorus. Mrs. Daniels was advised to maintain the same diet for several days after Tom arrived home.

Fortunately for Tom, the physicians were correct. Tom had no additional adverse reactions to the poison other than some GI tract irritation, and he was released from the hospital after five days of observation. Home again, Tom found that his dad had cleaned up the garage—the Daniels family wasn't taking any more chances with poisons.

After you've studied this chapter, you should be able to:

1. Distinguish between the common name of a compound and its systematic name.

2. Write the formula of a binary compound containing two nonmetals when you are given the name of the compound.

3. Write the formula of a binary compound containing a metal with a fixed oxidation number and a nonmetal when you are given the name of the compound.

4. Write the formula of a binary compound containing a metal with a variable oxidation number and a nonmetal when you are given the name of the compound.

5. Write the formula of a polyatomic ion when you are given its name, and write the name of a polyatomic ion when you are given its formula.

6. Write the formula of a ternary or higher compound when you are given the name of the compound.

7. Write the name of a binary compound containing two nonmetals when you are given the formula of the compound.

8. Write the name of a binary compound containing a metal with a fixed or variable oxidation number and a nonmetal when you are given the formula of the compound.

9. Find the oxidation numbers of less familiar elements when you are given chemical formulas of compounds containing these elements.

10. Write the name of a ternary or higher compound when you are given the formula of the compound.

11. Write the formula of an inorganic acid when you are given its name, and write the name of an inorganic acid when you are given its formula.

12. Write the formula of a compound when you are given its common name, and write the common name of a compound when you are given its formula or systematic name.

One of the most important topics to master in your study of chemistry is **chemical nomenclature,** which refers to the system for naming chemical compounds and writing their formulas. In any introductory college course, learning the basic vocabulary of the discipline is extremely important. This allows you to discuss the subject matter using the language of the discipline. Learning to write the names and formulas of chemical compounds will build your chemistry vocabulary. In fact, chemical nomenclature is a major part of the language of chemistry. In this chapter, you will learn how to write the formula of a compound when you are given its name. You will also learn how to name a compound when you are given its formula.

As we begin our study of chemical nomenclature, you should be aware that there are two ways of designating chemical compounds: the **systematic chemical name** and the **common name.** The rules governing the systematic names have been developed by the *International Union of Pure and Applied Chemistry (IUPAC)*. The IUPAC is made up of chemists from all over the world. Members of the IUPAC Commission on the Nomenclature of Inorganic Compounds first met in 1921 to develop a system for naming compounds. This commission meets periodically to refine this system as new types of compounds are synthesized or discovered.

There are no rules that govern the common names of compounds. Most common names, such as table salt (for sodium chloride, NaCl) and water (for H_2O), have been derived from common usage or simply handed down through chemical history. We'll have more to say about common names of compounds later in this chapter. Let's begin our study of nomenclature by learning how to write the formulas of compounds from their systematic chemical names.

■ LEARNING GOAL 1
Distinguishing between common name and systematic name

8.1 Writing the Formulas of Compounds from Their Systematic Names

In developing the rules for naming inorganic compounds, the IUPAC wanted to make sure that each chemical formula had a unique chemical name and vice versa. The commission also decided that the more *positive portion* of the compound should be written first, and the more *negative portion* last. Therefore, in the formula of a compound composed of two ions, the positive ion precedes the negative ion. That is why the formula for sodium chloride is always written NaCl, not ClNa. In the formula for a covalent compound, the element with the lower electronegativity (the more positive portion) generally precedes the element with the higher electronegativity (the more negative portion). Thus, the formula for water is written as H_2O, not OH_2.

These simple rules can be used to write the formulas of compounds from their systematic names. We'll begin with **binary compounds** (compounds made up of two different elements) and then move on to **ternary** and higher compounds (those made up of three or more elements).

8.2 Writing the Formulas of Binary Compounds Containing Two Nonmetals

The formula of a compound containing two nonmetals can easily be deduced from the compound's name, because the name of the compound will contain the name of both elements. For binary compounds, the ending of the second

TABLE 8.1

Greek Prefixes and Their Meanings

Mono-	= 1
Di-	= 2
Tri-	= 3
Tetra-	= 4
Penta-	= 5
Hexa-	= 6
Hepta-	= 7
Octa-	= 8
Nona-	= 9
Deca-	= 10

LEARNING GOAL 2

Formulas of binary compounds containing two nonmetals

element is always *-ide*. Therefore chlor*ine* becomes chlor*ide,* sul*fur* becomes sul*fide,* and ox*ygen* becomes ox*ide*. In addition, the number of atoms of each element can be deduced from Greek prefixes that are part of the compound's name. Table 8.1 lists the Greek prefixes used in naming compounds. Note that the Greek prefix *mono-* (meaning "one") is rarely used for the first-named element, though it is used for the second-named element. *The absence of a Greek prefix for the first-named element means that there is only one atom in the formula unit of that compound.* The following example will clarify this point.

EXAMPLE 8.1

Write the formulas of the following binary compounds composed of two nonmetals:
(a) carbon monoxide
(b) carbon dioxide
(c) phosphorus trichloride
(d) dinitrogen monoxide

Solution

(a) Carbon monoxide is a compound containing carbon and oxygen. Ox*ygen* becomes ox*ide* because it is the second-named element. Notice that there is no prefix with carbon, and the prefix *mono-* appears with oxide. There is one atom of each element in the formula unit: CO.

(b) Carbon dioxide is a compound containing carbon and oxygen. Ox*ygen* becomes ox*ide* because it is the second-named element. Notice that there is no prefix with carbon, and the prefix *di-* appears with oxide. There is one atom of carbon and two atoms of oxygen in the formula unit: CO_2.

(c) Phosphorus trichloride is a compound containing phosphorus and chlorine. Chlor*ine* becomes chlor*ide* because it is the second-named element. Notice that there is no prefix with phosphorus, and the prefix *tri-* appears with chloride. There is one atom of phosphorus and three atoms of chlorine in the formula unit: PCl_3.

(d) Dinitrogen monoxide is a compound containing nitrogen and oxygen. Ox*ygen* becomes ox*ide* because it is the second-named element. Notice the prefix *di-* on nitrogen and the prefix *mono-* on oxide. There are two atoms of nitrogen and one atom of oxygen in the formula unit: N_2O.

Practice Exercise 8.1   Write the formulas of the following binary compounds composed of two nonmetals:
(a) Sulfur dioxide
(b) Dinitrogen pentoxide
(c) Dinitrogen tetroxide
(d) Phosphorus pentabromide

8.3 Writing the Formulas of Binary Compounds Containing a Metal and a Nonmetal

Before we learn how to write the formulas of binary compounds containing a metal and a nonmetal, we must learn about *oxidation numbers.* Chemists who have studied how elements combine to form compounds have discovered certain trends. It seems that elements tend to form ions with specific charges or they tend to form only a certain number of covalent bonds. To describe this phenomenon, chemists have devised a system to indicate how elements combine to form compounds. This system involves the assignment of oxidation numbers to the substances involved in forming the compound. An **oxidation number** is *the positive or negative number that expresses the combining capacity of an element in a particular compound.* Oxidation numbers are not always real, but are "tools" created by chemists for bookkeeping purposes.

The oxidation number can be positive or negative, depending on whether the element tends to attract electrons strongly or give them up. In carbon tetrachloride (Figure 8.1), the carbon has an oxidation number of 4+ and the chlorine has an oxidation number of 1−. Elements with high electronegativity values usually have negative oxidation numbers, and elements with low electronegativity values usually have positive oxidation numbers.

Tables 8.2 and 8.3 list the oxidation numbers of some important ions. Note that for many monatomic ions, the oxidation number is the same as

■ **FIGURE 8.1**

Covalent bonding in carbon tetrachloride

TABLE 8.2

Oxidation Numbers of Positive Ions Frequently Used in Chemistry

1+		2+	
Hydrogen	H^{1+}	Calcium	Ca^{2+}
Lithium	Li^{1+}	Magnesium	Mg^{2+}
Sodium	Na^{1+}	Barium	Ba^{2+}
Potassium	K^{1+}	Zinc	Zn^{2+}
Mercury(I)*	Hg^{1+} (also called mercurous)	Mercury(II)	Hg^{2+} (also called mercuric)
Copper(I)	Cu^{1+} (also called cuprous)	Tin(II)	Sn^{2+} (also called stannous)
Silver	Ag^{1+}	Iron(II)	Fe^{2+} (also called ferrous)
Ammonium	$(NH_4)^{1+}$	Lead(II)	Pb^{2+} (also called plumbous)
Rubidium	Rb^{1+}	Copper(II)	Cu^{2+} (also called cupric)
Cesium	Cs^{1+}	Strontium	Sr^{2+}
		Nickel(II)	Ni^{2+}
		Chromium(II)	Cr^{2+} (also called chromous)
		Cobalt(II)	Co^{2+} (also called cobaltous)
		Manganese(II)	Mn^{2+} (also called manganous)

3+		4+	
Aluminum	Al^{3+}	Tin(IV)	Sn^{4+} (also called stannic)
Iron(III)	Fe^{3+} (also called ferric)	Lead(IV)	Pb^{4+} (also called plumbic)
Bismuth(III)	Bi^{3+}	Manganese(IV)	Mn^{4+}
Chromium(III)	Cr^{3+} (also called chromic)		
Cobalt(III)	Co^{3+} (also called cobaltic)		

*Note that the mercury(I) ion is a diatomic ion. In other words, you never find the Hg^{1+} ion alone, but always as Hg^{1+}—Hg^{1+}. The two Hg^{1+} ions are bonded to each other.

the charge on the ion. In other words, because the sodium atom tends to give up one electron when forming a chemical compound, it has a charge of 1+ as a sodium ion. Therefore its oxidation number is also 1+. Elements in Group IA, which tend to become ions with 1+ charges, all have oxidation numbers of 1+. Elements in Group IIA, which tend to become ions with 2+ charges, all have oxidation numbers of 2+. Elements in Group IIIA, which tend to become ions with 3+ charges, all have oxidation numbers of 3+.

Group IVA elements do not usually form ions. However, they still have oxidation numbers, because they do form chemical compounds. The most common oxidation numbers of the Group IVA elements are 4+ and 4−.

TABLE 8.3

Oxidation Numbers of Negative Ions Frequently Used in Chemistry

1−		2−		3−	
Fluoride	F^{1-}	Oxide	O^{2-}	Nitride	N^{3-}
Chloride	Cl^{1-}	Sulfide	S^{2-}	Phosphide	P^{3-}
Hydroxide	$(OH)^{1-}$	Sulfite	$(SO_3)^{2-}$	Phosphate	$(PO_4)^{3-}$
Nitrite	$(NO_2)^{1-}$	Sulfate	$(SO_4)^{2-}$	Arsenate	$(AsO_4)^{3-}$
Nitrate	$(NO_3)^{1-}$	Carbonate	$(CO_3)^{2-}$	Borate	$(BO_3)^{3-}$
Acetate	$(C_2H_3O_2)^{1-}$	Chromate	$(CrO_4)^{2-}$		
Bromide	Br^{1-}	Dichromate	$(Cr_2O_7)^{2-}$		
Iodide	I^{1-}	Oxalate	$(C_2O_4)^{2-}$		
Hypochlorite	$(ClO)^{1-}$				
Chlorite	$(ClO_2)^{1-}$				
Chlorate	$(ClO_3)^{1-}$				
Perchlorate	$(ClO_4)^{1-}$				
Permanganate	$(MnO_4)^{1-}$				
Cyanide	$(CN)^{1-}$				
Hydrogen sulfite	$(HSO_3)^{1-}$				
Hydrogen sulfate	$(HSO_4)^{1-}$				
Hydrogen carbonate	$(HCO_3)^{1-}$				

The Group VA elements are a bit more difficult to explain. When they form simple ions, they usually do so by gaining three electrons. Therefore they have a charge of 3−. In this case, their oxidation number is also 3−. However, in other compounds in which they bond covalently, their oxidation numbers can be anywhere from 5+ to 5−.

Elements in Group VIA, which tend to become ions with 2− charges, all have oxidation numbers of 2−. (Remember, Group VIA elements tend to gain two electrons to complete their octet.) Elements in Group VIIA, which tend to become ions with 1− charges, all have oxidation numbers of 1−.

Once you know the oxidation numbers of the various ions, you can predict the chemical formulas of any compounds they form. This is because all chemical compounds must be electrically neutral; in other words, they must have no net (overall) charge. If you want to become proficient in writing and naming compounds, you should memorize the names and symbols of the ions in Tables 8.2 and 8.3. A good idea is to use index cards to prepare a set of flash cards. Review these cards every day until you have memorized the names and symbols of the ions.

Binary Compounds Containing a Metal with a Fixed Oxidation Number and a Nonmetal

LEARNING GOAL 3

Formulas of binary compounds containing a metal with a fixed oxidation number and a nonmetal

In the examples that follow, we will consider only metals that have fixed oxidation numbers in combination with nonmetals. These metals are the A-group elements of Groups IA, IIA, and IIIA. The nonmetals we will consider are the A-group elements of Groups VA, VIA, and VIIA. Although you can find the oxidation number of many of these elements in Tables 8.2 and 8.3, it is easier to remember that Group IA elements take on a charge of 1+, Group IIA elements take on a charge of 2+, and Group IIIA elements take on a charge of 3+. (*Note:* Although hydrogen is a nonmetal, it is a Group IA element and takes on a charge of 1+ when it enters into most chemical combinations.) For the nonmetals, Group VA elements take on a charge of 3−, Group VIA elements take on a charge of 2−, and Group VIIA elements take on a charge of 1−.

Greek prefixes are not used in the names of binary compounds of this type. Therefore, to write the correct formula, you must choose the correct subscripts for the formula unit by noting the charges of the metal and non-metal portions. The subscripts you choose must cancel out the charges. The next example will show you how to do this.

EXAMPLE 8.2

Write the chemical formula for sodium chloride.

Solution

Sodium is a Group IA element, and therefore it takes on a charge of 1+ when it enters into chemical combination. Chlorine is a Group VIIA element, and therefore it takes on a charge of 1− when it enters into chemical combination. Thus the formula unit for sodium chloride is

$$Na_1^{1+} Cl_1^{1-}$$

The subscripts show that we need one ion of each element to obtain a neutral compound. The subscript 1 is usually not written. Likewise, the charges 1+ and 1− are left out, giving us the chemical formula NaCl.

Practice Exercise 8.2   Write the chemical formula for potassium fluoride.

Let's look more closely at how to choose the proper subscripts for a chemical formula. We know that a formula unit has a positive portion and a negative portion. The oxidation number tells us *how* positive or negative each portion is. Our job is to choose the correct subscripts to balance the charges, so that the formula unit is electrically neutral. Here are some guidelines to help you choose the correct subscripts.

1. Write the symbol of each element of the compound, along with its oxidation number. The element with the positive oxidation number is written first. For example, for the compound calcium chloride, write

$$Ca^{2+}Cl^{1-}$$

2. You want the positive side of the compound to balance the negative side. You can bring this about by crisscrossing the numbers.

$$Ca^{2+}Cl^{1-} \quad \text{or} \quad Ca_1{}^{2+}Cl_2{}^{1-} \quad \text{or} \quad CaCl_2$$

3. Note that the subscript numbers are written without charge (that is, without a plus or minus sign). For example, the formula for aluminum oxide is

$$Al^{3+}O^{2-} \quad \text{or} \quad Al_2{}^{3+}O_3{}^{2-} \quad \text{or} \quad Al_2O_3$$

4. Also note that subscript numbers should be in least-common-denominator form. For example, the formula for barium oxide is

$$Ba^{2+}O^{2-} \quad \text{or} \quad Ba_2{}^{2+}O_2{}^{2-} \quad \text{or} \quad Ba_1{}^{2+}O_1{}^{2-} \quad \text{or} \quad BaO$$

EXAMPLE 8.3

Write the formula for barium chloride.

Solution

The periodic table helps us to obtain the charges of barium and chlorine when they enter into chemical combination. We begin by writing $Ba^{2+}Cl^{1-}$. Then we choose the right subscripts to balance the charges:

$$Ba_1{}^{2+}Cl_2{}^{1-} \quad \text{or} \quad BaCl_2$$

Practice Exercise 8.3   Write the formula for magnesium oxide.

EXAMPLE 8.4

Write the formula for aluminum sulfide.

Solution

First we write $Al^{3+}S^{2-}$. Then we choose the right subscripts to balance the charges:

$$Al_2{}^{3+}S_3{}^{2-} \quad \text{or} \quad Al_2S_3$$

Practice Exercise 8.4   Write the formula for strontium sulfide.

EXAMPLE 8.5

Write the chemical formula for sodium phosphide.

Solution
First we write $Na^{1+}P^{3-}$. Then we choose the right subscripts to balance the charges:

$$Na_3^{1+}P_1^{3-} \quad \text{or} \quad Na_3P$$

Practice Exercise 8.5   Write the formula for potassium oxide.

EXAMPLE 8.6

Write the chemical formula for aluminum nitride.

Solution
First we write $Al^{3+}N^{3-}$. Then we choose the right subscripts to balance the charges:

$$Al_1^{3+}N_1^{3-} \quad \text{or} \quad AlN$$

Practice Exercise 8.6   Write the formula for aluminum bromide.

Binary Compounds Containing a Metal with a Variable Oxidation Number and a Nonmetal

LEARNING GOAL 4

Formulas of binary compounds containing a metal with a variable oxidation number and a nonmetal

Table 8.2 shows that some atoms have variable oxidation numbers; that is, they can form more than one kind of ion when they enter into chemical combination. This is especially true of the transition metals (the ones in the middle of the periodic table—the B-group elements), because of their unique electron configurations. An element such as copper can combine with other elements in two different ways. It can combine as Cu^{1+} or Cu^{2+}.

There are two methods to name these ions. In the IUPAC *Stock system,* the oxidation number of the metal ion, in the form of a Roman numeral, is used as part of the name. The Roman numeral is placed in parentheses and immediately follows the name of the metal. Therefore Cu^{1+} is called copper(I) (read as "copper-one"), and Cu^{2+} is called copper(II) (read as "copper-two"). Under the Stock system, the oxidation number of a metal that has a variable oxidation number will *always* be part of the compound's name.

The Stock system is now the system of choice, but the older *-ous* and *-ic* system for naming these ions is still used. In the older system, the Latin name of the metal is used along with the suffix *-ous* or *-ic.* The ion with the lower oxidation number is given the suffix *-ous,* and the ion with the higher oxidation number is given the suffix *-ic.* For example, Fe^{2+} is called the ferr*ous* ion, and Fe^{3+} is called the ferr*ic* ion. (Note the suffixes and the use of *ferr-,* not *iron,* as the main stem of the element.) Another example is copper: Cu^{1+} is called the cupr*ous* ion, and Cu^{2+} is called the cupr*ic* ion. The disadvantage of this system is that you have to memorize which ion of a given element has the higher or lower oxidation number to apply the correct suffix. Table 8.2 lists some additional examples of positive ions and their names, using the rules for both systems.

Although you may find the Stock system easier to use, you must also become familiar with the older system, because it is still in use. In the examples that follow, you will practice writing the formulas of compounds when you are given the Stock name or the older *-ous* or *-ic* name.

EXAMPLE 8.7

Write the formula of each of the following compounds whose metal ions have variable oxidation numbers:
(a) copper(I) chloride
(b) copper(II) chloride
(c) iron(III) sulfide
(d) vanadium(V) oxide
(e) mercuric bromide
(f) ferrous nitride

Solution
(a) The (I) tells us that copper is Cu^{1+}. We already know that chloride is Cl^{1-}, because chlorine is a Group VIIA element. Therefore we have

$$Cu^{1+}Cl^{1-} \quad \text{or} \quad CuCl$$

(b) The (II) tells us that copper is Cu^{2+}. We already know that chloride is Cl^{1-}. Therefore we have

$$Cu_1{}^{2+}Cl_2{}^{1-} \quad \text{or} \quad CuCl_2$$

(c) The (III) tells us that iron is Fe^{3+}. We already know that sulfide is S^{2-}, because sulfur is a Group VIA element. Therefore we have

$$Fe_2{}^{3+}S_3{}^{2-} \quad \text{or} \quad Fe_2S_3$$

(d) The (V) tells us that vanadium is V^{5+}. We already know that oxide is O^{2-}, because oxygen is a Group VIA element. Therefore we have

$$V_2{}^{5+}O_5{}^{2-} \quad \text{or} \quad V_2O_5$$

(e) Table 8.2 tells us that the mercuric ion is Hg^{2+}. We already know that bromide is Br^{1-}, because bromine is a Group VIIA element. Therefore we have

$$Hg_1{}^{2+}Br_2{}^{1-} \quad \text{or} \quad HgBr_2$$

(f) Table 8.2 tells us that the ferrous ion is Fe^{2+}. We already know that nitride is N^{3-}, because nitrogen is a Group VA element. Therefore we have

$$Fe_3{}^{2+}N_2{}^{3-} \quad \text{or} \quad Fe_3N_2$$

Practice Exercise 8.7   Write the formula of each of the following compounds whose metal ions have variable oxidation numbers:

(a) mercury(II) oxide
(b) iron(III) bromide
(c) cobalt(II) iodide
(d) manganese(IV) oxide
(e) cuprous sulfide
(f) ferric sulfide

Before we leave this topic, we should point out that some transition metals do not have variable oxidation numbers. Therefore neither the name of the compound nor the group that it is in will tell you the charge of the ion in that compound. Two important examples are zinc and silver. You should *memorize* that zinc has an oxidation number of 2+ (Zn^{2+}) and silver has an oxidation number of 1+ (Ag^{1+}).

EXAMPLE 8.8

Write the formulas of the following compounds: (a) silver sulfide
(b) zinc phosphide

Solution

(a) We know that silver is Ag^{1+}. We already know that sulfide is S^{2-}, because sulfur is a Group VIA element. Therefore we have

$$Ag_2^{1+}S_1^{2-} \qquad \text{or} \qquad Ag_2S$$

(b) We know that zinc is Zn^{2+}. We already know that phosphide is P^{3-}, because phosphorus is a Group VA element. Therefore we have

$$Zn_3^{2+}P_2^{3-} \qquad \text{or} \qquad Zn_3P_2$$

Practice Exercise 8.8   Write the formulas of the following compounds: (a) silver chloride   (b) zinc oxide

---

8.4 Polyatomic Ions

LEARNING GOAL 5

Formulas and names of polyatomic ions

Over the years, in studying the composition of many compounds, chemists have found that certain groups of covalently bonded atoms appear over and over again. Because these groups are electrically charged, they are not molecules but ions. More specifically, they are *polyatomic ions* (ions that are combinations of many atoms); the prefix *poly-* is from the Greek word for "many." A **polyatomic ion** is a *charged group of covalently bonded atoms.*

Examples of polyatomic ions are nitrate, $(NO_3)^{1-}$; sulfite, $(SO_3)^{2-}$; sulfate, $(SO_4)^{2-}$; carbonate, $(CO_3)^{2-}$; and phosphate, $(PO_4)^{3-}$. Polyatomic ions are common in minerals, plants, animals, and human beings. *They never exist in an uncombined state* but are always part of a chemical compound. For example, one can't isolate nitrate ions and put them in a bottle, but there are hundreds of chemical compounds made up, in part, of nitrate ions. Some examples of nitrate compounds are potassium nitrate, KNO_3 (used in chemical fertilizers); calcium nitrate, $Ca(NO_3)_2$ (used in chemical fertilizers and also in matches); and silver nitrate, $AgNO_3$ (used in photography, mirror manufacturing, hair dyeing, and silver plating, and as an external medicine).

Tables 8.2 and 8.3 list the more common polyatomic ions. You should memorize the name, symbol, and charge of each ion. That information will be useful as you learn to write the formulas of compounds containing these ions in the next section.

8.5   Writing the Formulas of Ternary and Higher Compounds

LEARNING GOAL 6

Formulas of ternary and higher compounds

In writing the formulas of compounds containing three or more elements (usually in the form of polyatomic ions), we follow much the same procedure we used to write formulas of binary compounds. The only difference is that we must use the formula and charge of the polyatomic ion in deriving the formulas of ternary and higher compounds. Most of the polyatomic ions that we'll encounter have negative charges and are listed in Table 8.3. The only common polyatomic ion with a positive charge is the ammonium ion, $(NH_4)^{1+}$, which is listed in Table 8.2.

The names of most polyatomic ions end in *-ite* or *-ate*. For example, $(SO_3)^{2-}$ is called the sulf*ite* ion, and $(SO_4)^{2-}$ is called the sulf*ate* ion. Notice that the *-ate* ion has one more oxygen than the corresponding *-ite* ion. There are two polyatomic ions in Table 8.3 that do not end in *-ite* or *-ate*. They are the hydrox*ide* ion, $(OH)^{1-}$, and the cyan*ide* ion, $(CN)^{1-}$.

In the next example, you will practice writing the formulas of compounds containing polyatomic ions. If you've already memorized the polyatomic ions, try writing the formulas using only the periodic table as a reference. Otherwise, look up the formulas and charges for the polyatomic ions in Tables 8.2 and 8.3.

EXAMPLE 8.9

Write the formula of each of the following ternary compounds:
(a) aluminum sulfate
(b) calcium arsenate
(c) copper(II) phosphate
(d) sodium nitrate

Solution

(a) Aluminum is a Group IIIA element, and therefore it takes on a charge of 3+ when it enters into chemical combination. Sulfate ion has the formula $(SO_4)^{2-}$. Thus the formula for aluminum sulfate is

$$Al_2^{3+}(SO_4)^{2-}   \text{or}   Al_2(SO_4)_3$$

(b) Calcium is a Group IIA element, and therefore it takes on a charge of 2+ when it enters into chemical combination. Arsenate ion has the formula $(AsO_4)^{3-}$. Thus the formula for calcium arsenate is

$$Ca_3^{2+}(AsO_4)_2^{3-} \quad \text{or} \quad Ca_3(AsO_4)_2$$

(c) Copper(II) has a charge of 2+. Phosphate ion has the formula $(PO_4)^{3-}$. Therefore the formula for copper(II) phosphate is

$$Cu_3^{2+}(PO_4)_2^{3-} \quad \text{or} \quad Cu_3(PO_4)_2$$

(d) Sodium is a Group IA element, and therefore it takes on a charge of 1+ when it enters into chemical combination. Nitrate ion has the formula $(NO_3)^{1-}$. Thus the formula for sodium nitrate is

$$Na_1^{1+}(NO_3)_1^{1-} \quad \text{or} \quad NaNO_3$$

Notice that the parentheses are dropped when the polyatomic ion has a subscript of 1.

Practice Exercise 8.9   Write the formula of each of the following ternary compounds:
(a) ferric nitrite
(b) barium phosphate
(c) copper(II) sulfate
(d) potassium chromate

8.6 Writing the Names of Binary Compounds Containing Two Nonmetals

Now let's turn our attention to writing the names of compounds when we are given their formulas. We will begin by learning how to name binary compounds composed of two nonmetals.

In Section 8.2, you learned how to write the formulas of these compounds from their names. You need only to reverse the process to write the name of the compound from the formula. Remember that Greek prefixes are used to designate the number of atoms in a formula unit of the compound.

EXAMPLE 8.10

Write the names of the following binary compounds composed of two nonmetals: (a) SO_3 (b) NO (c) N_2O (d) CO_2

Solution
(a) SO_3 is called sulfur trioxide. This compound contains one sulfur atom and three oxygen atoms. There is no prefix used for the sulfur portion of the compound. The prefix *tri-* is used for the oxygen

portion of the compound. (Also, remember that in naming a binary compound, you change ox*ygen* to ox*ide* because it is the second-named element.)

(b) NO is called nitrogen monoxide. This compound contains one nitrogen atom and one oxygen atom. There is no prefix used for the nitrogen portion of the compound. The prefix *mono-* is used for the oxygen portion of the compound. (Remember the rule for *mono-*: The prefix *mono-* is used for the second-named element, but rarely appears with the first-named element.)

(c) N_2O is called dinitrogen monoxide. This compound contains two nitrogen atoms and one oxygen atom. The prefix *di-* is used for the nitrogen portion of the compound. The prefix *mono-* is used for the oxygen portion of the compound.

(d) CO_2 is called carbon dioxide. This compound contains one carbon atom and two oxygen atoms. There is no prefix used for the carbon portion of the compound. The prefix *di-* is used for the oxygen portion of the compound.

Practice Exercise 8.10   Write the names of the following binary compounds composed of two nonmetals: (a) P_2O_5 (b) OF_2 (c) SO_2 (d) CO

8.7

Writing the Names of Binary Compounds Containing a Metal and a Nonmetal

LEARNING GOAL 8

Names of binary compounds containing a metal and a nonmetal

LEARNING GOAL 9

Determining oxidation numbers

We already know how to write the formulas of binary compounds composed of a metal and a nonmetal when we are given their names (Section 8.3). Once again, we can simply reverse this process to write the name of such a compound from its formula. Remember that for these compounds, *no Greek prefixes are used.* For compounds that contain metals with fixed oxidation numbers, just name the elements. (Don't forget to change the ending of the second-named element to *-ide.*) For compounds that contain metals with variable oxidation numbers, the Roman numeral, which represents the oxidation number of the metal, must be given as part of the compound's name (or the *-ous* and *-ic* suffix system may be used).

Finding Oxidation Numbers

Tables 8.2 and 8.3 list oxidation numbers only for some of the more common ions. However, once you know these, you can use them to find the oxidation numbers of less common ions from the compounds they form. You need only remember that every compound must be neutral. That is, the sum of the oxidation numbers of all the atoms in the compound must be zero.

EXAMPLE 8.11

What is the oxidation number of cobalt in $CoCl_3$?

Solution

We know that each chloride ion has a charge of $1-$. The molecule contains three chloride ions, for a total charge of $3-$. For the compound to be neutral, the single cobalt ion must have a charge of $3+$: $Co_1^{3+}Cl_3^{1-}$.

Practice Exercise 8.11   What is the oxidation number of copper in $CuCl_2$?

EXAMPLE 8.12

Find the oxidation number of each underlined element:
(a) $\underline{Mn}O_2$ (b) $K\underline{Mn}O_4$ (c) $\underline{Cs}_2SO_3$ (d) $\underline{Ga}_2O_3$
We follow the same procedure used in the previous example.

Solution

(a) We know that the oxidation number of oxygen is $2-$: $Mn_1^?O_2^{2-}$
    For the compound to be electrically neutral, the manganese must have a charge of $4+$: $Mn_1^{4+}O_2^{2-}$
(b) We know that the oxidation number of potassium is $1+$, and that of oxygen is $2-$: $K_1^{1+}Mn_1^?O_4^{2-}$
    For the compound to be electrically neutral, the manganese must have a charge of $7+$: $K_1^{1+}Mn_1^{7+}O_4^{2-}$
(c) We know that the oxidation number of the sulfite group is $2-$: $Cs_2^?(SO_3)_1^{2-}$
    For the compound to be electrically neutral, the cesium must have a charge of $1+$: $Cs_2^{1+}(SO_3)_1^{2-}$
(d) We know that the oxidation number of oxygen is $2-$: $Ga_2^?O_3^{2-}$
    For the compound to be electrically neutral, the gallium must have a charge of $3+$: $Ga_2^{3+}O_3^{2-}$

Practice Exercise 8.12   Find the oxidation number of each underlined element: (a) $\underline{Sn}F_4$ (b) $\underline{W}Cl_5$

EXAMPLE 8.13

Write the names of the following binary compounds composed of a metal and a nonmetal: (a) $AlCl_3$ (b) Na_2O (c) Mg_3N_2 (d) K_3P
(e) $CoCl_3$ (f) Fe_2O_3 (g) FeO (h) MnO_2

Solution

(a) $AlCl_3$ is called aluminum chloride. All we have to do is name the elements and remember to change the name of chlor*ine* to chlor*ide*.

(b) Na_2O is called sodium oxide. All we have to do is name the elements and remember to change the name of ox*ygen* to ox*ide.*

(c) Mg_3N_2 is called magnesium nitride. All we have to do is name the elements and remember to change the name of nitro*gen* to nit*ride.*

(d) K_3P is called potassium phosphide. All we have to do is name the elements and remember to change the name of phosph*orus* to phosph*ide.*

(e) $CoCl_3$ is called cobalt(III) chloride (or cobaltic chloride). Cobalt is a metal with a variable oxidation number, and a Roman numeral representing that oxidation number must be part of the compound's name. To determine the oxidation number of cobalt in this compound, we go through the following reasoning: We know that each chloride ion has a charge of $1-$. This molecule contains three chloride ions, for a total charge of $3-$. For the compound to be neutral, the single cobalt ion must have a charge of $3+$:

$$Co_1^{3+}Cl_3^{1-}$$

(f) Fe_2O_3 is called iron(III) oxide (or ferric oxide). Iron is a metal with a variable oxidation number, and a Roman numeral representing that oxidation number must be part of the compound's name. To determine the oxidation number of iron in this compound, we go through the following reasoning: We know that each oxide ion has a charge of $2-$. This molecule contains three oxide ions, for a total charge of $6-$. For the compound to be neutral, the two iron ions must have a total charge of $6+$. Therefore each iron ion must have a charge of $3+$:

$$Fe_2^{3+}O_3^{2-}$$

(g) FeO is called iron(II) oxide (or ferrous oxide). Iron is a metal with a variable oxidation number, and a Roman numeral representing that oxidation number must be part of the compound's name. To determine the oxidation number of iron in this compound, we go through the following reasoning: We know that each oxide ion has a charge of $2-$. This molecule contains one oxide ion, for a total charge of $2-$. For the compound to be neutral, the single iron ion must have a charge of $2+$:

$$Fe_1^{2+}O_1^{2-}$$

(h) MnO_2 is called manganese(IV) oxide. Manganese is a metal with a variable oxidation number, and a Roman numeral representing that oxidation number must be part of the compound's name. To determine the oxidation number of manganese in this compound, we go through the following reasoning: We know that each oxide ion has a charge of $2-$. This molecule contains two oxide ions, for a total charge of $4-$. For the compound to be neutral, the single manganese ion must have a charge of $4+$:

$$Mn_1^{4+}O_2^{2-}$$

Practice Exercise 8.13   Write the names of the following binary compounds composed of a metal and a nonmetal: (a) RbCl   (b) Ga_2S_3 (c) SrO   (d) ZnI_2 (e) $NiCl_2$ (f) FeI_3 (g) HgS   (h) Cu_2O

8.8   Writing the Names of Ternary and Higher Compounds

You learned how to write the formulas of ternary and higher compounds from their names in Section 8.5. You can just reverse the process to write the name of this type of compound from the formula. Remember that for these compounds, *no Greek prefixes are used,* but you must include the name of the polyatomic ion. For compounds that contain metals with fixed oxidation numbers, simply name the element and the polyatomic ion. For compounds that contain metals with variable oxidation numbers, the Roman numeral, which represents the oxidation number of the metal, must be given as part of the compound's name (or the *-ous* and *-ic* suffix system may be used).

EXAMPLE 8.14

Write the names of the following ternary and higher compounds:
(a) $Al_2(CrO_4)_3$ (b) Li_2SO_3 (c) $Mg(NO_2)_2$ (d) $(NH_4)_3PO_4$
(e) $Cr(OH)_2$ (f) $Fe_2(Cr_2O_7)_3$

Solution

(a) $Al_2(CrO_4)_3$ is called aluminum chromate. All we have to do is name the metal, aluminum, and the polyatomic ion, chromate.

(b) Li_2SO_3 is called lithium sulfite. All we have to do is name the metal, lithium, and the polyatomic ion, sulfite.

(c) $Mg(NO_2)_2$ is called magnesium nitrite. All we have to do is name the metal, magnesium, and the polyatomic ion, nitrite.

(d) $(NH_4)_3PO_4$ is called ammonium phosphate. All we have to do is name two polyatomic ions, ammonium and phosphate.

(e) $Cr(OH)_2$ is called chromium(II) hydroxide (or chromous hydroxide). Chromium is a metal with a variable oxidation number, and a Roman numeral representing that oxidation number must be part of the compound's name. Therefore we have chromium(II). The polyatomic ion is hydroxide.

(f) $Fe_2(Cr_2O_7)_3$ is called iron(III) dichromate (or ferric dichromate). Iron is a metal with a variable oxidation number, and a Roman numeral representing that oxidation number must be part of the compound's name. Therefore we have iron(III). The polyatomic ion is dichromate.

Practice Exercise 8.14   Write the names of the following ternary and higher compounds: (a) $(NH_4)_2O$ (b) $Mg(CN)_2$ (c) $Al(NO_2)_3$ (d) $Zn_3(PO_4)_2$ (e) CaC_2O_4 (f) $Ni_3(BO_3)_2$

8.9   Writing the Names and Formulas of Inorganic Acids

LEARNING GOAL 11

Formulas and names of inorganic acids

Svante Arrhenius, a well-known Swedish chemist of the late 1800s and early 1900s, developed a classical definition of an acid. According to Arrhenius, an *acid* is a substance that releases hydrogen ions, H^{1+}, in aqueous solutions. Two major classes of inorganic acids exist, and the naming rules differ for the two classes. First we'll examine the rules for naming the non-oxygen-containing acids. Then we'll examine the rules for naming the oxygen-containing acids. We'll have more to say about acids in Chapter 16.

Non-Oxygen-Containing Acids

These acids usually consist of hydrogen plus a nonmetal ion. For example, HCl, which is hydrogen chloride gas in its pure form, becomes hydrochloric acid when dissolved in water. The covalently bonded HCl molecules ionize in water to form hydrogen ions and chloride ions. We derive the names of acids such as HCl by adding the prefix *hydro-* to the name of the nonmetal, which is given an *-ic* suffix. For example,

$$hydro\text{——}chlor\text{——}ic\text{——}acid$$

Prefix   Name of   Suffix
           nonmetal

EXAMPLE 8.15

Name each of the following non-oxygen-containing acids: (a) HF (b) HBr   (c) H_2S (d) HCN

Solution

We'll give the name of each substance first as a covalent compound and then as an acid in aqueous solution.

NAME AS COVALENT COMPOUND	NAME AS ACID IN AQUEOUS SOLUTION
(a) Hydrogen fluoride	Hydrofluoric acid
(b) Hydrogen bromide	Hydrobromic acid
(c) Hydrogen sulfide	Hydrosulfuric acid
(d) Hydrogen cyanide	Hydrocyanic acid

Practice Exercise 8.15   Name each of the following non-oxygen-containing acids: (a) HCl   (b) H_2Se (c) HI

Oxygen-Containing Acids

Oxygen-containing acids formed from hydrogen and polyatomic ions (containing oxygen) are named as follows:

   1. If the polyatomic ion has an *-ate* suffix, we name the acid by replacing the *-ate* suffix with an *-ic* suffix and adding the word *acid*. For example, HNO_3, composed of a hydrogen ion and nitr*ate* ion, is called nitr*ic* acid in aqueous solution.

EXAMPLE 8.16

Name each of the following oxygen-containing acids: (a) H_2SO_4
(b) HC_2H_3O (c) $HBrO_3$

Solution

These three acids are composed of hydrogen plus a polyatomic ion that has the suffix *-ate*.
(a) H_2SO_4 is called sulfur*ic* acid. (SO_4^{2-} is the sulf*ate* ion.)
(b) $HC_2H_3O_2$ is called acet*ic* acid. ($C_2H_3O_2^{1-}$ is the acet*ate* ion.)
(c) $HBrO_3$ is called brom*ic* acid. (BrO_3^{1-} is the brom*ate* ion.)
Note that H_2SO_4 is an exception to our rule, in that the whole name of the element is used rather than the name of the sulfate group.

Practice Exercise 8.16   Name each of the following oxygen-containing acids: (a) H_3PO_4 (b) $HClO_3$

---

   2. If the polyatomic ion has an *-ite* suffix, we name the acid by replacing the *-ite* suffix with an *-ous* suffix and adding the word *acid*. For example, HNO_2, composed of a hydrogen ion and a nitr*ite* ion, is called nitr*ous* acid.

EXAMPLE 8.17

Name each of the following oxygen-containing acids: (a) H_2SO_3
(b) H_3PO_3 (c) $HBrO_2$

Solution

These three acids are composed of hydrogen plus a polyatomic ion that has the suffix *-ite*.
(a) H_2SO_3 is called sulfur*ous* acid. (SO_3^{2-} is the sulf*ite* ion.)
(b) H_3PO_3 is called phosphor*ous* acid. (PO_3^{3-} is the phosph*ite* ion.)
(c) $HBrO_2$ is called brom*ous* acid. (BrO_2^{1-} is the brom*ite* ion.)
Note that H_2SO_3 is an exception to our rule, in that the whole name of the element is used rather than the name of the sulfite group.

Practice Exercise 8.17   Name each of the following oxygen-containing acids: (a) $HClO_2$ (b) HNO_2

3. Some elements form more than two oxygen-containing acids. Most nota-
   ble are the acids formed by the elements chlorine, bromine, and iodine.
   This occurs because these elements can form four polyatomic ions. For
   example, bromine forms the following ions:

   BrO_4^{1-}, perbrom*ate* ion BrO_3^{1-}, brom*ate* ion
   BrO_2^{1-}, brom*ite* ion BrO^{1-}, hypobrom*ite* ion

   Note that the first two ions listed have the suffix *-ate* and that the last
   two ions listed have the suffix *-ite*. Therefore we use rules 1 and 2. Keep
   the name of the polyatomic ion, but change the *-ate* suffix to *-ic* or the
   -ite suffix to *-ous*.

EXAMPLE 8.18

Name each of the following oxygen-containing acids: (a) $HBrO_4$
(b) $HBrO_3$ (c) $HBrO_2$ (d) HBrO

Solution
(a) $HBrO_4$ is called perbrom*ic* acid. (BrO_4^{1-} is the perbrom*ate* ion.)
(b) $HBrO_3$ is called brom*ic* acid. (BrO_3^{1-} is the brom*ate* ion.)
(c) $HBrO_2$ is called brom*ous* acid. (BrO_2^{1-} is the brom*ite* ion.)
(d) HBrO is called hypobrom*ous* acid. (BrO^{1-} is the hypobrom*ite* ion.)

Practice Exercise 8.18   Name each of the following oxygen-containing
acids: (a) $HClO_4$ (b) $HClO_3$ (c) $HClO_2$ (d) HClO

8.10   Common Names of Compounds

LEARNING GOAL 12

**Common names and
formulas of compounds**

As we stated at the beginning of this chapter, there are no rules governing
the common names of compounds. Most common names have been derived
from common usage or simply handed down through chemical history. Many
common names are still in use, mostly because the systematic names are too
complex for everyday use. For example, baking soda is a common substance
that has a variety of uses around the house, but hardly anyone would call
it sodium hydrogen carbonate. The same could be said of Epsom salts,
whose systematic name is magnesium sulfate heptahydrate. Table 8.4 lists
the common names, systematic chemical names, and formulas of several
familiar substances.

SUMMARY

The system for naming chemical compounds and designating their formulas
is called chemical nomenclature. Chemical compounds may be referred to

TABLE 8.4

Common Names of Some Chemical Compounds

COMMON NAME	FORMULA	SYSTEMATIC CHEMICAL NAME
Baking soda	$NaHCO_3$	Sodium hydrogen carbonate
Borax	$Na_2B_4O_7 \cdot 10H_2O$	Sodium tetraborate decahydrate
Dry ice	CO_2	Carbon dioxide
Epsom salts	$MgSO_4 \cdot 7H_2O$	Magnesium sulfate heptahydrate
Gypsum	$CaSO_4 \cdot 2H_2O$	Calcium sulfate dihydrate
Laughing gas	N_2O	Nitrous oxide
Marble	$CaCO_3$	Calcium carbonate
Milk of magnesia	$Mg(OH)_2$	Magnesium hydroxide
Muriatic acid	HCl	Hydrochloric acid
Oil of vitriol	H_2SO_4	Sulfuric acid
Quickline	CaO	Calcium oxide
Saltpeter	$NaNO_3$	Sodium nitrate

by their systematic chemical names or their common names. The rules for the systematic names have been developed by the International Union of Pure and Applied Chemistry (IUPAC). There are no rules that govern the common names. These names have simply been handed down through chemical history.

According to IUPAC rules for writing the formula of a compound, the more positive portion of the compound is written first. In a compound composed of two ions, the positive ion precedes the negative ion. In a covalent compound, the element with the lower electronegativity (the more positive portion) precedes the element with the higher electronegativity (the more negative portion).

Greek prefixes are used to write the names of binary compounds composed of two nonmetals. For binary compounds, the ending of the second element is always *-ide*. Oxidation numbers are used to write the formulas of binary compounds containing a metal and a nonmetal. An oxidation number is the positive or negative number that expresses the combining capacity of an element in a particular compound. Some metals (typically the A-group metals) have fixed oxidation numbers. Other metals (typically the B-group, or transition, metals) have variable oxidation numbers and can combine with a nonmetal element in more than one way. The periodic table can be used to determine the oxidation number of an A-group metal: The group number of the A-group metal is its common oxidation number. The oxidation number of metals with variable oxidation numbers can be determined from the name of the compound, either through the Roman numeral attached to the name of the metal or through the *-ous* or *-ic* suffix attached to the Latin name of the metal.

Ternary and higher compounds contain polyatomic ions. A polyatomic ion is a charged group of covalently bonded atoms. The names of most polyatomic ions end in *-ite* or *-ate*. We use the formula and charge of the polyatomic ion and the rules for writing the formulas of binary compounds in deriving the formulas of ternary and higher compounds.

A number of chemicals are known by their common names, either because of tradition or because the systematic names are too complex for convenient reference. Because there are no rules that govern the common names of compounds, the formulas of these compounds may be associated with their names only through memorization.

KEY TERMS

binary compound **(8.1)**

chemical nomenclature
  (Introduction)

common name **(Introduction)**

oxidation number **(8.3)**

polyatomic ion **(8.4)**

systematic chemical name
  (Introduction)

ternary compound **(8.1)**

CAREER SKETCH

Research and Development Careers

Many of the careers explored throughout this book are in research and development (R&D). R&D chemists work on methods for storing and retrieving information, develop applications for artificial intelligence systems, and develop hypertext for chemical information systems.

A broad range of skills is helpful, and a team-oriented person will be most successful. Strong oral and written communication skills are necessary.

For the chemist interested in information technology (IT), the ability to conceptualize is necessary. Knowledge of current software programs and of the business for which the product is being developed is also a plus.

Research and development chemists work on creating new molecules.
(Terry Vine/Tony Stone Images)

CHEMICAL FRONTIERS

*Airplane Food?**

Airplane food has never been a favorite of passengers, but frequent flyers may soon be praising a new "food" substance that will be on board. Researchers at the National Aeronautics & Space Administration's Ames Research Center in California have developed an anti-icing agent so environmentally safe that all of its ingredients have a government "food-grade" rating.

Almost all anti-icing agents in use today have propylene glycol and water as their base. These chemicals are relatively inert and do not harm the environment. The problem with these agents, however, is that highly toxic additives are mixed with the inert base to achieve the desired de-icing effect for a prolonged period of time.

The newly developed agent, nicknamed "green goop" because of its color (achieved by nothing more than food coloring), also contains

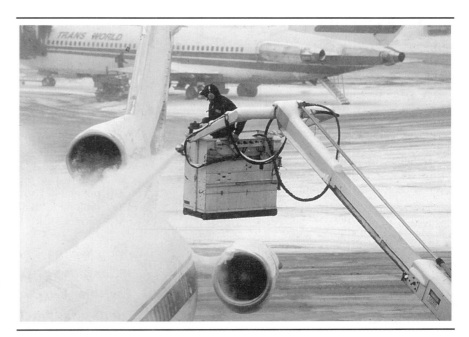

"Green goop," a new deicing agent for aircraft, does not harm the environment.
(Alvis Upitis/The Image Bank)

harmless propylene glycol and water as its base. Unlike in most commercially available fluids, however, the other ingredients are also safe—nonreactive and nonconductive.

The researchers chose to use a xanthan polysaccharide as the thickening agent in the green goop. This allows the fluid to cling to the surface of an aircraft, acting as a barrier to ice and snow. As the plane starts down the runway, the material slides off the plane just before it takes off. Another harmless additive, an aliphatic alcohol, prevents the fluid from being diluted.

The innocuous fluid may also be safer than conventional anti-icing agents. The green goop is effective for almost two hours, surpassing government specifications nearly four times over. Furthermore the material has passed corrosion resistance and humidity tests.

The material, which is inexpensive to manufacture and simple to apply, may have uses beyond de-icing aircraft. Engineers at the Army Corp of Engineers' Cold Regions Research and Engineering Laboratory in Hanover, New Hampshire, are testing the material to see how well it performs when used to de-ice highways, bridges, and railways.

*Adapted from *Chemical & Engineering News,* June 23, 1997, pp. 11–12

SELF-TEST EXERCISES

LEARNING GOAL 1: **Distinguishing Between Common Name and Systematic Name**

◀ 1. Explain why chemists developed systematic names for compounds, when common names were already in existence.

◀ 2. Define or explain the following terms: (a) IUPAC (b) Stock system (c) binary compound (d) ternary compound

3. The common name for H_2O is water. What is its systematic name?

4. What is the common name and the systematic name of solid CO_2?

LEARNING GOAL 2: **Formulas of Binary Compounds Containing Two Nonmetals**

5. Write the number that corresponds to each of the following prefixes: (a) *hepta-* (b) *di-* (c) *tri-* (d) *octa-* (e) *mono-*

6. Write the Greek prefix for each of the following numbers: (a) 3 (b) 7 (c) 2 (d) 9

7. Write the formulas of the following binary compounds composed of two nonmetals:
(a) diphosphorus pentasulfide
(b) chlorine dioxide
(c) dinitrogen tetroxide
(d) dichlorine heptoxide

8. Write the formulas of the following binary compounds composed of two nonmetals:
(a) carbon tetrachloride
(b) tetraphosphorus decaoxide
(c) phosphorus pentabromide
(d) selenium dioxide

LEARNING GOAL 3: **Formulas of Binary Compounds Containing a Metal with a Fixed Oxidation Number and a Nonmetal**

9. Write the formulas of the following binary compounds composed of a metal with a fixed oxidation number and a nonmetal:

(a) aluminum sulfide
(b) lithium oxide
(c) sodium nitride
(d) strontium phosphide
(e) silver bromide
(f) zinc oxide

10. Write the formulas of the following binary compounds composed of a metal with a fixed oxidation number and a nonmetal:
(a) lithium iodide
(b) calcium oxide
(c) strontium bromide
(d) potassium phosphide
(e) rubidium sulfide
(f) barium nitride

LEARNING GOAL 4: **Formulas of Binary Compounds Containing a Metal with a Variable Oxidation Number and a Nonmetal**

11. Give the Roman numerals that correspond to the following Arabic numerals: (a) 9 (b) 8 (c) 7 (d) 6

12. Give the Arabic numerals that correspond to the following Roman numerals: (a) III (b) VIII (c) VI (d) IV

◀ **13.** Give a definition or an explanation of the following suffixes: (a) *-ic* (b) *-ous* (c) *-ide*

◀ 14. Explain the difference between (a) ferrous and ferric (b) cuprous and cupric (c) cobaltous and cobaltic (d) stannous and stannic

15. Write the formulas of the following binary compounds composed of a metal with a variable oxidation number and a nonmetal:
(a) copper(I) sulfide
(b) mercury(II) chloride
(c) iron(II) oxide
(d) tin(II) iodide
(e) cobaltic bromide
(f) mercuric nitride

16. Write the formulas of the following binary compounds composed of a metal with a variable oxidation number and a nonmetal:
(a) uranium(VI) fluoride
(b) iron(III) nitride
(c) manganous chloride
(d) ferrous sulfide

(e) cuprous oxide
(f) mercurous chloride

LEARNING GOAL 5: *Formulas and Names of Polyatomic Ions*

◀ *17.* Write the formulas of the following polyatomic ions:
(a) ammonium ion
(b) hydroxide ion
(c) borate ion
(d) hydrogen carbonate ion
(e) oxalate ion
(f) sulfite ion

◀ 18. Write the formulas of the following polyatomic ions:
(a) cyanide ion
(b) chlorite ion
(c) chlorate ion
(d) chromate ion
(e) dichromate ion
(f) phosphate ion

19. Write the names of the following polyatomic ions: (a) $(AsO_4)^{3-}$ (b) $(MnO_4)^{1-}$ (c) $(SO_3)^{2-}$ (d) $(NH_4)^{1+}$ (e) $(HSO_4)^{1-}$ (f) $(C_2H_3O_2)^{1-}$

20. Write the names of the following polyatomic ions: (a) $(NO_2)^{1-}$ (b) $(NO_3)^{1-}$ (c) $(ClO_4)^{1-}$ (d) $(CN)^{1-}$ (e) $(BO_3)^{3-}$ (f) $(ClO)^{1-}$

LEARNING GOAL 6: *Formulas of Ternary and Higher Compounds*

21. Write the formulas of the following ternary and higher compounds:
(a) mercury(II) phosphate
(b) tin(II) arsenate
(c) iron(III) acetate
(d) lithium phosphate
(e) aluminum sulfite
(f) zinc nitrite

22. Write the formulas of the following ternary and higher compounds:
(a) cesium hydroxide
(b) copper(I) arsenate
(c) ammonium sulfate
(d) potassium carbonate
(e) ferric cyanide
(f) cuprous sulfate

◀ *23.* Write the formulas of the following ternary and higher compounds:
(a) lead(II) sulfate
(b) cobalt(II) phosphate

(c) ammonium dichromate
(d) calcium oxalate
(e) stannous nitrate
(f) magnesium hydrogen sulfate

◀ 24. Write the formulas of the following ternary and higher compounds:
(a) potassium chlorate
(b) zinc hydroxide
(c) zinc phosphate
(d) silver nitrite
(e) mercuric nitrate
(f) cuprous sulfite

LEARNING GOAL 7: *Names of Binary Compounds Containing Two Nonmetals*

25. Write the names of the following binary compounds composed of two nonmetals: (a) P_2S_5 (b) CO (c) SiO_2 (d) ClO_2

26. Write the names of the following binary compounds composed of two nonmetals: (a) N_2O (b) NO_2 (c) SO_3 (d) N_2O_5

LEARNING GOAL 8: *Names of Binary Compounds Containing a Metal and a Nonmetal*

◀ *27.* Write the names of the following binary compounds containing metals with fixed oxidation numbers:
(a) Al_2O_3 (b) NaI (c) $ZnCl_2$ (d) Mg_3N_2 (e) Ag_2S (f) LiI

◀ 28. Write the names of the following binary compounds containing metals with fixed oxidation numbers:
(a) Cs_2O (b) Al_2S_3 (c) BaI_2 (d) $GaCl_3$ (e) K_2O (f) MgS

LEARNING GOAL 9: *Determining Oxidation Numbers*

29. Find the oxidation number of the underlined element or polyatomic ion: (a) $Na\underline{Cl}O_4$ (b) $\underline{Pb}_3O_4$ (c) $Ge\underline{S}_2$ (d) $\underline{V}OCl_3$ (e) $Ca(\underline{HCO_3})_2$

◀ 30. Find the oxidation number of the underlined element: (a) $H\underline{Cl}O$ (b) $H\underline{Cl}O_2$ (c) $H\underline{Cl}O_3$ (d) $H\underline{Cl}O_4$

◀ *31.* Find the oxidation number of the underlined element or polyatomic ion: (a) $\underline{V}_2O_5$ (b) $\underline{In}_2O$ (c) $\underline{N}_2O$ (d) $Pb\underline{P}_5$ (e) $Mg_3(\underline{BO_3})_2$ (f) $Al(\underline{ClO_3})_3$

32. Find the oxidation number of the underlined element or polyatomic ion: (a) $H_3\underline{B}O_3$ (b) $H\underline{Br}O_3$ (c) $HI\underline{O}_3$ (d) $\underline{Cr}PO_3$ (e) $Al(\underline{HCO_3})_3$ (f) $Li\underline{Cl}O$

◀ **33.** Write the names of the following binary compounds containing metals with variable oxidation numbers: (a) OsO_4 (b) Hg_6P_2 (c) FeS (d) $CoCl_2$ (e) Cu_3N (f) Cu_2O

◀ 34. Write the names of the following binary compounds containing metals with variable oxidation numbers: (a) CuS (b) $AuBr_3$ (c) FeO (d) Cu_3P_2 (e) V_2O_5 (f) MnO_2

LEARNING GOAL 10: *Names of Ternary and Higher Compounds*

35. Write the names of the following ternary and higher compounds: (a) Ag_2CO_3 (b) $Hg_3(PO_4)_2$ (c) $Fe_2(SO_4)_3$ (d) $NaNO_3$ (e) $CuCrO_4$ (f) $Zn(OH)_2$

36. Write the names of the following ternary and higher compounds: (a) $CaCO_3$ (b) $NaNO_2$ (c) $NaOH$ (d) $Mg(OH)_2$ (e) $K_2Cr_2O_7$ (f) NH_4I

◀ **37.** Write the names of the following ternary and higher compounds: (a) $CaSO_4$ (b) KCN (c) $AlPO_4$ (d) $Cu_2C_2O_4$ (e) $Fe_2(CrO_4)_3$ (f) $Cu(NO_2)_2$

◀ 38. Write the names of the following ternary and higher compounds: (a) Rb_2SO_4 (b) $Fe(C_2H_3O_2)_2$ (c) $Mg_3(BO_3)_2$ (d) $KMnO_4$ (e) $Bi_2(SO_4)_3$ (f) $(NH_4)_2C_2O_4$

LEARNING GOAL 11: *Formulas and Names of Inorganic Acids*

39. Write the formulas of the following inorganic acids:
(a) hypochlorous acid
(b) hydrobromic acid
(c) nitric acid
(d) chlorous acid
(e) perbromic acid
(f) sulfuric acid

40. Write the formulas of the following inorganic acids:
(a) acetic acid
(b) sulfurous acid
(c) chloric acid
(d) hydrofluoric acid
(e) phosphoric acid
(f) nitrous acid

◀ **41.** Name the following compounds as inorganic acids:
(a) HCN (b) H_2S (c) $HBrO_3$ (d) H_2SO_4 (e) $HClO$ (f) $HBrO_2$

◀ 42. Name the following compounds as inorganic acids:
(a) HIO_4 (b) $HClO_3$ (c) HBr (d) HF (e) $HClO_2$ (f) HIO

LEARNING GOAL 12: *Common Names and Formulas of Compounds*

43. Write the formula of each of the following compounds designated by their common names:
(a) milk of magnesia
(b) oil of vitriol
(c) saltpeter
(d) laughing gas

44. Write the formula of each of the following compounds designated by their common names:
(a) Epsom salts
(b) muriatic acid
(c) marble
(d) table salt

45. Write the common name of the compound designated by each formula: (a) CaO (b) $CaCO_3$ (c) $CaSO_4 \cdot 2H_2O$ (d) H_2SO_4

46. Write the systematic chemical name of each of the following substances: (a) dry ice (b) saltpeter (c) baking soda (d) borax

EXTRA EXERCISES

◀ **47.** Predict the compound that is *most likely* to be formed from each of the following pairs of elements: (a) Sr and O (b) Al and N (c) Rb and S

48. Write the formula of the compound formed from each of the following pairs of ions:
(a) Al^{3+} and N^{3-}
(b) V^{5+} and O^{2-}
(c) Fe^{3+} and $(OH)^{1-}$
(d) $(NH_4)^{1+}$ and S^{2-}

◀ **49.** Name each of the compounds in Exercise 48.

50. Write the formulas of the following compounds:
(a) tin(II) ion plus sulfide ion
(b) copper(I) ion plus phosphate ion
(c) iron(III) ion plus nitrite ion
(d) magnesium ion plus acetate ion

51. Complete the table following Exercise 53 by writing the correct formula of the compound formed by each metal ion in combination with each of the given nonmetal ions.

52. Write the correct name of each of the following compounds: (a) $KClO_2$ (b) $Fe(CN)_3$ (c) P_2S_5 (d) CCl_4 (e) $HC_2H_3O_2$ (in water) (f) SnO_2

53. Write the correct formula of each of the following compounds:
    (a) gallium fluoride
    (b) palladium(II) nitrate
    (c) gold(III) phosphide
    (d) lanthanum(III) acetate
    (e) plutonium(IV) oxide
    (f) ruthenium(VIII) oxide

	NONMETAL ION		
Metal ion	*Bromide*	*Sulfate*	*Phosphate*
Sodium	_____	_____	_____
Calcium	_____	_____	_____
Aluminum	_____	_____	_____

1. Write the electron-dot structure for (a) a calcium atom, (b) a calcium ion.

2. Write the electron-dot structure for (a) a chlorine atom, (b) a chloride ion.

3. Distinguish among an ionic bond, a covalent bond, and a coordinate covalent bond.

4. Draw the electron-dot diagram for C_3H_8.

◀ **5.** List the following compounds in order of decreasing covalence: HCl, $BaCl_2$, CaF_2, and $CaCl_2$.

6. Which of the following compounds is most ionic? LiF, NaF, RbF, or CsF

7. Write the chemical formula of each of the following compounds:
(a) copper(I) sulfite
(b) osmium tetroxide
(c) ammonium carbonate
(d) dinitrogen monoxide

◀ **8.** Find the oxidation number of each underlined element: (a) $\underline{W}Cl_5$ (b) $\underline{Sn}F_4$ (c) $\underline{V}_2O_5$ (d) $\underline{In}I_3$

9. Write the name of each of the following compounds: (a) NH_4I (b) CaC_2 (c) Cu_2CrO_4

◀ **10.** Determine the formula of a hypothetical compound that would be formed between elements X and Y if X has three electrons in the outer shell and Y has six electrons in the outer shell.

11. The electron-dot structure for carbon dioxide is

$$:\overset{\times}{\underset{\times}{\ddot{O}}} \overset{\times}{\underset{\times}{C}} \overset{\times}{\underset{\times}{\ddot{O}}}:$$

True or false?

12. An atom having the electron configuration $1s^2 2s^2 2p^6 3s^2$ would form an ion with a 2+ charge. True or false?

13. A Br^{1-} ion, an atom of Kr, and a Rb^{1+} ion have the same electron structure. True or false?

14. The bonds in a molecule of

are nonpolar. True or false?

15. A molecule of $CHCl_3$ is polar. True or false?

16. Write the chemical formula for the combination of atoms that have the following electron configuration:

$$1s^2 2s^2 2p^4 \quad \text{and} \quad 1s^2 2s^2 2p^6 3s^2$$

◀ **17.** Which of the following bonds is most covalent? C—H, C—Cl, or C—O

18. Write the formula of the compound formed from Na^{1+} and SO_4^{2-}.

◀ **19.** Predict the oxidation number that the element with atomic number 16 is most likely to have.

20. Draw the electron-dot diagram for each of the following: (a) NO_3^{1-} (b) NH_4^{1+}

◀ **21.** Calculate the oxidation number for the element indicated in the following compounds or ions: (a) S in $(SO_4)^{2-}$ (b) Mn in MnO_2 (c) Mn in $(MnO_4)^{-1}$ (d) N in HNO_2

22. Use the symbol δ^+ to show a partial positive charge and the symbol δ^- to show a partial negative charge for each of the atoms in the following compounds: (a) H_2O (b) HI (c) Cl_2O (d) PCl_5

23. Write the electron-dot structure for each of the following compounds: (a) F_2 (b) C_2Cl_2 (c) PCl_5 (d) C_2Cl_4

◀ **24.** State the common oxidation number for each of the following elements when they enter into chemical combination. Base your answer on the element's position in the periodic table. (a) cesium (b) astatine (c) barium (d) gallium

25. Write the formulas of the compounds formed by the following ions: (a) ferrous, Fe^{2+}, and phosphate, $(PO_4)^{3-}$ (b) mercuric, Hg^{2+}, and cyanide, $(CN)^{1-}$ (c) zinc, Zn^{2+}, and bicarbonate, $(HCO_3)^{1-}$ (d) ferric, Fe^{3+}, and dichromate, $(Cr_2O_7)^{2-}$

26. Write the formulas of the following compounds:
(a) indium sulfide
(b) magnesium arsenate
(c) gallium oxide
(d) rubidium selenide
(e) iron(II) oxide
(f) cobalt(II) chloride
(g) copper(I) nitride
(h) copper(I) oxide

27. Write the formulas of the following compounds:
  (a) periodic acid
  (b) chloric acid
  (c) hydrobromic acid
  (d) chlorous acid

◀ **28.** What is a coordinate covalent bond? Explain.

◀ **29.** Is it possible for nonpolar molecules to contain polar bonds? Explain.

◀ **30.** Element number 116 has not yet been discovered. But using what you know about periodic trends, answer the following questions:
  (a) In what family (group) would this element be placed?
  (b) How many electrons would be in its outermost energy level?
  (c) What would be its most common oxidation number?

CALCULATIONS INVOLVING CHEMICAL FORMULAS

Let's journey to the beginning of the nineteenth century. John Dalton had just published his atomic theory. A belief that matter was made up of discrete particles called atoms was beginning to gain favor among many chemists, believers of Dalton's theory, who were called atomists. However, there were also many respected chemists at the time who did not believe in the existence of atoms. Both the atomists and non-atomists continued to debate this issue as new chemical research was published.

It was about this time (1808) that the French chemist Joseph Gay-Lussac performed experiments that indicated an amazing regularity in the way that (elemental) gases combined to form gaseous products. In one of these experiments, Gay-Lussac discovered that the volume ratio of hydrogen gas to oxygen gas that was required to form water (in the form of steam) was 1.9989 to 1.0000. Because this result was very close to a simple 2-to-1 ratio, Gay-Lussac attributed the slight difference in his experimental value to experimental error. Gay-Lussac discovered that simple numerical ratios held between reactants and products, as long as both gases were measured under identical conditions of temperature and pressure. For example,

2 volumes of hydrogen gas + 1 volume of oxygen gas →

2 volumes of steam

or 1 volume of nitrogen gas + 3 volumes of hydrogen gas →

2 volumes of ammonia gas

These experiments gave rise to the *Law of Combining Volumes*. This law stated that the volumes of gases in chemical reactions are related by simple whole numbers. To the modern-day chemist these experiments would be viewed as adding evidence to the existence of atoms, but to Dalton these observations appeared to be in conflict with his atomic theory. Dalton held that particles in gaseous elements must be atoms. Therefore, according to Dalton 1 atom of hydrogen should unite with 1 atom of oxygen to form 1 molecule of steam.

$$H + O \rightarrow HO$$

The Law of Combining Volumes appeared to be in conflict with atomic theory, yet there was much evidence to support both. Chemists of the day were asking how both Dalton's atomic theory and Gay-Lussac's Law of Combining Volumes could be valid. It was the Italian chemist Amedeo Avogadro in 1811 who resolved the apparent conflict when he presented his two-part hypothesis:

1. Equal volumes of gases under the same conditions of temperature and pressure contain an equal number of particles.
2. The ultimate *physical* units of elemental substances may be different from their ultimate *chemical* units.

Avogadro's first assumption, although important to resolving the conflict between Dalton and Gay-Lussac, had been stated by others. His second assumption, however, was original and showed miraculous insight into what was happening on a molecular level. Avogadro proposed that the ultimate physical units of both hydrogen and oxygen were not *atoms* but *molecules,* each composed of two atoms. According to Avogadro's hypothesis:

$$2H_2 + 1O_2 \rightarrow 2H_2O$$

where H_2 represents a molecule of hydrogen gas, and O_2 represents a molecule of oxygen gas. Therefore, 2 molecules of hydrogen gas plus 1 molecule of oxygen gas produces 2 molecules of water vapor (steam).

Avogadro's insight into the formulas of gases coupled with what was occurring on the molecular level made a significant contribution both to the understanding of atomic theory and to the advancement of modern-day chemistry.

LEARNING GOALS

After you've studied this chapter, you should be able to:
1. Calculate the number of moles in a sample of an element when you are given the mass of the sample.
2. Calculate the mass, in grams, of a sample of an element when you are given the number of moles.

3. Calculate the number of atoms of an element in a sample when you are given the mass of the sample.

4. Calculate the empirical formula of a compound when you are given its percentage composition.

5. Calculate the number of moles in a sample of a compound when you are given the mass of the sample.

6. Calculate the mass, in grams, of a sample of a compound when you are given the number of moles.

7. Calculate the number of formula units of a compound in a sample when you are given the mass of the sample.

8. Determine the molecular formula of a compound from its molecular mass and its empirical formula or percentage composition data.

9. Calculate the percentage composition by mass of a compound when you are given its chemical formula.

Amedeo Avogadro
(The Granger Collection, New York)

▐▐▐ INTRODUCTION

We began our study of chemistry by reviewing the basic structure of matter. We learned about elements, compounds, atoms, and molecules. We discussed the basic structure of the atom and the concept of chemical bonding. We also learned how to name and write the formulas of chemical compounds. We did all of this in a qualitative fashion and didn't get very involved with quantitative calculations. In this chapter, we will revisit the topic of elements and compounds. However, this time we will discuss some quantitative aspects involving these substances.

We will begin our study of calculations involving elements and compounds by learning about the *mole,* a useful concept for expressing the amount of a chemical substance. Before proceeding with this chapter, you may find it helpful to review the discussion of atomic mass and formula mass in Chapter 3. Also, because we will use the factor-unit method in our calculations, you may find it helpful to review the discussion of the factor-unit method in Chapter 2.

9.1 Gram-Atomic Mass and the Mole

The **gram-atomic mass** of an element is *its atomic mass expressed in grams.* For example, the atomic mass of gold is 197.0 amu, so gold's gram-atomic mass is 197.0 g. And the gram-atomic mass of carbon is 12.0 g.

One gram-atomic mass of any element contains the same number of atoms as one gram-atomic mass of any other element. For example, 197.0

1 mole of silver (Ag) atoms has a mass of 107.9 g.

1 mole of lead (Pb) atoms has a mass of 207.2 g.

1 mole of neon (Ne) atoms has a mass of 20.2 g.

■ **FIGURE 9.1**

1 mole of atoms of some common substances

g of gold contain the same number of atoms as 12.0 g of carbon and the same number of atoms as 1.0 g of hydrogen. In about 1870, scientists discovered that this number is 6.02×10^{23}. It is called **Avogadro's number,** in honor of the scientist whose thinking led to its discovery.

Later scientists took this idea one step further and defined 6.02×10^{23} atoms as 1 **mole** of atoms. Thus 1 mole is equal to 6.02×10^{23}; it is simply a particular number of items. We could just as easily talk about 1 mole of people, 1 mole of apples, or 1 mole of dollars. (By the way, if we had 1 mole of dollar bills, we would have \$602,000,000,000,000,000,000,000. If we distributed our money equally among the 5 billion people on the earth, each person would get more than 120 trillion dollars:

$$\frac{\$602{,}000{,}000{,}000{,}000{,}000{,}000{,}000}{5{,}000{,}000{,}000} = 1.20 \times 10^{14}$$

A person could spend a million dollars every day, each day of the year, and never run out of money for 330,000 years.)

A mole of hydrogen atoms weighs 1.0 g; a mole of oxygen atoms weighs 16.0 g; a mole of carbon atoms weighs 12.0 g; and a mole of uranium atoms weighs 238.0 g. (To simplify the mathematics, we are rounding off the atomic masses found in the periodic table to one decimal place. We will do this throughout the text.) Figure 9.1 shows the mass of a mole of some other kinds of atoms.

The mole is very important in chemistry. It gives the chemist a convenient way to describe a large number of atoms or molecules and to relate this number of atoms or molecules to a mass (usually in grams). In fact, the mole has replaced the gram-atomic mass, which is seldom used today. But you do have to be able to convert from moles of atoms of an element to mass in grams, and vice versa. This is like converting inches to feet and feet to inches.

LEARNING GOAL 1

Moles from mass of an element

EXAMPLE 9.1

Do the following conversions: (a) $6\overline{0}$ in = ? ft   (b) 8.0 ft = ? in

Solution

(a) There are 12 in in 1 ft, so

$$? \text{ ft} = 6\overline{0} \text{ in} \times \frac{1 \text{ ft}}{12 \text{ in}} = 5.0 \text{ ft}$$

(b) In the same way,

$$? \text{ in} = 8.0 \text{ ft} \times \frac{12 \text{ in}}{1 \text{ ft}} = 96 \text{ in}$$

Practice Exercise 9.1   Do the following conversions:
(a) 72.0 in = ? ft   (b) 10.0 ft = ? in

EXAMPLE 9.2

(a) How many moles of oxygen atoms are there in $8\overline{0}$ g of oxygen?
(b) What is the mass in grams of 0.50 mole of oxygen atoms?

Solution

UNDERSTAND THE PROBLEM

(a) In this example, a specific mass of oxygen atoms is given. This mass must be converted into moles.

DEVISE A PLAN

We use the factor-unit method as described in Chapter 2. (For convenience, we will round all atomic masses to one decimal place.)

CARRY OUT THE PLAN

$$\text{Moles of O atoms} = \text{g} \times \frac{\text{moles of O atoms}}{\text{g}}$$

Then, because the atomic mass of oxygen is 16.0 amu,

$$? \text{ moles of O atoms} = 8\overline{0} \text{ g} \times \frac{1 \text{ mole of O atoms}}{16.0 \text{ g}}$$

$$= 5.0 \text{ moles of O atoms}$$

We find that $8\overline{0}$ g of oxygen atoms equals 5.0 moles.

LOOK BACK

We reason that if 1 mole of oxygen atoms contains 16.0 g, then 5 moles should have five times as many grams. Since 16.0 × 5.0 equals $8\overline{0}$, the answer does make sense.

UNDERSTAND THE PROBLEM

(b) In this case, the number of moles of oxygen is given and it must be converted into a mass (in grams).

DEVISE A PLAN

We use the factor-unit method.

CARRY OUT THE PLAN

Using the atomic mass of oxygen, we write

$$g = \text{moles of O atoms} \times \frac{g}{\text{moles of O atoms}}$$

$$? g = 0.50 \text{ mole of O atoms} \times \frac{16.0 \text{ g}}{1 \text{ mole of O atoms}}$$

$$= 8.0 \text{ g}$$

LOOK BACK

We know that 1 mole of oxygen has a mass of 16.0 g; therefore 0.50 mole of oxygen should have half that mass, or 8.0 g. We conclude that our answer is sensible.

Practice Exercise 9.2   (a) How many moles of oxygen atoms are there in $16\bar{0}$ g of oxygen?   (b) What is the mass in grams of 2.00 moles of oxygen atoms?

EXAMPLE 9.3

Find the number of moles of atoms in each of the following samples: (a) 46 g of Na   (b) 5.4 g of Al   (c) 0.12 g of C   (d) 23.8 g of U

Solution

We need to look up the atomic mass of each element in the periodic table inside the front cover.

(a) The atomic mass of Na is 23.0 amu. Therefore

$$? \text{ moles of Na atoms} = 46 \text{ g} \times \frac{1 \text{ mole of Na atoms}}{23.0 \text{ g}}$$

$$= 2.0 \text{ moles of Na atoms}$$

(b) The atomic mass of Al is 27.0 amu. Therefore

$$? \text{ moles of Al atoms} = 5.4 \text{ g} \times \frac{1 \text{ mole of Al atoms}}{27.0 \text{ g}}$$

$$= 0.20 \text{ mole of Al atoms}$$

(c) The atomic mass of C is 12.0 amu. Therefore

$$? \text{ moles of C atoms} = 0.12 \text{ g} \times \frac{1 \text{ mole of C atoms}}{12.0 \text{ g}}$$

$$= 0.010 \text{ mole of C atoms}$$

(d) The atomic mass of U is 238.0 amu. Therefore

$$? \text{ moles of U atoms} = 23.8 \text{ g} \times \frac{1 \text{ mole of U atoms}}{238.0 \text{ g}}$$

$$= 0.100 \text{ mole of U atoms}$$

Practice Exercise 9.3   Find the number of moles of atoms in each of the following samples: (a) 11.50 g of Na   (b) 2.70 g of Al   (c) $28\overline{0}$ g of N   (d) 0.0238 g of U

EXAMPLE 9.4

Find the mass in grams of each of the following: (a) 0.20 mole of Zn atoms   (b) 4.0 moles of Br atoms   (c) 1.50 moles of Ca atoms

Solution
We need to look up the atomic mass of each element in the periodic table.
(a) The atomic mass of Zn is 65.4 amu. Therefore

LEARNING GOAL 2

Mass, in grams, from moles of an element

$$? \text{ g} = 0.20 \text{ mole of Zn atoms} \times \frac{65.4 \text{ g}}{1 \text{ mole of Zn atoms}} = 13 \text{ g}$$

(b) The atomic mass of Br is 79.9 amu. Therefore

$$? \text{ g} = 4.0 \text{ moles of Br atoms} \times \frac{79.9 \text{ g}}{1 \text{ mole of Br atoms}} = 320 \text{ g}$$

(c) The atomic mass of Ca is 40.1 amu. Therefore

$$? \text{ g} = 1.50 \text{ moles of Ca atoms} \times \frac{40.1 \text{ g}}{1 \text{ mole of Ca atoms}} = 60.2 \text{ g}$$

Practice Exercise 9.4   Find the mass in grams of each of the following: (a) 15.0 moles of Zn atoms   (b) 0.200 mole of Br atoms   (c) 3.00 moles of Ca atoms

---

EXAMPLE 9.5

How many *atoms* of each element are there in the samples given in Example 9.4?

LEARNING GOAL 3

Number of atoms of an element from mass of a sample

Solution
We know that 1 mole of atoms of any element is 6.02×10^{23} atoms of that element; in other words, there are

$$\frac{6.02 \times 10^{23} \text{ atoms}}{1 \text{ mole of atoms}}$$

for any element. But for this example, we will round off Avogadro's number to 6.0×10^{23}.

(a) For the zinc,

$$? \text{ Zn atoms} = 0.20 \text{ ~~mole of Zn atoms~~} \times \frac{6.0 \times 10^{23} \text{ Zn atoms}}{1 \text{ ~~mole of Zn atoms~~}}$$

$$= 1.2 \times 10^{23} \text{ Zn atoms}$$

(b) For the bromine,

$$? \text{ Br atoms} = 4.0 \text{ ~~moles of Br atoms~~} \times \frac{6.0 \times 10^{23} \text{ Br atoms}}{1 \text{ ~~mole of Br atoms~~}}$$

$$= 24 \times 10^{23} \text{ Br atoms (or } 2.4 \times 10^{24})$$

(c) For the calcium,

$$? \text{ Ca atoms} = 1.50 \text{ ~~moles of Ca atoms~~} \times \frac{6.0 \times 10^{23} \text{ Ca atoms}}{1 \text{ ~~mole of Ca atoms~~}}$$

$$= 9.0 \times 10^{23} \text{ Ca atoms}$$

Practice Exercise 9.5   How many *atoms* of each element are there in the samples given in Practice Exercise 9.4?

9.2   Empirical Formulas

Chemists prepare hundreds of new compounds in their search for substances that may be beneficial in medicine, agriculture, industry, and the home. A first step in determining the nature of such a new or unknown compound is to obtain its empirical formula. The **empirical formula** of a compound is *the simplest whole-number ratio of the atoms that make up a formula unit of the compound.* For example, the empirical formula of water is H_2O. The subscripts indicate that the ratio of hydrogen atoms to oxygen atoms in this molecule is 2 to 1, often written $2:1$. (The lack of a subscript on the O is taken to mean a subscript of 1.)

To find the empirical formula of a compound, the chemist measures the percentage by mass of each element in the compound. From the percentage of each element, the chemist then determines (1) the number of moles of atoms of each element in 100 g of the compound and (2) the ratio of the moles of atoms. (We take 100 g of compound merely as a convenience.)

EXAMPLE 9.6

Determine the empirical formula of a compound whose composition is 50.05% S and 49.95% O by mass.

Solution

If we had 100 g of the compound, 50.05 g would be sulfur and 49.95 g would be oxygen. All we have to do now is convert these masses to moles of atoms and then find their whole-number ratio.

$$? \text{ moles of S atoms} = 50.05 \text{ g} \times \frac{1 \text{ mole of S atoms}}{32.1 \text{ g}}$$

$$= 1.56 \text{ moles of S atoms}$$

$$? \text{ moles of O atoms} = 49.95 \text{ g} \times \frac{1 \text{ mole of O atoms}}{16.0 \text{ g}}$$

$$= 3.12 \text{ moles of O atoms}$$

Therefore the formula may be written as $S_{1.56}O_{3.12}$, but this formula does not have whole-number subscripts. One way to obtain a formula with whole-number subscripts is to *divide* all the subscripts by the *smallest* subscript. This gives us

$$S_{1.56/1.56}O_{3.12/1.56} \quad \text{or} \quad SO_2$$

This is an empirical formula, because it has the lowest possible ratio of whole-number subscripts. (The compound is called sulfur dioxide.)

Practice Exercise 9.6   Determine the empirical formula of a compound whose composition is 88.9% oxygen and 11.1% hydrogen by mass.

EXAMPLE 9.7

Find the empirical formula of a compound whose composition is 3.1% H, 31.5% P, and 65.4% O by mass.

Solution

In 100 g of this compound, there are 3.1 g of H, 31.5 g of P, and 65.4 g of O. Therefore there are

$$? \text{ moles of H atoms} = 3.1 \text{ g} \times \frac{1 \text{ mole of H atoms}}{1.0 \text{ g}}$$

$$= 3.1 \text{ moles of H atoms}$$

$$? \text{ moles of P atoms} = 31.5 \text{ g} \times \frac{1 \text{ mole of P atoms}}{31.0 \text{ g}}$$

$$= 1.02 \text{ moles of P atoms}$$

$$? \text{ moles of O atoms} = 65.4 \text{ g} \times \frac{1 \text{ mole of O atoms}}{16.0 \text{ g}}$$

$$= 4.09 \text{ moles of O atoms}$$

This formula may be written $H_{3.1}P_{1.02}O_{4.09}$, but this formula does not have whole-number subscripts. Therefore we divide each subscript by the smallest subscript to obtain

$$H_{3.1/1.02}P_{1.02/1.02}O_{4.09/1.02} \quad \text{or} \quad H_3PO_4$$

Practice Exercise 9.7   Find the empirical formula of a compound whose composition is 51.9% Cr and 48.1% S by mass.

EXAMPLE 9.8

Determine the empirical formula of a compound whose composition is 23.8% C, 5.9% H, and 70.3% Cl by mass.

Solution

In 100 g of this compound, there are 23.8 g of C, 5.9 g of H, and 70.3 g of Cl. Therefore there are

$$? \text{ moles of C atoms} = 23.8 \text{ g} \times \frac{1 \text{ mole of C atoms}}{12.0 \text{ g}}$$

$$= 1.98 \text{ moles of C atoms}$$

$$? \text{ moles of H atoms} = 5.9 \text{ g} \times \frac{1 \text{ mole of H atoms}}{1.0 \text{ g}}$$

$$= 5.9 \text{ moles of H atoms}$$

$$? \text{ moles of Cl atoms} = 70.3 \text{ g} \times \frac{1 \text{ mole of Cl atoms}}{35.5 \text{ g}}$$

$$= 1.98 \text{ moles of Cl atoms}$$

One possible formula is $C_{1.98}H_{5.9}Cl_{1.98}$, but this does not have whole-number subscripts. Again we divide by the smallest subscript to obtain

$$C_{1.98/1.98}H_{5.9/1.98}Cl_{1.98/1.98} \qquad \text{or} \qquad CH_3Cl$$

Practice Exercise 9.8   Determine the empirical formula of a compound whose composition is 41.5% Zn, 17.8% N, and 40.7% O by mass.

After you divide the subscripts in determining an empirical formula, you may round subscripts to the nearest whole number if they are within 0.1 of the whole number. Otherwise you must search for a factor that will give you whole numbers. For example, suppose a compound contains 68.4% Cr and 31.6% O by mass. This means that 100 g of the compound contains 68.4 g (or 1.32 moles) of Cr and 31.6 g (or 1.98 moles) of O. Therefore the formula is

$$Cr_{1.32}O_{1.98}$$

Following our usual procedure, we divide by the smallest number of moles (the smallest subscript) and obtain

$$Cr_{1.32/1.32}O_{1.98/1.32} \qquad \text{or} \qquad CrO_{1.5}$$

The formula $CrO_{1.5}$ does not have *whole*-number subscripts. What do we do now? Should we simply round off the 1.5 to the number 2 and make the empirical formula CrO_2? The answer is NO! We may round off only when the number is within 0.1 of the whole number. Therefore we must search for a factor that will give us a whole-number ratio. When we look carefully at $Cr_1O_{1.5}$, we see that if we multiply each subscript by 2, we can get whole-number subscripts. $Cr_1O_{1.5}$ becomes Cr_2O_3, and the empirical formula is Cr_2O_3.

9.3   Gram-Formula Mass and the Mole

We're now going to learn how to apply the concept of the *mole* to compounds. Because some compounds are composed of molecules and other compounds are composed of ions, we'll use the term *formula mass* to help us with this application. Recall from Chapter 3 that the formula mass of a compound is *the sum of the atomic masses of all the atoms or ions that make up a formula unit of the compound.* The formula mass of a compound in grams is called the **gram-formula mass.*** This is the mass of a collection of 6.02×10^{23} formula units of the compound. Therefore *1 gram-formula mass of a compound is the mass in grams of 1 mole of molecules for a molecular compound or 1 mole of formula units for an ionic compound.*

Consider some examples. Water (H_2O) is a compound composed of molecules, each containing two hydrogen atoms and one oxygen atom. The formula mass of water is 18.0 amu—two times the atomic mass of hydrogen (2×1.0) plus the atomic mass of oxygen (16.0). So the mass of 1 mole of water molecules is 18.0 g (Figure 9.2). Calcium chloride ($CaCl_2$) is a compound

1 mole of water (H_2O) molecules has a mass of 18.0 g.

1 mole of glucose sugar ($C_6H_{12}O_6$) molecules has a mass of 180.0 g.

1 mole of TNT ($C_7H_5N_3O_6$) molecules has a mass of 227.0 g and can make a lot of noise.

1 *Scalopus aquaticus*— alias 1 mole—has varying mass and can make very little noise.

■ **FIGURE 9.2**

One mole of molecules of some common substances

*The term **molar mass** is used in some texts as a general term to describe gram-formula mass and gram-atomic mass of a substance. In this text, we will continue to use the more traditional terms, *gram-formula mass* and *gram-atomic mass.*

composed of ions; the formula unit of the compound consists of one calcium ion and two chloride ions. The formula mass of calcium chloride is 111.1 amu—the atomic mass of calcium (40.1) plus two times the atomic mass of chlorine (2×35.5). So the mass of 1 mole of calcium chloride formula units is 111.1 g.

It is important to be able to convert the number of grams of a compound into the number of moles, and vice versa. You will need to perform these calculations when we study the topic of chemical stoichiometry in Chapter 11. The following examples will give you some practice.

EXAMPLE 9.9

Find the number of moles in each of the following: (a) 32 g of CH_4
(b) 0.32 g of CH_4 (c) 81 g of H_2O (d) 37 g of $Ca(OH)_2$

Solution
We first find the formula mass of each compound and then use the factor-units method to determine the number of moles.
(a) The formula mass of CH_4 is 16.0 amu. Therefore we have

$$? \text{ moles of } CH_4 \text{ molecules} = 32 \text{ g} \times \frac{1 \text{ mole of } CH_4 \text{ molecules}}{16.0 \text{ g}}$$

$$= 2.0 \text{ moles of } CH_4 \text{ molecules}$$

(b) The formula mass of CH_4 is 16.0 amu. Therefore we have

$$? \text{ moles of } CH_4 \text{ molecules} = 0.32 \text{ g} \times \frac{1 \text{ mole of } CH_4 \text{ molecules}}{16.0 \text{ g}}$$

$$= 0.020 \text{ mole of } CH_4 \text{ molecules}$$

(c) The formula mass of H_2O is 18.0 amu. Therefore

$$? \text{ moles of } H_2O \text{ molecules} = 81 \text{ g} \times \frac{1 \text{ mole of } H_2O \text{ molecules}}{18.0 \text{ g}}$$

$$= 4.5 \text{ moles of } H_2O \text{ molecules}$$

(d) The formula mass of $Ca(OH)_2$ is 74.1 amu. Therefore

$? \text{ moles of } Ca(OH)_2 \text{ formula units}$

$$= 37 \text{ g} \times \frac{1 \text{ mole of } Ca(OH)_2 \text{ formula units}}{74.1 \text{ g}}$$

$$= 0.50 \text{ mole of } Ca(OH)_2 \text{ formula units}$$

Practice Exercise 9.9   Find the number of moles in each of the following: (a) $9\overline{0}$ g of H_2O (b) 0.016 g CH_4

EXAMPLE 9.10

■ LEARNING GOAL 6

Mass, in grams, from moles of a compound

Determine the number of grams in each of the following: (a) 6.00 moles of butane (C_4H_{10}) (b) 0.025 mole of CO_2 (c) 7.00 moles of Al_2O_3 (d) 0.400 mole of Cu_3N

Solution

We first find the formula mass of each compound and then use the factor-unit method to determine the number of grams.

(a) The formula mass of C_4H_{10} is 58.0 amu. Therefore we have

$$? \text{ g } C_4H_{10} = 6.00 \text{ moles} \times \frac{58.0 \text{ g}}{1 \text{ mole}} = 348 \text{ g}$$

(b) The formula mass of CO_2 is 44.0 amu. Therefore we have

$$? \text{ g } CO_2 = 0.025 \text{ mole} \times \frac{44.0 \text{ g}}{1 \text{ mole}} = 1.1 \text{ g}$$

(c) The formula mass of Al_2O_3 is 102.0 amu. Therefore we have

$$? \text{ g } Al_2O_3 = 7.00 \text{ moles} \times \frac{102.0 \text{ g}}{1 \text{ mole}} = 714 \text{ g}$$

(d) The formula mass of Cu_3N is 204.5 amu. Therefore we have

$$? \text{ g } Cu_3N = 0.400 \text{ mole} \times \frac{204.5 \text{ g}}{1 \text{ mole}} = 81.8 \text{ g}$$

Practice Exercise 9.10   Determine the number of grams in each of the following: (a) 4.00 moles of NO_2 (b) 0.050 mole of $CaCO_3$ (c) 2.50 moles of $Ca(C_2H_3O_2)_2$ (d) 0.060 mole of $(NH_4)_2SO_4$

EXAMPLE 9.11

In Example 9.9, how many formula units are present in each sample?

Solution

To obtain our factor unit, we recall that there are 6.0×10^{23} formula units in 1 mole of a compound.

(a) For the first sample of CH_4, composed of molecules,

■ LEARNING GOAL 7

Number of formula units of a compound from mass of a sample

$$? \text{ molecules of } CH_4 = 2.0 \text{ moles} \times \frac{6.0 \times 10^{23} \text{ molecules of } CH_4}{1 \text{ mole}}$$

$$= 12 \times 10^{23} \text{ molecules of } CH_4$$

(b) For the second sample of CH_4, composed of molecules,

$$? \text{ molecules of } CH_4 = 0.020 \text{ mole} \times \frac{6.0 \times 10^{23} \text{ molecules of } CH_4}{1 \text{ mole}}$$

$$= 1.2 \times 10^{22} \text{ molecules of } CH_4$$

(c) For the H_2O, composed of molecules,

$$? \text{ molecules of } H_2O = 4.5 \text{ moles } \times \frac{6.0 \times 10^{23} \text{ molecules of } H_1O}{1 \text{ mole}}$$

$$= 27 \times 10^{23} \text{ molecules of } H_2O$$

(d) For the $Ca(OH)_2$, composed of ions,

? formula units of $Ca(OH)_2$

$$= 0.50 \text{ mole } \times \frac{6.0 \times 10^{23} \text{ formula units of } Ca(OH)_2}{1 \text{ mole}}$$

$$= 3.0 \times 10^{23} \text{ formula units of } Ca(OH)_2$$

Practice Exercise 9.11   In Practice Exercise 9.10, how many *formula units* are present in each sample?

9.4 Molecular Formulas

LEARNING GOAL 8

Molecular formula of a compound

The **molecular formula** of a compound is *a formula that shows the actual number of atoms of each element that are in one molecule of that compound.* The molecular formula is found from the percentages by mass of the elements *and* the molecular mass of the compound.

EXAMPLE 9.12

A compound is composed of $4\bar{0}\%$ C, 6.6% H, and 53.4% O by mass. The molecular mass of the compound is 180.0. Determine its empirical and molecular formulas.

Solution
Find the empirical formula first, by the usual method.

$$? \text{ moles of C atoms} = 4\bar{0} \text{ g } \times \frac{1 \text{ mole of C atoms}}{12.0 \text{ g}}$$

$$= 3.3 \text{ moles of C atoms}$$

$$? \text{ moles of H atoms} = 6.6 \text{ g } \times \frac{1 \text{ mole of H atoms}}{1.0 \text{ g}}$$

$$= 6.6 \text{ moles of H atoms}$$

$$? \text{ moles of O atoms} = 53.4 \text{ g } \times \frac{1 \text{ mole of O atoms}}{16.0 \text{ g}}$$

$$= 3.3 \text{ moles of O atoms}$$

Hence one possible formula is $C_{3.3}H_{6.6}O_{3.3}$, but this does not have whole-number subscripts. Therefore we divide by the smallest number of moles to get

$$C_{3.3/3.3}H_{6.6/3.3}O_{3.3/3.3} \quad \text{or} \quad CH_2O$$

This is the empirical formula. But it certainly is not the molecular formula of the compound, because CH_2O does not have a molecular mass of 180.0. What is the molecular formula of the compound? It could be any formula wherein the ratio of $C:H:O$ is $1:2:1$. Is it $C_2H_4O_2$, $C_3H_6O_3$, $C_4H_8O_4$, $C_5H_{10}O_5$, $C_6H_{12}O_6$, or what? Remember, the correct formula is the one that has a molecular mass of 180.0. Let's determine the molecular mass of each proposed formula.

MOLECULAR FORMULA	MOLECULAR MASS (AMU)
CH_2O	30.0
$C_2H_4O_2$	60.0
$C_3H_6O_3$	90.0
$C_4H_8O_4$	120.0
$C_5H_{10}O_5$	150.0
$C_6H_{12}O_6$	180.0

It is clear from our trial-and-error method that the molecular formula is $C_6H_{12}O_6$. It's the only one that has a molecular mass of 180.0. But is there an easier way to do this? Yes. We simply ask ourselves, "How many times must we take the molecular mass of the empirical formula (30.0 in this problem) to get the true molecular mass (180.0 in this problem)?" The answer is obviously *six* times. It takes six times the mass of the empirical formula, CH_2O, to give the true molecular mass, so the true molecular formula must be $C_6H_{12}O_6$.

Practice Exercise 9.12   A compound has a molecular mass of 180.0 and is composed of 60.0% C, 4.48% H, and 35.5% O by mass. Calculate the empirical and molecular formulas of this common compound, which is called aspirin.

9.5 Percentage Composition by Mass

LEARNING GOAL 9

Percentage composition from chemical formula

You have seen how to determine the empirical formula of a compound from its percentage composition. There will also be times when you will have to do the opposite—find the percentage by mass of each element in a compound. To do so, you need to know the empirical or molecular formula of the compound and the atomic mass of its constituent elements.

EXAMPLE 9.13

Find the percentage by mass of H and O in water.

Solution

In each mole of water (H_2O), there are 2 moles of H, having a mass of 1.0 g/mole, and 1 mole of O, having a mass of 16.0 g/mole. Therefore

$$\text{Percent H} = \frac{\text{mass of hydrogen in 1 mole of water}}{\text{mass of 1 mole of water}} \times 100$$

$$= \frac{2.0 \text{ g of H}}{18.0 \text{ g of } H_2O} \times 100 = 11\%$$

$$\text{Percent O} = \frac{\text{mass of oxygen in 1 mole of water}}{\text{mass of 1 mole of water}} \times 100$$

$$= \frac{16.0 \text{ g of O}}{18.0 \text{ g of } H_2O} \times 100 = 89\%$$

Note that the sum of the percentages must be 100.

Practice Exercise 9.13   Find the percentage by mass of K and Br in potassium bromide (KBr).

EXAMPLE 9.14

Find the percentage by mass of each element in the following compounds: (a) C_8H_{18} (b) NH_3 (c) FeO   (d) Fe_2O_3

Solution

We must find the formula mass of each compound and the percent contribution of each element to that mass.
(a) C_8H_{18} contains 96.0 g of C + 18.0 g of H = 114.0 g in each mole.
    Therefore

$$\text{Percent C} = \frac{96.0 \text{ g of C}}{114.0 \text{ g of } C_8H_{18}} \times 100 = 84.2\%$$

$$\text{Percent H} = \frac{18.0 \text{ g of H}}{114.0 \text{ g of } C_8H_{18}} \times 100 = 15.8\%$$

(b) NH_3 contains 14.0 g of N + 3.0 g of H = 17.0 g in each mole.
    Therefore

$$\text{Percent N} = \frac{14.0 \text{ g of N}}{17.0 \text{ g of } NH_3} \times 100 = 82.4\%$$

$$\text{Percent H} = \frac{3.0 \text{ g of H}}{17.0 \text{ g of } NH_3} \times 100 = 17.6\% \text{ (round to 18\%)}$$

(c) FeO contains 55.8 g of Fe + 16.0 g of O = 71.8 g in each mole. Therefore

$$\text{Percent Fe} = \frac{55.8 \text{ g of Fe}}{71.8 \text{ g of FeO}} \times 100 = 77.7\%$$

$$\text{Percent O} = \frac{16.0 \text{ g of O}}{71.8 \text{ g of FeO}} \times 100 = 22.3\%$$

(d) Fe_2O_3 contains 111.6 g of Fe + 48.0 g of O = 159.6 g in each mole. Therefore

$$\text{Percent Fe} = \frac{111.6 \text{ g of Fe}}{159.6 \text{ g of Fe}_2\text{O}_3} \times 100 = 69.92\%$$

$$\text{Percent O} = \frac{48.0 \text{ g of O}}{159.6 \text{ g of Fe}_2\text{O}_3} \times 100 = 30.1\%$$

Practice Exercise 9.14 Find the percentage by mass of each element in the following compounds: (a) $C_{22}H_{44}$ (b) NaCl (c) C_6H_6

EXAMPLE 9.15

How many grams of sulfur are there in 256 g of SO_3?

Solution

Determine the percentage of sulfur in SO_3 and multiply this by the amount of the sample (256 g). SO_3 contains 32.1 g of S + 48.0 g of O = 80.1 g/mole. Therefore

$$\text{Percent S} = \frac{32.1 \text{ g of S}}{80.1 \text{ g of SO}_3} \times 100 = 40.1\%$$

So 40.1% of the 256 g of SO_3 is sulfur (S); there are 0.401 × 256 g = 102.7 g of sulfur (or 103 g of sulfur to three significant figures).

Practice Exercise 9.15 How many grams of carbon are there in $\overline{100}$ g of CH_4?

▌▌▌ SUMMARY

Each element has its own symbol, and every compound has its own chemical formula. Each element has a unique atomic mass. One mole of anything is equal to 6.02 × 10²³ things (Avogadro's number of things): 1 mole of an element contains 6.02 × 10²³ atoms of that element, and 1 mole of a compound contains 6.02 × 10²³ formula units of that compound.

One mole of atoms of an element has a mass equal to the atomic mass of that element in grams. One mole of formula units of a compound has a mass equal to the formula mass of that compound in grams.

The empirical formula of a compound is the simplest whole-number ratio of the atoms that make up a formula unit of the compound. The molecular formula of a compound is the actual number of atoms of each element that are in one molecule (or formula unit) of the compound. The empirical formula of a compound can be found from percentage composition data. The molecular formula of a compound can be found from its empirical formula and its molecular mass. The percentage composition of a compound may be determined from its empirical or molecular formula.

▍▍▍ KEY TERMS

Avogadro's number **(9.1)**

empirical formula **(9.2)**

gram-atomic mass **(9.1)**

gram-formula mass **(9.3)**

molar mass **(9.3)**

mole **(9.1)**

molecular formula **(9.4)**

CAREER SKETCH

Computer Careers

Over the past few decades, computers have become an increasingly important aspect of chemistry. While knowledge of computers is important for all chemists, many have made careers out of specializing in the relationship between chemistry and computers.

Molecular chemistry, for example, relies heavily on the use of computers and on those familiar with the use of computers. Computer programs are able to develop models of different molecules, allowing chemists to determine how to make new molecules or what molecules to make. Chemists can also use computer models to predict the properties of a given molecule, or even compare different molecules.

Chemists are often called on to design new software. This could involve programming algorithms, acquiring data, or analyzing the results of an experiment. Another software application could involve the development of data management or data analysis programs. Chemists in the computer field can also be called upon when one is preparing to publish findings, for help designing reports that are clear and accurate.

Computer technology helps chemists design and test the properties of new molecules before they are synthesized in the laboratory.
(Lawrence Migdale)

In the lab, computer skills are also in high demand. A chemist could be called upon to develop a program in a highly technical computer language. Or a computer expert could be asked to solve a statistics problem that a chemist encounters. Knowledge of the major computer operating systems, as well as of the Internet, is often needed by chemists.

There are computer jobs in chemistry for people with varying educational backgrounds. Opportunities exist for B.S., M.S., and Ph.D. chemists. Besides a solid chemistry background, strong computer and mathematics skills are necessary for jobs in this field. Problem-solving skills, good oral and written communication skills, and an orientation toward customer service are all important to bridge the gap between chemistry and computers.

CHEMICAL FRONTIERS

*TV from Your Wrist to the Wall**

The technology that makes the digital numbers appear on your wristwatch may someday turn a wall in your home into a giant television set.

Chemist Stephen Moratti and physicist Neil Greenham have produced a light-generating polymer with high enough efficiency to be useful as a light-emitting diode (LED). These new plastic light sources apparently work as well as the LEDs currently in use in the displays of the electronic equipment that has become part of our everyday life—from wrist-watches to control panels in automobiles and computers.

Here's how a typical LED works: Positive and negative charges are emitted by electrodes on opposite sides of an inorganic material. Where the charges meet, a photon is emitted in the visible wavelength. If you put a variety of these photons together, you can create a word, picture, or whatever visual image you need on your LED screen.

In order for this technology to be useful for bigger products (like a wall-sized television, for example), the process needs to be more efficient than in traditional LEDs. Moratti and Greenham were able to do this by sandwiching poly (*p*-phenylene vinylene), or PPV, between layers of a new light-generating thin plastic polymer. The combination efficiently accepts negative charges. Further, to be a viable option for television, the technology needs to produce vibrant colors. Chemists are able to do this by altering the shapes and positions of the polymer's building blocks—altering the polymer's physical makeup slightly—almost as if it were a dye.

Much more research and development is necessary before this technology is perfected or even fine-tuned enough to be useful in commercial products. Besides television, other possible uses for this new LED may be light-emitting clothing for police or rescue workers, glowing tools, or radiating vehicles.

*Adapted from *Science News,* October 16, 1993, p. 246

SELF-TEST EXERCISES

LEARNING GOAL 1: *Moles from Mass of an Element*

1. Explain the importance of the mole concept in chemistry.
2. (a) Using scientific notation, write the number that represents 1 mole.
   (b) Write this number without the use of scientific notation.
3. A sample of gold (Au) has a mass of 19.7 g. How many moles of gold atoms is this?
4. How many moles are there in 5.62 g of silicon?
5. The atomic department store is selling uranium atoms at the reduced price of 100,000,000 atoms for 1 cent. What would be the cost in dollars of 1 mg of uranium?

◄ 6. A sample of sulfur has a mass of 0.963 g. How many moles of sulfur atoms are there in this sample?

7. Determine the number of *moles* of atoms in each of the following samples:
   (a) 16.2 g of Al
   (b) 239.5 g of Ti
   (c) 0.06075 g of Mg
   (d) 3,570.0 g of U

◄ 8. Determine the number of *moles* of atoms in each of the following samples:
   (a) 2397.0 g of Cu
   (b) 13.79 g of W
   (c) 0.00399 g of Ar
   (d) 16.19 g of Ag

LEARNING GOAL 2: **Mass, in Grams, from Moles of an Element**

◄ 9. Mercury (Hg), element number 80 in the periodic table, has an atomic mass of 200.6. What mass (in grams) of Hg would you need to obtain 0.250 mole?

10. What is the mass in grams of 1.50 moles of rubidium (Rb) atoms?

11. What is the mass in grams of 0.750 mole of cobalt (Co) atoms?

◄ 12. How many grams of Cr are contained in 0.0250 mole of Cr?

13. Determine the number of grams in each of the following:
   (a) 4.600 moles of Ni
   (b) 0.00300 mole of Br_2
   (c) 200.0 moles of Ca
   (d) 0.04000 mole of S

14. Determine the number of grams in each of the following:
   (a) 0.400 mole of Rn
   (b) 19.0 moles of F_2
   (c) 0.0350 mole of Hg
   (d) 7.200 moles of Yb

LEARNING GOAL 3: **Number of Atoms of an Element from Mass of a Sample**

◄ 15. After you've finished eating your lunch, you recycle the aluminum (Al) foil in which it was wrapped. If the aluminum foil weighed 2.00 g, how many atoms of aluminum did you recycle?

◄ 16. At the end of an experiment, you find that there are 13.8 g of sodium left in the reaction vessel. How many atoms of sodium are left?

*17. Diamond is made up of carbon atoms. A 2-carat diamond costs $1,800. What is the cost of 1 carbon atom in the diamond? (Assume that 1 carat weighs 0.2 g.)

*18. How many atoms are present in a sample of water molecules that has a mass of 1,800 g?

19. The radius of a silver atom is 1.34 angstroms, or 1.34 Å (1 Å = 10^{-8} cm). If the atoms of 0.1 mole of silver atoms were placed in a straight line, touching each other, what distance (in miles) would they cover?

*20. How many *atoms* of Cl_2 are present in a sample of Cl_2 that has a mass of 350 g?

LEARNING GOAL 4: **Empirical Formula from Percentage Composition**

21. Sodium, the eleventh element in the periodic table (whose name means "headache" in medieval Latin), is a very dangerous chemical when uncombined. But when 39.6% sodium is combined with 60.4% chlorine by mass, a common compound is formed. What is the empirical formula of this compound, and what is its common name?

22. When 15.8% Al is combined with 28.2% S and 56.1% O by mass, a compound is formed. Determine the empirical formula of this compound.

◄ 23. Cholesterol, a compound suspected of causing hardening of the arteries, has a molecular mass of 386 and the following percentage composition by mass: 84.0% C, 11.9% H, and 4.1% O. What is the molecular formula of cholesterol?

24. When 24.2% Ca is combined with 17.1% N and 58.5% O by mass, a common compound is formed. What are its empirical formula and name?

25. Determine the empirical formula of a compound that is 80.0% C and 20.0% H by mass. The molecular mass of this compound is 30.0. What is its molecular formula?

◄ 26. When 60.56% C is combined with 11.18% H and 28.26% N by mass, the compound spherophysine, which is used for treating high blood pressure, is formed. What is the empirical formula of this compound?

27. Determine the empirical formula of a compound that is 54.55% C, 9.09% H, and 36.36% O by mass. The molecular mass of this compound is 88. What is its molecular formula?
28. When 64.81% C is combined with 13.60% H and 21.59% O by mass, a compound that is used as a surgical anesthetic is formed. The name of this compound is diethyl ether. What is its empirical formula?

LEARNING GOAL 5: *Moles from Mass of a Compound*

29. How many moles of $Ba_3(PO_4)_2$ are there in 451.4 g of $Ba_3(PO_4)_2$?
30. A sample of $CaSO_4$ has a mass of 0.1705 g. How many moles of $CaSO_4$ are there in this sample?
31. How many moles of HCl are there in 100.0 g of this compound?
32. How many moles of O_2 molecules are there in 6.40 g of oxygen gas?
◄ 33. A sample of laughing gas (N_2O) has a mass of 2.20 g. How many moles of N_2O is this?
◄ 34. Milk of magnesia, $Mg(OH)_2$, is used as an antacid in small doses and as a laxative in large doses. The typical laxative dose for an adult is about 2.00 g. How many moles of $Mg(OH)_2$ is this?

LEARNING GOAL 6: *Mass, in Grams, from Moles of a Compound*

35. Determine the number of grams in each of the following:
   (a) 2.50 moles of CH_4
   (b) 0.0400 mole of SO_3
   (c) 0.00600 mole of $Al_2(SO_4)_3$
   (d) 0.020 mole of $(NH_4)_3PO_4$
36. What is the mass in grams of 10.0 moles of H_2O?
37. How many grams of SO_2 are contained in 0.20 mole of SO_2?
38. How many grams of H_2SO_4 are there in 4.00 moles of this compound?
◄ 39. Portable gas grills use propane (C_3H_8) as a fuel. A cylinder of propane contains 51.6 moles of C_3H_8. How many grams of propane are in the cylinder? How many pounds of propane are in the cylinder?
◄ 40. Baking soda ($NaHCO_3$) has many uses around the house. A box of baking soda contains 5.40 moles of $NaHCO_3$. How many grams is this? How many pounds is this?

LEARNING GOAL 7: *Number of Formula Units of a Compound from Mass of a Sample*

◄ 41. How many *molecules* of CO_2 are there in 198.0 g of CO_2?
42. How many *molecules* of SO_2 are there in 6.40 g of SO_2?
◄ 43. How many formula units of NaCl are there in 14.6 g of NaCl?
◄ 44. How many molecules of C_3H_8 are contained in the tank of propane described in Exercise 39?
◄ 45. How many formula units of $NaHCO_3$ are contained in the box of baking soda described in Exercise 40?
◄ 46. How many formula units of magnesium nitride are contained in 20.2 g of Mg_3N_2?

LEARNING GOAL 8: *Molecular Formula of a Compound*

47. According to Nobel Prize winner Linus Pauling, vitamin C, also known as ascorbic acid, is a cure and preventative for colds. The empirical formula of vitamin C is $C_3H_4O_3$, and its molecular mass is 176.0. What is its molecular formula?
48. A compound is found to have the empirical formula HF and a molecular mass of 40.0. What is its molecular formula?
49. The empirical formula of a very important compound, called benzene, is CH. Its molecular mass is 78.0. What is its molecular formula?
50. The compound fumaric acid has the empirical formula CHO. Its molecular mass is 116.0. What is its molecular formula?
51. The plastic melamine, used to make Melmac® dishes, has the empirical formula CH_2N_2. Its molecular mass is 126.0. What is its molecular formula?
52. The compound benzidine, used in the manufacture of dyes, has the empirical formula C_6H_6N. Its molecular mass is 184.0. What is its molecular formula?
◄ 53. Dextrose, a type of sugar, has the empirical formula CH_2O. Acetic acid, which is found in vinegar, shares the same empirical formula as dextrose. The molecular mass of dextrose is 180.0, whereas the molecular

mass of acetic acid is 60.0. Write the molecular formula of each compound.

54. A compound of iron and sulfur with a formula mass of 208.0 has an empirical formula of Fe_2S_3. What is its molecular formula?

LEARNING GOAL 9: *Percentage Composition from Chemical Formula*

◄ **55.** Find the percentages of K, Cr, and O by mass in potassium chromate (K_2CrO_4).

56. Find the percentages by mass of the elements in each of the following compounds: (a) C_2H_6
(b) NO_2 (c) CO_2

57. Determine the percentages by mass of the elements in the following compounds: (a) BF_3
(b) UF_6 (c) $C_3H_8O_3$

58. Determine the percentage by mass of oxygen in each of the following compounds: (a) H_2O
(b) CO   (c) H_2SO_4

*59. Some sodium chloride (NaCl) and some sugar ($C_{12}H_{22}O_{11}$) are mixed together. The mixture is analyzed and found to contain 20% chlorine by mass. The mixture has a mass of 50 g. How much sugar is present?

60. How many moles of nitrogen atoms can be obtained from 1,700 grams of NH_3?

61. Determine the percentages by mass of the elements in the following compounds: (a) SO_2
(b) CH_4 (c) $C_{12}H_{22}O_{11}$ (d) $CaCO_3$

◄ 62. How many grams of hydrogen can be obtained from 180 g of water?

63. How many grams of copper can be obtained from 1,600 g of CuO?

64. How many grams of oxygen can be obtained from 320.0 g of SO_2?

65. How many grams of iron can be obtained from 350.0 g of Fe_2O_3?

66. Determine the percentage by mass of sulfur in H_2S.

67. How many grams of oxygen combine with 200 g of sulfur in the compound SO_2? (*Hint:* Do Exercise 61 (a) to obtain the percentages by mass of S and O in SO_2.)

68. Determine the percentage by mass of hydrogen in CH_3OH.

*69. An unknown mixture contains sodium sulfide (Na_2S) and iron(III) oxide (Fe_2O_3). The mixture is analyzed and found to contain 25.0% sulfur by mass. The mixture has a mass of 500 g. How many grams of each compound are present?

70. Determine the percentage by mass of N in HNO_3.

EXTRA EXERCISES

71. If you had Avogadro's number of dollars and spent 1 dollar every second of every day, how long would it take you to run out of money?

72. What is the simplest formula for the compound that has the following percentage composition by mass: 24.3% C, 4.1% H, and 71.6% Cl?

73. If the molecular mass of the compound in Exercise 72 is determined to be 99.0, what is its molecular formula?

74. What is the mass of an Fe_3O_4 sample that contains 100 g of oxygen?

◄ **75.** How many carbon atoms are there in a 1.75-carat diamond? (*Hint:* 1.00 carat = 200 mg.)

76. Two ores of iron are hematite (Fe_2O_3) and magnetite (Fe_3O_4). Calculate the percentage by mass of iron in each ore to determine the richer source of iron.

◄ **77.** Two ores of copper have the following mass percentage analyses. What is the empirical formula of each ore?

ORE A	ORE B
Cu, 63.3%	Cu, 34.6%
Fe, 11.1%	Fe, 30.4%
S, 25.6%	S, 34.9%

78. The percentages by mass of oxygen and sulfur are almost the same in the compound SO_2. Calculate the percentage by mass of each.

79. What is the mass in grams of 10 atoms of silver?

80. How many moles of phosphorus are there in 392 g of H_3PO_4?

81. In Exercise 80, how many grams of phosphorus are there?

82. How much is 1 atomic mass unit in grams?

83. Ethyl alcohol, which is found is some popular beverages, has the molecular formula C_2H_6O. You have 9.2 g of ethyl alcohol.
 (a) How many moles of ethyl alcohol do you have?
 (b) How many molecules of ethyl alcohol?
 (c) How many moles of carbon atoms?
 (d) How many moles of hydrogen atoms?
 (e) How many moles of oxygen atoms?
 (f) How many atoms of carbon?
 (g) How many atoms of hydrogen?
 (h) How many atoms of oxygen?

84. The formula for methanol, also known as wood alcohol, is CH_3OH. You have 3.2 grams of methanol.
 (a) How many moles of methanol do you have?
 (b) How many molecules of methanol?
 (c) How many moles of carbon atoms?
 (d) How many moles of hydrogen atoms?
 (e) How many moles of oxygen atoms?
 (f) How many atoms of carbon?
 (g) How many atoms of hydrogen?
 (h) How many atoms of oxygen?

85. Determine the number of *moles* of molecules or formula units in each of the following samples:
 (a) 7.20 g of H_2O (c) 0.070 g of N_2
 (b) 260.7 g of MnO_2 (d) 1,980.0 g of $(NH_4)_2SO_4$

86. Determine the number of moles of molecules in each of the following: (a) 0.34 g of NH_3
 (b) 1,335.0 g of MnO_2

◀ **87.** Glycine is the simplest amino acid. It has the molecular formula $C_2H_5O_2N$. You have 300.0 g of glycine.
 (a) How many moles of glycine do you have?
 (b) How many molecules of glycine?
 (c) How many moles of carbon atoms?
 (d) How many moles of hydrogen atoms?
 (e) How many moles of oxygen atoms?
 (f) How many moles of nitrogen atoms?
 (g) How many grams of carbon?
 (h) How many grams of hydrogen?
 (i) How many grams of oxygen?
 (j) How many grams of nitrogen?
 (k) What should be true about the sum of the numbers of grams of each element in (g), (h), (i), and (j) and the number of grams of glycine we started with?

◀ 88. Determine the number of moles of molecules in each of the following: (a) $72\overline{0}$ g of H_2O
 (b) 13.35 g of MnO_2

THE CHEMICAL EQUATION:
Recipe for a Reaction

10

This is the story of Anne Williams, wife, mother, and teacher, and how she saved her husband, Tony, from serious injury when he was charging his car battery. Although Tony was the mechanic in the family, it was something that Anne remembered from her college chemistry course that saved the day.

Our story begins on a Saturday afternoon. Tony was having trouble starting his car, which was in the garage, so he decided to charge the battery with a battery charger that operated off the house current. After plugging in the unit, he attached the charger to his car battery and the battery began to charge.

Tony decided to wait in the garage as the battery charged, have a cigarette, and read the newspaper. After reading for about fifteen minutes, Tony thought he would check the battery. He was smoking his third cigarette as he approached the car and was about to lean over the battery when Anne came walking into the garage. Seeing what he was about to do she screamed, "Tony! No!" Startled by this outburst, Tony's mouth dropped open as he jerked his head up, turning away from the car toward Anne. A second later, as Tony gazed at Anne, still surprised by her warning, the battery exploded.

The lit cigarette that fell from Tony's mouth when Anne startled him had landed on the battery casing—hydrogen gas that was venting from the charging battery had ignited and exploded. Fortunately, Tony had turned away from the battery in time and was not hurt. Anne explained to Tony that flammable hydrogen gas is produced when a car battery is being charged and that the battery could have exploded in his face had he leaned over it with a lit cigarette. That's why she had screamed at him. Thanks to Anne's quick action, only the car battery had to be replaced, and Tony was saved from serious injury.

■■■ **LEARNING GOALS**

After you've studied this chapter, you should be able to:

1. Write formula equations from word equations, and vice versa.
2. Balance formula equations.
3. Recognize and give examples of some general types of chemical reactions (combination, decomposition, single-replacement, and double-replacement).
4. Predict the products for the various types of reactions.
5. Define the terms *acid, base,* and *salt* and recognize each from its formula.
6. Use the activity series and solubility table to predict reactions.
7. Define the terms *oxidation, reduction, oxidizing agent,* and *reducing agent.*
8. Determine which substance has been oxidized and which substance has been reduced in a redox reaction.
9. Determine which substance is the reducing agent and which substance is the oxidizing agent in a redox reaction.

■■■ **INTRODUCTION**

Combine 4 cups of sifted flour and 1 cake of yeast soaked in $1\frac{1}{3}$ cups of 35°C water. Add 2 tablespoons of olive oil and 1 teaspoon of salt. Roll out the dough, add tomato sauce and mozzarella cheese, and bake. When these ingredients are baked, has a chemical reaction occurred? Yes. What's more, this recipe is a set of directions that tells you how to mix certain *reactants* to yield a delectable *product,* a pizza.

In a sense, we can think of a chemical reaction as a set of directions. When we follow these directions and mix the proper chemicals, we get the products that we want. In this chapter, you will study some important classes of chemical reactions. You will also learn how to write chemical equations and how to interpret the information they provide.

10.1 Word Equations

LEARNING GOAL 1

Formula and word equations

One way to represent a chemical reaction is with words. For example, we can say

Sulfur trioxide + water → sulfuric acid

In a chemical equation such as this one, the reacting substances, or **reactants,** are shown on the left-hand side. The **products** are shown on the right-hand side. The plus sign means *and,* and the arrow means *yields* or *gives.* Our

Sulfur burns in oxygen with a pale blue flame to form gaseous SO_2 and SO_3, which are used to make sulfuric acid. An older name for sulfur is *brimstone*, which means "burning stone." (Rich Treptow/Photo Researchers, Inc.)

word equation, then, means that "sulfur trioxide and water react to yield sulfuric acid." (By the way, this equation represents part of the process by which "acid rain" develops from one byproduct of fuel combustion.)

Our word equation tells us what happens when sulfur trioxide combines with water. But it tells us nothing about the chemical formulas of the reactants and products. Nor does it tell us anything about *how* the reactants combine to form the products. That's why chemists prefer to use formula notation to write chemical reactions. And, like other people, chemists prefer to use the simpler shorthand notation whenever possible.

10.2 The Formula Equation

Let us substitute the formulas for the reactants and product in our word equation:

$$\text{Sulfur trioxide} + \text{water} \rightarrow \text{sulfuric acid}$$
$$SO_3 \qquad + H_2O \rightarrow \qquad H_2SO_4$$

LEARNING GOAL 2
Balancing equations

Now let's do some "atomic bookeping" to see whether we have complied with the Law of Conservation of Mass. On the left-hand side of this equation are one sulfur atom, four oxygen atoms, and two hydrogen atoms. And there are exactly the same numbers of these atoms on the right-hand side. So we have not "created" or "destroyed" atoms in writing the formula equation (see Figure 10.1). The equation is *balanced*. A chemical equation is **balanced** when *the total number of atoms of each element on the left-hand side is exactly equal to the total number of atoms of each element on the right-hand side.*

Let's now look at the chemical reaction in which hydrogen and oxygen form water. Because hydrogen and oxygen are both diatomic elements, we must write the equation for this chemical reaction as follows:

$$H_2 + O_2 \xrightarrow[\text{spark}]{\text{Electric}} H_2O$$

$$SO_3 \quad + \quad H_2O \longrightarrow \quad H_2SO_4$$

■ FIGURE 10.1

Models show how sulfur trioxide and water produce sulfuric acid.

(Many reactions occur only under special conditions. We specify these conditions by writing them above or around the arrow.) Now for the bookkeeping: There are two hydrogens on the left side and two hydrogens on the right side of the equation. However, there are two oxygens on the left side but only *one* oxygen on the right side. This equation is *not balanced*, because the number of atoms of each element on the left *does not* equal the number of atoms of each element on the right.

We can't let the equation stand this way. Every chemical equation must be balanced, because chemical reactions satisfy the Law of Conservation of Mass. As it stands, our reaction falsely implies that an atom of oxygen was lost or destroyed.

10.3   Balancing a Chemical Equation

To balance a chemical equation, we determine how many molecules of each substance are needed to satisfy the Law of Conservation of Mass. We indicate this by placing **coefficients,** or small whole numbers, before the formulas in the equation. These coefficients are usually obtained by trial and error. For example, in our equation for the formation of water, we have

$$H_2 + O_2 \xrightarrow[\text{spark}]{\text{Electric}} H_2O \qquad \text{(not balanced)}$$

Because this unbalanced equation lacks one oxygen atom on the right, we place the coefficient 2 before the formula for water on the right:

$$H_2 + O_2 \xrightarrow[\text{spark}]{\text{Electric}} 2H_2O \qquad \text{(not balanced)}$$

Now we have two oxygen atoms on the left and two on the right. But the coefficient 2 has unbalanced the hydrogen atoms. There are now four on the right but only two on the left. To remedy this, we use two molecules of H_2:

$$2H_2 + O_2 \xrightarrow[\text{spark}]{\text{Electric}} 2H_2O \qquad \text{(balanced)}$$

Now the equation is balanced: there are four hydrogen atoms and two oxygen atoms on the left and the same numbers on the right. The balanced equation

■ **FIGURE 10.2**

The formation of water from
hydrogen and oxygen

$$2H_2 \quad + \quad O_2 \quad \longrightarrow \quad 2H_2O$$

states that two hydrogen molecules and one oxygen molecule react to form
two water molecules (Figure 10.2).

There are two rules to keep in mind when you balance a chemical
equation:

RULE 1   The equation must be accurate. Correct formulas for all reactants and
products must be shown.

RULE 2   The Law of Conservation of Mass must be observed. The same number
of atoms of each element must appear on both sides of the balanced
equation.

EXAMPLE 10.1

Write a balanced chemical equation for the following reaction:

$$\text{Mercury(II) oxide} \xrightarrow{\text{Heat}} \text{mercury metal} + \text{oxygen gas}$$

Solution

We begin by writing the formulas for the reactants and products. Remember that oxygen gas is a diatomic element, so we must write O_2.

$$\text{HgO} \xrightarrow{\text{Heat}} \text{Hg} + O_2 \quad \text{(unbalanced)}$$

The mercury is balanced, but there is only one oxygen atom on the left
and two oxygens on the right. We balance the oxygens by placing the
coefficient 2 in front of HgO:

$$2\text{HgO} \xrightarrow{\text{Heat}} \text{Hg} + O_2 \quad \text{(unbalanced)}$$

Now the oxygens are balanced, but there are two mercury atoms on
the left and only one on the right. We balance the mercury atoms by
placing a 2 in front of the Hg:

$$2\text{HgO} \xrightarrow{\text{Heat}} 2\text{Hg} + O_2 \quad \text{(balanced)}$$

The balanced chemical equation shows two mercury atoms and two oxygen atoms on each side of the equation.

Practice Exercise 10.1   Write a balanced chemical equation for the following reaction:

$$\text{Copper(II) oxide} \rightarrow \text{copper metal} + \text{oxygen gas}$$

Here are two important points to remember.

1. You *cannot* balance an equation using *subscripts*, because this would change the formulas of the compounds. For example,

$$\text{HgO} \xrightarrow{\text{Heat}} \text{Hg} + \text{O}_2$$

cannot be balanced by doing this:

$$\text{HgO}_2 \xrightarrow{\text{Heat}} \text{Hg} + \text{O}_2$$

because HgO_2 is not the formula for mercury(II) oxide.

2. You *cannot* insert a coefficient between elements in a chemical formula. For example,

$$\text{HgO} \xrightarrow{\text{Heat}} \text{Hg} + \text{O}_2$$

cannot be balanced by doing this:

$$\text{Hg2O} \xrightarrow{\text{Heat}} \text{Hg} + \text{O}_2$$

because such an expression has no meaning.

After you have balanced a number of equations, you will find it easy to choose the proper coefficients. Here are some suggestions that may help:

1. Balance metal and nonmetal atoms first.
2. Balance hydrogen and oxygen atoms last.
3. Balance polyatomic ions that contain more than one atom *as a group* if they appear unchanged on both sides of the equation.

EXAMPLE 10.2

Write a balanced chemical equation for the following reaction:

$$\text{Aluminum} + \text{sulfuric acid} \rightarrow \text{aluminum sulfate} + \text{hydrogen gas}$$

Solution

We first write the correct formula for each compound:

$$\text{Al} + \text{H}_2\text{SO}_4 \rightarrow \text{Al}_2(\text{SO}_4)_3 + \text{H}_2 \qquad \text{(unbalanced)}$$

There is one aluminum atom on the left-hand side but two aluminum atoms on the right-hand side. We balance the aluminums by placing the coefficient 2 in front of the Al on the left-hand side:

$$2\text{Al} + \text{H}_2\text{SO}_4 \rightarrow \text{Al}_2(\text{SO}_4)_3 + \text{H}_2 \qquad \text{(unbalanced)}$$

There is one SO_4 group on the left-hand side and three SO_4 groups on the right-hand side. We balance the SO_4 groups by placing a 3 in front of the H_2SO_4:

$$2Al + 3H_2SO_4 \rightarrow Al_2(SO_4)_3 + H_2 \quad \text{(unbalanced)}$$

Now there are six hydrogens on the left-hand side and two hydrogens on the right-hand side. We balance the hydrogens by placing a 3 in front of the H_2:

$$2Al + 3H_2SO_4 \rightarrow Al_2(SO_4)_3 + 3H_2 \quad \text{(balanced)}$$

There are now two aluminum atoms, three sulfate groups, and six hydrogen atoms on each side of the equation. And so we have a balanced chemical equation.

Practice Exercise 10.2   Write a balanced chemical equation for the following reaction:

$$\text{Zinc} + \text{sulfuric acid} \rightarrow \text{zinc sulfate} + \text{hydrogen gas}$$

EXAMPLE 10.3

Write a balanced equation for each of the following reactions:

(a) Hydrogen gas + nitrogen gas $\xrightarrow[\text{Pressure}]{\text{Heat}}$ ammonia (NH_3)

(b) Sodium oxide + water $\rightarrow$ sodium hydroxide

(c) Calcium bromide + chlorine $\rightarrow$ calcium chloride + bromine

(d) Sodium hydroxide + phosphoric acid (H_3PO_4) $\rightarrow$ sodium phosphate + water

(e) Iron(III) oxide + carbon monoxide $\xrightarrow{\text{Heat}}$ iron + carbon dioxide

Solution

We use the same procedure as in the previous examples: Write the formula for each substance and then balance the equation.

(a) $H_2 + N_2 \xrightarrow[\text{Pressure}]{\text{Heat}} NH_3 \quad \text{(unbalanced)}$

$3H_2 + N_2 \xrightarrow[\text{Pressure}]{\text{Heat}} 2NH_3 \quad \text{(balanced)}$

(b) $Na_2O + H_2O \rightarrow NaOH \quad \text{(unbalanced)}$
$Na_2O + H_2O \rightarrow 2NaOH \quad \text{(balanced)}$

(c) $CaBr_2 + Cl_2 \rightarrow CaCl_2 + Br_2 \quad \text{(balanced)}$

(d) $NaOH + H_3PO_4 \rightarrow Na_3PO_4 + H_2O \quad \text{(unbalanced)}$
$3NaOH + H_3PO_4 \rightarrow Na_3PO_4 + 3H_2O \quad \text{(balanced)}$

(e) $Fe_2O_3 + CO \xrightarrow{\text{Heat}} Fe + CO_2 \quad \text{(unbalanced)}$

$Fe_2O_3 + 3CO \xrightarrow{\text{Heat}} 2Fe + 3CO_2 \quad \text{(balanced)}$

Practice Exercise 10.3   Write a balanced equation for each of the following word equations:
(a) Aluminum + chlorine gas → aluminum chloride
(b) Sodium + water → sodium hydroxide + hydrogen gas
(c) Potassium nitrate → potassium nitrite + oxygen gas
(d) Nitric acid + barium hydroxide → barium nitrate + water

10.4   Types of Chemical Reactions

The millions of known chemical reactions may be classified in many ways. We will use a classification scheme that is helpful in predicting the products of many chemical processes. In this scheme, there are four basic classes:

> LEARNING GOAL 3
> **Recognizing types of chemical reactions**

1. Combination reactions
2. Decomposition reactions
3. Single-replacement reactions
4. Double-replacement (or double-exchange) reactions

All of these chemical reactions, with the exception of double-replacement reactions, may also be classified as redox reactions. (*Redox* is short for "oxidation–reduction.") We have more to say about these reactions later in the chapter.

Combination Reactions

Combination reactions are *reactions in which two or more substances combine to form a more complex substance* (Figure 10.3). The general formula is

$$A + B \rightarrow AB$$

In some combination reactions, two *elements* join to form a compound. For example,

$$4Fe + 3O_2 \rightarrow 2Fe_2O_3$$
Iron   Oxygen   Iron(III)
        gas   oxide

Iron(III) oxide is iron rust. When a product made of iron contacts moist air, the iron rusts, forming a crust of iron(III) oxide. The crust flakes off, a new

■ **FIGURE 10.3**

A combination reaction: iron + sulfur → iron(II) sulfide

Iron   Sulfur   Iron(II) sulfide

A vivid example of a combination reaction! The space shuttles are launched by liquid-fuel rockets that contain liquid hydrogen and liquid oxygen, which combine to form water. The shuttle shown here is Endeavour. (NASA)

surface of iron becomes visible, and it too begins to rust. The iron can keep rusting until the entire piece of metal is eaten away.

Another example of a combination reaction that joins two elements is

$$\underset{\text{Carbon}}{C} \ + \ \underset{\text{Sulfur}}{2S} \ \rightarrow \ \underset{\substack{\text{Carbon} \\ \text{disulfide}}}{CS_2}$$

Carbon disulfide is an important chemical used in manufacturing rayon and in electronic vacuum tubes. However, it is a highly poisonous substance and should be used only in well-ventilated areas.

In some combination reactions, two *compounds* join to form a more complex compound. For example,

$$\underset{\substack{\text{Sulfur} \\ \text{trioxide}}}{SO_3} \ + \ \underset{\text{Water}}{H_2O} \rightarrow \underset{\substack{\text{Sulfuric} \\ \text{acid}}}{H_2SO_4}$$

$$\underset{\substack{\text{Sodium} \\ \text{oxide}}}{Na_2O} \ + \ \underset{\text{Water}}{H_2O} \rightarrow \underset{\substack{\text{Sodium} \\ \text{hydroxide}}}{2NaOH}$$

EXAMPLE 10.4

Complete and balance the following equations for combination reactions:

LEARNING GOAL 4

Predicting products of chemical reactions

(a) $H_2 + Cl_2 \rightarrow$?
(b) $K + Br_2 \rightarrow$?
(c) $Ba + O_2 \rightarrow$?

Solution

We are told that these are combination reactions, so we know that the products are formed by combination of the elements involved. We first write the correct formula for the product. Then we balance the equation.

(a) $H_2 + Cl_2 \rightarrow HCl$ (unbalanced)
    $H_2 + Cl_2 \rightarrow 2HCl$ (balanced)

(b) $K + Br_2 \rightarrow KBr$ (unbalanced)
    $2K + Br_2 \rightarrow 2KBr$ (balanced)

(c) $Ba + O_2 \rightarrow BaO$ (unbalanced)
    $2Ba + O_2 \rightarrow 2BaO$ (balanced)

Practice Exercise 10.4   Complete and balance the following equations for combination reactions:

(a) $K + O_2 \rightarrow$?
(b) $Ca + O_2 \rightarrow$?
(c) $H_2 + Br_2 \rightarrow$?

Decomposition Reactions

Decomposition reactions are *reactions in which a complex substance is broken down into simpler substances.* They are thus the reverse of combination reactions. The general formula for a decomposition reaction is

$$AB \rightarrow A + B$$

Some examples of common decomposition reactions follow.

1. The decomposition of a metal oxide (Figure 10.4):

$$2HgO \xrightarrow{\text{Heat}} 2Hg + O_2$$

Mercury(II)   Mercury   Oxygen
oxide   gas

2. The decomposition of a nonmetal oxide:

$$2H_2O \xrightarrow{\text{Electricity}} 2H_2 + O_2$$

Left: The decomposition of ammonium dichromate occurs spontaneously once it has been started by heat. *Right:* The voluminous powder (green in color) is chromium(III) oxide, which is used as an abrasive and as a pigment, especially for coloring glass and in printing fabrics and paper money. (Ken O'Donoghue)

3. The decomposition of a metal chlorate:

$$2KClO_3 \xrightarrow{\text{Heat}} 2KCl + 3O_2$$

Potassium   Potassium
chlorate   chloride

4. The decomposition of a metal carbonate:

$$MgCO_3 \xrightarrow{\text{Heat}} MgO + CO_2$$

Magnesium   Magnesium
carbonate   oxide

5. The decomposition of a metal hydroxide:

$$Ca(OH)_2 \xrightarrow{\text{Heat}} CaO + H_2O$$

Calcium   Calcium
hydroxide   oxide

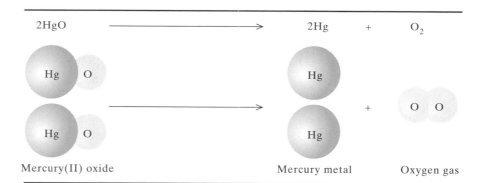

■ **FIGURE 10.4**

The decomposition of mercury(II) oxide molecules

Decomposition reactions are a very important class of reactions. They are the ones that produce pure samples of various elements. For example, very pure hydrogen and oxygen can be obtained from the decomposition of water. Pure copper can be produced by the decomposition of copper(II) oxide.

EXAMPLE 10.5

Complete and balance the following equations for decomposition reactions:

(a) $Mg(OH)_2 \xrightarrow{\text{Heat}} ? + ?$

(b) $CaCO_3 \xrightarrow{\text{Heat}} ? + ?$

(c) $NaClO_3 \xrightarrow{\text{Heat}} ? + ?$

(d) $HCl \rightarrow ? + ?$

Solution

UNDERSTAND THE PROBLEM

We must predict the products and balance the equations for these decomposition reactions.

DEVISE A PLAN

We are told that these are decomposition reactions, so we know that the products are formed by a breakdown of the elements in the original substance. We want to balance the equation according to the Law of Conservation of Mass. Look back in the chapter for similar kinds of compounds undergoing decomposition.

CARRY OUT THE PLAN

We first write the correct formulas for the products and then balance the equation.

LOOK BACK

We look back and check to be sure we have written an accurate equation.

(a) $Mg(OH)_2 \xrightarrow{\text{Heat}} MgO + H_2O$ (balanced)

(b) $CaCO_3 \xrightarrow{\text{Heat}} CaO + CO_2$ (balanced)

(c) $NaClO_3 \xrightarrow{\text{Heat}} NaCl + O_2$ (unbalanced)

   $2NaClO_3 \xrightarrow{\text{Heat}} 2NaCl + 3O_2$ (balanced)

(d) $HCl \rightarrow H_2 + Cl_2$ (unbalanced)
   $2HCl \rightarrow H_2 + Cl_2$ (balanced)

Practice Exercise 10.5   Complete and balance the following equations for decomposition reactions:
(a) $Sr(OH)_2 \rightarrow ? + ?$
(b) $SrCO_3 \rightarrow ? + ?$
(c) $KClO_3 \rightarrow ? + ?$
(d) $KCl \rightarrow ? + ?$

Single-Replacement Reactions

A **single-replacement reaction** is *one in which an uncombined element replaces another element that is in a compound* (Figure 10.5). The general formula for this type of reaction is

$$A + BC \rightarrow AC + B$$

Examples of single-replacement reactions follow.

1. The replacement in aqueous solution of one metal by another metal that is more active (the activity of metals is discussed in Section 10.5):

$$Mg + Cu(NO_3)_2 \rightarrow Mg(NO_3)_2 + Cu$$

2. The replacement of hydrogen in water by a Group IA metal:

$$2Na + 2H_2O \rightarrow 2NaOH + H_2$$

3. The replacement of hydrogen in an acid by an active metal:

$$Zn + 2HCl \rightarrow ZnCl_2 + H_2$$

EXAMPLE 10.6

Complete and balance the following equations for single-replacement reactions:
(a) $K + H_2O \rightarrow ? + ?$
(b) $Mg + H_2SO_4 \rightarrow ? + ?$
(c) $Zn + NiCl_2 \rightarrow ? + ?$

■ **FIGURE 10.5**

A single-replacement reaction: zinc + copper(II) sulfate → copper + zinc sulfate

Zinc   Copper(II)sulfate   Copper   Zinc sulfate

Solution

We first write the correct formulas for the products and then balance the equation.

(a) K + H$_2$O → KOH + H$_2$ (unbalanced)
    2K + 2H$_2$O → 2KOH + H$_2$ (balanced)
(b) Mg + H$_2$SO$_4$ → MgSO$_4$ + H$_2$ (balanced)
(c) Zn + NiCl$_2$ + ZnCl$_2$ + Ni   (balanced)

Practice Exercise 10.6   Complete and balance the following equations for single-replacement reactions:

(a) Li + H$_2$O → ? + ?
(b) Al + H$_2$SO$_4$ → ? + ?
(c) Mg + Al(NO$_3$)$_3$ → ? + ?

Single-replacement reactions are also a very important class of reactions. Chemists often use single-replacement reactions to obtain various metals from compounds containing these metals. For example, you can obtain pure silver metal from the compound silver nitrate by causing a solution of silver nitrate to react with copper metal.

$$Cu + 2AgNO_3 → Cu(NO_3)_2 + 2Ag$$

The thermite reaction between iron(III) oxide and aluminum metal is an example of a single-replacement reaction. (Richard Megna/Fundamental Photographs)

■ FIGURE 10.6

A double-replacement
reaction: potassium
chloride + silver nitrate →
potassium nitrate + silver
chloride

| Potassium chloride | Silver nitrate | | Potassium nitrate | Silver chloride |

Double-Replacement Reactions

A **double-replacement** (or *double-exchange*) **reaction** is *one in which two compounds exchange ions with each other* (Figure 10.6). The general formula is

$$A^+B^- + C^+D^- \rightarrow A^+D^- + C^+B^-$$

Double-replacement reactions usually take place in water. To show that they do, we use the notation (*aq*), which stands for *aqueous*, or in water.

⬛ LEARNING GOAL 5
Acids, bases, and salts

There are two important kinds of double-replacement reactions. One kind takes place between two salts. A **salt** is a compound composed of a positive and a negative ion. The positive ion may be a monatomic ion of a metal (for example, Na^{1+} or Ca^{2+}) or a polyatomic ion such as $(NH_4)^{1+}$. The negative ion may be a monatomic ion of a nonmetal (for example, Cl^{1-}) or a polyatomic ion such as $(SO_4)^{2-}$ or $(NO_3)^{1-}$.

A double-replacement reaction between two salts can occur if one of the *products* separates from the solution—in other words, if it is insoluble in the solution (or if a gas is formed). When a product is insoluble, it separates out of the solution in solid form (called a **precipitate**). To show this, we use

Soluble barium chloride
reacts with soluble sodium
sulfate to form insoluble
barium sulfate and soluble
sodium chloride.
(Yoav Levy/Phototake)

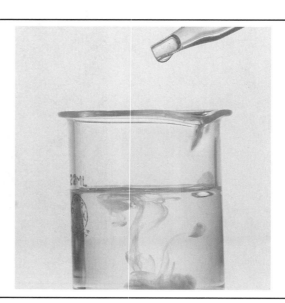

the notation (s), meaning *solid*. Table B.4 in Appendix B tells you whether a particular product is soluble in water.

When a salt dissolves in water it **dissociates**, or breaks up into its ions. For example, when we dissolve silver nitrate ($AgNO_3$) in water, there are Ag^+ and $(NO_3)^-$ ions in the solution. There are no $AgNO_3$ units as such in the water.

Now let's see what happens when we mix an aqueous solution of silver nitrate and an aqueous solution of potassium chloride. Both salts are soluble in water, so both solutions are clear. But when we mix the two solutions and allow them to react, we get a solid precipitate. This precipitate is silver chloride, which is *not* soluble in water. The reaction is

$$KCl(aq) + AgNO_3(aq) \rightarrow KNO_3(aq) + AgCl(s)$$

The other product, potassium nitrate, remains in solution [as K^+ and $(NO_3)^-$ ions] because it *is* soluble in water.

The second important kind of double-replacement reaction takes place between an acid and a base. The simplest **acids** are *compounds that dissociate in water into hydrogen ions and negatively charged ions of a nonmetal (or negatively charged polyatomic ions)*. Acids release hydrogen ions in water. The simplest **bases** are *composed of metallic ions (or positively charged polyatomic ions) combined with hydroxide ions*. Bases release hydroxide ions in water.

In an acid–base reaction, the acid and base react to form a salt and water. This type of reaction is also called a **neutralization reaction.** Some examples of neutralization reactions are

$$\underset{\substack{\text{Hydrochloric} \\ \text{acid}}}{HCl} + \underset{\substack{\text{Sodium} \\ \text{hydroxide}}}{NaOH} \rightarrow NaCl + H_2O$$

$$\underset{\substack{\text{Sulfuric} \\ \text{acid}}}{H_2SO_4} + 2NaOH \rightarrow Na_2SO_4 + 2H_2O$$

$$\underset{\substack{\text{Phosphoric} \\ \text{acid}}}{H_3PO_4} + \underset{\substack{\text{Lithium} \\ \text{hydroxide}}}{3LiOH} \rightarrow Li_3PO_4 + 3H_2O$$

We make much use of neutralization reactions in our daily lives. For example, if you have an acid stomach (caused by excess hydrochloric acid), you take a product such as milk of magnesia to neutralize the excess acidity. Milk of magnesia contains a base—magnesium hydroxide, $Mg(OH)_2$.

EXAMPLE 10.7

Complete and balance the following equations for double-replacement reactions:
(a) $BaCl_2(aq) + Na_2SO_4(aq) \rightarrow$? + ?
(b) $H_3PO_4 + Ca(OH)_2 \rightarrow$? + ?

Solution

As usual, we first write the correct formula for each product and then balance the equation.

(a) $BaCl_2(aq) + Na_2SO_4(aq) \rightarrow$
$$NaCl(aq) + BaSO_4(s) \qquad \text{(unbalanced)}$$
$BaCl_2(aq) + Na_2SO_4(aq) \rightarrow$
$$2NaCl(aq) + BaSO_4(s) \qquad \text{(balanced)}$$

(b) $H_3PO_4 + Ca(OH)_2 \rightarrow Ca_3(PO_4)_2 + H_2O$ (unbalanced)
$2H_3PO_4 + 3Ca(OH)_2 \rightarrow Ca_3(PO_4)_2 + 6H_2O$ (balanced)

Practice Exercise 10.7   Complete and balance the following equations for double-replacement reactions:

(a) $BiCl_3 + H_2S \rightarrow ? + ?$

(b) $Fe(OH)_3 + H_2SO_4 \rightarrow ? + ?$

10.5   The Activity Series

■ LEARNING GOAL 6

Using the activity series and solubility table

The **activity series** (Table 10.1) is *a list of elements in decreasing order of their reactivity, or ability to react chemically in aqueous solution.* The most active elements (those that react most readily) appear at the top of the table, and reactivity decreases as we move down the list. Under ordinary conditions, each element in the table can replace any element that appears below it by taking part in a single-replacement reaction. For example, aluminum is higher in Table 10.1 than zinc, so it replaces zinc in the compound zinc nitrate:

$$2Al + 3Zn(NO_3)_2 \rightarrow 2Al(NO_3)_3 + 3Zn$$

The activity series is very helpful in determining whether a particular reaction—especially a single-replacement reaction—can take place under ordinary conditions.

EXAMPLE 10.8

Predict whether each of the following reactions can occur:

(a) $Cu + 2HCl \xrightarrow{\text{Heat}} CuCl_2 + H_2$

(b) $H_2 + CuO \xrightarrow{\text{Heat}} H_2O + Cu$

Solution

(a) Table 10.1 shows that copper is *below* hydrogen in the activity series. Therefore copper cannot replace hydrogen in the compound HCl, and

$$Cu + HCl \rightarrow \text{no reaction}$$

(b) Hydrogen is above copper in the activity series, so it can replace copper in the compound CuO. This reaction can occur:

$$H_2 + CuO \rightarrow H_2O + Cu$$

TABLE 10.1

The Activity Series

METALS	NONMETALS
Lithium	Fluorine
Potassium	Chlorine
Calcium	Bromine
Sodium	Iodine
Magnesium	
Aluminum	
Zinc	
Chromium	
Iron	
Nickel	
Tin	
Lead	
HYDROGEN*	
Copper	
Mercury	
Silver	
Platinum	
Gold	

*Hydrogen is in capital letters because the activities of the other metals were calculated relative to the activity of hydrogen. The most active elements are at the top of the table. The activity of the elements decreases going down the table.

Practice Exercise 10.8   Predict whether each of the following reactions can occur:
(a) $Cu + H_2SO_4 \rightarrow CuSO_4 + H_2$
(b) $H_2 + CuCl_2 \rightarrow 2HCl + Cu$

10.6 Oxidation–Reduction (Redox) Reactions

We've just finished classifying reactions according to one scheme—as combination, decomposition, single-replacement, or double-replacement reactions. Now we will introduce another scheme for classifying reactions—as either oxidation–reduction (redox, for short) reactions or nonredox reactions. Let's begin with some definitions.

Oxidation and Reduction

In the past, chemists defined oxidation as the combining of a substance with oxygen. (That's how the process got its name.) Now, however, the concept of oxidation is expanded to include reactions that do not involve oxygen at all. **Oxidation** is *the loss of electrons by a substance undergoing a chemical reaction. (Alternatively, it is an increase in the oxidation number of a substance.)* (If you don't recall the meaning of *oxidation number*, refer to Section 8.3, and note that the oxidation number of an element in the uncombined state is 0.) As an example, look at the following equation, in which the oxidation number of the iron is shown over its symbol:

$$\overset{0}{4\text{Fe}} + 3\text{O}_2 \longrightarrow 2\overset{3+}{\text{Fe}_2}\text{O}_3$$

The oxidation number of the iron changes from 0 to 3+. By our definition, this means that the iron is oxidized. But what happens to the oxygen? We say that the oxygen has been *reduced*, which leads us to our definition of reduction. **Reduction** is *the gain of electrons by a substance undergoing a chemical reaction. (Alternatively, it is a decrease in the oxidation number of a substance.)*

In the reaction between iron and oxygen, the oxidation number of the oxygen changes from 0 to 2−, because each oxygen atom gains two electrons from the iron:

$$4\text{Fe} + 3\overset{0}{\text{O}_2} \longrightarrow 2\text{Fe}_2\overset{2-}{\text{O}_3}$$

Note that in this reaction, one substance (iron) is oxidized, while the other substance (oxygen) is reduced. This is not a coincidence: The processes of oxidation and reduction always occur together. In other words, oxidation cannot take place in a reaction unless there is also reduction. And reduction can't take place in a reaction unless there is also oxidation. A reaction in which oxidation and reduction take place is called an **oxidation–reduction** (or **redox**) **reaction.** All other reactions are **nonredox reactions.**

EXAMPLE 10.9

In each of the following reactions, determine which substance is oxidized and which substance is reduced:
(a) $2\text{K} + \text{Br}_2 \rightarrow 2\text{KBr}$
(b) $\text{S} + \text{O}_2 \rightarrow \text{SO}_2$
(c) $\text{Mg} + \text{H}_2\text{SO}_4 \rightarrow \text{MgSO}_4 + \text{H}_2$
(d) $2\text{H}_2\text{O} \rightarrow 2\text{H}_2 + \text{O}_2$

Solution

In each case, we use the ideas of Section 8.3 (and the fact that the oxidation number of an element in its uncombined state is 0) to write the oxidation numbers for the elements.

(a)
$$2K + Br_2 \longrightarrow 2KBr$$

The oxidation number of the potassium changes from 0 to 1+. This increase means that the potassium is oxidized. The oxidation number of the bromine changes from 0 to 1−. This decrease means that the bromine is reduced.

(b)
$$S + O_2 \longrightarrow SO_2$$

The oxidation number of the sulfur changes from 0 to 4+, so the sulfur is oxidized. The oxidation number of the oxygen changes from 0 to 2−, so the oxygen is reduced.

(c)
$$Mg + H_2(SO_4) \longrightarrow Mg(SO_4) + H_2$$

The oxidation number of the magnesium changes from 0 to 2+, so the magnesium is oxidized. The oxidation number of the hydrogen changes from 1+ to 0. This decrease means that the hydrogen is reduced. The oxidation number of the sulfate group does not change, so it is neither oxidized nor reduced.

(d)
$$2H_2O \longrightarrow 2H_2 + O_2$$

The oxidation number of the oxygen changes from 2− to 0, so the oxygen is oxidized. The oxidation number of the hydrogen changes from 1+ to 0, so the hydrogen is reduced.

Practice Exercise 10.9   In each of the following reactions, determine which substance is oxidized and which substance is reduced:
(a) $Cu + 2AgNO_3 \rightarrow 2Ag + Cu(NO_3)_2$
(b) $2Na + Cl_2 \rightarrow 2NaCl$
(c) $Cl_2 + 2NaBr \rightarrow 2NaCl + Br_2$
(d) $Zn + S \rightarrow ZnS$

Oxidizing and Reducing Agents

LEARNING GOAL 9

Determining oxidizing and reducing agents

An **oxidizing agent** is *a substance that causes something else to be oxidized.* A **reducing agent** is *a substance that causes something else to be reduced.* As things work out, the oxidizing agent is the substance being reduced, and the reducing agent is the substance being oxidized. Let's take another look at the reaction in which iron and oxygen form iron(III) oxide:

$$4Fe + 3O_2 \rightarrow 2Fe_2O_3$$

The iron is the substance being oxidized, and the oxygen is the substance being reduced. Therefore the iron is the *reducing agent* and the oxygen is the *oxidizing agent*.

$$\underset{\text{(reducing agent)}}{\underset{\text{Substance oxidized}}{\overset{0}{4Fe}}} + \underset{\text{(oxidizing agent)}}{\underset{\text{Substance reduced}}{\overset{0}{3O_2}}} \longrightarrow \overset{3+ \; 2-}{2Fe_2O_3}$$

EXAMPLE 10.10

For each reaction in Example 10.9, state which substance is the oxidizing agent and which substance is the reducing agent.

Solution

Just remember that the substance being oxidized is the reducing agent and the substance being reduced is the oxidizing agent.

(a)
$$\underset{\text{(reducing agent)}}{\underset{\text{Substance oxidized}}{\overset{0}{2K}}} + \underset{\text{(oxidizing agent)}}{\underset{\text{Substance reduced}}{\overset{0}{Br_2}}} \longrightarrow \overset{1+ \; 1-}{2KBr}$$

(b)
$$\underset{\text{(reducing agent)}}{\underset{\text{Substance oxidized}}{\overset{0}{S}}} + \underset{\text{(oxidizing agent)}}{\underset{\text{Substance reduced}}{\overset{0}{O_2}}} \longrightarrow \overset{4+ \; 2-}{S \, O_2}$$

(c)
$$\underset{\text{(reducing agent)}}{\underset{\text{Substance oxidized}}{\overset{0}{Mg}}} + \underset{\text{(oxidizing agent)}}{\underset{\text{Substance reduced}}{\overset{1+ \; 2-}{H_2(SO_4)}}} \longrightarrow \overset{2+ \; 2-}{Mg(SO_4)} + \overset{0}{H_2}$$

Actually, in this case, the whole compound, H_2SO_4, is considered the oxidizing agent.

(d)
$$\overset{1+ \; 2-}{2H_2O} \longrightarrow \overset{0}{2H_2} + \overset{0}{O_2}$$

The hydrogen is reduced and the oxygen is oxidized. The compound water acts as both oxidizing agent and reducing agent.

Practice Exercise 10.10   For each reaction in Practice Exercise 10.9, state which substance is the oxidizing agent and which substance is the reducing agent.

SUMMARY

A chemical equation is a representation of a chemical reaction. A word equation indicates (in words) which reactants combine (or decompose) and which products are formed as a result of the reaction. In a formula equation, the words are replaced with chemical formulas. In addition, the formula equation must be balanced, so that the number of atoms of each element on the left side of the equation equals the number of atoms of each element on the right side. We balance an equation by placing the proper coefficients before elements or compounds in the equation. Coefficients are generally chosen by trial and error.

Chemical reactions may be classified as combination reactions, decomposition reactions, single-replacement reactions, or double-replacement reactions. Combination reactions are those in which two or more substances combine to form a more complex substance. Decomposition reactions (the reverse of combination reactions) involve the breakdown of a complex substance into simpler substances. A single-replacement reaction is one in which an uncombined element replaces another element in a compound. A double-replacement reaction is one in which two compounds exchange ions with each other.

To predict whether or not a reaction will occur under ordinary conditions, it is important to know the relative activities of the elements involved in the reaction. The activity series is a list of elements in decreasing order of activity or reactivity. It can be used to predict whether or not a single-replacement reaction will occur. That is, each element in the list can replace any element below it in a single-replacement reaction.

Another way of classifying chemical reactions is by dividing them into oxidation–reduction (redox) reactions and non-oxidation–reduction (nonredox) reactions. Oxidation is the loss of electrons by a substance undergoing a chemical reaction (in other words, an increase in the oxidation number of the substance). Reduction is the gain of electrons by a substance undergoing a chemical reaction (in other words, a decrease in the oxidation number of the substance). Oxidation and reduction always occur together. An oxidizing agent is a substance that causes something else to be oxidized, and a reducing agent is a substance that causes something else to be reduced. A redox reaction is a reaction in which oxidation and reduction take place. In such a reaction, the oxidizing agent is reduced, and the reducing agent is oxidized.

KEY TERMS

acid **(10.4)**

activity series **(10.5)**

balanced equation **(10.2)**

base **(10.4)**

coefficient **(10.3)**

combination reaction **(10.4)**

decomposition reaction **(10.4)**

dissociate **(10.4)**

double-replacement (double-exchange) reaction **(10.4)**

neutralization reaction **(10.4)**

nonredox reaction **(10.6)**

oxidation **(10.6)**

oxidation–reduction (redox) reaction **(10.6)**

oxidizing agent **(10.6)**

precipitate **(10.4)**

product **(10.1)**

reactant **(10.1)**

reducing agent **(10.6)**

reduction **(10.6)**

salt **(10.4)**

single-replacement reaction **(10.4)**

CAREER SKETCH

Physical Chemist

Are you interested in using the latest technology in your job or developing advanced instrumentation? Does bridging the gap between fundamental phenomena and engineering practice sound fascinating? If so, the job of *physical chemist* may be a career option for you.

Physical chemists most often work in chemical, instrument, or computer companies. In these fields, a physical chemist may be responsible for designing an instrument to be used in the laboratory or creating a software program that will help a chemist analyze the makeup of a sample. Many physical chemists are also employed by academic institutions, in either teaching or research roles. The government, too, hires physical chemists to do research and development. While physical chemists can perform a number of tasks, often they are asked to conduct work in such areas as laser science, surface science, molecular spectroscopy, high-temperature chemistry, nuclear chemistry, or superconductor chemistry.

The most important skills for a physical chemist are excellent problem-solving and analytical skills. A physical chemist should also have a strong background in computer programming. Creativity, high-level math skills, and mechanical skills are also highly valuable to employers looking to hire physical chemists. Course work should include a strong background in theoretical chemistry and physics, a broad overview of chemistry and its applications, and calculus and electronics.

CHEMICAL FRONTIERS

*Using Coral to Fix Your Body**

If you are injured in a serious accident, the following chemical equation could become very important to you: $5CaCO_3 + 3(NH_4)_2HPO_4 + H_2O \rightarrow Ca_5(PO_4)_3OH + 3(NH_4)_2CO_3 + 2H_2CO_3$. This is the process incorporated by one company to turn coral from the sea ($CaCO_3$) into a synthetic material that mimics natural bone.

The concept for this process began in the 1960s when Marshall Urist, a professor of orthopedic surgery at the University of California, Los Angeles, demonstrated the phenomenon of osteoinduction—the process of bone desorption and formation. Later, researchers built upon Urist's work and learned more about bone structure and the process of bone formation. This led to the development of calcium-based synthetic bone products that not only act like natural bone, but are actually able to be reabsorbed into the body.

The method of creating the bone is simple. Coral, collected from the sea, is heated with ammonium phosphate at more than 200 degrees Celsius for 24 to 60 hours. This creates about 95% hydroxyapatite [$Ca_5(PO_4)_3OH$], which is processed into block or granular form and sterilized by gamma radiation.

Calcium-based synthetic bone products act like natural bone.
(Tony Stone Images)

The substance can then be used to fill in for natural bone that may have been damaged or destroyed in an accident. A surgeon can shape a block of the material to fit into virtually any place natural bone is needed. After a metal plate and screws are added to stabilize the area (both plate and screws can later be removed), the process is complete. Within a year of the surgery, full use of the bone can be achieved.

Adapted from Chemical & Engineering News, August 25, 1997, pp. 27–32

SELF-TEST EXERCISES

LEARNING GOALS 1 AND 2: *Formula and Word Equations; Balancing Equations*

◀ *1.* Write balanced equations for each of the following reactions:
(a) Potassium + water →
      potassium hydroxide + hydrogen gas
(b) Acetic acid + calcium hydroxide →
      calcium acetate + water
(c) Magnesium + copper(II) nitrate →
      magnesium nitrate + copper
(d) Sodium oxide + water → sodium hydroxide
(e) Zinc sulfide + oxygen gas →
      zinc oxide + sulfur dioxide
(f) Potassium hydroxide + aluminum nitrate →
      aluminum hydroxide + potassium nitrate

◀ 2. Write a balanced chemical equation for each of the following reactions:
(a) Calcium hydroxide + phosphoric acid →
      calcium phosphate + water
(b) Strontium + oxygen gas → strontium oxide
(c) Magnesium chlorate →
      magnesium chloride + oxygen gas
(d) Iron(III) hydroxide + hydrochloric acid →
      iron(III) chloride + water
(e) Potassium + oxygen gas → potassium oxide
(f) Phosphorus + bromine →
      phosphorous tribromide

3. Write a balanced chemical equation for each of the following reactions:
(a) Iron + oxygen gas → iron(III) oxide
(b) Sulfuric acid + aluminum hydroxide →
      aluminum sulfate + water

(c) Silver nitrate + barium chloride →
      silver chloride + barium nitrate
(d) Copper(I) sulfide + oxygen gas →
      copper + sulfur dioxide

4. Write a balanced chemical equation for each of the following reactions:
(a) Hydrogen gas + oxygen gas → water
(b) Potassium hydroxide + acetic acid →
      potassium acetate + water
(c) Nitrogen gas + hydrogen gas → ammonia
(d) Potassium chlorate →
      potassium chloride + oxygen gas

LEARNING GOAL 3: *Recognizing Types of Chemical Reactions*

◀ *5.* Identify each of the following reactions as a combination, decomposition, single-replacement, or double-replacement reaction:
(a) $4Fe + 3O_2 \rightarrow 2Fe_2O_3$
(b) $2HBr \rightarrow H_2 + Br_2$
(c) $AgNO_3 + KCl \rightarrow AgCl + KNO_3$
(d) $Zn + H_2SO_4 \rightarrow ZnSO_4 + H_2$
(e) $Mg(CN)_2 + 2HCl \rightarrow MgCl_2 + 2HCN$

6. Identify each of the following reactions as a combination, decomposition, single-replacement, or double-replacement reaction:
(a) $H_2 + Br_2 \rightarrow 2HBr$
(b) $2PbO_2 \xrightarrow{\text{Heat}} 2PbO + O_2$
(c) $2KI + Br_2 \rightarrow 2KBr + I_2$
(d) $2Fe(OH)_3 + 3H_2SO_4 \rightarrow Fe_2(SO_4)_3 + 6H_2O$
(e) $2Al + 3Cl_2 \rightarrow 2AlCl_3$
(f) $2K_3PO_4 + 3BaCl_2 \rightarrow 6KCl + Ba_3(PO_4)_2$

_7._ Identify the type (combination, decomposition, single-replacement, or double-replacement) of each reaction in Exercise 13.

8. Identify the type (combination, decomposition, single-replacement, or double-replacement) of each reaction in Exercise 14.

_9._ Identify each of the reactions in Exercise 3 as a combination, decomposition, single-replacement, or double-replacement reaction.

◀ 10. Identify each of the reactions in Exercise 2 as a combination, decomposition, single-replacement, or double-replacement reaction.

LEARNING GOAL 4: *Predicting Products of Chemical Reactions*

◀ _11._ Complete and balance each of the following equations. Use the activity series and solubility table when necessary.

COMBINATION REACTIONS

(a) $H_2 + Br_2 \rightarrow$?
(b) $BaO + H_2O \rightarrow$?
(c) $Na + Cl_2 \rightarrow$?
(d) Nitrogen gas + oxygen gas $\rightarrow$ nitrogen dioxide

DECOMPOSITION REACTIONS

(e) $CaCO_3 \xrightarrow{\text{Heat}}$? + ?
(f) $KOH \rightarrow$? + ?
(g) $Hg(ClO_3)_2 \xrightarrow{\text{Heat}}$? + ?
(h) $PbCl_2 \rightarrow$? + ?

SINGLE-REPLACEMENT REACTIONS

(i) $Li + H_2O \rightarrow$? + ?
(j) Zinc + sulfuric acid $\rightarrow$? + ?
(k) $Ni + Al(NO_3)_3 \rightarrow$? + ?
(l) $Al + Hg(C_2H_3O_2)_2 \rightarrow$? + ?

DOUBLE-REPLACEMENT REACTIONS

(m) Sulfuric acid + ammonium hydroxide $\rightarrow$? + ?
(n) Silver nitrate + barium chloride $\rightarrow$? + ?
(o) H_2SO_3 (sulfurous acid) + $Al(OH)_3 \rightarrow$? + ?
(p) $NaNO_3(aq) + KCl(aq) \rightarrow$? + ?

12. Complete and balance each of the following equations. Use the activity series and solubility table when necessary.

COMBINATION REACTIONS

(a) $Cu + S \rightarrow$?
(b) $CaO + H_2O \rightarrow$?
(c) $Na + F_2 \rightarrow$?
(d) Phosphorus + iodine $\rightarrow$?

DECOMPOSITION REACTIONS

(e) $MgCO_3 \xrightarrow{\text{Heat}}$? + ?
(f) $LiOH \rightarrow$? + ?
(g) $Mg(ClO_3)_2 \rightarrow$? + ?
(h) $Mg_3N_2 \rightarrow$? + ?

SINGLE-REPLACEMENT REACTIONS

(i) $Al + H_3PO_4 \rightarrow$? + ?
(j) $Zn + Pb(NO_3)_2 \rightarrow$? + ?
(k) $Cl_2 + KBr \rightarrow$? + ?
(l) $Br_2 + NaCl \rightarrow$? + ?

DOUBLE-REPLACEMENT REACTIONS

(m) $AgNO_3 + H_2S \rightarrow$? + ?
(n) Barium chloride + ammonium carbonate $\rightarrow$? + ?
(o) $CaCO_3 + H_3PO_4 \rightarrow$? + ?
(p) Silver nitrate + magnesium chloride $\rightarrow$? + ?

_13._ Complete and balance each of the following equations. Use the activity series and solubility table when necessary.

(a) $N_2O \xrightarrow{\text{Heat}}$? + ?
(b) $H_2CO_3 \xrightarrow{\text{Heat}}$? + ?
(c) Sodium nitrate $\xrightarrow{\text{Heat}}$ sodium nitrite + oxygen gas
(d) $H_2 + F_2 \rightarrow$?
(e) $N_2 + H_2 \xrightarrow[\text{High pressure}]{\text{Heat}}$?
(f) $Ca + HCl \rightarrow$? + ?
(g) $Cu + NiCl_2 \rightarrow$? + ?
(h) $AgNO_3(aq) + K_3AsO_4(aq) \rightarrow$? + ?

◀ 14. Complete and balance each of the following equations. Use the activity series and solubility table when necessary.

(a) $Na_2O \rightarrow$? + ?
(b) $H_2 + I_2 \rightarrow$?

(c) Phosphorus + oxygen gas →
diphosphorus trioxide

(d) $K_3PO_4 + BaCl_2 →$? + ?

(e) $KClO_3 →$? + ?

(f) $Ca + O_2 →$?

(g) $Cl_2 + NH_4I →$? + ?

(h) $Ag_2O \xrightarrow{\text{Heat}}$? + ?

15. Complete and balance each of the following equations:

COMBINATION REACTIONS

(a) $K + Cl_2 →$?

(b) $K_2O + H_2O →$?

(c) $MgO + H_2O →$?

(d) $CaO + CO_2 →$?

DECOMPOSITION REACTIONS

(e) $NH_3 →$? + ?

(f) $SrCO_3 →$? + ?

(g) $NaClO_3 →$? + ?

(h) $Mg(OH)_2 →$? + ?

16. Complete and balance each of the following equations:

COMBINATION REACTIONS

(a) $K + Br_2 →$?

(b) $Li_2O + H_2O →$?

(c) $BaO + H_2O →$?

(d) $MgO + CO_2 →$?

DECOMPOSITION REACTIONS

(e) $MgCO_3 →$? + ?

(f) Copper(I) sulfide → ? + ?

(g) Chromium(III) carbonate → ? + ?

(h) $Al(ClO_3)_3 \xrightarrow{\text{Heat}}$? + ?

LEARNING GOAL 5: *Acids, Bases, and Salts*

◀ **17.** The reaction of an acid and a base produces a salt and water. Find the acid and base needed to produce each of the following salts: (a) NaCl (b) K_2SO_4 (c) $Al(NO_3)_3$

18. The reaction of an acid and a base produces a salt and water. Find the acid and base needed to produce each of the following salts: (a) KCl (b) $MgSO_4$ (c) $Fe_2(SO_4)_3$

LEARNING GOAL 6: *Using the Activity Series and Solubility Table*

19. Complete and balance each of the following equations. Check the activity series and solubility table to be sure that the reaction can take place.

SINGLE-REPLACEMENT REACTIONS

(a) $Zn + HNO_2 →$? + ?

(b) $Ag + NiCl_2 →$? + ?

(c) $Zn + AgNO_3 →$? + ?

(d) $Cs + H_2O →$? + ?

DOUBLE-REPLACEMENT REACTIONS

(e) $HCl + Al(OH)_3 →$? + ?

(f) $KNO_3 + ZnCl_2 →$? + ?

(g) $Al(NO_3)_3 + NaOH →$? + ?

(h) $K_2CrO_4 + Pb(NO_3)_2 →$? + ?

◀ 20. Complete and balance each of the following equations. Check the activity series and solubility table to be sure that the reaction can take place.

SINGLE-REPLACEMENT REACTIONS

(a) $Cu + AgCl →$? + ?

(b) $Cl_2 + KF →$? + ?

(c) $Zn + HNO_3 →$? + ?

(d) $Cu + HCl →$? + ?

DOUBLE-REPLACEMENT REACTIONS

(e) $NH_4NO_3 + H_3PO_4 →$? + ?

(f) $ZnCl_2 + KOH →$? + ?

(g) $Ni_3(PO_4)_2 + HCl →$? + ?

(h) $AgNO_3 + KCl →$? + ?

LEARNING GOAL 7: *Oxidation and Reduction; Oxidizing and Reducing Agents*

◀ 21. Define the following terms and give an example of each: (a) oxidation (b) reduction (c) oxidizing agent (d) reducing agent

22. Complete each of the following statements:

(a) A loss of electrons is called _____ .

(b) A gain of electrons is called _____ .

(c) A substance that causes something else to be oxidized is called a(n) _____ .

(d) A substance that causes something else to be reduced is called a(n) _____ .

LEARNING GOAL 8: *Determining Oxidized and Reduced Substances*

◀ <u>23.</u> Balance each of the following reactions and determine which substance is oxidized and which is reduced:
   (a) $K + Cl_2 \rightarrow KCl$
   (b) $NH_3 \rightarrow N_2 + H_2$
   (c) $CuO + H_2 \rightarrow Cu + H_2O$
   (d) $Sn + Cl_2 \rightarrow SnCl_4$

◀ 24. Balance each of the following reactions and determine which substance is oxidized and which is reduced:
   (a) $Na + F_2 \rightarrow NaF$
   (b) $Al + C \rightarrow Al_4C_3$
   (c) $Cu + S \rightarrow CuS$
   (d) $Ag_2O \rightarrow Ag + O_2$

LEARNING GOAL 9: *Determining Oxidizing and Reducing Agents*

◀ <u>25.</u> For each of the reactions in Exercise 23, determine the oxidizing agent and the reducing agent.

◀ 26. For each of the reactions in Exercise 24, determine the oxidizing agent and the reducing agent.

<u>27.</u> For each single-replacement reaction in Exercise 19, determine the oxidizing agent and the reducing agent.

28. For each single-replacement reaction in Exercise 20, determine the oxidizing agent and the reducing agent.

EXTRA EXERCISES

◀ <u>29.</u> For the reaction

$$5Zn + V_2O_5 \rightarrow 5ZnO + 2V$$

list (a) the oxidizing agent, (b) the reducing agent, (c) the element oxidized, (d) the element reduced

<u>30.</u> Magnesium carbonate and calcium carbonate, when heated, both decompose to form carbon dioxide and the metal oxides. Write these two reactions.

<u>31.</u> Determine whether the following equations are balanced. If they are not, balance them.
   (a) $N_2 + 3H_2 \rightarrow NH_3$

   (b) $Fe + H_2O \rightarrow Fe_3O_4 + H_2$
   (c) $PCl_3 + Cl_2 \rightarrow PCl_5$

◀ 32. A student is asked to balance the following:

Calcium hydroxide + hydrochloric acid → calcium chloride + water

The student works the problem and gets the following solution:

$$CaOH + HCl \rightarrow CaCl + H_2O$$

This attempt is marked incorrect. Can you explain why the answer is wrong?

<u>33.</u> Write a balanced equation for each of the following. If no reaction occurs, write *no reaction*.
   (a) Carbon burning in oxygen gas to form carbon monoxide.
   (b) Hydrogen gas and chlorine gas reacting to form hydrogen chloride gas
   (c) Metallic sodium dropped into water to create sodium hydroxide and hydrogen gas
   (d) Silver metal and hydrochloric acid reacting to produce _____

◀ <u>34.</u> Consider an experiment in which metal A is immersed in a water solution of ions of metal B. For each of the following pairs A and B, determine whether or not a reaction occurs. If a reaction does occur, write the balanced equation, assuming that the B ions are part of nitrate salts.

	A	B
(a)	Cu	Zn
(b)	Mg	Zn
(c)	Cu	Ni
(d)	Zn	Cu

35. Explain why an oxidizing agent is reduced.

<u>36.</u> Write balanced formula equations for the following word equations and state what type of reaction each is:
   (a) Calcium hydroxide + hydrochloric acid →
   (b) Zinc metal + sulfuric acid →
   (c) Barium metal + oxygen gas →
   (d) Cesium oxide + water →

37. Balance the equation

$$NH_3(g) + O_2(g) \rightarrow N_2O_4(g) + H_2O(g)$$

38. Consider the redox reaction

$$(Cr_2O_7)^{2-} + 14H^{1+} + 6Fe^{2+} \rightarrow$$
$$2Cr^{3+} + 7H_2O + 6Fe^{3+}$$

(a) Determine which substance is oxidized.
(b) Determine which substance is reduced.
(c) Determine the oxidizing agent.
(d) Determine the reducing agent.

39. Balance the following equation and write the chemical name of each product and reactant in the balanced equation.

$$Na + P_4 \rightarrow Na_3P$$

◀ 40. Write a balanced equation for a reaction that produces magnesium chloride. (*Hint:* In other words, magnesium chloride has to be at least one of the products.)

41. Write a balanced equation for an acid–base reaction that produces lithium chloride as one of the products.

42. Are single-replacement reactions also redox reactions? Explain with examples.

43. Write a balanced equation for any reaction that produces carbon dioxide as one of the products.

44. A double-replacement reaction produces zinc phosphate. What are the reactants? Write a balanced equation for the reaction.

STOICHIOMETRY:
The Quantities in Reactions

11

One of the most important contributors to the advancement of modern-day chemistry is the French chemist Antoine Laurent Lavoisier (1743–1794). As the 1800s drew to a close, Lavoisier melded together some of the most important discoveries about gases, thereby establishing the quantitative aspects of chemistry.

From the very beginning Lavoisier recognized the importance of accurate measurement. In some of his most important work, around 1772, Lavoisier heated metals, such as tin and lead, in closed containers with a limited supply of air. Both metals formed a layer of *calx* on the surface and then rusted no further. A popular (but soon-to-be-proven incorrect) theory at the time, called the *phlogiston theory*, had been used by many chemists to explain these observations. According to the phlogiston theory, anything that burned contained the substance phlogiston, and as the material burned the phlogiston was released into the air. The phlogistonists explained the results of Lavoisier's experiments by stating that the air had absorbed all of the phlogiston that it could from the metal. However, experiments had shown that the calx weighed more, not less, than the metal itself. More important, when Lavoisier weighed the entire container (metal, calx, air, and all) after heating, the container weighed exactly the same as it weighed before the heating.

Based on the results of these experiments, Lavoisier conjectured that if the metal had gained weight by being partially turned into calx, then something else in the vessel must have lost an equivalent amount of weight. That something else, it seemed, would have to be air (or something that was part of air). If this were true, then a partial vacuum should exist in the container. Sure enough, when Lavoisier opened the container, air rushed in. Once that had occurred, the container and its contents weighed more. Lavoisier had shown that the conversion of a metal into its calx was not the result of the loss of the mysterious substance phlogiston, but the gain of something from the air. Now Lavoisier had to determine what in the air combined with the metal.

Fortunately, about the same time that Lavoisier was performing his experiments, another famous chemist, Joseph Priestly, had made a startling discovery. In 1774 Priestly decomposed a calx of mercury (known today as mercuric oxide) and obtained an unknown gas that he called *dephlogisticated* air (known today as oxygen gas). When he visited Lavoisier in Paris later that year, Priestly described his discovery. Lavoisier immediately saw the significance of this gas! The following year, Lavoisier published a paper in which he stated that air is not a simple substance, but a mixture of at least two substances, one of them being Priestly's dephlogisticated air (oxygen gas).

It wasn't long before Lavoisier had determined that air consisted of about 20% oxygen and 80% nitrogen. The importance of measurement had made its initial impact on the chemical sciences.

LEARNING GOALS

After you've studied this chapter, you should be able to:

1. Calculate the quantities of reactants needed or products yielded in a chemical reaction.
2. Determine which starting material is the limiting reactant in a chemical reaction.

INTRODUCTION

In Chapter 10, we wrote balanced equations showing how many atoms or molecules of each substance take part in a reaction. For example, we considered the reaction

$$N_2 + 3H_2 \xrightarrow[\text{Pressure}]{\text{Heat}} 2NH_3$$

In this reaction, one molecule of nitrogen combines with three molecules of hydrogen to form two molecules of ammonia (Figure 11.1). Industry uses this reaction to make ammonia for fertilizers, explosives, and plastics, among many other products. The industrial process based on this reaction is called the **Haber process** for the synthesis of ammonia. (To **synthesize** means to combine the parts into the whole.)

Suppose we also wanted to produce ammonia in large quantities for commercial use. To prevent waste and thereby minimize costs, we would

■ FIGURE 11.1

The Haber process

$3H_2$ + N_2 ⟶ $2NH_3$

like to combine just enough of each reactant. But how much is "just enough of each reactant"? And how much ammonia would we obtain? Or suppose we wanted to produce just 100 kilograms of ammonia. How much nitrogen and hydrogen would we need?

To answer such questions, we need to "translate" balanced chemical equations into quantities of reactants and products. In this chapter, you will learn how to do so by using *stoichiometry*, which is pronounced "stoy-key-OM-eh-tree."

11.1 The Mole Method

Stoichiometry is defined as the *calculation of the quantities of substances involved in chemical reactions.* An important idea in stoichiometry is that a balanced equation tells us how many *moles* of the various substances take part in the reaction. For example, the balanced equation for the Haber process,

$$N_2 + 3H_2 \xrightarrow[\text{Pressure}]{\text{Heat}} 2NH_3$$

tells us that 1 *mole* of nitrogen and 3 *moles* of hydrogen react to form 2 *moles* of ammonia. (If you've forgotten what a mole is, refer back to Chapter 9.)

The reasoning behind this idea is as follows: The balanced chemical equation tells us the number of molecules or atoms of each substance that is involved in a single instance of the reaction. It also tells us in what *ratio* these molecules or atoms are involved. In the Haber process, the molecules are involved in the ratio one N_2 to three H_2 to two NH_3, or $1N_2 : 3H_2 : 2NH_3$. Using this ratio, we can also say that 1 *Avogadro's number* of nitrogen molecules and 3 *Avogadro's numbers* of hydrogen molecules produce 2 *Avogadro's numbers* of ammonia molecules. Because 1 Avogadro's number of molecules is 1 mole, we can say that 1 mole of nitrogen and 3 moles of hydrogen react to produce 2 moles of ammonia.

This same sort of reasoning holds true for every balanced equation. *The coefficients in a balanced chemical equation indicate the number of moles of each substance that takes part in the reaction.*

Once we know the number of moles of each substance in a reaction, we can find out how many grams (or kilograms or pounds) of each substance are involved. For example, we know that the atomic mass of nitrogen is 14.0 and that the atomic mass of hydrogen is 1.0. Then we can say that since 1 mole of N_2 has a mass of 28.0 grams, 1 mole of H_2 has a mass of 2.0 grams, and 1 mole of NH_3 has a mass of 17.0 grams, 28.0 grams of nitrogen react with 6.0 grams of hydrogen to produce 34.0 grams of ammonia:

$$N_2 \quad + \quad 3H_2 \quad \xrightarrow[\text{Pressure}]{\text{Heat}} \quad 2NH_3$$

1 mole	3 moles	2 moles
28.0 grams	6.0 grams	34.0 grams

Note that we have satisfied the Law of Conservation of Mass. The mass of the product is equal to the sum of the masses of the reactants. In the remainder of this chapter, we shall apply this basic idea to several types of stoichiometric problems.

11.2   Quantities of Reactants and Products

■ LEARNING GOAL 1

Simple stoichiometry

The best way to learn how to solve stoichiometric problems is to work through a number of them. As we do this, remember that the balanced equation for a reaction contains the basic information about the relative quantities of substances involved. Also note that if we know the quantity of even a single reactant or product, we can obtain the quantities of all other reactants or products.

We can begin by asking how many moles of N_2 are needed to produce 5 moles of NH_3. The balanced equation is

$$N_2 + 3H_2 \rightarrow 2NH_3$$

1 mole	2 moles
?	5 moles

Set up the conversion:

$$5 \text{ moles } NH_3 \times \frac{1 \text{ mole } N_2}{2 \text{ moles } NH_3} = 2.5 \text{ moles } N_2$$

Next we can ask how many moles of H_2 are needed to produce 5 moles of NH_3. The balanced equation is

$$N_2 + 3H_2 \rightarrow 2NH_3$$

3 moles	2 moles
?	5 moles

Set up the conversion:

$$5 \text{ moles } NH_3 \times \frac{3 \text{ moles } H_2}{2 \text{ moles } NH_3} = 7.5 \text{ moles } H_2$$

A balance is used to weigh precise quantities of materials in the laboratory.
(Courtesy Mettler–Toledo)

Let us assume that we want to synthesize 136 grams of ammonia. How many grams of nitrogen and hydrogen do we need? The first thing we do is set up the balanced chemical equation.

$$N_2 + 3H_2 \rightarrow 2NH_3$$

Next we record what is given and what must be found.

$$N_2 + 3H_2 \rightarrow 2NH_3$$
$$? \text{ g} \quad ? \text{ g} \quad 136 \text{ g}$$

The next step is to convert grams of ammonia into moles (because the chemical equation relates substances by moles, not grams).

$$? \text{ moles of } NH_3 = (136 \cancel{\text{ g}}) \left(\frac{1 \text{ mole}}{17.0 \cancel{\text{ g}}} \right) = 8.00 \text{ moles of } NH_3$$

$$N_2 + 3H_2 \rightarrow \qquad 2NH_3$$
$$? \text{ g} \quad ? \text{ g} \quad 136 \text{ g (which is 8.00 moles)}$$

Now we use the coefficients of the balanced equation to calculate the number of moles of hydrogen and nitrogen necessary to form 8 moles of ammonia. We do it in three steps. First, we determine the number of moles of nitrogen:

$$? \text{ moles of } N_2 = (8.00 \cancel{\text{ moles } NH_3}) \left(\frac{1 \text{ mole } N_2}{2 \cancel{\text{ moles } NH_3}} \right) = 4.00 \text{ moles of } N_2$$

We obtained the factor

$$\left(\frac{1 \text{ mole } N_2}{2 \text{ moles } NH_3} \right)$$

from the coefficients of the balanced equation. This factor tells us that 1 mole of nitrogen is necessary to produce 2 moles of ammonia. Now we are ready for the second step.

$$? \text{ moles of } H_2 = (8.00 \cancel{\text{ moles } NH_3}) \left(\frac{3 \text{ moles } H_2}{2 \cancel{\text{ moles } NH_3}} \right) = 12.0 \text{ moles of } H_2$$

We also obtained the factor

$$\left(\frac{3 \text{ moles } H_2}{2 \text{ moles } NH_3} \right)$$

from the balanced equation, which tells us that 3 moles of hydrogen are necessary to produce 2 moles of ammonia. We now have

$$N_2 \quad + \quad 3H_2 \quad \rightarrow \quad 2NH_3$$
$$? \text{ g} \qquad ? \text{ g} \qquad 136 \text{ g}$$
$$4.00 \text{ moles} \quad 12.0 \text{ moles} \quad 8.00 \text{ moles}$$

Another way to obtain the same information is to remember that the moles of reactants and products must be in the ratio of

$$N_2 : H_2 : NH_3$$
$$1 : 3 : 2$$

We could ask ourselves the question,

$$1:3:2 \quad \text{is the same as} \quad ?:?:8$$

To answer this, we would note that

$$1:3:2$$
$$\downarrow \times 4$$
$$?:?:8$$

or

$$1:3:2 \quad \text{is the same as} \quad 4:12:8$$

Our final step is to change *moles* of nitrogen and hydrogen into *grams*.

$$? \text{ grams of } N_2 = (4.00 \text{ moles}) \left(\frac{28.0 \text{ g}}{1 \text{ mole}} \right) = 112 \text{ g of } N_2$$

$$? \text{ grams of } H_2 = (12.0 \text{ moles}) \left(\frac{2.0 \text{ g}}{1 \text{ mole}} \right) = 24 \text{ g of } H_2$$

The problem is solved.

$$N_2 \quad + \quad 3H_2 \quad \rightarrow \quad 2NH_3$$

112 g	24 g	136 g
4.00 moles	12.0 moles	8.00 moles

We can check our mathematics by seeing whether this reaction obeys the Law of Conservation of Mass. The grams of nitrogen and hydrogen we start with should equal the grams of ammonia produced.

MASS OF REACTANTS	MASS OF PRODUCTS
Grams of N_2 = 112 g	Grams of NH_3 = 136 g
Grams of H_2 = 24 g	
Total mass = 136 g	Total mass = 136 g

To practice our stoichiometry, let's solve another problem. Many scientists have suggested using propane gas (C_3H_8) in motor vehicles. Because propane burns cleaner than today's gasolines, these scientists feel that the use of propane would drastically reduce the amount of automobile pollution. When propane reacts with sufficient oxygen, the resulting products are carbon dioxide and water, in the form of water vapor. Our problem is this: How many grams of oxygen are needed for combustion of 22 grams of propane? How many grams of carbon dioxide would be produced? How many grams of water would be produced? (Remember that C_3H_8 is the formula for propane.)

STEP 1

Write a balanced equation, listing what is given and what we want to find.

$$C_3H_8 + 5O_2 \rightarrow 3CO_2 + 4H_2O$$

22 g   ? g   ? g   ? g

STEP 2

Change the grams of C_3H_8 into moles.

$$? \text{ moles of } C_3H_8 = (22 \text{ g}) \left(\frac{1 \text{ mole}}{44.0 \text{ g}}\right) = 0.50 \text{ mole of } C_3H_8$$

$$C_3H_8 + 5O_2 \rightarrow 3CO_2 + 4H_2O$$

22g   ? g   ? g   ? g

0.50 mole

STEP 3

Use the coefficients of the equation to determine the moles of oxygen needed, as well as the moles of carbon dioxide and water produced.

$$? \text{ moles of } O_2 \text{ needed} = (0.50 \text{ mole } C_3H_8) \left(\frac{5 \text{ moles } O_2}{1 \text{ mole } C_3H_8}\right)$$

$$= 2.5 \text{ moles of } O_2 \text{ needed}$$

$$? \text{ moles of } CO_2 \text{ produced} = (0.50 \text{ mole } C_3H_8) \left(\frac{3 \text{ moles } CO_2}{1 \text{ mole } C_3H_8}\right)$$

$$= 1.5 \text{ moles of } CO_2 \text{ produced}$$

$$? \text{ moles of } H_2O \text{ produced} = (0.50 \text{ mole } C_3H_8) \left(\frac{4 \text{ moles } H_2O}{1 \text{ mole } C_3H_8}\right)$$

$$= 2.0 \text{ moles of } H_2O \text{ produced}$$

$$C_3H_8 + 5O_2 \rightarrow 3CO_2 + 4H_2O$$

22g   ? g   ? g   ? g

0.50 mole   2.5 moles   1.5 moles   2.0 moles

Note that you can obtain the same result by the other method.

1 : 5 : 3 : 4

0.50 : ? : ? : ?

or

1 : 5 : 3 : 4

0.50 : 2.5 : 1.5 : 2.0

STEP 4

Change moles of O_2, CO_2, and H_2O into grams.

$$? \text{ grams of } O_2 = (2.5 \text{ moles}) \left(\frac{32.0 \text{ g}}{1 \text{ mole}} \right) = 8\overline{0} \text{ g of } O_2$$

$$? \text{ grams of } CO_2 = (1.5 \text{ moles}) \left(\frac{44.0 \text{ g}}{1 \text{ mole}} \right) = 66 \text{ g of } CO_2$$

$$? \text{ grams of } H_2O = (2.0 \text{ moles}) \left(\frac{18.0 \text{ g}}{1 \text{ mole}} \right) = 36 \text{ g of } H_2O$$

$$C_3H_8 \quad + \quad 5O_2 \quad \rightarrow \quad 3CO_2 \quad + \quad 4H_2O$$

22 g	8$\overline{0}$ g	66 g	36 g
0.50 mole	2.5 moles	1.5 moles	2.0 moles

STEP 5

Check your results.

MASS OF REACTANTS		MASS OF PRODUCTS	
Grams of C_3H_8 =	22 g	Grams of CO_2 =	66 g
Grams of O_2 =	8$\overline{0}$ g	Grams of H_2O =	36 g
Total mass =	102 g	Total mass =	102 g

Let's summarize the procedure for solving stoichiometric problems:

1. Write a balanced chemical equation for the reaction. Record the given number of grams under the chemical formula of that substance. Place question marks for the unknown numbers of grams under the corresponding chemical formulas.
2. Change the given number of grams into moles.
3. Use the coefficients of the balanced equation and the moles of the given amount of substance to determine the moles of the other substances involved in the reaction.
4. Change the moles of the other substances into grams.
5. Check the results to see whether the mass of the reactants equals the mass of the products.

Another way of summarizing this procedure is to think of it in the following sequence:

Grams known → moles known → moles unknown → grams unknown

Before we continue we would like to show one more way to approach this problem. Some students find it easier to solve a problem such as this one using the mass-mass ratio from the balanced equation. We will find the grams of carbon dioxide and the grams of water produced using this method.

$$C_3H_8 + 5O_2 \rightarrow 3CO_2 + 4H_2O$$

1 mole	3 moles	4 moles
44.0 g	3(44.0 g)	4(18.0 g)
44.0 g	132.0 g	72.0 g

$$22 \; \cancel{g} \; C_3H_8 \times \frac{132.0 \text{ g } CO_2}{44.0 \; \cancel{g} \; C_3H_8} = 66 \text{ g } CO_2$$

$$22 \; \cancel{g} \; C_3H_8 \times \frac{72.0 \text{ g } H_2O}{44.0 \; \cancel{g} \; C_3H_8} = 36 \text{ g } H_2O$$

This procedure for solving mass-mass stoichiometric problems can be summarized as follows:

1. First write a balanced chemical equation for the reaction.
2. Underneath the chemical formulas, write the number of moles of the substance given and the number of moles of the substance asked for as indicated by the coefficients.
3. Convert these mole quantities into grams. This may be written below the formulas in the equation.
4. Use the gram-gram relationship as a conversion factor with the mass of the given substance to find the mass of the substance in question.

Practice solving stoichiometric problems by doing the following exercises. Work each problem in steps according to the procedure shown. You may also use the alternate strategy.

EXAMPLE 11.1

The *thermite* reaction is a chemical reaction that is important to railroads, because railroad workers use it to weld railroad tracks. In this reaction, a mixture of aluminum metal and iron(III) oxide (known as thermite) is heated to yield aluminum oxide and iron metal. Once begun, the reaction releases a tremendous amount of heat—so much that the molten iron produced fuses the railroad tracks (Figure 11.2).

How many grams of aluminum and iron(III) oxide are needed to produce 279 grams of iron metal? How many grams of aluminum oxide will be produced?

Solution

UNDERSTAND THE PROBLEM

We are told how many grams of product (iron metal) are formed, and we can use this information to calculate the grams of reactants [aluminum and iron(III) oxide] necessary to produce this amount of product.

DEVISE A PLAN

Use the strategy that was described previously.

Magnesium ribbon acts as a starting fuse

Starting mixture (Mg powder)

Asbestos board with a hole in the center

Thermite mixture

Cone made of filter paper

Pan containing dry sand

■ FIGURE 11.2

The thermite reaction in the lab

CARRY OUT THE PLAN

The procedure may be followed in steps. [*Note:* Fe_2O_3 stands for iron(III) oxide, and Al_2O_3 stands for aluminum oxide.]

STEP 1

$$2Al + Fe_2O_3 \rightarrow Al_2O_3 + 2Fe$$
$$\text{? g} \qquad \text{? g} \qquad \text{? g} \qquad 279 \text{ g}$$

STEP 2

$$\text{? moles of Fe} = (279 \, \cancel{g}) \left(\frac{1 \text{ mole}}{55.8 \, \cancel{g}} \right) = 5.00 \text{ moles of Fe}$$

$$2Al + Fe_2O_3 \rightarrow Al_2O_3 + 2Fe$$
$$\text{? g} \qquad \text{? g} \qquad \text{? g} \qquad 279 \text{ g}$$
$$\qquad\qquad\qquad\qquad\qquad\qquad 5.00 \text{ moles}$$

STEP 3

$$\text{? moles of Al} = (5.00 \, \cancel{\text{moles Fe}}) \left(\frac{2 \text{ moles Al}}{2 \, \cancel{\text{moles Fe}}} \right)$$
$$= 5.00 \text{ moles of Al}$$

$$\text{? moles of Fe}_2O_3 = (5.00 \, \cancel{\text{moles Fe}}) \left(\frac{1 \text{ mole Fe}_2O_3}{2 \, \cancel{\text{moles Fe}}} \right)$$
$$= 2.50 \text{ moles of Fe}_2O_3$$

$$\text{? moles of Al}_2O_3 = (5.00 \, \cancel{\text{moles Fe}}) \left(\frac{1 \text{ mole Al}_2O_3}{2 \, \cancel{\text{moles Fe}}} \right)$$
$$= 2.50 \text{ moles of Al}_2O_3$$

$$2Al \quad + \quad Fe_2O_3 \quad \rightarrow \quad Al_2O_3 \quad + \quad 2Fe$$

$$?\text{ g} \qquad\qquad ?\text{ g} \qquad\qquad ?\text{ g} \qquad\qquad 279\text{ g}$$

5.00 moles   2.50 moles   2.50 moles   5.00 moles

STEP 4

$$? \text{ grams of Al} = (5.00 \text{ moles}) \left(\frac{27.0 \text{ g}}{1 \text{ mole}} \right) = 135 \text{ g of Al}$$

$$? \text{ grams of Fe}_2O_3 = (2.50 \text{ moles}) \left(\frac{159.6}{1 \text{ mole}} \right) = 399 \text{ g of Fe}_2O_3$$

$$? \text{ grams of Al}_2O_3 = (2.50 \text{ moles}) \left(\frac{102.0 \text{ g}}{1 \text{ mole}} \right) = 255 \text{ g of Al}_2O_3$$

$$2 \text{ Al} \quad + \quad Fe_2O_3 \quad \rightarrow \quad Al_2O_3 \quad + \quad 2Fe$$

135 g   399 g   255 g   279 g

5.00 moles   2.50 moles   2.50 moles   5.00 moles

LOOK BACK

We check our results to be sure they satisfy the Law of Conservation of Mass.

MASS OF REACTANTS		MASS OF PRODUCTS	
Grams of Al	= 135 g	Grams of Al_2O_3 =	255 g
Grams of Fe_2O_3 =	399 g	Grams of Fe	= 279 g
Total mass	= 534 g	Total mass	= 534 g

Practice Exercise 11.1   Magnesium oxide is a compound that occurs in nature as the mineral periclase. It is composed of the elements magnesium and oxygen. In industry, magnesium oxide is used in the manufacture of refractory crucibles, firebricks, and casein glue. It is also used as a reflector in optical instruments. It combines with water to form magnesium hydroxide, which acts as an antacid in low concentrations. How many grams of magnesium and oxygen gas are needed to produce 4.032 grams of magnesium oxide?

EXAMPLE 11.2

Polyvinyl chloride (PVC) is one of the best-known and most widely used plastics in the world today. PVC is used to make food wrappings, luggage, plastic garbage bags, phonograph records, and automobile upholstery. It was originally thought that PVC was an extremely safe compound. But studies now indicate that there may be some hidden hazards. The problem is not with the PVC itself but with the compound vinyl chloride (C_2H_3Cl), from which the PVC is made. Vinyl chloride is a colorless gas, and studies show that it may cause liver damage and some forms of cancer in people who are in constant contact with it. Another problem with PVC and vinyl chloride is that when the compounds are burned, an extremely poisonous gas (hydrogen chloride, HCl) can be produced.

Our problem is this: Assume that the following reaction occurs when we burn vinyl chloride (C_2H_3Cl).

$$2C_2H_3Cl + 5O_2 \rightarrow 4CO_2 + 2H_2O + 2HCl$$

How many grams of hydrogen chloride (HCl) will be produced if 6.25 g of vinyl chloride are burned? How many grams of O_2 are needed, and how many grams of CO_2 and water are produced?

Solution

Follow the procedure we used before. Do it in steps.

STEP 1

$$2C_2H_3Cl + 5O_2 \rightarrow 4CO_2 + 2H_2O + 2HCl$$
$$\text{6.25 g} \qquad \text{? g} \qquad \text{? g} \qquad \text{? g} \qquad \text{? g}$$

STEP 2

$$? \text{ moles of } C_2H_3Cl = (6.25\text{ g})\left(\frac{1 \text{ mole}}{62.5\text{ g}}\right) = 0.100 \text{ mole of } C_2H_3Cl$$

$$2C_2H_3Cl + 5O_2 \rightarrow 4CO_2 + 2H_2O + 2HCl$$
$$\text{6.25 g} \qquad \text{? g} \qquad \text{? g} \qquad \text{? g} \qquad \text{? g}$$
$$\text{0.100 mole}$$

STEP 3

$$? \text{ moles of } O_2 = (0.100 \text{ mole } C_2H_3Cl)\left(\frac{5 \text{ moles } O_2}{2 \text{ moles } C_2H_3Cl}\right)$$
$$= 0.250 \text{ mole of } O_2$$

$$? \text{ moles of } CO_2 = (0.100 \text{ mole } C_2H_3Cl)\left(\frac{4 \text{ moles } CO_2}{2 \text{ moles } C_2H_3Cl}\right)$$
$$= 0.200 \text{ mole of } CO_2$$

$$? \text{ moles of } H_2O = (0.100 \text{ mole } C_2H_3Cl)\left(\frac{2 \text{ moles } H_2O}{2 \text{ moles } C_2H_3Cl}\right)$$
$$= 0.100 \text{ mole of } H_2O$$

$$? \text{ moles of HCl} = (0.100 \text{ mole } C_2H_3Cl) \left(\frac{2 \text{ moles HCl}}{2 \text{ moles } C_2H_3Cl} \right)$$

$$= 0.100 \text{ mole of HCl}$$

$2C_2H_3Cl$	$+$	$5O_2$	$\rightarrow$	$4CO_2$	$+$	$2H_2O$	$+$	$2HCl$
6.25 g		? g		? g		? g		? g
0.100 mole		0.250 mole		0.200 mole		0.100 mole		0.100 mole

STEP 4

$$? \text{ grams of } O_2 = (0.250 \text{ mole}) \left(\frac{32.0 \text{ g}}{1 \text{ mole}} \right) = 8.00 \text{ g of } O_2$$

$$? \text{ grams of } CO_2 = (0.200 \text{ mole}) \left(\frac{44.0 \text{ g}}{1 \text{ mole}} \right) = 8.80 \text{ g of } CO_2$$

$$? \text{ grams of } H_2O = (0.100 \text{ mole}) \left(\frac{18.0 \text{ g}}{1 \text{ mole}} \right) = 1.80 \text{ g of } H_2O$$

$$? \text{ grams of } HCl = (0.100 \text{ mole}) \left(\frac{36.5 \text{ g}}{1 \text{ mole}} \right) = 3.65 \text{ g of } HCl$$

STEP 5

MASS OF REACTANTS		MASS OF PRODUCTS	
Grams of C_2H_3Cl =	6.25 g	Grams of CO_2 =	8.80 g
Grams of O_2 =	8.00 g	Grams of H_2O =	1.80 g
		Grams of HCl =	3.65 g
Total mass =	14.25 g	Total mass =	14.25 g

Practice Exercise 11.2   Aluminum nitrate can be produced in a single-replacement reaction. Aluminum replaces zinc in the compound zinc nitrate, because aluminum is higher than zinc in the activity series. The reaction is as follows:

$$2Al + 3Zn(NO_3)_2 \rightarrow 2Al(NO_3)_3 + 3Zn$$

Aluminum nitrate is used in tanning leather, in antiperspirant preparations, and as a corrosion inhibitor. How many grams of aluminum nitrate are produced when 9.469 g of zinc nitrate react with aluminum? How many grams of aluminum are needed, and how many grams of zinc are produced?

11.3   The Limiting-Reactant Problem

Sometimes a chemist who wants to synthesize a compound finds that the quantity of product that can be formed is limited by the amount of one of the starting materials. To find out how much product can be formed, the chemist must first determine which one of the starting materials will be completely used up when the reaction is completed. The other will be in *excess*, so not all of it will be used up to form the product. In other words, the chemist must determine which one of the starting materials is the *limiting reactant.*

To clarify this, let's imagine an automobile manufacturer, the Ajax Automobile Company, which makes a model of a car that is called the Ajax. In order to make just *one* of these automobiles, the assembly line needs (among other things) the following parts: *one* body and *four* wheels. We can represent this statement as follows:

$$1 \text{ body} + 4 \text{ wheels} \rightarrow 1 \text{ automobile}$$

On a particular day, the assembly line has 50 bodies and 160 wheels. How many automobiles can be produced? If you use the number of bodies as the basis for your answer, you might think Ajax could produce 50 automobiles. However, if you use the number of wheels as the basis for your answer, you'll find that Ajax can produce only 40 automobiles. (Remember, it takes 4 wheels to produce an automobile—not counting spare tires. Therefore $160 \div 4 = 40$.) So the company has 10 extra bodies but not enough wheels to use them. As a result, the most automobiles it can produce is 40. The "limiting reactant" in this case is the wheels, and the bodies are the "reactant" that is in excess.

Let's go through this entire procedure again, this time treating it as if it were a chemical problem.

STEP 1

We follow our usual procedure.

$$1 \text{ body} \quad + \quad 4 \text{ wheels} \quad \rightarrow \quad 1 \text{ automobile}$$
$$50 \text{ bodies} \quad \quad 160 \text{ wheels} \quad \quad ? \text{ automobiles}$$

STEP 2

In a chemical problem of this type, we would have to change grams into moles at this point. However, in our car problem, we already know the relationship of bodies to wheels. Therefore all we have to do is determine which substance gets used up first (bodies or wheels).

a. Determine how many bodies would be needed to use up 160 wheels.

$$? \text{ bodies needed} = (160 \text{ wheels}) \left(\frac{1 \text{ body}}{4 \text{ wheels}} \right) = 40 \text{ bodies}$$

b. Determine how many wheels would be needed to use up 50 bodies.

$$? \text{ wheels needed} = (50 \text{ bodies}) \left(\frac{4 \text{ wheels}}{1 \text{ body}} \right) = 200 \text{ wheels}$$

But the Ajax Company doesn't have 200 wheels—only 160. Therefore the wheels limit the number of automobiles it can produce.

STEP 3

We may use the number of wheels to calculate the number of automobiles Ajax can produce and the number of bodies it will use. (Note that we've already calculated the number of bodies that the company will use in step 2a.)

$$? \text{ automobiles} = (160 \text{ wheels}) \left(\frac{1 \text{ automobile}}{4 \text{ wheels}} \right) = 40 \text{ automobiles}$$

Now let's do a chemical problem. Assume that a chemist has 203 grams of $Mg(OH)_2$ (magnesium hydroxide) and 164 grams of HCl (hydrogen chloride). The chemist wants to make these substances react to form $MgCl_2$ (magnesium chloride), which can be used for fireproofing wood or in a disinfectant. How much $MgCl_2$ can the chemist produce, and how much water is formed?

STEP 1

We follow our usual procedure.

$$Mg(OH)_2 + 2HCl \rightarrow MgCl_2 + 2H_2O$$
$$203 \text{ g} \qquad 164 \text{ g} \qquad ? \text{ g} \qquad ? \text{ g}$$

STEP 2

Determine the number of moles of each starting substance.

$$? \text{ moles of } Mg(OH)_2 = (203 \text{ g}) \left(\frac{1 \text{ mole}}{58.3 \text{ g}} \right) = 3.48 \text{ moles of } Mg(OH)_2$$

$$? \text{ moles of HCl} = (164 \text{ g}) \left(\frac{1 \text{ mole}}{36.5 \text{ g}} \right) = 4.49 \text{ moles of HCl}$$

STEP 3

To find which starting substance is used up, use the following procedure.

a. Write the balanced equation, then place the number of moles of each substance below its formula.

$$Mg(OH)_2 + \quad 2HCl \quad \rightarrow MgCl_2 + 2H_2O$$
$$203 \text{ g} \qquad 164 \text{ g} \qquad ? \text{ g} \qquad ? \text{ g}$$
$$3.48 \text{ moles} \quad 4.49 \text{ moles}$$

b. Determine how much HCl would be needed to use up all the $Mg(OH)_2$.

$$? \text{ moles of HCl needed} = (3.48 \text{ moles } Mg(OH)_2) \left(\frac{2 \text{ moles HCl}}{1 \text{ mole } Mg(OH)_2} \right)$$
$$= 6.96 \text{ moles of HCl needed}$$

But we have only 4.49 moles of HCl. Therefore:

c. Our limiting reactant must be HCl, and the $Mg(OH)_2$ must be the reactant in excess. We can check this by calculating the amount of $Mg(OH)_2$ needed to react with all the HCl.

$$? \text{ moles of } Mg(OH)_2 \text{ needed} = (4.49 \text{ moles HCl})\left(\frac{1 \text{ mole } Mg(OH)_2}{2 \text{ moles HCl}}\right)$$

$$= 2.25 \text{ moles of } Mg(OH)_2 \text{ needed (or 131 g)}$$

We have 3.48 moles of $Mg(OH)_2$, so we can expect to have $3.48 - 2.25 = 1.23$ moles of $Mg(OH)_2$ left over, or in excess.

STEP 4

Using the number of moles of HCl that we have, calculate the moles of $MgCl_2$ and H_2O that will be formed.

$$? \text{ moles of } MgCl_2 = (4.49 \text{ moles HCl})\left(\frac{1 \text{ mole } MgCl_2}{2 \text{ moles HCl}}\right)$$

$$= 2.25 \text{ moles of } MgCl_2$$

$$? \text{ moles of } H_2O = (4.49 \text{ moles HCl})\left(\frac{2 \text{ moles } H_2O}{2 \text{ moles HCl}}\right)$$

$$= 4.49 \text{ moles of } H_2O$$

STEP 5

Calculate the number of grams of $MgCl_2$ and H_2O formed.

$$? \text{ grams of } MgCl_2 = (2.25 \text{ moles})\left(\frac{95.3 \text{ g}}{1 \text{ mole}}\right) = 214 \text{ g of } MgCl_2$$

$$? \text{ grams of } H_2O = (4.49 \text{ moles})\left(\frac{18.0 \text{ g}}{1 \text{ mole}}\right) = 80.8 \text{ g of } H_2O$$

$Mg(OH)_2$	+	$2HCl$	$\rightarrow$	$MgCl_2$	+	$2H_2O$
131 g		164 g		214 g		80.8 g
2.25 moles		4.50 moles		2.25 moles		4.50 moles

STEP 6

Check your mathematics by seeing whether the reaction obeys the Law of Conservation of Mass.

MASS OF REACTANTS		MASS OF PRODUCTS	
Grams of $Mg(OH)_2 = 131$ g		Grams of $MgCl_2$	$= 214$ g
Grams of HCl used $= 164$ g		Grams of H_2O	$= 80.8$ g
Total mass used $= 295$ g		Total mass produced $= 294.8$ g (or 295)

As we expected, we had 72 g of extra $Mg(OH)_2$. This material remained unchanged during the course of the reaction.

EXAMPLE 11.3

Chloroform ($CHCl_3$), a quick-acting powerful anesthetic (often used in spy stories to incapacitate the victim), can decompose when it reacts with oxygen. The products formed are HCl and deadly phosgene gas ($COCl_2$), a substance so harmful when inhaled that it was used as a poison gas against enemy troops in World War I. How many grams of $COCl_2$ and HCl can be formed from 35.9 grams of $CHCl_3$ and 6.40 grams of O_2?

Solution

Follow the procedure we used before.

STEP 1

$$2CHCl_3 + O_2 \rightarrow 2COCl_2 + 2HCl$$
$$35.9 \text{ g} \quad 6.40 \text{ g} \quad \text{? g} \quad \text{? g}$$

STEP 2

$$\text{? moles of } CHCl_3 = (35.9 \text{ g})\left(\frac{1 \text{ mole}}{119.5 \text{ g}}\right) = 0.300 \text{ mole of } CHCl_3$$

$$\text{? moles of } O_2 = (6.40 \text{ g})\left(\frac{1 \text{ mole}}{32.0 \text{ g}}\right) = 0.200 \text{ mole of } O_2$$

STEP 3

Determine which of the substances is the limiting reactant.

$$2CHCl_3 + O_2 \rightarrow 2COCl_2 + 2HCl$$
$$35.9 \text{ g} \quad 6.40 \text{ g} \quad \text{? g} \quad \text{? g}$$
$$0.300 \text{ mole} \quad 0.200 \text{ mole}$$

$$\text{? moles of } CHCl_3 \text{ needed} = (0.200 \text{ mole } O_2)\left(\frac{2 \text{ moles } CHCl_3}{1 \text{ mole } O_2}\right)$$

$$= 0.400 \text{ mole of } CHCl_3 \text{ needed}$$

We would need 0.400 mole of $CHCl_3$ to use all the oxygen. However, we do not have 0.400 mole of $CHCl_3$. Therefore the $CHCl_3$ must be the limiting reactant, and the O_2 must be the reactant in excess. Let's check this.

$$\text{? moles of } O_2 \text{ needed} = (0.300 \text{ mole } CHCl_3)\left(\frac{1 \text{ mole } O_2}{2 \text{ moles } CHCl_3}\right)$$

$$= 0.150 \text{ mole of } O_2 \text{ needed (or 4.80 g)}$$

We have more than enough oxygen present to use all the chloroform because we have 0.200 mole of O_2 and we need only 0.150 mole of it.

STEP 4

Using the moles of $CHCl_3$ that are present, calculate the moles of $COCl_2$ and HCl that will be formed.

$$? \text{ moles of } COCl_2 = (0.300 \text{ mole } CHCl_3)\left(\frac{2 \text{ moles } COCl_2}{2 \text{ moles } CHCl_3}\right)$$

$$= 0.300 \text{ mole of } COCl_2$$

$$? \text{ moles of } HCl = (0.300 \text{ mole } CHCl_3)\left(\frac{2 \text{ moles } HCl}{2 \text{ moles } CHCl_3}\right)$$

$$= 0.300 \text{ mole of } HCl$$

STEP 5

$$? \text{ grams of } COCl_2 = (0.300 \text{ mole})\left(\frac{99.0 \text{ g}}{1 \text{ mole}}\right) = 29.7 \text{ g of } COCl_2$$

$$? \text{ grams of } HCl = (0.300 \text{ mole})\left(\frac{36.5 \text{ g}}{1 \text{ mole}}\right) = 11.0 \text{ g of } HCl$$

$2CHCl_3$	+	O_2	→	$2COCl_2$	+	$2HCl$
35.9 g		4.80 g		29.7 g		11.0 g
0.300 mole		0.150 mole		0.300 mole		0.300 mole

STEP 6

Check your mathematics by seeing whether the reaction obeys the Law of Conservation of Mass.

MASS OF REACTANTS		MASS OF PRODUCTS	
Grams of $CHCl_3$ used =	35.9 g	Grams of $COCl_2$	= 29.7 g
Grams of O_2 used =	4.8 g	Grams of HCl	= 11.0 g
Total mass used =	40.7 g	Total mass produced =	40.7 g

Note that in this reaction we had an excess of 0.050 mole (1.6 g) of O_2. This excess oxygen remained unchanged during the course of the reaction.

Practice Exercise 11.3   Potassium acetate, $KC_2H_3O_2$, is a salt with a molecular mass of 98.1. It can be produced by the reaction of potassium hydroxide with acetic acid. Potassium acetate has been used in veterinary medicine, in combating cardiac arrhythmia (irregular heartbeat), and as an expectorant. It acts as a diuretic in animals. How many grams of potassium acetate can be produced from 28.0 g of potassium hydroxide and 120.0 g of acetic acid?

SUMMARY

Stoichiometry is the calculation of the quantities of elements or compounds involved in chemical reactions. Knowledge of stoichiometry enables chemists to determine the quantities of reactants needed and the quantities of products that are formed. It also permits chemists to produce industrial and commercial chemicals with little or no waste.

The coefficients in a balanced equation indicate the ratio in which molecules (and thus moles) of substances take part in a chemical reaction. The mole method for computing the quantities of reactants and products makes use of this fact. First we find the number of moles of each substance involved in the reaction from this ratio. Then, from the number of moles, we find the number of grams of each substance. The same method may be used to determine which of several reactants is the limiting reactant. It may also be used to determine the quantity of product that can be formed with the available amount of the limiting reactant.

KEY TERMS

Haber process **(Introduction)**
stoichiometry **(11.1)**

synthesize **(Introduction)**

CAREER SKETCH

Law Careers

Careers in law are often pursued by scientists with and without law degrees. There are numerous directions that chemists can explore in the legal field, each requiring its own special skills.

All interested in a career in the legal field need excellent writing skills, strong analytical abilities, good problem-solving and decision-making skills, and attention to detail. A degree in chemistry is almost always a requirement for someone pursuing a career that links chemistry and the law. Certain subfields also require that special exams be taken. Someone interested in becoming a patent agent, for example, must take the patent bar exam even though he or she does not have to go to law school.

If you are interested in intellectual property law, you should have a strong background in the physical sciences. This will allow you to deal with such high-profile matters as patents, copyrights, and trademarks related to biotechnology companies. Environmental law affords you the opportunity to work with product liability cases, pollution issues, and contamination litigation. Food and drug law requires a strong background in chemistry or toxicology.

Scientists specializing in legal issues are often employed by the federal government. Others work for law firms, corporations, or as private consultants. Some even serve as expert witnesses at trials, a privilege that accompanies the distinguished reputation that can be achieved by a scientist after years in the field and a significant amount of publishing.

CHEMICAL FRONTIERS

*Daily Cup of Tea for Better Health**

Almost every child can remember being told that "an apple a day keeps the doctor away." Recently, scientists have discovered that there might be an even better way of keeping the doctor away—drinking a cup of tea each day.

The news comes from a Dutch study that focused on elderly men. The researchers examined 805 people, collecting detailed information about the dietary patterns of the subjects. Five years later, the researchers re-evaluated the surviving subjects and collected data on those who had died. They found that participants who had diets rich in flavonoids, antioxidants that occur naturally in plants and drinks derived from plants, appeared to significantly cut their chance of dying from coronary heart disease.

Here's why: Research has found that heart disease is likely caused by oxidation of low-density lipoprotein (LDL), or "bad" cholesterol. Dietary agents (such as flavonoids) prevent LDLs from undergoing the oxidation process, in turn minimizing the chance that heart disease will occur.

So the researchers decided to begin a chemical study of different foods to find out which are highest in flavonoid content. They found a high amount of the chemical in tea. In fact, when they went back and looked at their study participants, they found that the men received 61 percent of the flavonoids in their diets from tea, compared with only 13 percent from apples.

Future research will likely find other sources for this important antioxidant. Until then, the next time you eat an apple for your health, you might consider following it with a cup of tea.

*Adapted from *Science News*, October 30, 1993, p. 278

Flavonoids, chemicals found in tea, may reduce coronary artery disease.
(Fundamental Photographs)

SELF-TEST EXERCISES

LEARNING GOAL 1: *Simple Stoichiometry*

1. Sodium thiosulfate, $Na_2S_2O_3$, known as "hypo" in photography, is used to remove excess silver bromide, AgBr, in the film-developing process. The reaction is

$$Na_2S_2O_3 + AgBr \rightarrow$$
$$Na_3Ag(S_2O_3)_2 + NaBr \text{ (unbalanced)}$$

(a) Balance the equation.
(b) How many grams of sodium thiosulfate are needed to react with 46.95 g of silver bromide?
(c) How many grams of $Na_3Ag(S_2O_3)_2$ are produced? How many grams of NaBr are produced?

2. A spectacular laboratory demonstration is the synthesis of sodium chloride (table salt) from its elements. In this reaction, pure sodium metal is heated in an enclosed glass container in the presence of green-colored chlorine gas. A violent reaction begins producing heat, light, and a white cloud of sodium chloride. The reaction can be represented by the following chemical equation:

$$Na + Cl_2 \rightarrow NaCl \text{ (unbalanced)}$$

(a) Balance the equation.
(b) How many grams of NaCl can be produced from 4.60 g of sodium metal?
(c) How many grams of chlorine gas are needed to react with the sodium metal?

3. Nitrous oxide (N_2O), commonly called laughing gas, can be prepared from the decomposition of ammonium nitrate:

$$NH_4NO_3 \xrightarrow{\text{Heat}} N_2O + H_2O$$

(a) Balance the equation.
(b) How many grams of ammonium nitrate are needed to prepare 2.2 g of nitrous oxide?
(c) How many grams of water are produced?

4. Lithium phosphate is a white crystalline powder that is soluble in dilute acids. It can be prepared by the reaction

$$H_3PO_4 + LiOH \rightarrow Li_3PO_4 + H_2O$$

(a) Balance the equation.
(b) How many grams of lithium phosphate can be produced from 168.0 g of lithium hydroxide?
(c) How many grams of phosphoric acid are needed, and how many grams of water are produced?

5. We can prepare the chemical explosive trinitrotoluene (TNT) by the reaction of toluene (C_7H_8) with nitric acid:

$$C_7H_8 + HNO_3 \rightarrow \underset{\text{TNT}}{C_7H_5N_3O_6} + H_2O$$

(a) Balance the equation.
(b) How many grams of toluene and nitric acid are necessary to manufacture $1,\overline{0}00$ g of TNT?
(c) How many grams of water will be produced in the reaction?

6. Calcium hydroxide, also known as slaked lime, is used in mortar, plaster, cement, and other building and paving materials. When it reacts with phosphoric acid, it produces calcium phosphate and water.

(a) Write a balanced chemical equation for this re-action.

(b) How many grams of calcium phosphate can be produced from 148 g of calcium hydroxide?

(c) How many grams of phosphoric acid are needed, and how many grams of water are produced?

◀ **7.** The compound ammonium sulfate can be used as a local analgesic (pain reliever). This compound can be prepared by a reaction of ammonium chloride and sulfuric acid.

(a) Write a balanced equation for this reaction.

(b) How many grams of sulfuric acid are needed to react with 15.9 g of ammonium chloride?

(c) How many grams of ammonium sulfate and hydrogen chloride are produced?

◀ 8. Barium chloride is used in manufacturing pigments and paints, in weighting and dyeing textile fabrics, and in tanning and finishing leather. When it reacts with potassium phosphate, the products formed are potassium chloride and barium phosphate.

(a) Write a balanced equation for this reaction.

(b) How many grams of potassium chloride can be produced from 0.208 g of barium chloride?

(c) How many grams of potassium phosphate are needed, and how many grams of barium phosphate are produced?

* **9.** Suppose you are a chemist who has been asked to analyze a sample containing potassium chlorate ($KClO_3$) and sodium chloride. You are asked to determine the percentage of potassium chlorate in the sample. You can do this by heating a weighed amount of the sample. The potassium chlorate will decompose to potassium chloride and oxygen gas. You can reweigh the sample after heating, to tell how much oxygen was lost. Then you can backtrack to find the number of grams of potassium chorate that decomposed to yield this mass of oxygen. From this information, you can calculate the percentage of potassium chlorate in the sample.

Assume that the masses of your sample are as follows, and find the percentage of potassium chlorate in the mixture of potassium chlorate and sodium chloride:

Grams of sample before heating = 5.00 g

Grams of sample after heating = 4.00 g

10. A sample containing $NaClO_3$ and KCl must be analyzed to determine the percentage of $NaClO_3$. The sample is weighed and then heated, and the $NaClO_3$ decomposes to NaCl and O_2. The sample is weighed after heating to determine the amount of O_2 lost during decomposition. The number of grams of $NaClO_3$ that decomposed to yield the mass of oxygen is calculated. The percentage of $NaClO_3$ in the sample is then calculated. Perform this analysis using the following information:

Grams of sample before heating = 10.00 g

Grams of sample after heating = 8.00 g

11. Hydrogen peroxide (H_2O_2) is a compound with many uses. In dilute solutions (3% H_2O_2 in water), it can be used as an antiseptic for cuts. In a solution of 90% concentration, it can be used as a fuel for rockets. H_2O_2 is prepared in the following way:

$BaO_2 + H_3PO_4 \rightarrow$

Barium peroxide $\qquad H_2O_2 + Ba_3(PO_4)_2$ (unbalanced)

(a) Balance the equation.

(b) How many grams of hydrogen peroxide can be made from 338 g of barium peroxide?

(c) How much phosphoric acid is needed to react with 338 g of barium peroxide?

12. Lead(II) iodide is a chemical used in bronzing, printing, and photography, as well as in making gold pencils. It is composed of lead and iodine.

(a) Write a balanced equation for the formation of lead(II) iodide from its elements.

(b) How many grams of lead and iodine are needed to produce 1.38 g of lead(II) iodide?

*13. Suppose you have a solution containing 588 g of sulfuric acid, and you put 200 g of magnesium into it. A reaction occurs. After the reaction is complete, 56 g of the original magnesium remain. How many grams of hydrogen were produced?

*14. Suppose you have a solution containing 216.8 g of HCl, and you put 100.0 g of magnesium into it. A reaction occurs. After the reaction is complete, 28.00 g of the original magnesium remain. How many grams of hydrogen were produced?

*15. How many grams of *air* are necessary for a complete combustion of 90.0 g of ethane (C_2H_6), producing carbon dioxide and water vapor? Assume that the air contains 23.0% oxygen by weight.

*16. How many grams of *air* are necessary for the complete combustion of 64.0 g of methane (CH_4), producing carbon dioxide and water vapor? Assume that the air contains 23.0% oxygen by weight.

◀ **17.** Butane, C_4H_{10}, is an organic compound used as a fuel in some types of cigarette lighters. When butane is burned with oxygen, the following reaction takes place.

$$C_4H_{10} + O_2 \rightarrow CO_2 + H_2O \text{ (unbalanced)}$$

(a) Balance the equation.
(b) How many grams of oxygen are needed to burn 23.2 g of butane?
(c) How many grams of carbon dioxide and water are produced?

◀ 18. Ethane, C_2H_6, is an organic compound that together with 90% propane and 5% butane composes bottled gas. When ethane is burned with oxygen, the following reaction takes place:

$$C_2H_6 + O_2 \rightarrow CO_2 + H_2O \text{ (unbalanced)}$$

(a) Balance the equation.
(b) How many grams of oxygen are needed to burn $18\overline{0}$ g of ethane?
(c) How many grams of carbon dioxide and water are produced?

19. Ozone, O_3, is a pale-blue gas that can be formed in the atmosphere from O_2. The reaction is

$$3O_2 \rightarrow 2O_3$$

How many grams of ozone can be formed from 64.0 g of O_2?

◀ 20. Ozone (O_3) is formed from oxygen gas in the reaction

$$3O_2 \rightarrow 2O_3$$

How many grams of oxygen gas are needed to form 0.048 g of ozone?

21. Aluminum sulfate has been used to tan leather, to treat sewage, and as an antiperspirant. The compound can be formed by the acid–base reaction

$$Al(OH)_3 + H_2SO_4 \rightarrow$$
$$Al_2(SO_4)_3 + H_2O \text{ (unbalanced)}$$

(a) Balance the equation.
(b) How many grams of aluminum hydroxide and sulfuric acid are needed to produce $50\overline{0}$ g of aluminum sulfate?
(c) How many grams of water are produced?

22. Zinc sulfide is used as a pigment for paints, oilcloths, linoleum, and leather. It is composed of the elements zinc and sulfur.
(a) Write a balanced equation for the formation of zinc sulfide from its elements.

(b) How many grams of zinc and sulfur are needed to produce 2.44 g of zinc sulfide?

23. In a blast furnace, iron(III) oxide is converted to iron metal. The overall reaction for this process is

$$Fe_2O_3 + CO \rightarrow Fe + CO_2 \text{ (unbalanced)}$$

(a) Balance the equation.
(b) How many grams of iron(III) oxide are needed to produce 454.0 g of iron metal (about one pound)?
(c) How many grams of CO are needed, and how many grams of CO_2 are produced?

24. Silver nitrate, a topical anti-infective agent, reacts with copper to form copper(II) nitrate and silver metal.
(a) Write a balanced equation for this reaction.
(b) How many grams of copper are needed to react with 1.70 g of silver nitrate?
(c) How many grams of silver metal and copper(II) nitrate are formed from the reaction?

25. Nitric acid is prepared commercially by the *Ostwald process*. The first part of this process involves the reaction of ammonia with oxygen to produce nitric oxide, using platinum as a catalyst:

$$NH_3 + O_2 \xrightarrow{\text{Pt}} NO + H_2O \text{ (unbalanced)}$$

(a) Balance the equation.
(b) How many grams of oxygen are needed to react with 425 g of ammonia?
(c) How many grams of nitric oxide and water are produced?

26. Glucose, $C_6H_{12}O_6$, reacts with oxygen to produce carbon dioxide and water.
(a) Write a balanced equation for this reaction.
(b) How many grams of glucose are needed to react with 3.20 g of oxygen?
(c) How many grams of carbon dioxide and water are produced?

*27. Suppose that you have a solution of $50\overline{0}$ g of dissolved $Cu(NO_3)_2$ and you add $30\overline{0}$ g of Zn metal to it. A reaction occurs. After the reaction is complete, 127 g of Zn remain. How many grams of $Zn(NO_3)_2$ are produced?

28. Hydrogen gas reacts with iodine to produce hydrogen iodide. Hydrogen iodide is a strong irritant that is used in the manufacture of hydroiodic acid.
(a) Write a balanced equation for the reaction between hydrogen and iodine.

(b) How many grams of hydrogen gas and iodine are needed to produce $32\overline{0}$ g of hydrogen iodide?

LEARNING GOAL 2: *Limiting-Reactant Problems*

◀ **29.** The compound acetylene (C_2H_2) is used as a fuel for welding (for example, in the oxyacetylene torch). In years gone by, acetylene was used as a surgical anesthetic. The compound results from the reaction

$$CaC_2 + H_2O \rightarrow Ca(OH)_2 + C_2H_2 \text{ (unbalanced)}$$
Calcium   Acetylene
carbide

   (a) Balance this equation.
   (b) If you have 128 g of calcium carbide and 144 g of water, how many grams of calcium hydroxide and acetylene can you produce?

◀ 30. In the Ostwald process, which is used to produce nitric acid, ammonia gas (NH_3) is initially converted to nitric oxide (NO) according to the following reaction:

$$NH_3 + O_2 \rightarrow NO + H_2O \text{ (unbalanced)}$$

   (a) Balance the equation.
   (b) How many grams of NO and how many grams of H_2O are produced from the reaction of 102.0 g of NH_3 and 320.0 g of O_2?

31. Pure iron is obtained by extracting it from its most abundant ore, hematite, Fe_2O_3. The process is carried out in a device known as a blast furnace. The chemical reaction can be represented by the following equation:

$$Fe_2O_3 + CO \rightarrow Fe + CO_2$$

   (a) Balance the equation.
   (b) How many grams of iron and how many grams of carbon dioxide can be obtained from the reaction of 399.8 g of Fe_2O_3 and 168.0 g of CO?

32. Nitrogen gas reacts with hydrogen gas to form ammonia gas (NH_3). How many grams of ammonia gas can be formed from 0.014 g of N_2 and 0.020 g of H_2?

◀ **33.** Water can be produced from its elements by the reaction

$$H_2 + O_2 \xrightarrow{\text{Electric spark}} H_2O \text{ (unbalanced)}$$

   (a) Balance the equation.
   (b) How many grams of water can be produced from $1\overline{0}$ g of H_2 and 64 g of O_2?

34. Propane, C_3H_8, reacts with oxygen gas to form carbon dioxide and water according to the equation

$$C_3H_8 + 5O_2 \rightarrow 3CO_2 + 4H_2O$$

How many grams of carbon dioxide and water can be formed from 11.0 g of propane and 32 g of oxygen gas?

35. When calcium phosphide is placed in water, it forms phosphine gas (PH_3). In air, PH_3 usually bursts into flames. In fact, this substance (PH_3) may be responsible for people observing faint flickers of light in marshes. The PH_3 could be produced by the reduction of naturally occuring phosphorus compounds. The reaction for the formation of PH_3 from calcium phosphide is

$$Ca_3P_2 + H_2O \rightarrow PH_3 + Ca(OH)_2 \text{ (unbalanced)}$$

   (a) Balance the equation.
   (b) How many grams of PH_3 can be produced from 515 g of calcium phosphide and 216 g of water?
   (c) How many grams of calcium hydroxide are produced?

◀ 36. Aluminum reacts with iron(III) oxide to produce iron and aluminum oxide.
   (a) Write a balanced equation for this reaction.
   (b) How many grams of aluminum oxide and iron can be produced from 135 g of Al and $80\overline{0}$ g of iron(III) oxide?

37. Elementary phosphorus can be produced by the reaction

$$Ca_3(PO_4)_2 + 3SiO_2 + 5C \rightarrow$$
$$3CaSiO_3 + 5CO + P_2$$

Determine the number of moles of each product formed from 8.0 moles of $Ca_3(PO_4)_2$, $2\overline{0}$ moles of SiO_2, and 45 moles of C.

38. Barium metal reacts with oxygen gas to form barium oxide, a chemical used for drying gases and solvents.
   (a) Write a balanced equation for this reaction.
   (b) How many grams of barium oxide can be produced from $10\overline{0}$ g of barium and $10\overline{0}$ g of oxygen gas?

39. An inexpensive way of preparing pure hydrogen gas involves the following reaction between iron and steam:

$$Fe + H_2O \rightarrow Fe_3O_4 + H_2 \text{ (unbalanced)}$$

   (a) Balance the equation.
   (b) How many grams of hydrogen gas can be prepared from 225 g of Fe and 225 g of H_2O?
   (c) How many grams of Fe_3O_4 are produced?

40. Sulfur dioxide is a chemical used in preserving fruits and vegetables and as a disinfectant in breweries and food factories. It is produced from the reaction of sulfur and oxygen gas.
    (a) Write a balanced equation for this reaction.
    (b) How many grams of sulfur dioxide can be prepared from 100.0 g of sulfur and 200.0 g of oxygen gas?

EXTRA EXERCISES

◀ **41.** (a) Write the balanced equation for the reaction in which chlorine, Cl_2, reacts with potassium metal, K, to produce potassium chloride. How many moles of chlorine are needed to react with 10.0 g of potassium?
    (b) How many grams of potassium chloride are produced?

◀ **42.** Consider the reaction

$$Cu(NO_3)_2 \rightarrow CuO + NO_2 + O_2 \text{ (unbalanced)}$$

    (a) Balance the equation.
    (b) What mass of O_2 is produced if $1\overline{00}$ g of copper(II) nitrate react?
    (c) How much copper(II) nitrate must react if 54.0 g of NO_2 are formed?

43. Assume that $1\overline{00}$ g of carbon and $15\overline{0}$ g of oxygen are available for the reaction

$$C + O_2 \rightarrow CO_2$$

    (a) What is the limiting reactant?
    (b) How much CO_2 is formed?

44. A reaction is carried out with 8.70 g of hydrogen and 64.0 g of oxygen to produce water.
    (a) Write a balanced equation for this reaction.

(b) What amount of each substance will be on hand when the reaction is complete?

45. Calculate the number of grams of oxygen gas produced by the decomposition of 1.00 kg of potassium chlorate. [*Hint:* $KClO_3 \rightarrow KCl + O_2$ (unbalanced)]

46. Write the equation for the reaction between zinc metal and sulfuric acid. What mass of zinc is required to produce $6\overline{00}$ g of hydrogen gas?

47. Two experiments are performed. In the first, a 50.0-g plate of silver is immersed in a solution of copper(II) chloride. In the second experiment, a 50.0-g plate of lead is immersed in the same solution.
    (a) Determine in which experiment a reaction occurs, and write a balanced equation for this reaction.
    (b) Assuming that the metal is the limiting reactant, calculate the masses of all products formed in the reaction that occurs.

48. Aluminum reacts with sulfuric acid to produce aluminum sulfate and hydrogen gas. How many grams of hydrogen are produced from the reaction of 60.0 g of aluminum with excess sulfuric acid?

49. Hydrochloric acid reacts with sodium hydroxide to produce sodium chloride and water. Suppose that 36.5 g of HCl react with 80.0 g of NaOH. How many grams of sodium chloride are produced? How many grams of the excess reactant are left at the end of the reaction?

50. Carbon reacts with oxygen under certain conditions to produce carbon monoxide. Suppose that 48.0 g of carbon react with 64.0 g of oxygen gas to produce *only* carbon monoxide. How many grams of carbon monoxide are produced?

1. How many moles are there in 3.43 g of $Al_2(SO_4)_3$?

2. How many moles are there in 4.8 g of C atoms?

3. Determine the number of grams in each of the following:
(a) 0.750 mole of H_2SO_4
(b) 5.00 moles of H_2O
(c) 0.00700 mole of H_2CO_3
(d) 50.0 moles of $HC_2H_3O_2$

4. After completing an experiment, you find 4.6 g of unreacted Na in the vessel. How many atoms does this represent?

5. After a reaction is complete, it is determined that 5.328 g of Sn combined with 1.436 g of O. Calculate the empirical formula of this compound.

6. The empirical formula for the antibiotic nonactin is $C_{10}H_{16}O_3$. If the molecular mass of nonactin is 736, what is its molecular formula?

7. The empirical formula for fructose, the natural sugar found in fruit juice, fruits, and honey, is CH_2O. If the molecular mass of fructose is $18\overline{0}$, what is its molecular formula?

8. The molecular formula for aspirin is $C_9H_8O_4$. You have 36.0 g of aspirin.
(a) How many moles of aspirin are there?
(b) How many molecules of aspirin are there?
(c) How many moles of carbon atoms are there?
(d) How many moles of hydrogen atoms are there?
(e) How many moles of oxygen atoms are there?

9. A 6.08-g sample of nitrogen combines with 13.90 g of oxygen to produce a compound whose molecular mass is 92.0. What is the molecular formula of this compound?

10. Calculate the molecular mass of each of the following compounds: (a) $Ca_3(PO_4)_2$ (b) CH_4 (c) NO_2 (d) $Ba(C_2H_3O_2)_2$

11. Calculate the percentage composition of each of the following compounds: (a) H_2S (b) NaCl (c) $Fe(C_2H_3O_2)_3$ (d) K_2SO_4

12. Determine the molecular formula of a compound that has a molecular mass of 86.0 and is composed of 83.7% carbon and 16.3% hydrogen by mass.

13. Find the percentage by mass of carbon, hydrogen, and nitrogen in the compound nicotine, given that its empirical formula is C_5H_7N and that its molecular mass is 162.0.

14. How many water molecules are there in a 0.720-g sample of H_2O?

15. A patient has been told to take a $1,\overline{0}00$-mg supplement of elemental calcium each day. How many grams of $CaCO_3$ must she take in order to receive $1,\overline{0}00$ mg of elemental calcium?

16. Write a balanced chemical equation for the reaction in which calcium bromide + sulfuric acid forms hydrogen bromide + calcium sulfate.

17. When magnesium metal reacts with oxygen, magnesium oxide is formed. Write a balanced equation for this reaction.

18. A flask found in a chemistry laboratory was coated with a thin film that had a cloudy appearance. Chemical analysis of the film showed it to be ammonium chloride formed by reaction of ammonia gas and hydrogen chloride gas. Write the balanced equation for this reaction.

19. Identify each of the following reactions as a combination, decomposition, single-replacement, or double-replacement reaction:
(a) $C_{12}H_{22}O_{11} \rightarrow 12C + 11H_2O$
(b) $H_2 + Br_2 \rightarrow 2HBr$
(c) $2KBr + Cl_2 \rightarrow 2KCl + Br_2$
(d) $2HCl + Ca(OH)_2 \rightarrow CaCl_2 + 2H_2O$

20. Complete and balance each of the following equations. Use the activity series and solubility table when necessary.
(a) $H_2 + Cl_2 \rightarrow$?
(b) $NaCl(aq) + Pb(NO_3)_2(aq) \rightarrow$? + ?
(c) $Al + H_2SO_4 \rightarrow$? + ?
(d) $SrCO_3 \xrightarrow{\text{Heat}}$? + ?

21. Complete and balance the following equation:
$$CaCO_3 \xrightarrow{\text{Heat}} ? + ?$$

22. Balance the following equations:
(a) $TiCl_4 + H_2O \rightarrow TiO_2 + HCl$
(b) $P_4O_{10} + H_2O \rightarrow H_3PO_4$

23. Complete and balance the following equations. If necessary, check the solubility table to be sure that the reaction can take place.
(a) $Fe(OH)_3 + H_3PO_4 \rightarrow$? + ?
(b) $Pb(OH)_2 + HNO_3 \rightarrow$? + ?

24. Acid rain is formed in the atmosphere by the reaction of sulfur trioxide (an air pollutant) and water, forming sulfuric acid. Complete and balance the equation describing this process.

25. In an automobile engine, fuel that contains a chemical called octane (C_8H_{18}) reacts with oxygen to form carbon dioxide and water. Write a balanced equation for this reaction.

26. Isopropyl alcohol (C_3H_7OH), or rubbing alcohol, reacts with oxygen to produce carbon dioxide and water. Write a balanced equation for this reaction.

27. Hard water containing $CaSO_4$ can be softened by adding washing soda (Na_2CO_3). Write a balanced chemical equation for this reaction. Which chemical precipitates out, causing the water to soften?

28. When the sugar sucrose ($C_{12}H_{22}O_{11}$) is heated, it decomposes to form carbon and water. Write a balanced chemical equation for this reaction.

29. In the following reaction, the sum of the coefficients is 9. True or false? (*Hint:* Balance the equation before answering.)

$$Al + H_2SO_4 \rightarrow H_2 + Al_2(SO_4)_3$$

30. In the following reaction, the sum of the coefficients is 4. True or false? (*Hint:* Balance the equation before answering.)

$$NaNO_3 \rightarrow NaNO_2 + O_2$$

31. In the following reaction, metallic zinc is oxidized and hydrogen ions are reduced. True or false?

$$Zn + H_2SO_4 \rightarrow ZnSO_4 + H_2$$

32. In the following reaction, sodium is oxidized and chlorine is reduced. True or false?

$$Na + Cl_2 \rightarrow NaCl$$

33. In the following reaction, tin is the reducing agent and HNO_3 is the oxidizing agent. True or false?

$$Sn + HNO_3 \rightarrow SnO_2 + NO_2 + H_2O$$

34. The sum of the coefficients for the following oxidation–reduction reaction is 38. True or false?

$$K_2Cr_2O_7 + FeCl_2 + HCl \rightarrow$$
$$CrCl_3 + KCl + FeCl_3 + H_2O \text{ (unbalanced)}$$

35. Balance the following redox reaction:

$$C + H_2SO_4 \rightarrow CO_2 + SO_2 + H_2O$$

36. When copper(II) oxide reacts with hydrogen, copper and water are produced. How many moles of copper oxide are needed to completely react with 5 moles of hydrogen?

37. Consider the following chemical equation:

$$2Na + 2H_2O \rightarrow 2NaOH + H_2$$

How many moles of sodium hydroxide are produced when 0.46 mole of sodium reacts with 0.20 mole of water?

38. Calculate the number of grams of propane (C_3H_8) needed to produce 6.0 moles of carbon dioxide by the following reaction:

$$C_3H_8 + 5O_2 \rightarrow 3CO_2 + 4H_2O$$

39. How many grams of hydrochloric acid are needed to produce 340 g of hydrogen sulfide by the following reaction?

$$FeS + HCl \rightarrow FeCl_2 + H_2S \text{ (unbalanced)}$$

40. A 30.0-g sample of ethane (C_2H_6) is allowed to react with 16.0 g of oxygen to produce carbon dioxide and water. How many grams of carbon dioxide will this reaction produce?

41. How many grams of phosphoric acid are needed to react with 1,360 g of lithium hydroxide in the following reaction?

$$H_3PO_4 + 3LiOH \rightarrow Li_3PO_4 + 3H_2O$$

42. Calcium hydroxide reacts with phosphoric acid to produce calcium phosphate and water. How many grams of calcium phosphate can be produced from 49.0 g of phosphoric acid? How many grams of calcium hydroxide are needed?

43. When the compound Epsom salts (magnesium sulfate heptahydrate, $MgSO_4 \cdot 7H_2O$) is heated, 1 mole of magnesium sulfate and 7 moles of water are produced. How many moles of magnesium sulfate are produced when 15.0 moles of Epsom salts are heated? How many grams is this?

44. When hydrogen peroxide (H_2O_2) decomposes, water and oxygen are produced. How many grams of hydrogen peroxide are needed to produce 63.0 g of water?

45. When sodium chloride and lead(II) nitrate react, lead(II) chloride solid and sodium nitrate result. When 66.4 g of lead(II) nitrate react with 5.85 g of sodium chloride, how many grams of lead(II) chloride form?

46. Carbon dioxide gas reacts with lithium hydroxide to form lithium carbonate and water. How many grams of lithium hydroxide are needed to produce 148 g of lithium carbonate?

47. Magnesium and oxygen react to form magnesium oxide. How many grams of magnesium are required to react with 48.0 g of oxygen?

48. Barium and water form barium hydroxide and hydrogen gas upon reaction. When 137.3 g of barium are mixed with 72.0 g of water, how much barium hydroxide and how much hydrogen are produced? Which reactant is in excess and by how much?

HEATS OF REACTION:
Chemistry and Energy

12

It is the year 1840. Experiments performed by the English physicist James Joule and the German physicists Julius von Mayer and Hermann von Helmholtz prove that when changes in energy occur in a physical process, the energy is neither created nor destroyed, just converted from one form to another. This principle is called the *Law of Conservation of Energy* and is known today as the *First Law of Thermodynamics*.

Around the same time, research performed by several other scientists, most notably Nicholas Camot, Lord Kelvin, and Rudolf Clausius, shows that, when left to itself, heat flows spontaneously from a point of high temperature to one of lower temperature. Research results also show that work can be obtained from heat only when such a flow across temperature differences exists.

It is at this point that the worlds of chemistry and physics finally meet. The Swiss-Russian chemist Henri Hess has been busy performing numerous experiments in which he carefully measures the heat evolved in chemical reactions of known quantities of materials. Now, in 1840, he publishes his work and shows that the heat produced or absorbed in going from one substance to another is the same, no matter what chemical route is chosen, and no matter how many steps are involved. Because of the importance of this discovery, Hess becomes known as the *father of thermochemistry,* and the results of his experiments become known as *Hess's Law.*

Hess's Law supplies the evidence necessary to show that the Law of Conservation of Energy applies to chemical changes as well as physical changes. In other words, the laws of thermodynamics apply to chemistry as well as physics. Over the next two decades several other chemists and physicists expand on Hess's work, producing a wealth of knowledge and understanding regarding the importance of energy in chemical reactions.

■ ■ ■   **LEARNING GOALS**

After you have studied this chapter, you should be able to:

1. Distinguish between exothermic reactions and endothermic reactions.
2. Define the terms *heat of reaction*, *heat of formation*, and *enthalpy*.
3. Define the terms *calorie* and *specific heat*.
4. Calculate the amount of heat gained or lost by a substance when its temperature changes.
5. Write a thermochemical equation.
6. Explain the significance of positive and negative values of ΔH.
7. Calculate the heat of formation or heat of reaction when you are given a balanced equation and the appropriate information.

■ ■ ■   **INTRODUCTION**

Turn on a *light*, take a ride in your *car*, cook a roast in your *oven*, or just turn on your portable *radio*. All these devices use energy, and this energy usually comes from chemical reactions. The entire world operates on energy. In fact, the use of energy from chemical reactions is the basis of life itself. In this chapter, we discuss the changes in energy that occur in chemical reactions, and we examine some of the practical uses of energy obtained from various reactions.

12.1   The Release or Absorption of Energy

LEARNING GOAL 1
Exothermic and endothermic reactions

Associated with every substance is a quantity of energy (a sort of chemical potential energy). When a substance reacts to form a new substance, it can either lose energy to the surroundings or gain energy from the surroundings. This is because in chemical reactions, old bonds between atoms are broken and new bonds are formed. Breaking a bond usually requires energy, while forming a bond usually releases energy. Depending on the number of bonds broken and the number formed, and depending on the strengths of the bonds, energy is either released to the surroundings or absorbed from them. This energy is usually in the form of heat.

Reactions that release energy to the surroundings are called **exothermic reactions**. For example, the explosion of a stick of dynamite is a chemical reaction that releases a tremendous quantity of heat energy to the environment. *Reactions that absorb energy from the surroundings* are called **endothermic reactions.** For example, plants live and grow by a process called photosynthesis. In this process, plants use carbon dioxide and water and they absorb *energy from the sun* (sunlight) to make glucose and oxygen.

Solar energy is our most powerful energy source. Solar panels on the roof of this experimental house provide the occupants with most of the heat and hot water they need.
(Courtesy Dr. K. W. Böer)

12.2 Heat and the Chemical Reaction

LEARNING GOAL 2

Definitions of energy terms

Any collection of molecules contains an energy that, if released by chemical reaction, can produce heat: a tangible energy effect you can feel with your hands. *The heat content of a specific amount of a particular substance* is called its **enthalpy**. (The symbol for enthalpy is *H*.) One mole of any substance at a given temperature and pressure has a specific enthalpy, which may be thought of as its heat content. We cannot measure the heat content (enthalpy) of a substance itself. However, we can measure the *change* in heat content that takes place during a chemical reaction. That is, the heat that is absorbed or released during a chemical reaction can be measured. *The change in heat content that results from a chemical reaction* is defined as the **heat of reaction.**

The heat of reaction is usually measured at 25°C and 1 atmosphere (atm) pressure. (The temperature 25°C was probably chosen because it is close to room temperature. The pressure 1 atmosphere—normal atmospheric pressure at sea level—is the pressure in most laboratories. This temperature and pressure are called *standard-state conditions.*) The reactants are kept at a fixed temperature before the reaction, and the products are brought back to the same temperature after the reaction. This allows us to measure only the heat given up or absorbed during the reaction, and not the heat involved in temperature or pressure changes. We measure this heat in units of calories.

LEARNING GOAL 3

Definitions of calorie and specific heat

One **calorie** is *the quantity of heat needed to raise the temperature of 1 gram of water by 1 Celsius degree.* If we were to heat 1 gram of water from 20°C to 21°C we might ask how much energy is absorbed. With our knowledge of the definition of a calorie, the answer is 1 calorie. Next we might ask how much energy 1 gram of water absorbs if its temperature changes from 20°C to 25°C. The answer is 5 calories, or 1 calorie for each degree Celsius rise in temperature. Large heat changes are measured in kilocalories (1,000 cal = 1 kcal). (It is the kilocalorie, also known among nutritionists as the large calorie or Calorie, that dieters count when they are trying to lose weight.)

With the adoption of the SI metric system (the revised metric system) in 1960, scientists recognized the *joule* as the basic unit of energy, replacing the calorie. However, the calorie is still used by many scientists to express the quantity of energy something contains, and it may be some time before it disappears from use. In case you find it necessary to convert from calories to joules or vice versa, remember that 1 calorie equals 4.18 joules.

$$1 \text{ calorie} = 4.18 \text{ joules}$$

EXAMPLE 12.1

One liter of water ($1,\overline{000}$ g) is heated from $2\overline{0}°C$ to $8\overline{0}°C$. How much energy does the water absorb?

Solution

UNDERSTAND THE PROBLEM

We want to find out how many calories are absorbed by a liter of water when its temperature is raised by 60 Celsius degrees.

DEVISE A PLAN

Given our definition of the calorie, we can say that every gram of water that we heat by 1 Celsius degree absorbs 1 calorie of heat energy. This can be stated as

Heat, in calories = (mass, in grams) $\times$ (temperature change, in °C)

(for water only)

We can abbreviate this relationship as

$$Q_{cal} = (m)(\Delta t) \quad \text{(for water only)}$$

The symbol Δ (the Greek capital letter delta) means "change in." The mass of water heated and the number of degrees that the water was heated can be substituted into this relationship to obtain the number of calories.

CARRY OUT THE PLAN

Substitute the given values into the abbreviated relationship.

$$Q_{cal} = (m)(\Delta t) \quad \text{(for water only)}$$

$$= (1,\overline{000} \text{ g})(8\overline{0}°C - 2\overline{0}°C) = (1\overline{000} \text{ g})(6\overline{0}°C)$$

$$= 6\overline{0},000 \text{ cal} \quad \text{or} \quad 6\overline{0} \text{ kcal}$$

LOOK BACK

We check to see whether the solution makes sense. If 1 calorie is required to raise the temperature of 1 gram of water by 1 Celsius degree, it makes sense that 60,000 times more heat would be required to raise the temperature of 1,000 g of water by 60 Celsius degrees.

Practice Exercise 12.1   You heat $2,\overline{000}$ mL of water ($2,\overline{000}$ g) from $5\overline{0}°C$ to $7\overline{0}°C$. How much energy does the water absorb?

TABLE 12.1

Specific Heat of Some Substances at 25°C

SUBSTANCE	SPECIFIC HEAT (c) IN cal/(g)(°C)
Aluminum	0.215
Copper	0.0920
Ethyl alcohol	0.581
Gold	0.0308
Iron	0.107
Nickel	0.106
Zinc	0.0922
Water (liquid)	1.000

12.3   Specific Heat

The formula used in Example 12.1,

$$Q_{cal} = (m)(\Delta t)$$

works only for water. However, we can use this formula to perform calculations on other substances once we understand the concept of *specific heat.* The **specific heat** of a substance is *the number of calories needed to raise the temperature of 1 gram of the substance by 1 Celsius degree.* The specific heat of water is taken to be 1.00 calorie per gram per degree Celsius, or 1.00 cal/(g)(°C). Table 12.1 gives the specific heats of some other substances.

LEARNING GOAL 4

Calculating calories

The general formula, then, for calculating the heat (in calories) gained or lost by any substance is

$$Q_{cal} = (c)(m)(\Delta t)$$

where c is the symbol for specific heat. Let's work some examples using this formula.

EXAMPLE 12.2

How much heat is absorbed when $5\overline{00}$ g of water are heated from 25.0°C to 75.0°C?

Solution

We write the general formula and substitute known values. Remember that the specific heat, c, for water is 1.00 cal/(g)(°C).

$$Q_{cal} = (c)(m)(\Delta t)$$

$$= \left(1.00 \frac{cal}{(\cancel{g})(\cancel{°C})}\right)(5\overline{00}\,\cancel{g})(50.0\,\cancel{°C})$$

$$= 25{,}\overline{0}00\,cal \quad \text{or} \quad 25.0\,kcal$$

Practice Exercise 12.2   How much heat is absorbed when 250.0 g of water are heated from 10.0°C to 60.0°C?

EXAMPLE 12.3

How much heat is needed to raise the temperature of a $2\overline{0}0$-g sample of aluminum from 20.0°C to 50.0°C?

Solution

Again we write the general formula and then substitute known values. Table 12.1 tells us that c for aluminum is 0.215 cal/(g)(°C). And the change in temperature is 50.0°C − 20.0°C, or 30.0°C. Then,

$$Q_{cal} = (c)(m)(\Delta t)$$

$$= \left(0.215 \frac{cal}{(g)(°C)}\right)(2\overline{0}0 \, g)(30.0°C)$$

$$= 1,290 \, cal \quad \text{or} \quad 1.29 \, kcal$$

Practice Exercise 12.3   Find the amount of heat needed to raise the temperature of a 500.0-g sample of aluminum from 25.0°C to 75.0°C.

EXAMPLE 12.4

If we had heated iron in place of aluminum in Example 12.3, how much heat would we have needed?

Solution

We solve this problem exactly as we did Example 12.3. However, the specific heat of iron is 0.107 cal/(g)(°C).

$$Q_{cal} = (c)(m)(\Delta t)$$

$$= \left(0.107 \frac{cal}{(g)(°C)}\right)(2\overline{0}0 \, g)(30.0°C)$$

$$= 642 \, cal \quad \text{or} \quad 0.642 \, kcal$$

(Iron requires about half as much heat as aluminum requires.)

Practice Exercise 12.4   Repeat Practice Exercise 12.3, substituting iron for aluminum.

12.4 The Calorimeter

An instrument called a *calorimeter* can be used to measure changes in the heat content of substances when they undergo chemical reactions. For example, the *bomb calorimeter* shown in Figure 12.1 is frequently used to measure heat-content changes in substances reacting with oxygen.

Suppose we want to measure the heat of reaction of glucose as it reacts with oxygen. We obtain a sample of glucose, say 1.80 g. We place it inside a special heat-transferring container that fits into the calorimeter. Then we position the container in the calorimeter along with an ignition device that will cause a spark, which will allow the sugar and oxygen to react. The calorimeter is filled with a known amount of water, say 1,000 g. We measure the temperature of the water and find it to be 25.0°C. We press the ignition

■ FIGURE 12.1

A bomb calorimeter to determine the heat of combustion of C(s)

switch, and a large quantity of heat is liberated by the reaction between glucose and oxygen.

$$C_6H_{12}O_6(s) + 6O_2(g) \rightarrow 6CO_2(g) + 6H_2O(l) + \text{energy}$$

The heat passes through the heat-transferring container to the water, the temperature of which rises to 31.7°C. Now we can calculate how much heat is required to raise the temperature of $1,\overline{000}$ g of water from 25.0°C to 31.7°C. This is exactly the amount of heat that was liberated by the glucose–oxygen reaction. We proceed as follows:

$$Q_{cal} \text{ (absorbed by water)} = (c_{water})(m_{water})(\Delta t_{water})$$

$$= \left(1.00 \frac{\text{cal}}{(\text{g})(°C)}\right)(1,\overline{000} \text{ g})(31.7°C - 25.0°C)$$

$$= \left(1.00 \frac{\text{cal}}{(\text{g})(°C)}\right)(1,\overline{000} \text{ g})(6.7°C)$$

$$= 6,700 \text{ cal} \quad \text{or} \quad 6.7 \text{ kcal}$$

Heat gained by water = Heat produced by reaction

Thus 6.7 kcal of heat energy must have been available to raise the temperature of the water. But this energy could have come only from the glucose–oxygen reaction. We may therefore say that the reaction of 1.80 g of glucose with oxygen released 6.7 kcal of energy.

It is customary to express heats of reaction in units of kilocalories per mole. To do so for the glucose–oxygen reaction, we first note that the molecular mass of glucose is $18\overline{0}$ g/mole. Then,

$$? \frac{\text{kcal}}{\text{mole}} = \left(\frac{6.7 \text{ kcal}}{1.80 \text{ g}}\right)\left(\frac{18\overline{0} \text{ g}}{1 \text{ mole}}\right) = 670 \frac{\text{kcal}}{\text{mole}}$$

Now, in the equation for the glucose–oxygen reaction, we can indicate the exact heat of reaction:

$$C_6H_{12}O_6(s) + 6O_2(g) \rightarrow 6CO_2(g) + 6H_2O(l) + 670 \text{ kcal}$$

An equation in this form is called a *thermochemical equation*. A **thermochemical equation** is a *chemical equation that is balanced, indicates the physical state of each substance (solid s, liquid l, or gas g), and shows the amount of heat liberated or absorbed during the reaction*. The foregoing equation states that 1 mole of glucose in solid form reacts with 6 moles of gaseous oxygen to form 6 moles of carbon dioxide gas and 6 moles of liquid water, releasing 670 kcal for each mole of glucose that reacts. Note that this also means that 670 kcal are released for every 6 moles of oxygen that react with the glucose, or for every 6 moles of CO_2 or every 6 moles of water that are formed.

12.5   Writing Thermochemical Equations

▌LEARNING GOAL 5
Writing thermochemical equations

Writing a thermochemical equation is simply a matter of obtaining the necessary information and putting it in the proper form. Here are some examples that illustrate the procedure and a few of the "fine points."

EXAMPLE 12.5

When 1 mole of carbon (graphite) reacts with oxygen at 25°C and 1 atm pressure, carbon dioxide is formed. In addition, 94.1 kcal of heat are released per mole of carbon dioxide formed. Write the thermochemical equation for this reaction.

Solution

$$C(\text{graphite}) + O_2(g) \rightarrow CO_2(g) + 94.1 \text{ kcal}$$

The reactants (C and O_2) have a total heat content that is 94.1 kcal greater than the product (CO_2). The heat is given up to the surroundings when the product is formed.

Practice Exercise 12.5   When ethane, $C_2H_6(g)$, reacts with oxygen gas, the products formed are carbon dioxide gas and liquid water. When this reaction is carried out at 25°C and 1 atm pressure, 186 kcal of heat are released per mole of carbon dioxide formed. Write the thermochemical equation for this reaction.

Heats of reaction are usually expressed in kilocalories per mole of product formed. However, this is not always the case, so each equation must be examined carefully. Example 12.6 will help explain this point.

EXAMPLE 12.6

When hydrogen gas reacts with oxygen gas, water is produced and heat is given off. Experiments have shown that 136.6 kcal of heat are produced when *2* moles of water are formed. Write the thermochemical equation for this reaction.

Solution

$$2H_2(g) + O_2(g) \rightarrow 2H_2O(l) + 136.6 \text{ kcal}$$

This equation tells us that 136.6 kcal are released to the environment when 2 moles of water are formed. However, you may sometimes see the equation written in the following way:

$$H_2(g) + \tfrac{1}{2}O_2(g) \rightarrow H_2O(l) + 68.3 \text{ kcal}$$

This equation tells us that 68.3 kcal are released to the environment when 1 mole of water is formed. Because the quantity of heat produced is proportional to the amount of water formed, both equations are correct.

Practice Exercise 12.6   When 2.00 moles of hexane, $C_6H_{14}(l)$, burn in sufficient oxygen gas, the products are carbon dioxide gas and water. In addition, 1,980 kcal of heat are given up to the environment. Write a balanced thermochemical equation for this reaction.

Once you have found the heat of reaction when 1 mole of a substance is formed, you can use this information to find the heat produced or absorbed when *any* amount of that substance is formed.

EXAMPLE 12.7

How much heat is produced when 144 g of $H_2O(l)$ are formed from the reaction between hydrogen and oxygen?

Solution

Because the formation of 1 mole of water produces 68.3 kcal, we must find out how many moles of water there are in 144 g.

$$? \text{ moles of } H_2O = (144 \text{ g}) \left(\frac{1 \text{ mole}}{18.0 \text{ g}} \right) = 8.00 \text{ moles of } H_2O$$

$$? \text{ kcal of heat produced} = (8.00 \text{ moles}) \left(\frac{68.3 \text{ kcal}}{1 \text{ mole}} \right)$$

$$= 546 \text{ kcal of heat produced}$$

Practice Exercise 12.7   How much heat is produced when 54.0 g of $H_2O(l)$ are formed from the reaction between hydrogen gas and oxygen gas?

So far we have neglected endothermic reactions (reactions that absorb heat from the surroundings). The next example shows you how to write a thermochemical equation for this type of reaction.

EXAMPLE 12.8

When hydrogen iodide is formed from hydrogen and iodine, heat is *absorbed* from the surroundings. Experiments have shown that 6.2 kcal of heat are absorbed per mole of hydrogen iodide produced. Write the thermochemical equation for the reaction.

Solution

There are two ways to write this type of equation.

(a) Write it with the heat of reaction on the right-hand side of the equation. This shows that the reactants (H_2 and I_2) require heat to yield the product (HI).

$$6.2 \text{ kcal} + \tfrac{1}{2}H_2(g) + \tfrac{1}{2}I_2(s) \rightarrow HI(g)$$

(b) Write it with the heat of reaction on the right-hand side of the equation, as a negative number. Think of this as indicating that the product must *lose* heat for the reaction to be completed.

$$\tfrac{1}{2}H_2(g) + \tfrac{1}{2}I_2(s) \rightarrow HI(g) - 6.2 \text{ kcal}$$

Chemists usually prefer writing heats of reaction on the right-hand side of the equation.

Practice Exercise 12.8   When $PCl_5(g)$ breaks down to form $PCl_3(g)$ and $Cl_2(g)$, heat is absorbed from the surroundings. Experiments have shown that 22.2 kcal of heat are absorbed per mole of $PCl_5(g)$ broken down. Write the thermochemical equation for this reaction.

12.6   Another Way of Writing Heats of Reaction

LEARNING GOAL 6

Significance of ΔH

The symbol for heat content (enthalpy) is H, so we can represent a *change* in heat content with the symbol ΔH. (Again we are using the Greek letter Δ to mean "change in.") Because a heat of reaction is a change in heat content during a reaction, we represent it with ΔH.

Chemists define ΔH as the sum of the heat contents of the products less the sum of the heat contents of the reactants:

$$\Delta H = H(\text{products}) - H(\text{reactants})$$

With this notation, ΔH is negative for an exothermic reaction and positive for an endothermic reaction (Figure 12.2).

In the formation of 1 mole of $CO_2(g)$ from its elements in their standard states, 94.1 kcal of heat are released to the environment. This means that the products have a lower heat content than the reactants, and $\Delta H = -94.1$ kcal. This may be indicated by writing

$$C(\text{graphite}) + O_2(g) \rightarrow CO_2(g) \qquad \Delta H = -94.1 \text{ kcal}$$

or by placing the value of ΔH in the equation and writing

$$C(\text{graphite}) + O_2(g) \rightarrow CO_2(g) + 94.1 \text{ kcal}$$

In the formation of 1 mole of $HI(g)$ under standard conditions, heat is absorbed, so $\Delta H = +6.2$ kcal. This may be shown as

$$\tfrac{1}{2}H_2(g) + \tfrac{1}{2}I_2(s) \rightarrow HI(g) \qquad \Delta H = +6.2 \text{ kcal}$$

or by placing the value of ΔH in the equation and writing

$$\tfrac{1}{2}H_2(g) + \tfrac{1}{2}I_2(s) \rightarrow HI(g) - 6.2 \text{ kcal}$$

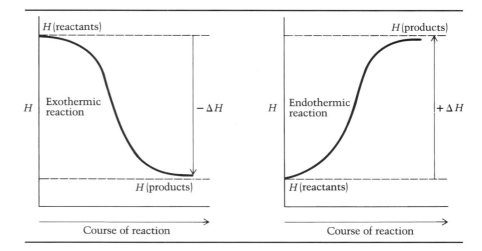

12.7   Heats of Formation

LEARNING GOAL 7

Heats of formation and heats of reaction

Up to this point, we have used the term *heat of reaction* for the change in heat content (or enthalpy) during the course of a reaction. However, the examples we used all involved the formation of a compound from its elements:

$$\tfrac{1}{2}H_2(g) + \tfrac{1}{2}I_2(g) \rightarrow HI(g) \qquad \Delta H = +6.2 \text{ kcal}$$

$$H_2(g) + \tfrac{1}{2}O_2(g) \rightarrow H_2O(l) \qquad \Delta H = -68.3 \text{ kcal}$$

$$C(\text{graphite}) + O_2(g) \rightarrow CO_2(g) \qquad \Delta H = -94.1 \text{ kcal}$$

When 1 mole of a compound is formed from its elements in their usual states at 1 atm pressure, the heat of reaction is called the **heat of formation** of the compound and is given the symbol ΔH_f^0. Heats of formation have been determined experimentally for many compounds, and Table 12.2 lists some of them. They can be used to determine ΔH for complex reactions.

12.8   Determining Heats of Reaction

You will find that Table 12.2 makes it easy to determine heats of reaction. However, you must keep the following rules in mind.

RULE 1

The heat of reaction is equal to the heat of formation of the products less the heat of formation of the reactants:

$$\Delta H = \Delta H_f^0(\text{products}) - \Delta H_f^0(\text{reactants})$$

RULE 2

The heat of formation of an element at standard-state conditions is zero:

$$\Delta H_f^0(\text{element}) = 0$$

TABLE 12.2

Heats of Formation at 25°C and 1 atm Pressure

SUBSTANCE	ΔH_f^0 (kcal/mole)	SUBSTANCE	ΔH_f^0 (kcal/mole)
$BaCl_2(s)$	-205.56	$H_2O(g)$	-57.8
$BaO_2(s)$	-150.5	$H_2O(l)$	-68.3
$CH_4(g)$	-17.9	$H_2O_2(l)$	-44.8
$C_2H_2(g)$	$+54.2$	$H_2S(g)$	-4.8
$C_2H_6(g)$	-20.2	$H_2SO_4(aq)$	-217
$C_8H_{18}(l)$	-59.97	$H_3PO_4(aq)$	-308.2
$C_6H_{12}O_6(s)$	-304.4	$SO_2(g)$	-70.96
$CCl_4(l)$	-33.4	$SO_3(g)$	-94.45
$CO(g)$	-26.4	$NH_3(g)$	-11.04
$CO_2(g)$	-94.1	$NO(g)$	$+21.5$
$CaO(s)$	-151.9	$NO_2(g)$	$+8.1$
$CaCO_3(s)$	-288.5	$NaCl(aq)$	-97.3
$HF(g)$	-64.2	$NaOH(aq)$	-112
$HCl(g)$	-22.1	$Fe_2O_3(s)$	-197
$HBr(g)$	-8.7	$KOH(aq)$	-115
$HI(g)$	$+6.2$	$K_2O(s)$	-86.4
$HCl(aq)$	-39.6		

Also keep in mind that the values in Table 12.2 are *per mole of compound.* This is important when more or less than 1 mole is involved in the balanced equation.

EXAMPLE 12.9

Determine ΔH for the reaction

$$CaO(s) + CO_2(g) \rightarrow CaCO_3(s)$$

Solution

We obtain the ΔH_f^0 value for each compound from Table 12.2 and then use Rule 1 above.

SUBSTANCE	ΔH_f^0(kcal/mole)
$CaO(s)$	-151.9
$CO_2(g)$	-94.1
$CaCO_3(s)$	-288.5

$$\Delta H = \Delta H_f^0(\text{products}) - \Delta H_f^0(\text{reactants})$$
$$= (\Delta H_{CaCO_3}^0) - (\Delta H_{CaO}^0 + \Delta H_{CO_2}^0)$$
$$= (-288.5 \text{ kcal}) - (-151.9 \text{ kcal} - 94.1 \text{ kcal})$$
$$= (-288.5 \text{ kcal}) - (-246.0 \text{ kcal})$$
$$= -288.5 \text{ kcal} + 246.0 \text{ kcal}$$
$$= -42.5 \text{ kcal}$$

This reaction is exothermic. It produces 42.5 kcal of heat per mole of $CaCO_3$ formed.

Practice Exercise 12.9   Use Table 12.2 to determine ΔH for the reaction

$$K_2O(s) + H_2O(l) \rightarrow 2KOH(aq)$$

EXAMPLE 12.10

Determine ΔH for the reaction

$$2C_2H_6(g) + 7O_2(g) \rightarrow 4CO_2(g) + 6H_2O(l)$$
$$\text{Ethane}$$

Solution
Again we obtain ΔH_f^0 values from Table 12.2:

SUBSTANCE	ΔH_f^0(kcal/mole)
$C_2H_6(g)$	-20.2
$O_2(g)$	0.0
$CO_2(g)$	-94.1
$H_2O(l)$	-68.3

Now, because 6 moles of water and 4 moles of CO_2 are formed, we have to multiply ΔH_f^0 for water by 6 and ΔH_f^0 for CO_2 by 4. And because 2 moles of ethane and 7 moles of oxygen are used, we have to multiply ΔH_f^0 for ethane by 2 and ΔH_f^0 for oxygen by 7.

$$\Delta H = \Delta H_f^0(\text{products}) - \Delta H_f^0(\text{reactants})$$

$$= [6(\Delta H_{H_2O}^0) + 4(\Delta H_{CO_2}^0)] - [2(\Delta H_{C_2H_6}^0) + 7(\Delta H_{O_2}^0)]$$

$$= [6(-68.3 \text{ kcal}) + 4(-94.1 \text{ kcal})]$$

$$- [2(-20.2 \text{ kcal}) + 7(0.0 \text{ kcal})]$$

$$= -786.2 \text{ kcal} + 40.4 \text{ kcal} = -745.8 \text{ kcal}$$

The reaction is exothermic. The combustion of 2 moles of ethane releases 745.8 kcal of heat to the environment.

Practice Exercise 12.10   Use Table 12.2 to determine ΔH for the reaction

$$2CO(g) + O_2(g) \rightarrow 2CO_2(g)$$

12.9   Practical Applications of Heat Energy from Chemical Reactions

At the beginning of this chapter, we mentioned some uses of energy in our everyday lives. In most cases, this energy is obtained from fuels, substances that are rich in chemical potential energy—for example, oil, natural gas, and coal (often referred to as *fossil fuels*). We usually obtain heat energy from fuels by making them react in the presence of oxygen; in everyday language, we burn them. Then we use the heat energy they produce to do work.

EXAMPLE 12.11

Modern gas ranges use methane (CH_4) for fuel. Calculate ΔH for the combustion of 1 mole of methane:

$$CH_4(g) + 2O_2(g) \rightarrow CO_2(g) + 2H_2O(l)$$

Solution
Obtain ΔH_f^0 for each compound from Table 12.2 and then use Rule 1:

SUBSTANCE	ΔH_f^0(kcal/mole)
$CH_4(g)$	−17.9
$O_2(g)$	0.0
$CO_2(g)$	−94.1
$H_2O(l)$	−68.3

$$\Delta H = \Delta H_f^0(\text{products}) - \Delta H_f^0(\text{reactants})$$
$$= [2(\Delta H_{H_2O}^0) + 1(\Delta H_{CO_2}^0)] - [1(\Delta H_{CH_4}^0) + 2(\Delta H_{O_2}^0)]$$
$$= [2(-68.3 \text{ kcal}) + 1(-94.1 \text{ kcal})]$$
$$- [1(-17.9 \text{ kcal}) + 2(0.0 \text{ kcal})]$$
$$= -230.7 \text{ kcal} + 17.9 \text{ kcal} = -212.8 \text{ kcal}$$

This reaction is exothermic. When 1 mole of methane burns, 212.8 kcal of heat are released to the environment.

Practice Exercise 12.11 The first step in the synthesis of nitric acid is the reaction of ammonia gas and oxygen gas to form nitric oxide gas and steam. Calculate ΔH for this reaction:

$$4NH_3(g) + 5O_2(g) \rightarrow 4NO(g) + 6H_2O(g)$$

EXAMPLE 12.12

The combustion of methane releases considerable energy to do work. Let's compare this energy with the energy released by the combustion of 1 mole of octane, C_8H_{18} (used in automobile fuels):

$$C_8H_{18}(l) + 12\tfrac{1}{2}O_2(g) \rightarrow 8CO_2(g) + 9H_2O(l)$$

Solution

As usual, obtain ΔH_f^0 for each compound from Table 12.2 and then use Rule 1.

SUBSTANCE	ΔH_f^0(kcal/mole)
$C_8H_{18}(l)$	−59.97
$O_2(g)$	0.0
$CO_2(g)$	−94.1
$H_2O(l)$	−68.3

$$\Delta H = \Delta H_f^0(\text{products}) - \Delta H_f^0(\text{reactants})$$
$$= [8(\Delta H_{CO_2}^0) + 9(\Delta H_{H_2O}^0)] - [1(\Delta H_{C_8H_{18}}^0) + 12\tfrac{1}{2}(\Delta H_{O_2}^0)]$$
$$= [8(-94.1 \text{ kcal}) + 9(-68.3 \text{ kcal})]$$
$$- [1(-59.97 \text{ kcal}) + 12\tfrac{1}{2}(0.0 \text{ kcal})]$$
$$= -1,367.5 \text{ kcal} + 59.97 \text{ kcal} = -1,307.5 \text{ kcal}$$

The reaction is exothermic. When 1 mole of octane burns, 1,307.5 kcal of heat are released to the environment. This is more than six times the energy produced by the combustion of 1 mole of methane. Note, however, that because 1 mole of octane weighs seven times as much as 1 mole of methane, we actually get a little more heat from 1 g of *methane* than we do from 1 g of octane.

Practice Exercise 12.12   Iron ore (Fe_2O_3) is reduced in the reaction

$$Fe_2O_3(s) + 3CO(g) \rightarrow 3CO_2(g) + 2Fe(s)$$

Using Table 12.2, determine ΔH for this reaction.

EXAMPLE 12.13

We all know that sugar is a source of quick energy. In our bodies, sucrose (sugar) is broken down into carbon dioxide and water by the very complicated process of cellular respiration. Sucrose itself is made up of two simpler sugars, glucose and fructose. Let's determine how much energy our bodies obtain from the oxidation of 1 mole of the simpler sugar, glucose ($C_6H_{12}O_6$), in the reaction

$$C_6H_{12}O_6(s) + 6O_2(g) \rightarrow 6CO_2(g) + 6H_2O(l)$$

Solution

Again we use Table 12.2 and Rule 1.

SUBSTANCE	ΔH_f^0(kcal/mole)
$C_6H_{12}O_6(s)$	-304.4
$O_2(g)$	0.0
$CO_2(g)$	-94.1
$H_2O(l)$	-68.3

$$\Delta H = \Delta H_f^0(\text{products}) - \Delta H_f^0(\text{reactants})$$
$$= [6(\Delta H_{CO_2}^0) + 6(\Delta H_{H_2O}^0)] - [1(\Delta H_{C_6H_{12}O_6}^0) + 6(\Delta H_{O_2}^0)]$$
$$= [6(-94.1 \text{ kcal}) + 6(-68.3 \text{ kcal})]$$
$$- [1(-304.4 \text{ kcal}) + 6(0.0 \text{ kcal})]$$
$$= -974.4 \text{ kcal} + 304.4 \text{ kcal} = -67\overline{0} \text{ kcal}$$

The reaction is exothermic. Our bodies obtain $67\overline{0}$ kcal of heat from the oxidation of 1 mole of glucose.

Practice Exercise 12.13   Acetylene gas, which is used for welding, reacts with hydrogen to produce ethane gas. Calculate ΔH for this reaction:

$$C_2H_2(g) + 2H_2(g) \rightarrow C_2H_6(g)$$

▌▌▌ ▐ SUMMARY

In a chemical reaction, heat is either absorbed from the environment or released to the environment as old bonds are broken and new bonds are formed. A reaction in which energy is absorbed from the surroundings is called endothermic, and a reaction in which energy is released is called exothermic.

Associated with every substance is a heat content called its enthalpy. Although we cannot measure enthalpy directly, we can measure the amount of heat released or absorbed during a reaction. This change in heat content that occurs during a chemical reaction is called the heat of reaction. Heats of reaction are usually measured at standard-state conditions: 25°C and 1 atm pressure. Heat (including heat of reaction) is measured in calories. One calorie is the quantity of heat needed to raise the temperature of 1 gram of water by 1 Celsius degree. The specific heat of a substance is defined as the number of calories needed to raise the temperature of 1 gram of the substance by 1 Celsius degree.

A chemical equation that indicates the heat of reaction and the physical state of each reactant and product is called a thermochemical equation. Heat of reaction, ΔH, may be defined as the sum of the heat contents of the products less the sum of the heat contents of the reactants. For an exothermic reaction, this difference is negative; for an endothermic reaction, it is positive. When a compound is formed from its elements, the heat of reaction is called the heat of formation ΔH_f^0 for the compound. The heat of formation for an element (at standard-state conditions) is zero.

▌▌▌ ▐ KEY TERMS

calorie **(12.2)**   heat of formation **(12.7)**
endothermic reaction **(12.1)**   heat of reaction **(12.2)**
enthalpy **(12.2)**   specific heat **(12.3)**
exothermic reaction **(12.1)**   thermochemical equation **(12.4)**

CAREER SKETCH

Regulatory Chemist

Ensuring that human health and environmental issues are dealt with properly are the two principle responsibilities of a *regulatory chemist*. To do this, a regulatory chemist must ensure compliance with laws and regulations that protect human health and the environment. A working knowledge of local, state, and federal legislation is a necessary skill of a regulatory chemist, as well as the ability to develop programs to comply with these laws. A strong background in chemistry—often including a chemistry degree—is necessary for someone interested in the regulatory field. An engineering degree and background in the law are also helpful.

Regulatory chemists can work in a variety of settings. Since they work most often in agencies that conduct research, the government is a principal employer. Many different government agencies, ranging from the Food and Drug Administration to the Federal Highway Administration, employ regulatory chemists. Industry is another site where you might find a regulatory chemist. In this type of setting, one's duties may include acting as an environmental health and safety officer, a chemical hygiene officer, or a regulatory compliance officer. Academic institutions also hire regulatory chemists for similar purposes.

CHEMICAL FRONTIERS

*Chemistry Helps Farmers Protect Land**

Farmers around the world have accomplished quite a few miraculous feats. Perhaps one of the most interesting can be found in southern Idaho's Magic Valley, where farmers have been able to grow alfalfa, wheat, barley, potatoes, sugar beets, beans, and corn in a high-altitude desert that receives just seven inches of precipitation a year. They've done this primarily with the help of furrows, or long dirt ditches that run along each planted row of crops.

The system is not perfect, however. Often, as water runs along a ditch, it erodes the valuable topsoil. This causes important fertilizers and pesticides to be moved from the soil where they are needed, reducing their effectiveness and leading to pollution of nearby lakes or streams. Just as damaging, the eroded soil can block pores at the end of the furrow, limiting the amount of water that is actually able to reach the plants.

The solution to this problem can be found in polyacrylamide (PAM), a long-chain molecule. The term PAM refers not to a single substance, but to a variety of different chemicals, each of which may have a different electrical charge, number of building blocks, or number of molecules of acrylamide linked together. PAM, which is being used to help farmers, contains about 30,000 acrylamide units in each molecule.

A new group of chemicals called PAMs, helps farmers stop the erosion of soil. (William H. Mullins/Photo Researchers, Inc.)

Once added to a farm's irrigation system, the PAM searches for and then binds to the broken edges of crystalline clay particles. Since the clay particles are positively charged and the 30,000-acrylamide PAM is negatively charged, the two attract each other. The process increases the cohesiveness of soil particles at the surface of the field, making the dirt more resistant to the forces of flowing water. The treated soil now remains in place, instead of eroding as it would if left untreated.

International studies have focused on other positive aspects of PAMs. One study by a researcher with the Israeli government's Institute of Soils and Water found that PAMs can reduce the compaction of soil after a rainstorm, as well as reducing the crust formation that follows. The study found that treating the soil with PAMs before the rain can reduce by 90 percent or more these damaging effects.

The major drawback to this process is cost. For many farmers, the price of purchasing the chemical could equal the profit that the crops would ordinarily yield. As the process is perfected, however, costs are likely to become lower. And some reports seem to predict that the higher cost now will result in higher profits in the long run. If we reduce the loss of topsoil today, future generations may be able to farm the land for many years.

*Adapted from *Science News*, October 30, 1993, pp. 280–281

SELF-TEST EXERCISES

LEARNING GOALS 1 AND 2: *Exothermic and Endothermic Reactions; Definitions of Energy Terms*

◀ 1. Match the term on the left with its definition on the right.

(a) Exothermic reaction	1. The heat term associated with the formation of a mole of a compound from its elements.
(b) Endothermic reaction	2. A reaction in which heat is released into the environment.
(c) Heat of formation	3. The heat term associated with the heat gained or lost in a chemical reaction.
(d) Heat of reaction	4. A reaction in which heat is absorbed from the environment.

◀ 2. Explain the meaning of the symbol H.

◀ 3. How does the sum of the heat contents of the products compare with that of the reactants in (a) an endothermic reaction, (b) an exothermic reaction?

◀ 4. The change in heat content during a chemical reaction is called the _____.

LEARNING GOAL 3: *Definitions of Calorie and Specific Heat*

5. Define the terms *calorie* and *specific heat*.

6. What are standard-state conditions, and why is it necessary to adhere to them when calculating changes in heat?

LEARNING GOAL 4: *Calculating Calories*

◀ 7. How many calories of heat are needed to raise the temperature of 50.0 g of water from 20.0°C to 80.0°C?

◀ 8. How much heat is needed to raise the temperature of 500.0 g of water from 0.00°C to 75.0°C?

9. Calculate the calories of heat needed to raise the temperature of 40.0 g of iron from $4\overline{0}$°C to $3\overline{00}$°C.

10. Calculate the amount of heat needed to raise the temperature of 100.0 g of copper from 45.0°C to 80.0°C.

*11. How much heat is required to raise the temperature of 100.0 mL of ethyl alcohol from 15.0°C to 60.0°C? The density of ethyl alcohol is 0.800 g/mL.

*12. How much heat is required to raise the temperature of 50.0 cm³ of ethyl alcohol from 25.0°C to 55.0°C? The density of ethyl alcohol is 0.800 g/cm³.

13. Suppose that 308 cal of heat are added to a sample of gold that has a mass of $1,\overline{000}$ g. What is the final temperature of the gold, given that its initial temperature is 20.0°C?

14. Suppose that 5.00 kcal of heat are added to a sample of gold that has a mass of 2.00 kg. What is the final temperature of the gold if its initial temperature is 10.0°C?

*15. A 0.146-g sample of H_2 gas is placed in a calorimeter, where it is allowed to react with oxygen gas. The heat generated by the reaction is transferred to the calorimeter water. The water has a mass of $5\overline{00}$ g. The temperature of the water increases from 20.0°C to 30.0°C. From this information, calculate the heat of reaction in kilocalories per mole.

16. How much heat is needed to raise the temperature of 1.000 kg of water from 32.0°C to 99.9°C?

LEARNING GOAL 5: *Writing Thermochemical Equations*

17. Using Table 12.2, write thermochemical equations for the formation of the following compounds from their elements: (a) SO_3 (b) HBr (c) CCl_4 (d) NO

18. Using Table 12.2, write thermochemical equations for the formation of the following compounds from their elements: (a) HF(g) (b) HCl(g) (c) HBr(g)

LEARNING GOAL 6: *Significance of ΔH*

◀ 19. Explain the significance of positive and negative values of ΔH. Use a chemical equation to illustrate your answer.

20. (a) What is meant by the term *enthalpy*? (b) What is the difference between the symbols H and ΔH?

LEARNING GOAL 7: *Heats of Formation and Heats of Reaction*

21. Calculate the heat released when 102 g of hydrogen sulfide (H_2S) are produced from its elements. *Note:* ΔH_f^0 for H_2S is −4.8 kcal/mole. (*Hint:* Calculate the number of moles of H_2S.)

22. Calculate the heat released when 15.3 g of barium oxide (BaO) are produced from its elements. (ΔH_f^0 for BaO is -133.4 kcal/mole.)

23. Rocket engines often use liquid hydrogen and liquid oxygen for fuel. These reactants are heated, mixed, and united to produce steam.

$$H_2(g) + \tfrac{1}{2}O_2(g) \rightarrow H_2O(g) + 57.8 \text{ kcal}$$

Calculate the amount of heat released when 1 kg of steam is produced.

24. Hexane is a chemical that is sometimes used in thermometers, instead of mercury. Hexane burns in oxygen to produce carbon dioxide and water according to the equation

$$2C_6H_{14}(l) + 19O_2(g) \rightarrow$$
$$12CO_2(g) + 14H_2O(l) + 1{,}980 \text{ kcal}$$

How many grams of hexane are needed to produce 19,800 kcal of heat?

◀ 25. (a) Using Table 12.2, calculate ΔH for the reaction

$$2C_2H_2(g) + 5O_2(g) \rightarrow 4CO_2(g) + 2H_2O(l)$$

(b) Calculate how much heat is produced when 5,200 g of C_2H_2 are burned.

26. Carbon dioxide is formed from its elements as follows:

$$C(graphite) + O_2(g) \rightarrow CO_2(g) + 94.1 \text{ kcal}$$

How much energy is produced when 880 g of $CO_2(g)$ are formed?

27. Barium peroxide can be used as a solid propellant for rockets. When it is added to hydrochloric acid, the following reaction takes place:

$$BaO_2(s) + 2HCl(g) \rightarrow BaCl_2(s) + H_2O_2(l)$$
Barium
peroxide

(a) Using Table 12.2, calculate ΔH for the reaction.
(b) Calculate the number of grams of BaO_2 that must react to produce 500 kcal of heat.

28. Liquid water is broken down into its elements by electrolysis. This is an endothermic reaction requiring 68.3 kcal per mole of liquid water decomposed. How much energy is needed to produce 640 g of oxygen gas?

29. Given the reaction

$$P_2O_5(s) + 3H_2O(l) \rightarrow 2H_3PO_4(aq) + 228.5 \text{ kcal}$$

calculate ΔH_f^0 for P_2O_5 (use Table 12.2).

◀ 30. When you burn 1.20 g of carbon (graphite) in the presence of oxygen, carbon dioxide gas is produced. The heat generated by this reaction can raise the temperature of 470.5 g of water from 20.0°C to 40.0°C. Calculate ΔH_f^0 for carbon dioxide from this information and compare it with the value given in Table 12.2.

◀ 31. When you burn 3.20 g of sulfur in the presence of oxygen, sulfur trioxide (SO_3) is produced. The heat generated by this reaction can raise the temperature of 1,890 g of water from 25.00°C to 30.00°C. Calculate ΔH_f^0 for sulfur trioxide from this information and compare it with the value given in Table 12.2.

32. Calculate the heat released when 188.4 g of $K_2O(s)$ are produced from its elements.

33. Eric Cottell, a British-born inventor, has found a method of increasing the burning efficiency of oil and gasoline. He does this simply by mixing the oil or gasoline with water. Cottell has built a device called the Cottell Ultrasonic Reactor, which can actually mix oil and water by breaking down the oil and water droplets into extremely fine particles. By using a mixture containing three parts oil to one part water, Cottell has decreased the fuel consumption in his home furnace by 25%. He claims that cars can be made to run almost pollution-free on a mixture of 18% water and 82% gasoline.

One of the ingredients of home heating oil is $C_{16}H_{34}$. Given the following equation, determine ΔH_f^0 for this ingredient.

$$C_{16}H_{34}(s) + 24\tfrac{1}{2}O_2(g) \rightarrow$$
$$16CO_2(g) + 17H_2O(l) + 2{,}560 \text{ kcal}$$

34. Calculate the heat released when 7.30 g of $HCl(g)$ are produced from its elements.

◀ 35. Calculate the heat released when 12.8 g of $SO_2(g)$ are produced from its elements.

36. Calculate the heat released when 1.618 kg of $HBr(g)$ are produced from its elements.

37. Calculate the heat absorbed when 51.2 g of $HI(g)$ are produced from its elements.

◀ 38. Acetylene gas, C_2H_2, burns in oxygen to produce carbon dioxide gas and liquid water. For every mole of acetylene burned, 312.0 kcal of heat are released. From this information, calculate ΔH_f^0 for acetylene and compare your answer with the value given in Table 12.2.

39. Using Table 12.2, calculate ΔH for the reaction

$$2SO_2(g) + O_2(g) \rightarrow 2SO_3(g)$$

40. Use the following equation and Table 12.2 to determine ΔH_f^0 for $Ca(OH)_2(s)$:

$$CaO(s) + H_2O(l) \rightarrow Ca(OH)_2(s) + 15.6 \text{ kcal}$$

41. Hydrogen peroxide (H_2O_2) is sometimes used in diluted form as an antiseptic. Upon standing, hydrogen peroxide decomposes according to the reaction

$$2H_2O_2(l) \rightarrow 2H_2O(l) + O_2(g)$$

(a) Calculate ΔH for this reaction.
(b) Calculate how much heat is produced when 204 g of H_2O_2 decompose.

42. Calculate the amount of heat released when 2.000 kg of $BaCl_2(s)$ are produced from $Ba(s)$ and $Cl_2(g)$. Use Table 12.2.

43. Calculate ΔH for the decomposition of potassium chlorate:

$$2KClO_3(s) \rightarrow 2KCl(s) + 3O_2(g)$$

[ΔH_f^0 for $KClO_3(s)$ is -93.50 kcal, and ΔH_f^0 for $KCl(s)$ is -104.18 kcal.]

44. When 0.230 g of $Na(s)$ reacts with $Cl_2(g)$, $NaCl(s)$ is produced. The heat generated by this reaction can raise the temperature of 24.55 g of water from $10.0°C$ to $50.0°C$. Using this information, calculate ΔH_f^0 for $NaCl(s)$.

45. The following equation represents the formation of phosphoric acid from phosphorus pentachloride:

$$PCl_5(s) + H_2O(l) \rightarrow H_3PO_4(aq) + HCl(aq)$$

(a) Balance the equation.
(b) Calculate ΔH for this reaction. [ΔH_f^0 for $PCl_5(s)$ is -95.35 kcal, and ΔH_f^0 for $HCl(aq)$ is -40.02 kcal.]
(c) How much heat is produced from this reaction if 41.70 g of $PCl_5(s)$ are used?

46. Calculate the amount of heat absorbed when 5.00 moles of $CO_2(g)$ are formed from the reaction

$$C(graphite) + 2H_2O(g) \rightarrow \\ CO_2(g) + 2H_2(g) - 31.39 \text{ kcal}$$

*47. Calculate ΔH_f^0 for $CuO(s)$ from the following information:

$$3CuO(s) + 2NH_3(g) \rightarrow \\ 3Cu(s) + 3H_2O(l) + N_2(g) \qquad \Delta H = -71.1 \text{ kcal}$$

[ΔH_f^0 for $NH_3(g)$ is -11.04 kcal.]

48. Calculate the amount of heat released when 5.00 moles of $CO_2(g)$ are formed from the reaction

$$CO(g) + \tfrac{1}{2}O_2(g) \rightarrow CO_2(g) + 67.64 \text{ kcal}$$

EXTRA EXERCISES

49. In a simple calorimeter, a reaction is carried out in a vessel submerged in water. After the reaction occurs, 420 g of water increase in temperature from $25.0°C$ to $40.0°C$. What is the amount of energy released by the reaction?

50. The heat of reaction for $CaCO_3(s) \rightarrow CaO(s) + CO_2(g)$ is $\Delta H = 42.5$ kcal. Given that $\Delta H_{CaO}^0 = -151.9$ kcal/mole and $\Delta H_{CO_2}^0 = -94.1$ kcal/mole, calculate $\Delta H_{CaCO_3}^0$. Compare your calculated value with the value given in Table 12.2.

51. Using Table 12.2, determine whether heat is absorbed or released when hydrogen gas reacts at standard-state conditions with the appropriate elements to form each of the following: (a) $C_2H_2(g)$ (b) $HCl(g)$ (c) $H_2O(l)$ (d) $HI(g)$

52. When burned with excess oxygen, does 1.00 mole of methane (CH_4) or 1.00 mole of acetylene (C_2H_2) produce the greater amount of energy? The reactions are

$$CH_4(g) + 2O_2(g) \rightarrow CO_2(g) + 2H_2O(g)$$
$$C_2H_2(g) + 2\tfrac{1}{2}O_2(g) \rightarrow 2CO_2(g) + H_2O(g)$$

53. Energy is required to break chemical bonds. Considering this, determine whether the following reaction would be endothermic or exothermic:

$$N_2O_4(g) \rightarrow 2NO_2(g)$$

54. The heat of reaction for the burning of carbon is 94 kcal/mole. The coal used in a coal-fired power plant is 70.0% carbon. How much energy does 1.00 kg of coal yield?

55. An energy source that is gaining more attention these days is hydrogen gas. When it combines with oxygen, it forms water and releases 58 kcal/mole. Compare the fuel value of hydrogen gas with that of natural gas, which can be approximated as methane (CH_4), having a heat of reaction of 213 kcal/mole, and that of coal, which would yield 94 kcal/mole if it were pure carbon. (*Hint:* The best way to make this comparison is to calculate the heat of formation per gram for each substance.)

◄ 56. Explain why an exothermic reaction has a negative ΔH and an endothermic reaction has a positive ΔH.

57. Calculate the heat released by the combustion of 50.0 moles of methane.

58. Using Table 12.2, calculate the heat needed to produce the following change in 1.00 mole of water:

$$H_2O(l) \rightarrow H_2O(g)$$

THE GASEOUS STATE:
Ideal Behavior

13

This is the story of Danielle Morgan, boat owner and investment counselor, and how she saved a fellow sailor from possible brain damage and death.

It was a Saturday in mid-October at the marina on the New Jersey shore where Danielle kept her 28-foot cabin cruiser. Danielle was winterizing her boat and performing all of the chores necessary to protect her craft from the long winter ahead.

This was a very busy time at the marina. Boat owners and boat enthusiasts used this time to inspect their boats and plan for purchases and trades for the next year. Once in dry dock a boat is easy to inspect and to board for a leisurely look inside. Bill King, a friend of Danielle's, was doing just that. Bill had a small ski boat and was looking to purchase one with a cabin.

There was one particular boat in the marina that caught Bill's eye. It wasn't for sale, but he nevertheless wanted to see the interior cabin. The boat was in dry dock and had its protective canopy in place. In order to get onboard, Bill would have to unzip the canopy and squeeze through the opening. There was a sign on this boat, however, that read, "The Interior of This Boat is Filled with Nitrogen Gas—**DO NOT ENTER**."

Although filling the interior of a boat with nitrogen gas is not a common practice, some boat owners do this to protect the interior metal parts of the boat from oxidizing. Bill read the sign but figured that nitrogen is a component of air and isn't harmful, so he decided to board the boat anyway and look around.

Bill unzipped the canopy just enough so that he could squeeze through. Danielle Morgan, about 50 feet away, saw Bill just as he disappeared into the boat. She knew the owner of that boat and was aware that it was filled with nitrogen. Dropping some tools she was holding, she ran to the boat. As she peeked through the canopy she saw

307

Bill lying on the deck, apparently unconscious. Danielle, while yelling for help, unzipped the rest of the canopy in order to ventilate the deck of the boat and shoved him out of the boat. Although he was inside the boat for less than a minute, his breathing was shallow.

The rescue squad soon arrived and administered oxygen to Bill. His breathing returned to normal after a short period of time and he appeared to be fine, which an exam later administered at the hospital confirmed. But what had happened to Bill, and how did his friend Danielle know to come after him?

Danielle had studied chemistry and knew that nitrogen is not a toxic gas. But she also knew that oxygen gas is necessary for respiration. The fact that the boat was filled with nitrogen gas meant that all of the oxygen gas had been displaced—there was no oxygen gas in the sealed and covered boat, so it didn't take long for Bill to be overcome. Danielle's knowledge of the situation and her quick thinking saved her friend's life.

LEARNING GOALS

After you have studied this chapter, you should be able to:

1. State and illustrate the kinetic theory of gases.
2. Convert pressures in atmospheres to pressures in torr, and vice versa.
3. Use Boyle's Law to calculate changes in pressure and volume at constant temperature.
4. Use Charles's Law to calculate changes in temperature and volume at constant pressure.
5. Convert from Celsius degrees to kelvins, and vice versa.
6. Use the Combined Gas Law to calculate changes in temperature, pressure, or volume.
7. Use Dalton's Law to calculate the partial pressure of a gas in a mixture.
8. Use Avogadro's Principle to find the number of moles of a gas when you are given the volume (at fixed temperature and pressure), and vice versa.
9. Use the Ideal Gas Law to calculate the pressure, volume, temperature, or number of moles of a gas when you are given the other three.
10. Use the Ideal Gas Law to calculate the molecular mass of a gas.

■■■ **INTRODUCTION**

To many people the word *gas* seems to imply mainly the gas we use for heating and cooking. However, every time we breathe we inhale gases; in fact, we inhale 35 pounds of gases each day. Every time we pass a neon sign, use an aerosol spray, or open a refrigerator, we see examples of gas under pressure. A flying airplane is supported by gases.

Gases have deadly as well as useful aspects. A person in a large city inhales in one day as much toxic, or poisonous, material as he or she would by smoking two packs of cigarettes. More than half of this unhealthy mixture is contributed by the combustion of fuel.

Maybe because many gases are colorless and therefore seem invisible, people took a long time to get around to studying them. Until the late 1500s, air was the only vapor-like substance that scientists had studied. Even though alchemists had come into contact with many vapor-like substances, they had found them so mysterious that they called them spirits and ignored them. They used the word *spirits* for all liquids that vaporized, or turned to gas easily.

Jan Baptista van Helmont (1577–1644), a Flemish physician, was the first person to study vapors other than air. He observed that various vapors behaved like air but had different physical and chemical properties. Like air, these other vapors had no definite shape or volume.

In this chapter, we discuss the basic physical properties of gases: pressure, volume, and temperature. Several formulas are presented that allow us to predict the behavior of gases under various conditions.

13.1   Kinetic Theory of Gases

LEARNING GOAL 1
Kinetic theory of gases

We can explain the properties and characteristics of gases with a series of statements called the **kinetic theory of gases.** The kinetic theory is an attempt to account for properties of gases in terms of the forces that exist between the gas molecules and the energy of these molecules. The kinetic theory states:

1.   Nearly all gases are composed of molecules. (The so-called noble gases are composed of atoms, not molecules. Remember, atoms of noble gases don't combine readily with other atoms.)
2.   The forces of attraction between gas molecules increase as the molecules move closer together. At normal atmospheric pressure and room temperature, the distances between gas molecules are large compared to the size of the molecules and so the forces of attraction are very small.
3.   Gas molecules are always in motion. They often collide with other gas molecules or with their container. The molecules do not stick together after collisions occur, and they do not lose any energy on account of the collisions.

4. When the temperature rises, gas molecules move faster. They move more slowly when the temperature drops.

5. All gas molecules (heavy as well as light) have the same average kinetic energy—that is, energy of motion—at the same temperature.

A gas that behaves exactly as predicted by the kinetic theory under all conditions of temperature and pressure is called an *ideal gas*. An **ideal gas** is a gas whose molecules have no attraction for each other. In nature there are no ideal gases, only real gases such as hydrogen (H_2), helium (He), oxygen (O_2), and others. A **real gas** is a gas that does not completely obey the assumptions of the kinetic theory. At very low temperatures and very high pressures, real gases do not exhibit ideal behavior, because the molecules of the gas are very close together, which increases the attractive forces between them. Under moderate conditions of temperature and pressure, the forces of attraction are minimal and real gases behave as ideal gases. This characteristic allows us to use the kinetic theory to predict the behavior of real gases.

13.2 Air Pressure and the Barometer

In the early 1600s, the Italian mathematician and physicist Evangelista Torricelli discovered that air exerts a measurable pressure on the objects it touches. Torricelli filled a glass tube, which was sealed at one end, with liquid mercury. He then turned the tube upside down and placed it in a bowl, also filled with mercury (Figure 13.1). Torricelli noticed that most of the mercury remained in the upright tube. He concluded that the surrounding air exerted pressure on the surface of the mercury in the bowl, which in turn supported the column of mercury in the tube.

At sea level, the height of the mercury supported in the tube averages 760 mm. In honor of Torricelli, the pressure that supports a column of mercury exactly 1 mm high is now called 1 **torr.** Therefore, at sea level, the air pressure averages 760 torr, depending on the weather. (Note that the plural of *torr* is *torr*.)

If we take Torricelli's pressure-measuring device, called a **barometer,** to the top of Pikes Peak (which is 14,000 feet above sea level), the height of the mercury column decreases, because the air pressure at the top of the peak is less than it is at sea level. If we take the same barometer to Death Valley (which is 280 feet below sea level), the height of the mercury column increases, because the air pressure below sea level is greater than it is at sea level. We can say that the height of the mercury column is proportional to the atmospheric pressure. In fact, the height of a mercury column supported by *any* gas is proportional to the pressure of that gas. We can thus express the pressure of any gas as a number of torr.

Because we know that the air pressure at sea level supports a column of mercury 760 mm high, we can also define another unit of pressure, called

LEARNING GOAL 2
Converting atmospheres to torr, and vice versa

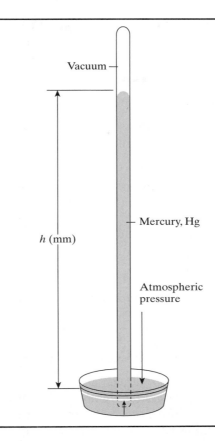

Vacuum

h (mm)

Mercury, Hg

Atmospheric
pressure

■ **FIGURE 13.1**

A simple barometer

the **atmosphere** (abbreviated *atm*). We have used this unit before, but now
we'll define it in terms of torr:

$$1 \text{ atmosphere} = 760 \text{ mm of mercury} = 760 \text{ torr}$$

EXAMPLE 13.1

Express each of the following pressures in torr: (a) 1.50 atm   (b)
0.150 atm

Solution

Use the factor-unit method and the fact that 1 atm = 760 torr.

(a) $? \text{ torr} = (1.50 \text{ atm}) \left(\dfrac{76\overline{0} \text{ torr}}{1 \text{ atm}} \right) = 1{,}140 \text{ torr}$

(b) $? \text{ torr} = (0.150 \text{ atm}) \left(\dfrac{76\overline{0} \text{ torr}}{1 \text{ atm}} \right) = 114 \text{ torr}$

Practice Exercise 13.1   Express each of the following pressures in torr: (a) 0.075 atm   (b) 9.700 atm

---

EXAMPLE 13.2

Express each of the following pressures in atmospheres: (a) 6,460 torr   (b) 152 torr

Solution

Use the factor-unit method and the fact that 1 atm = 760 torr.

(a) $? \text{ atm} = (6,460 \text{ torr}) \left(\dfrac{1 \text{ atm}}{760 \text{ torr}} \right) = 8.50 \text{ atm}$

(b) $? \text{ atm} = (152 \text{ torr}) \left(\dfrac{1 \text{ atm}}{760 \text{ torr}} \right) = 0.200 \text{ atm}$

Practice Exercise 13.2   Express each of the following pressures in atmospheres: (a) 7.6 torr   (b) 9,120 torr

---

Air Pressure: A Strong Force

The German physicist Otto von Guericke developed a pump that could *evacuate*, or empty, the air from a container. In 1657, intrigued by Torricelli's experiments, he performed an experiment in which he put together two tightly fitted metal hemispheres (Figure 13.2). He then used his pump to evacuate the air from inside them. The hemispheres held together. He attached a team of horses to each hemisphere by means of a rope. The horses pulled in opposite directions, but they could not separate the hemispheres, because air was pressing from the outside but *not* from the inside. The air pressure was stronger than the two teams of horses. When the hemispheres were again filled with air, they easily fell apart.

Gas inlet–outlet

■ **FIGURE 13.2**

von Guericke's hemispheres

13.3   Boyle's Law

LEARNING GOAL 3
Boyle's Law

After hearing of von Guericke's experiment, an English chemist named Robert Boyle decided to test the effect of pressure on the volume of gases. For his experiments in the mid-1600s, Boyle devised a J-shaped tube. He trapped air in the closed end of the tube by pouring mercury into the open end (Figure 13.3). He changed the pressure on the trapped air by changing the amount of mercury in the tube. He found that if he tripled the pressure on the air, the volume occupied by the air decreased to one-third the volume. And if he decreased the pressure on the trapped air to one-third the original pressure, the volume occupied by the air tripled.

■ **FIGURE 13.3**

Boyle's experiment

Boyle failed to realize that the temperature of the air affects this relationship. The French physicist Edme Mariotte, working on his own, also discovered the relationship, and he was the first to state what is now known as **Boyle's Law:** *The volume of a gas increases as the pressure on it decreases, if its temperature remains constant.* Another way to state this principle is to say that at constant temperature, the volume of a given mass of gas varies inversely with its pressure. This is what we mean by an *inverse relationship*.

Boyle's Law can also be stated in terms of the ratios of pressures and volumes:

$$\frac{P_i}{P_f} = \frac{V_f}{V_i} \qquad \text{(at constant temperature)}$$

or $P_i \times V_i = P_f \times V_f$ Equation 13.1

where P stands for pressure, V stands for volume, i indicates initial (or original) conditions, and f indicates final conditions. Equation 13.1 can be used to find the final pressure or volume of any gas, given a change in the other quantity.

EXAMPLE 13.3

Suppose a gas has an initial volume, V_i, of $5\overline{0}$ mL at a pressure of 1.0 atm. What will be its final volume, V_f, if the pressure is increased to 2.0 atm (P_f)? (Assume that the temperature remains constant.)

Solution

UNDERSTAND THE PROBLEM

A gas is described as having a particular volume at a certain pressure. We are asked to determine the volume of the gas if the pressure is increased with no change in temperature. We can use Boyle's Law, which states that at constant temperature, the volume of a gas decreases when its pressure increases. Since the pressure is increased, the final volume will decrease, that is, it will be less than the initial volume of $5\overline{0}$ mL.

CARRY OUT THE PLAN

First organize the data—that is, the facts that have been given.

$$V_i = 5\overline{0} \text{ mL} \qquad V_f = ?$$

$$P_i = 1.0 \text{ atm} \qquad P_f = 2.0 \text{ atm}$$

Then write Boyle's Law, solve for V_f, and enter the data in the result.

$$P_i \times V_i = P_f \times V_f \qquad \text{so} \qquad V_f = \frac{P_i V_i}{P_f}$$

$$V_f = \frac{(1.0 \text{ atm})(5\overline{0} \text{ mL})}{2.0 \text{ atm}} = 25 \text{ mL}$$

LOOK BACK

As expected, increasing the pressure on the gas decreases the volume of the gas. Once again, this is an *inverse relationship*.

Practice Exercise 13.3   A gas has an initial volume of 500.0 mL (V_i) at a pressure of 2.50 atm (P_i). What will be its final volume (V_f) if the pressure is increased to 5.00 atm (P_f)? (Assume that the temperature remains constant.)

EXAMPLE 13.4

Suppose a gas occupies a volume of $5\overline{00}$ mL under a pressure of 3.0 atm at $1\overline{00}°$ C. What will be the volume of the gas when the pressure is decreased to 1.0 atm? (Assume that the temperature remains constant.)

Solution

First organize the data:

$$V_i = 5\overline{00} \text{ mL} \qquad V_f = ?$$

$$P_i = 3.0 \text{ atm} \qquad P_f = 1.0 \text{ atm}$$

$$t = 1\overline{00}°\text{C and is constant}$$

Then write Boyle's Law, solve for V_f, and enter the data in the result.

$$P_i \times V_i = P_f \times V_f \qquad \text{so} \qquad V_f = \frac{P_i V_i}{P_f}$$

$$V_f = \frac{(3.0 \text{ atm})(5\overline{00} \text{ mL})}{1.0 \text{ atm}} = 1,500 \text{ mL}$$

When the pressure was decreased to one-third of its initial value, the volume increased to three times *its* initial value.

Practice Exercise 13.4   A gas occupies a volume of 250.0 mL under a pressure of 4.00 atm at 100.0° C. What will be the volume of the gas when the pressure is decreased to 2.00 atm? (Assume that the temperature remains constant.)

Boyle's Law can also be used to determine the pressure needed to cause a change in volume of a gas by solving the formula $P_iV_i = P_fV_f$ for P_f. See the next example.

EXAMPLE 13.5

A gas occupies a volume of 0.100 liter (L) under a pressure of $76\overline{0}$ torr at $2\overline{0}°C$. What pressure will cause the gas to occupy a volume of 50.0 mL at the same temperature?

Solution
Organize the data:

$$V_i = 0.100 \text{ L} = 1\overline{00} \text{ mL} \qquad V_f = 50.0 \text{ mL}$$

$$P_i = 76\overline{0} \text{ torr} \qquad\qquad P_f = ?$$

$$t = 2\overline{0}°C \text{ and is constant}$$

Write Boyle's Law, solve for P_f, and enter the data in the result.

$$P_i \times V_i = P_f \times V_f \quad \text{ so } \quad P_f = \frac{P_iV_i}{V_f}$$

$$P_f = \frac{(76\overline{0} \text{ torr})(1\overline{00} \text{ mL})}{50.0 \text{ mL}} = 1,520 \text{ torr}$$

Practice Exercise 13.5   A gas occupies a volume of 0.200 L under a pressure of $76\overline{0}$ torr at 50.0°C. What pressure will cause the gas to occupy a volume of 25.0 mL at the same temperature?

13.4 Charles's Law

LEARNING GOAL 4

Charles's Law

More than a hundred years after Boyle's work, the French physicist Jacques Charles discovered a relationship between the volume and the temperature of a gas. He found that a gas expands when its temperature is increased and that it shrinks when its temperature is decreased. A gas originally at 0°C expands 1/273 of its volume for each 1°C increase in temperature, and it shrinks 1/273 of its volume for each 1°C decrease (if the pressure of the gas is held constant). This is the basis of *Charles's Law*. It stands to reason: Most things do shrink as they get cooler and expand as they get hotter. The old tube-type tires used on automobiles are a good example. These tires

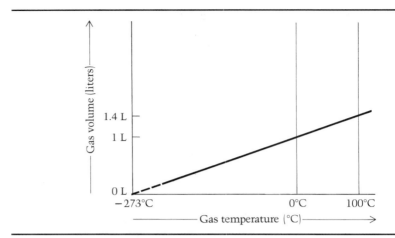

■ FIGURE 13.4

A theoretical plot of gas
volume versus temperature

expanded as the air inside them got warm, as often happened on a hot day
when the car was traveling at high speed. If the tires expanded enough, a
blowout would occur.

Theoretically, if we took 1 L (or any volume) of a gas at 0°C and kept
cooling it (at constant pressure), the volume of the gas would shrink to 0
mL at −273°C (Figure 13.4). (This, of course, has never been accomplished,
because all known gases become liquid before reaching that temperature.)
The temperature −273°C is called **absolute zero.** It is the basis of a temperature scale called the **Kelvin scale**, which was named after Lord Kelvin, the
British physicist who proposed it. The units of this scale are kelvins, K, not
degrees. One kelvin covers the same temperature interval as one Celsius
degree (Figure 13.5). The relationship between the two temperature scales is

LEARNING GOAL 5
Converting °C to K, and
vice versa

$$K = °C + 273 \qquad \text{Equation 13.2}$$

Using the Kelvin scale, we can state **Charles's Law** in the following way:
*The volume of a gas varies directly with its temperature in kelvins if the
pressure remains constant.* Charles's Law can also be stated mathematically:

$$\frac{V_i}{V_f} = \frac{T_i}{T_f} \qquad \text{(at constant pressure)} \qquad \text{Equation 13.3}$$

where T is the temperature in kelvins (a lower-case t stands for degrees
Celsius).

You must remember to change degrees Celsius to kelvins when you
work with the formula for Charles's Law. The direct relationship between
the temperature and the volume of a gas, as expressed in Equation 13.3,
applies *only* to temperatures measured in kelvins. To convert from one scale
to the other, use the formula

$$T = t + 273$$

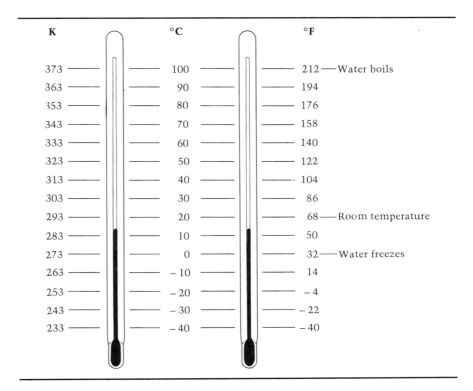

Comparison of kelvins,
degrees Celsius, and degrees
Fahrenheit

EXAMPLE 13.6

A gas at 27°C occupies a volume of $1,\overline{0}00$ mL. What will be its volume
at 127°C? (Assume that the pressure of the gas remains constant.)

Solution

First organize the data and change degrees Celsius to kelvins.

$$V_i = 1,\overline{0}00 \text{ mL} \qquad\qquad V_f = ?$$

$$T_i = t_i + 273 \qquad\qquad T_f = t_f + 273$$

$$= 27 + 273 = 3\overline{0}0 \text{ K} \qquad = 127 + 273 = 4\overline{0}0 \text{ K}$$

Then write Charles's Law, solve for V_f, and enter the data in the result.

$$\frac{V_i}{V_f} = \frac{T_i}{T_f} \qquad \text{so} \qquad V_f = \frac{V_i T_f}{T_i}$$

$$V_f = \frac{(1,\overline{0}00 \text{ mL})(4\overline{0}0 \cancel{\text{ K}})}{3\overline{0}0 \cancel{\text{ K}}}$$

$$= 1,333 \text{ mL or } 1,330 \text{ mL (three significant figures)}$$

As expected, increasing the temperature increases the volume of the gas. This is what we mean by a *direct relationship*.

You can also solve the problem using this alternate approach. Since Charles's Law predicts an increase in volume with an increase in temperature, the final volume will be greater than the initial volume of $1,\overline{000}$ mL. We may determine the final volume by using the following relationship:

Final volume = initial volume × temperature ratio

The correct temperature ratio is the one that will predict the expected volume increase, that is,

$$\text{Final volume} = 1,\overline{000} \text{ mL} \times \frac{400 \text{ K}}{300 \text{ K}}$$

$$= 1,330 \text{ mL (three significant figures)}$$

Practice Exercise 13.6   A gas at 127°C occupies a volume of $2,5\overline{00}$ mL. What will be its volume at 27.0°C? (Assume that the pressure of the gas remains constant.)

Charles's Law can also be used to determine the temperature of the gas when the change in volume is known. Solve the formula

$$\frac{V_i}{T_f} = \frac{T_i}{T_f}$$

for either T_i or T_f.

EXAMPLE 13.7

When a gas is cooled at constant pressure, it contracts from 50.0 mL to 45.0 mL. Its final temperature is 80°C. What was its initial temperature in degrees Celsius?

Solution

First organize the data and change degrees Celsius to kelvins.

$$V_i = 50.0 \text{ mL} \quad V_f = 45.0 \text{ mL}$$

$$t_i = ? \quad\quad\quad T_f = t_f + 273$$

$$T_i = ? \quad\quad\quad = 8\overline{0} + 273 = 353 \text{ K}$$

Then write Charles's Law, solve for T_i, and enter the data in the result.

$$\frac{V_i}{V_f} = \frac{T_i}{T_f} \quad \text{so} \quad T_i = \frac{V_i T_f}{V_f}$$

$$T_i = \frac{(50.0 \text{ mL})(353 \text{ K})}{45.0 \text{ mL}} = 392 \text{ K}$$

Finally, to find the answer in degrees Celsius, proceed as follows:

$$T_i = t_i + 273$$

$$t_i = T_i - 273$$

$$t_i = 392 - 273 = 119°C$$

Practice Exercise 13.7   When a certain gas is heated at constant pressure, it expands from 50.0 mL to 60.0 mL. Its final temperature is 75.0°C. What was its initial temperature in degrees Celsius?

13.5 The Combined Gas Law

Boyle's Law and Charles's Law can be combined into a single law that includes all three variables—pressure, volume, and temperature. The single law can be used to determine how changes in any two of these variables affect the third variable. This **Combined Gas Law** can be stated as follows:

LEARNING GOAL 6
Combined Gas Law

$$\frac{P_i V_i}{T_i} = \frac{P_f V_f}{T_f}$$
Equation 13.4

Equation 13.4 incorporates both Charles's Law and Boyle's Law, so it is the only equation you need to memorize. The following examples show why.

EXAMPLE 13.8

A certain gas at 2°C and 1.00 atm pressure fills a 4.00-L container. What volume will the gas occupy at $10\overline{0}$°C and $78\overline{0}$ torr pressure?

Solution

To begin, organize the data, change degrees Celsius to kelvins, and convert all the pressures to the same units.

$$P_i = 1.00 \text{ atm } (76\overline{0} \text{ torr}) \quad P_f = 78\overline{0} \text{ torr}$$

$$V_i = 4.00 \text{ L} \qquad\qquad V_f = ?$$

$$T_i = t_i + 273 \qquad\qquad T_f = t_f + 273$$

$$= 2 + 273 = 275 \text{ K} \qquad = 100 + 273 = 373 \text{ K}$$

Then write the Combined Gas Law, solve for V_f, and enter the data in the result.

$$\frac{P_i V_i}{T_i} = \frac{P_f V_f}{T_f} \quad \text{so} \quad V_f = \frac{P_i V_i T_f}{P_f T_i}$$

$$V_f = \frac{(76\overline{0} \text{ torr})(4.00 \text{ L})(373 \text{ K})}{(78\overline{0} \text{ torr})(275 \text{ K})} = 5.29 \text{ L}$$

(If you need help solving algebraic equations, see Appendix A, Section A.6.)

In this example the final volume depends on both the pressure change (Boyle's Law) and the temperature change (Charles's Law). The Combined Gas Law can be used to solve for the final volume, or the appropriate pressure ratio and temperature ratio can be used according to the following relationship:

Final volume = initial volume × pressure ratio × temperature ratio

From the data given, the pressure change would predict a decrease in the volume of the gas and the temperature change would predict an increase in the volume. Therefore:

$$\text{Final volume} = 4.00 \text{ L} \times \frac{760 \text{ torr}}{780 \text{ torr}} \times \frac{373 \text{ K}}{275 \text{ K}}$$

$$= 5.29 \text{ L}$$

Practice Exercise 13.8   A certain gas at 25.0°C and 1.00 atm pressure fills a 5.00-L container. What volume will the gas occupy at 50.0°C and 800.0 torr pressure?

The Combined Gas Law can be used to find the final temperature, as in Example 13.9.

EXAMPLE 13.9

Suppose a gas is kept in a container (that is, at constant volume), but the pressure of the gas is increased from $76\bar{0}$ torr to $86\bar{0}$ torr. This is done by changing the temperature of the gas from 0°C to some higher temperature. What is this higher temperature? (Report the answer in degrees Celsius.)

Solution

To begin, organize the data and change degrees Celsius to kelvins. Note that the volume remains constant.

$$P_i = 76\bar{0} \text{ torr} \qquad\qquad P_f = 86\bar{0} \text{ torr}$$
$$T_i = t_i + 273 \qquad\qquad\quad t_f = ?$$
$$= 0 + 273 = 273 \text{ K} \qquad T_f = ?$$
$$V \text{ is constant; this means} \quad V_i = V_f$$

Then write the Combined Gas Law:

$$\frac{P_i V_i}{T_i} = \frac{P_f V_f}{T_f}$$

But because V_i and V_f are equal, they cancel each other, and the equation becomes

$$\frac{P_i}{T_i} = \frac{P_f}{T_f}$$

Solve the equation for T_f and enter the data in the result.

$$T_f = \frac{P_f T_i}{P_i} = \frac{(860 \ \cancel{torr})(273 \ K)}{760 \ \cancel{torr}} = 309 \ K$$

Finally, change kelvins to degrees Celsius.

$$T_f = t_f + 273 \qquad so \qquad t_f = T_f - 273$$

$$t_f = 309 - 273 = 36°C$$

Practice Exercise 13.9 A gas is kept in a container at constant volume. The pressure of the gas is decreased from 850.0 torr to 750.0 torr. The temperature of the gas at the start of the experiment is 50.0°C. What is its final temperature in degrees Celsius?

EXAMPLE 13.10

A certain gas occupies a volume of 3 L at 25°C and 4 atm pressure. What will be its volume at 25°C and 2 atm pressure?

Solution

First organize the data and change degrees Celsius to kelvins. Note that the temperature remains constant.

$$P_i = 4 \ atm \qquad P_f = 2 \ atm$$

$$V_i = 3 \ L \qquad V_f = ?$$

$$t_i = t_f = 25°C \ (298 \ K)$$

Then write the Combined Gas Law.

$$\frac{P_i V_i}{T_i} = \frac{P_f V_f}{T_f}$$

T_i and T_f cancel each other because they are equal (remember that T is constant), and the equation becomes

$$P_i V_i = P_f V_f$$

Finally, solve this equation for V_f and enter the data in the result.

$$V_f = \frac{P_i V_i}{P_f} = \frac{(4 \ \cancel{atm})(3 \ L)}{2 \ \cancel{atm}} = 6 \ L$$

Practice Exercise 13.10 A gas occupies a volume of 5.00 L at 298 K and 1.50 atm pressure. What will be its volume at 25°C and 3.00 atm pressure?

13.6   Standard Temperature and Pressure (STP)

For work with gases, scientists have defined a set of standard reference conditions. **Standard temperature and pressure (STP)** are defined as 273 K (0°C) and 1 atm pressure (760 torr). Historically, these conditions were chosen because they were easy to create with the available technology. Whenever volumes of gases are reported, they may be corrected to STP so that comparisons may be made. That is, the Combined Gas Law is used to determine what the volume of the gas would be at STP.

EXAMPLE 13.11

A gas collected at 127°C has a volume of $1\overline{0}$ L and a pressure of $8\overline{00}$ torr. What would be its volume at STP?

Solution

To begin, organize the data and change degrees Celsius to kelvins.

$$P_i = 8\overline{00} \text{ torr} \qquad\qquad P_f = 7\overline{60} \text{ torr}$$

$$V_i = 1\overline{0} \text{ L} \qquad\qquad V_f = ?$$

$$T_i = t_i + 273 \qquad\qquad T_f = t_f + 273$$

$$= 127 + 273 = 4\overline{00} \text{ K} \qquad = 0 + 273 = 273 \text{ K}$$

Then write the Combined Gas Law, solve for V_f, and enter the data in the result.

$$\frac{P_i V_i}{T_i} = \frac{P_f V_f}{T_f} \qquad \text{so} \qquad V_f = \frac{P_i V_i T_f}{P_f T_i}$$

$$V_f = \frac{(8\overline{00} \text{ torr})\,(1\overline{0} \text{ L})(273 \text{ K})}{(7\overline{60} \text{ torr})(4\overline{00} \text{ K})} = 7.2 \text{ L}$$

Practice Exercise 13.11   A gas collected at 25.0°C has a volume of 250.0 mL and a pressure of 780.0 torr. What would be its volume at STP?

13.7   Dalton's Law of Partial Pressures

■ LEARNING GOAL 7

Dalton's Law

In the laboratory, gases are usually collected by the water-displacement method (Figure 13.6), because many gases are relatively insoluble in water. This is a convenient method for obtaining a sample of a gas, except for one thing: Gas collected by this method always has some water vapor in it. Therefore the pressure in the collection bottle (P_{total}) is due to both the pressure of the gas (P_{gas}) and the pressure of the water vapor (P_{water}). To find out how much gas has been collected, we need to know the pressure of

■ **FIGURE 13.6**

Collection of a gas over water. In this example, hydrochloric acid is added, through the thistle tube, to a piece of zinc in the bottle. Hydrogen gas travels out of the bottle through the rubber tube and is collected in the flask. As the hydrogen gas fills the flask, the water that was originally in the flask is pushed out.

Gas and water vapor

the gas alone—that is, its partial pressure P_{gas}. The **partial pressure** of any gas in a mixture of gases is the pressure that the gas *would* exert at the temperature of the mixture if it were the *only* gas in the container.

About 1800, John Dalton discovered a principle that is useful in this situation. **Dalton's Law of Partial Pressures** states that *the total pressure of a mixture of gases is the sum of the partial pressures of all the gases in the mixture.* In other words, each gas in a mixture exerts a pressure that is independent of the other gases present. This means that we can simply subtract the pressure of the water vapor from the total pressure to find the pressure exerted by the gas. This is stated mathematically as

$$P_{total} = P_{gas} + P_{water} \qquad \text{(Dalton's Law)}$$

so

$$P_{gas} = P_{total} - P_{water}$$

The partial pressure of water vapor is known at any given temperature. Table 13.1 gives this value for various temperatures.

EXAMPLE 13.12

Suppose we collect a sample of oxygen gas over water at 27°C and at a pressure of $75\overline{0}$ torr. The gas fills a $50\overline{0}$-mL container but has water vapor mixed in it. What would be the volume of the *dry* gas at STP?

Solution

First find the pressure of the dry gas (P_{O_2}). To do this, subtract the pressure of the water vapor (P_{H_2O}) from the total pressure in the collection bottle (P_{total}). Look up the vapor pressure of water at 27°C and find that it is 26.74 torr. Then

$$P_{O_2} = P_{total} - P_{H_2O}$$

$$P_{O_2} = 75\overline{0} \text{ torr} - 26.74 \text{ torr} = 723 \text{ torr}$$

Therefore the initial pressure of the dry oxygen gas (P_i) is 723 torr. The next step is to find its volume at STP.

Organize the data and change degrees Celsius to kelvins.

$$P_i = 723 \text{ torr} \qquad\qquad P_f = 76\overline{0} \text{ torr}$$

$$V_i = 50\overline{0} \text{ mL} \qquad\qquad V_f = ?$$

$$T_i = t_i + 273 \qquad\qquad T_f = t_f + 273$$

$$= 27 + 273 = 30\overline{0} \text{ K} \qquad = 0 + 273 = 273 \text{ K}$$

TABLE 13.1

Pressure of Water Vapor, P_{H_2O}, at Various Temperatures

TEMPERATURE (°C)	PRESSURE (torr)	TEMPERATURE (°C)	PRESSURE (torr)
0	4.58	33	37.73
5	6.54	34	39.90
10	9.21	35	42.18
15	12.79	36	44.56
16	13.63	37	47.07
17	14.53	38	49.69
18	15.48	39	52.44
19	16.48	40	55.32
20	17.54	45	71.88
21	18.65	50	92.51
22	19.83	55	118.04
23	21.07	60	149.38
24	22.38	65	187.54
25	23.76	70	233.7
26	25.21	75	289.1
27	26.74	80	355.1
28	28.35	85	433.6
29	30.04	90	525.8
30	31.82	95	633.9
31	33.70	100	760.0
32	35.66		

Then write the Combined Gas Law, solve for V_f, and enter the data in the result.

$$\frac{P_iV_i}{T_i} = \frac{P_fV_f}{T_f} \quad\text{so}\quad V_f = \frac{P_iV_iT_f}{P_fT_i}$$

$$V_f = \frac{(723 \text{ torr})(500 \text{ mL})(273 \text{ K})}{(760 \text{ torr})(300 \text{ K})} = 433 \text{ mL}$$

The dry oxygen would have a volume of 433 mL at STP.

Practice Exercise 13.12   A sample of oxygen gas is collected over water at 55.0°C and at a pressure of 800.0 torr. The gas occupies a volume of 250.0 mL. What would be the volume of the *dry* gas at STP?

EXAMPLE 13.13

Suppose 2.0 L of hydrogen gas are collected over water at 45°C and at a pressure of 700 torr. What would be the volume of the dry hydrogen gas at STP?

Solution

Again, first find the pressure of the dry hydrogen gas (P_{H_2}). To do this, subtract the pressure of the water vapor (P_{H_2O}) from the total pressure in the collection bottle (P_{total}). The vapor pressure of water a 45°C is 71.88 torr, so

$$P_{H_2} = P_{total} - P_{H_2O}$$

$$P_{H_2} = 700 \text{ torr} - 71.88 \text{ torr} = 628 \text{ torr}$$

Therefore the initial pressure of the dry hydrogen gas (P_i) is 628 torr. Next organize the data and change degrees Celsius to kelvins.

$P_i = 628$ torr	$P_f = 760$ torr
$V_i = 2.0$ L	$V_f = ?$
$T_i = t_i + 273$	$T_f = t_f + 273$
$= 45 + 273 = 318$ K	$= 0 + 273 = 273$ K

Write the Combined Gas Law, solve for V_f, and enter the data in the result.

$$\frac{P_iV_i}{T_i} = \frac{P_fV_f}{T_f} \quad\text{so}\quad V_f = \frac{P_iV_iT_f}{P_fT_i}$$

$$V_f = \frac{(628 \text{ torr})(2.0 \text{ L})(273 \text{ K})}{(760 \text{ torr})(318 \text{ K})} = 1.4 \text{ L}$$

Practice Exercise 13.13   A sample of hydrogen gas is collected over water at 20.0°C and 715 torr. It occupies a volume of 5.50 L. What would be the volume of the *dry* gas at STP?

13.8   Gay-Lussac's Law of Combining Volumes

Joseph Louis Gay-Lussac
(Archive Photos)

In the early 1800s, the French chemist Joseph Louis Gay-Lussac (pronounced "Gay-lew-SACK") investigated the reaction in which hydrogen gas and oxygen gas form water vapor. He found that if the gases were at the same temperature and pressure, 2 L of hydrogen gas plus 1 L of oxygen gas produced 2 L of water vapor. He also studied the reaction between hydrogen and chlorine. He found that 1 L of hydrogen gas plus 1 L of chlorine gas reacted to form 2 L of hydrogen chloride gas.

Hydrogen gas + oxygen gas → water vapor
2 L   1 L   2 L

Hydrogen gas + chlorine gas → hydrogen chloride gas
1 L   1 L   2 L

What Gay-Lussac had discovered is now called **Gay-Lussac's Law of Combining Volumes,** which states that *when gases at the same temperature and pressure combine, the volumes of the reactants and products are in the same ratio as the coefficients of the balanced equation for the reaction.*

The Law of Combining Volumes states, for example, that two volumes of hydrogen combine with one volume of oxygen to form two volumes of water vapor

$$2H_2(g) + O_2(g) \rightarrow 2H_2O(g)$$
2 volumes   1 volume   2 volumes

whether "volume" means liter, cubic inch, gallon, or even coffee-can-full, as long as all three gases are at the same temperature and pressure. This law provides an easy way to work stoichiometric problems involving gases.

EXAMPLE 13.14

How many liters of carbon dioxide can be produced with 10 L of carbon monoxide? How much oxygen is required (provided that all three gases are at the same temperature and pressure)?

Solution
First write the balanced equation and organize the data.

$$2CO(g) + O_2(g) \rightarrow 2CO_2(g)$$
10 L   ? L   ? L

Then use the coefficients of the balanced equation, just as in Chapter 11 (except that here liters can be substituted for moles), to calculate the amount of O_2 needed and the amount of CO_2 formed.

$$? \text{ L of } CO_2 = (10 \text{ L } CO) \left(\frac{2 \text{ L } CO_2}{2 \text{ L } CO} \right)$$

$$= 10 \text{ L of } CO_2$$

$$? \text{ L of } O_2 = (10 \text{ L } CO) \left(\frac{1 \text{ L } O_2}{2 \text{ L } CO} \right)$$

$$= 5 \text{ L of } O_2$$

Practice Exercise 13.14   How many liters of hydrogen fluoride gas can be produced from 20.0 L of hydrogen gas and 20.0 L of fluorine gas? (Assume that all gases are at the same temperature and pressure.)

13.9   Molar Gas Volume and Avogadro's Principle

In 1811, the Italian physicist Amedeo Avogadro, whose name we associate with Avogadro's number, advanced one of the most important hypotheses in chemistry. In an attempt to explain Gay-Lussac's Law of Combining Volumes, Avogadro said that *at the same temperature and pressure, equal volumes of all gases contain the same number of molecules.* This is called **Avogadro's Principle.**

LEARNING GOAL 8
Avogadro's Principle

To understand this principle, consider the following experiment: Suppose exactly 1 mole of hydrogen gas (H_2) is placed in a container whose volume can be changed and whose temperature can be kept at 0°C (273 K). The volume of the container is adjusted so that the pressure of the hydrogen gas is exactly 1 atm. We find that the volume of the container (and therefore of the gas) is 22.4 L.

We now remove the hydrogen gas and repeat the experiment with exactly 1 mole of oxygen (O_2). Again we find that the volume of the container (and the oxygen) is 22.4 L. And every time the experiment is repeated with exactly 1 mole of a different gas, we find that its volume is 22.4 L at standard temperature and pressure (Figure 13.7).

Our experiment enables us to state the following conclusion: *Exactly 1 mole of any gas at STP has a volume of 22.4 L.* This volume is called the **molar volume of a gas,** and this conclusion can be used to solve problems like those in the next two examples.

■ **FIGURE 13.7**
Molar gas volume

1 mole H_2 at STP
V = 22.4 liters

1 mole He at STP
V = 22.4 liters

1 mole N_2 at STP
V = 22.4 liters

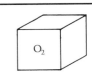

1 mole O_2 at STP
V = 22.4 liters

EXAMPLE 13.15

A certain gas occupies a volume of 179.2 L at STP. How many moles of gas are in this sample?

Solution

One mole of any gas has a volume of 22.4 L at STP. Therefore

$$? \text{ moles of gas} = (179.2 \text{ L}) \left(\frac{1 \text{ mole}}{22.4 \text{ L}} \right) = 8.00 \text{ moles of gas}$$

Practice Exercise 13.15   How many moles of gas occupy a volume of 448 mL at STP?

EXAMPLE 13.16

Determine the volume of 5.00 moles of a gas at STP.

Solution

Again, use the idea that 1 mole of a gas has a volume of 22.4 L at STP. Therefore

$$? \text{ L} = (5.00 \text{ moles}) \left(\frac{22.4 \text{ L}}{1 \text{ mole}} \right) = 112 \text{ L}$$

Practice Exercise 13.16   What is the volume of 0.200 mole of a gas at STP?

13.10   The Ideal Gas Law

We can now combine Boyle's Law, Charles's Law, and Avogadro's Principle into a relationship among the gas variables (P, V, T) and the number n of moles of gas. This relationship, called the **Ideal Gas Law**, is represented by the equation

LEARNING GOAL 9
Ideal Gas Law

Number of moles

$$\text{Pressure in atm} \longrightarrow PV = nRT \longleftarrow \text{Temperature in K} \qquad \text{Equation 13.5}$$

Volume   Ideal
in liters gas constant

If this equation is used with the pressure expressed in atmospheres, the volume in liters, the temperature in kelvins, and the amount of gas in moles, then the value for R, the ideal gas constant, is

$$0.0821 \frac{\text{(L) (atm)}}{\text{(K) (mole)}}$$

Notice the units of R. These units will remind you that when you use the Ideal Gas Law, pressure must be in atmospheres, volume in liters, temperature in kelvins, and gas quantity in moles. We can show that R has this value by solving the ideal gas equation for R and substituting known values for the other variables in the equation.

$$PV = nRT \qquad \text{so} \qquad R = \frac{PV}{nT}$$

We know that at STP (273 K, 1 atm), 1 mole of a gas has a volume of 22.4 L. These values may be substituted into the equation.

$$R = \frac{PV}{nT} = \frac{(1\text{ atm})\,(22.4\text{ L})}{(1\text{ mole})\,(273\text{ K})} = 0.0821\,\frac{\text{(L)}\,\text{(atm)}}{\text{(K)}\,\text{(mole)}}$$

The following examples show how the Ideal Gas Law is used.

EXAMPLE 13.17

Two moles of a gas exert a pressure of 1,520 torr at a temperature of 127°C. What volume does the gas occupy under these conditions?

Solution

First organize the data. Change the 1,520 torr to atmospheres and the degrees Celsius to kelvins. To convert torr to atmospheres, remember that 1 atm = $76\overline{0}$ torr.

$$? \text{ atm} = (1,520 \text{ torr}) \left(\frac{1 \text{ atm}}{76\overline{0} \text{ torr}} \right) = 2.00 \text{ atm}$$

$$P = 1,520 \text{ torr } (2.00 \text{ atm})$$

$$V = ?$$

$$n = 2.00 \text{ moles}$$

$$R = 0.0821\,\frac{\text{(L)}\,\text{(atm)}}{\text{(K)}\,\text{(mole)}}$$

$$T = t + 273 = 127 + 273 = 4\overline{00} \text{ K}$$

Next write the ideal gas equation, solve for V, and enter the data in the result.

$$PV = nRT \qquad \text{so} \qquad V = \frac{nRT}{P}$$

$$V = \frac{(2.00 \text{ moles}) \left(0.0821\,\dfrac{\text{(L)}\,\text{(atm)}}{\text{(K)}\,\text{(mole)}} \right) (4\overline{00} \text{ K})}{2.00 \text{ atm}} = 32.8 \text{ L}$$

Practice Exercise 13.17   Three moles of a gas exert a pressure of 1,555 torr at a temperature of 27.0°C. What volume does the gas occupy under these conditions?

EXAMPLE 13.18

A gas has a pressure of $38\overline{0}$ torr at a temperature of $5\overline{0}$°C. The volume of the gas is $8\overline{00}$ mL. How many moles of gas are present?

Solution

Organize the data. Change torr to atmospheres and degrees Celsius to kelvins.

$$? \text{ atm} = (38\overline{0} \text{ torr})\left(\frac{1 \text{ atm}}{76\overline{0} \text{ torr}}\right) = 0.500 \text{ atm}$$

$$P = 0.500 \text{ atm}$$

$$V = 8\overline{00} \text{ mL } (0.800 \text{ L})$$

$$n = ?$$

$$R = 0.0821 \frac{\text{(L) (atm)}}{\text{(K) (mole)}}$$

$$T = t + 273 = 5\overline{0} + 273 = 323 \text{ K}$$

Write the ideal gas equation, solve for n, and enter the data in the result.

$$PV = nRT \qquad \text{so} \qquad n = \frac{PV}{RT}$$

$$n = \frac{(0.500 \text{ atm}) (0.800 \text{ L})}{\left(0.0821 \dfrac{\text{(L) (atm)}}{\text{(K) (mole)}}\right)(323 \text{ K})} = 0.0151 \text{ mole}$$

Practice Exercise 13.18   A gas exerts a pressure of 444 torr at a temperature of 55.0°C. The volume of the gas is 1.00 L. How many moles of gas are present?

4 atm

■ **FIGURE 13.8**

A sample of gas in a small container. The pressure gauge shows the pressure that is exerted by the gas.

13.11   Determining Molecular Mass with the Ideal Gas Law

LEARNING GOAL 10

Ideal Gas Law and molecular mass

The number of moles n of a sample of a gas is equal to the mass of the sample (g) in grams divided by the molecular mass of the gas (MM).

$$n = \frac{\text{g}}{\text{MM}}$$

If g/MM is used in place of *n* in the Ideal Gas Law, we have a valuable tool for determining the molecular masses of gases:

$$PV = \frac{gRT}{MM} \qquad \text{Equation 13.6}$$

For an unknown gas, we can weigh a sample of it into a container, measure its pressure, volume, and temperature, and use this form of the Ideal Gas Law to calculate its molecular mass.

EXAMPLE 13.19

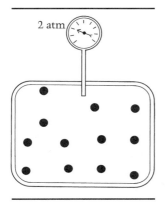

2 atm

■ FIGURE 13.9

The same sample of gas in a container twice the size of that shown in Figure 13.8

What is the molecular mass of the gas for which the following experimental data apply?

$$\text{Mass} = 1.00 \text{ g} \qquad V = 82.0 \text{ mL } (0.0820 \text{ L})$$

$$P = 1.50 \text{ atm} \qquad T = 3\overline{00} \text{ K}$$

Solution   Write Equation 13.6, solve for MM, and enter the data in the result.

$$PV = \frac{gRT}{MM} \quad \text{so} \quad MM = \frac{gRT}{PV}$$

$$MM = \frac{(1.00 \text{ g})\left(0.0821 \frac{(\cancel{L})(\cancel{atm})}{(\cancel{K})(\text{mole})}\right)(3\overline{00} \,\cancel{K})}{(1.50 \,\cancel{atm})(0.0820 \,\cancel{L})} = 2\overline{00} \text{ g/mole}$$

Practice Exercise 13.19   Determine the molecular mass of the gas to which the following hypothetical data apply: mass = 2.50 g, V = 100.0 mL, P = 2.00 atm, T = 400.0 K.

13.12 The Kinetic Theory Revisited

If the gas laws are considered in light of the kinetic theory presented at the beginning of this chapter, it is clear why gases behave as they do. For example, the kinetic theory explains Boyle's Law. Remember, Boyle's Law says that the volume of a gas increases as its pressure decreases and that the pressure of a gas increases as its volume decreases. When we assume that gases are composed of molecules that are constantly in motion, colliding with each other and with the walls of their container, it seems reasonable that:

1. Keeping these molecules in a smaller space causes them to strike the walls of their container more often; this means that more force is exerted against the walls of the container, and therefore the pressure is greater (Figure 13.8).

2. Allowing these molecules more space to move around in causes them to strike the walls less often; therefore the pressure is decreased (Figure 13.9).

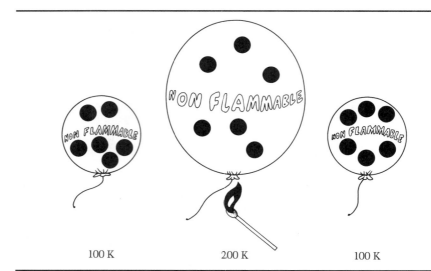

■ **FIGURE 13.10**

The balloon experiment shows the temperature–volume relationship expressed by Charles's Law.

The kinetic theory also explains Charles's Law, which says that the volume of a gas increases as its temperature increases and that the volume of a gas decreases as its temperature decreases. Consider a balloon filled with air (Figure 13.10). The molecules of air constantly move around, hitting the walls of the balloon. Suppose a candle is held near the balloon, which warms the air inside. The gas gets warmer because the molecules begin to move more rapidly, hitting the walls of the balloon with more force. Because the balloon is made of flexible material, it expands. As the gas cools to room temperature, the molecules of air move more slowly, hitting the walls of the balloon more gently, and the balloon contracts.

SUMMARY

The kinetic theory of gases is concerned with the motion of gas molecules and the forces of attraction between molecules. It can be used to explain the various physical properties of gases.

In the early 1600s, Torricelli devised a pressure-measuring device called a barometer and discovered that air exerts pressure on objects it touches. The units of pressure we use are atmospheres and torr, and air exerts a pressure of about 760 torr (1 atm) at sea level.

The experiments of Robert Boyle in the mid-1600s led to the formulation of Boyle's Law, which states that at constant temperature, the volume of a given mass of gas varies inversely with the pressure. In the 1780s, Jacques Charles discovered a relationship between the volume and the temperature of a gas. Charles's Law states that at constant pressure, the volume of a given amount of gas varies directly with its temperature in kelvins. On the Kelvin scale, the temperature in kelvins is equal to the Celsius temperature plus 273. Boyle's Law and Charles's Law can be combined into a single law, the Combined Gas Law, which shows the interrelationship of changes in the temperature, pressure, and volume of a gas.

Standard temperature and pressure (STP) for gases are defined as 273 K (0°C) and 1 atm pressure (760 torr). The Combined Gas Law may be used to determine what the volume of a gas would be at STP. Dalton's Law of Partial Pressures states that the total pressure exerted by a mixture of gases is the sum of the partial pressures that each of the gases in the mixture would exert on its own at the same temperature as the mixture.

In the early 1800s, Gay-Lussac discovered the Law of Combining Volumes. This law states that when gases at the same temperature and pressure combine, the volumes of the reactants and products are in the same ratio as the coefficients of the balanced equation for the reaction. In an attempt to explain the Law of Combining Volumes, Avogadro hypothesized that equal volumes of all gases at the same temperature and pressure contain the same number of molecules. From the hypothesis, called Avogadro's Principle, scientists determined that 1 mole of any gas at STP has a volume of 22.4 L. This is called the molar volume of a gas.

Boyle's Law, Charles's Law, and Avogadro's Principle can be combined into a single relationship among the pressure P, temperature T, volume V, and number of moles n of a gas. This relationship, called the Ideal Gas Law, is represented by the equation $PV = nRT$, where R is the ideal gas constant and is equal to 0.0821 (L) (atm)/(K) (mole). The Ideal Gas Law can be used to determine the molecular mass of a gas when n is replaced by the fraction that represents mass/molecular mass (g/MM) and the equation is rearranged.

KEY TERMS

absolute zero **(13.4)**

atmosphere **(13.2)**

Avogadro's Principle **(13.9)**

barometer **(13.2)**

Boyle's Law **(13.3)**

Charles's Law **(13.4)**

Combined Gas Law **(13.5)**

Dalton's Law of Partial Pressures **(13.7)**

Gay-Lussac's Law of Combining Volumes **(13.8)**

ideal gas **(13.1)**

Ideal Gas Law **(13.10)**

Kelvin scale **(13.4)**

kinetic theory of gases **(13.1)**

molar volume of a gas **(13.9)**

partial pressure **(13.7)**

real gas **(13.1)**

standard temperature and pressure (STP) **(13.6)**

torr **(13.2)**

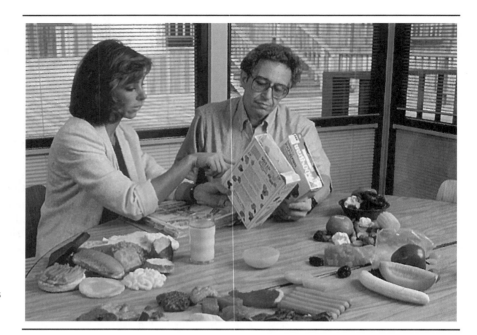

The dietitian recommends a nutritious eating plan for this patient.
(Nathan Benn/Stock, Boston)

CAREER SKETCH

Dietitian

As a dietitian, you will plan balanced and appetizing meals to ensure that healthy people stay healthy and sick people recover. You may choose to specialize in one of three areas: clinical (the largest area), administrative, or research.

As a *clinical dietitian*, sometimes called a *therapeutic dietitian*, you will plan diets and supervise the service of meals to patients in a hospital, nursing home, or clinic. You will also confer with physicians about patient's nutrition, instruct patients and their families on diet requirements, and suggest ways for patients to follow their diets after they return home.

As an *administrative dietitian*, you will supervise the planning and preparation of meals for large numbers of people in an institution such as a hospital, university, or school. You will also select, train, and direct food-service workers; purchase food, equipment, and supplies; enforce sanitary and safety regulations; and prepare records and reports. If you become the director of a dietetic department, you will also coordinate dietetic services with the activities of other departments. You will be responsible for your department's budget, which in a large organization may amount to millions of dollars annually.

As a *research dietitian*, you will conduct research aimed at improving the nutrition of both healthy and sick people. You may specialize in nutrition science, education, food management, or food-service systems and equipment. You may study how the body uses food, or you may investigate the nutritional needs of very young or very old people, of individuals with chronic diseases, or even of space travelers. Most research dietitians work in a medical center or at a college or university.

CHEMICAL FRONTIERS

*Measuring Temperature Using Light**

A typical thermometer works very well when you are trying to measure your body temperature, but what if you were trying to measure the temperature of just a few molecules? A regular thermometer would be too big to measure such a sample. Researchers at Purdue University in Indiana have invented a thermometer (Actually, they call it molecular fluorescence thermometry.) that measures temperature by using a beam of light.

The idea of this new thermometer is based on the work of astronomers, who use a similar method to study stars. Light is used by the astronomers to determine the elemental and structural makeup of the stars. The beam of light that is being used in the thermometer determines temperature, pressure, and viscosity.

The process of taking the temperature of a few molecules begins by mixing dye molecules into liquids or solids. Light is then passed through a fiber-optic cable to a tiny probe. The light is scattered by the dyes. The tiny probe gathers information based on how the light is scattered. A computer decodes the information from the probe and allows chemists to determine many characteristics about the molecules, including the temperature.

In the future, this method will allow scientists to work with living cells, circuit boards, computer chips, and other small samples without harming or touching them. This process may also be used in industry to measure the properties of lubricating oils between ball bearings, gears, and machine parts while they operate.

*Adapted from *Science News*, March 26, 1994, p. 206

SELF-TEST EXERCISES

LEARNING GOAL 1: *Kinetic Theory of Gases*

◀ 1. State the five basic principles of the kinetic theory of gases. Explain how each of these principles is related to the gas laws discussed in this chapter.

◀ 2. Explain the difference between a real gas and an ideal gas.

LEARNING GOAL 2: *Converting Atmospheres to Torr, and Vice Versa*

◀ <u>**3.**</u> Express the following pressures in atmospheres:
(a) 1,862 torr   (b) 471 torr   (c) 6,080 torr
(d) 125 torr

◀ 4. Express the following pressures in atmospheres:
(a) $35\overline{0}$ torr   (b) $2,50\overline{0}$ torr   (c) $1,96\overline{0}$ torr

<u>**5.**</u> Express the following pressures in torr:
(a) 4.00 atm   (b) 0.400 atm   (c) 0.00200 atm
(d) 5.25 atm

·6. Express the following pressures in torr:
(a) 2.00 atm   (b) 0.250 atm   (c) 0.123 atm

<u>**7.**</u> Express the following pressures in atmospheres:
(a) 76 torr   (b) 2,600 torr   (c) 3.80 torr

8. Express the following pressures in atmospheres:
(a) 152 torr   (b) 304 torr   (c) 0.152 torr

<u>**9.**</u> Express the following pressures in torr:
(a) 0.0100 atm   (b) 12.0 atm   (c) 5.50 atm

10. Express the following pressures in torr:
(a) 0.200 atm   (b) 15.2 atm   (c) 7.6 atm

LEARNING GOAL 3: *Boyle's Law*

<u>**11.**</u> Given the following data for a gas, use Boyle's Law to determine V_f: $P_i = 60.0$ torr, $V_i = 90.0$ mL, $P_f = 20.0$ torr

12. Given the following data for a gas, use Boyle's Law to determine V_f: $P_i = 0.500$ atm, $V_i = 40.0$ L, $P_f = 0.200$ atm.

◀ <u>**13.**</u> Given the following data for a gas, use Boyle's Law to determine V_f: $P_i = 3.00$ atm, $V_i = 50\overline{0}$ mL, $P_f = 38\overline{0}$ torr.

◀ 14. Given the following data for a gas, use Boyle's Law to determine V_f: $P_i = 2.00$ atm, $V_i = 25\overline{0}$ mL, $P_f = 40\overline{0}$ torr.

<u>**15.**</u> Given the following data for a gas, use Boyle's Law to determine P_f: $P_i = 1,90\overline{0}$ torr, $V_i = 25\overline{0}$ mL, $V_f = 50.0$ mL.

16. Given the following data for a gas, use Boyle's Law to determine P_f: $P_i = 1,52\overline{0}$ torr, $V_i = 455$ cm³, $V_f = 75.0$ cm³.

LEARNING GOAL 4: *Charles's Law*

◀ <u>**17.**</u> Given the following data for a gas, use Charles's Law to determine t_f (that is, express your answer in °C): $t_i = 127$°C, $V_i = 0.500$ L, $V_f = 2.00$ L.

◀ 18. Given the following data for a gas, use Charles's Law to determine t_f (that is, express your answer in °C): $t_i = 90.0$°C, $V_i = 4.00$ L, $V_f = 5.50$ L.

<u>**19.**</u> Given the following data for a gas, use Charles's Law to determine t_f (that is, express your answer in °C): $t_i = -73$°C, $V_i = 8.00$ L, $V_f = 32.0$ L.

20. Given the following data for a gas, use Charles's Law to determine t_f (that is, express your answer in °C): $t_i = 85.0$°C, $V_i = 7.50$ L, $V_f = 4.50$ L.

<u>**21.**</u> Given the following data for a gas, use Charles's Law to determine V_f: $t_i = 327$°C, $V_i = 20\overline{0}$ L, $t_f = -173$°C.

22. Given the following data for a gas, use Charles's Law to determine V_f: $t_i = 400.0$°C, $V_i = 45\overline{0}$ L, $t_f = -73$°C.

LEARNING GOAL 5: *Converting °C to K, and Vice Versa*

◀ <u>**23.**</u> Change the following temperatures to °C:
(a) $30\overline{0}$ K   (b) $2\overline{0}$ K

◀ 24. Change the following temperatures to °C:
(a) 353K   (b) 3 K

<u>**25.**</u> Change the following temperatures to kelvins:
(a) 10°C   (b) −10°C   (c) −73°C

26. Change the following temperatures to kelvins:
(a) 15°C   (b) −15°C   (c) −50°C

<u>**27.**</u> Change the following temperatures to °C:
(a) 0 K   (b) $10\overline{0}$ K   (c) $20\overline{0}$ K   (d) $30\overline{0}$ K

28. Change the following temperatures to °C:
(a) 5 K   (b) 150 K   (c) 250 K   (d) 350 K

29. Change the following temperatures to kelvins:
(a) 20°C (b) 0°C (c) −1$\overline{0}$0°C (d) 212°F

30. Change the following temperatures to kelvins:
(a) 25°C (b) −5°C (c) −75°C (d) 424°F

LEARNING GOAL 6: *Combined Gas Law*

◀ **31.** Given the following data for the gas, use the Combined Gas Law to determine P_f.

$P_i = 4.00$ atm $V_f = 24.0$ L
$V_i = 8.00$ L $t_f = 227°$
$t_i = 27°C$

32. Given the following data for a gas, use the Combined Gas Law to determine P_f:

$P_i = 0.20$ atm $V_f = 10.0$ L
$V_i = 8.0$ L $t_f = 35\overline{0}°C$
$t_i = 43°C$

33. A tire is filled with air at a pressure of 1,520 torr at 20.0°C. What will be the pressure in the tire after the car is driven and the temperature of the air in the tire rises to 34.0°C? (Assume that the volume of the tire remains constant, which is the case in a tubeless tire.)

◀ 34. A tubeless tire (whose volume remains constant) is filled with air at a pressure of 76$\overline{0}$ torr at 30.0°C. What will be the pressure in the tire after the car is driven and the temperature of the air in the tire rises to 35.0°C?

35. A balloon is filled with a gas to a volume of 4$\overline{00}$ mL, at a pressure of 1.00 atm and a temperature of 25.0 °C. The temperature is increased to 50.0°C, and the pressure is increased to 78$\overline{0}$ torr. What will be the volume of the balloon?

36. A balloon is filled with air to a volume of 350.0 mL, at a pressure of 1.20 atm and a temperature of 28.0°C. The temperature is increased to 55.0°C, and the pressure is increased to 2.00 atm. What is the new volume of the balloon?

37. Suppose that you collect a gas over water at 30.0°C. The volume of the gas is 8$\overline{00}$ mL, and the total pressure is 78$\overline{0}$ torr. What is the volume of *dry* gas at STP?

38. A gas is collected over water at 28.0°C. The volume of the gas is 750.0 mL, and the total pressure is 790.0 torr. What is the volume of the *dry* gas at STP?

39. Assume that your lungs can hold about 5$\overline{00}$ mL of air at STP. You take a deep breath and hold it, then dive into the ocean to a depth at which the temperature is 10.0°C and the pressure is 1,$\overline{000}$ torr. What is the volume of the air in your lungs now?

40. A gas is collected over water at 35.0°C. The volume of the gas is 1.00 L, and the total pressure is 800.0 torr. What is the volume of the *dry* gas at STP?

41. The Gulf of Mexico puffer fish can live in deep water, where the water pressures are enormous. These fish have the ability to compress and expand their bodies to compensate for changes in pressure. They do this by holding in or letting out air. If a fish of this type decided to move from an area in which the water pressure was 4.0 atm to an area in which it was 1.5 atm, by what ratio would the volume of its body increase?

42. Assume that the Gulf of Mexico puffer fish described in Exercise 41 moves from an area in which the water pressure is 2.00 atm to an area in which it is 4.50 atm. What happens to the volume of its body?

43. An airtight chamber is filled with a gas at a pressure of 25.0 atm and a temperature of 50.0°C. The temperature inside the chamber is increased to 5$\overline{00}$°C. What happens to the pressure of the gas? (The volume of the chamber remains constant.)

44. An airtight chamber is filled with a gas to a pressure of 15.0 atm and a temperature of 350.0°C. If the temperature inside the chamber is decreased to 50.0°C, what will happen to the pressure of the gas? (The volume of the chamber remains constant.)

45. A gas has a volume of 4$\overline{00}$ mL at a pressure of 3.00 atm and a temperature of 127°C. What would its volume be at STP?

46. A gas has a volume of 40.0 mL at a pressure of 2.50 atm and a temperature of 100.0°C. What would its volume be at STP?

47. A sample of oxygen gas is collected over water. The volume of gas collected is 3$\overline{00}$ mL, at a pressure of 75$\overline{0}$ torr and a temperature of 15.0°C. What is the volume of the *dry* gas at STP?

48. A sample of hydrogen gas is collected over water. The volume of gas collected is 250.0 mL, at a pressure of 815 torr and a temperature of 27.0°C. What is the volume of the *dry* gas at STP?

49. Complete the following table for each data set:

DATA SET

	P	V	n	T
1	constant	increase	constant	?
2	?	decrease	constant	constant
3	decrease	?	increase	constant

50. Complete the following table for each data set:

DATA SET

	P	V	n	T
1	decrease	?	constant	constant
2	?	increase	constant	constant
3	increase	constant	constant	?

LEARNING GOAL 7: *Dalton's Law*

◄ **51.** Find the total pressure (in torr) in an enclosed vessel that contains the following gases: argon at 125 torr, helium at 255 torr, neon at 0.200 atm, and xenon at 0.400 atm.

◄ 52. Find the total pressure in an enclosed vessel that contains the following gases: helium at $20\overline{0}$ torr, hydrogen at $15\overline{0}$ torr, nitrogen at $9\overline{0}$ torr, and oxygen at $15\overline{0}$ torr.

53. Nitrogen gas makes up about $8\overline{0}\%$ by volume percent of the atmosphere. What is the pressure exerted by this nitrogen gas when the total atmospheric pressure is $76\overline{0}$ torr?

54. Suppose we collect a $75\overline{0}$-mL sample of oxygen gas over water at 22.0°C and $77\overline{0}$ torr. What would be the volume of the *dry* gas at STP?

LEARNING GOAL 8: *Avogadro's Principle*

55. Butane burns in air to produce carbon dioxide and water vapor. How many liters of water vapor are produced from 5.00 L of butane? How many liters of oxygen are required to completely burn the butane?

56. Ethane (C_2H_6) burns in air to produce carbon dioxide and water vapor. How many liters of water vapor are produced from 15.0 L of ethane?

◄ **57.** Determine the volume of 4.00 moles of oxygen gas at STP.

58. What is the volume of 3.00 moles of hydrogen gas at STP?

59. You are given the reaction

$$2Na + 2H_2O \rightarrow 2NaOH + H_2$$

How many grams of sodium would be needed to release 10.0 L of hydrogen gas at STP?

*60. Consider the reaction

$$2Li + 2H_2O \rightarrow 2LiOH + H_2$$

How many grams of lithium are needed to release 10.0 L of hydrogen gas at STP?

*61. Suppose that $8\overline{0}$ mL of hydrogen gas react with $8\overline{0}$ mL of oxygen gas. How many milliliters of oxygen gas remain uncombined? How many milliliters of water vapor are formed?

*62. Suppose that 1.00 L of hydrogen gas reacts with 1.00 L of oxygen gas at STP. How many grams of oxygen gas remain uncombined? How many grams of water vapor are formed?

63. Sulfur dioxide can combine with oxygen in the air to produce sulfur trioxide:

$$SO_2(g) + O_2(g) \rightarrow SO_3(g)$$

(a) Balance the equation.
(b) How many liters of oxygen gas are needed to combine with $15\overline{0}$ L of SO_2?
(c) How many liters of SO_3 are produced?

64. Nitric oxide can combine with oxygen gas in the air to produce nitrogen dioxide:

$$NO(g) + O_2(g) \rightarrow NO_2(g)$$

(a) Balance the equation.
(b) How many liters of oxygen gas are needed to combine with 10.0 L of NO?
(c) How many liters of NO_2 are formed?

65. Butane burns in air to produce carbon dioxide and water vapor:

$$C_4H_{10}(g) + O_2(g) \rightarrow CO_2(g) + H_2O(g)$$

(a) Balance the equation.
(b) How many liters of oxygen gas are needed to burn 15.0 L of butane?
(c) How many liters of carbon dioxide and water are produced?

66. Propane burns in air to produce carbon dioxide and water vapor:

$$C_3H_8(g) + O_2(g) \rightarrow CO_2(g) + H_2O(g)$$

(a) Balance the equation.
(b) How many liters of oxygen gas are needed to burn 25.0 L of propane?
(c) How many liters of carbon dioxide gas are formed?

*67. Nitrogen gas can combine with oxygen gas to produce nitrogen dioxide:

$$N_2(g) + O_2(g) \rightarrow NO_2(g)$$

(a) Balance the equation.
(b) Suppose that 75.0 mL of nitrogen gas react with 75.0 mL of oxygen gas. How many milliliters of nitrogen gas remain uncombined?
(c) How many milliliters of $NO_2(g)$ are formed?

*68. Nitrogen gas can combine with hydrogen gas to produce ammonia:

$$N_2(g) + H_2(g) \rightarrow NH_3(g)$$

(a) Balance the equation.
(b) Suppose that 120.0 mL of N_2 can react with 120.0 mL of H_2 gas. How many milliliters of nitrogen gas remain uncombined?
(c) How many milliliters of ammonia gas are formed?

LEARNING GOAL 9: *Ideal Gas Law*

69. Determine the number of grams of nitrogen gas present in a 3.00-L container at a temperature of 27°C and a pressure of 1,520 torr.

70. Determine the number of grams of hydrogen gas present in a 2.00-L container at a temperature of 127°C and a pressure of 2,280 torr.

◄ 71. Part of Avogadro's Principle is that 1.00 mole of any gas at STP has a volume of 22.4 L. Using this information, solve the ideal gas equation for R.

72. What volume will 0.75 mole of hydrogen gas occupy at STP?

73. Determine the volume of 0.500 mole of $N_2(g)$ at STP.

◄ 74. How many grams of chlorine gas are present in a 1,500.0-mL flask at a temperature of 150.0°C and a pressure of 1.50 atm?

75. Determine the number of grams of NO_2 gas present in a 500-mL flask at a temperature of 227°C and a pressure of 3,040 torr.

76. How many moles of fluorine gas are present in a 2.00-L flask at a temperature of 127°C and a pressure of 1.00 atm?

77. Determine the pressure of 12.20 moles of H_2 gas occupying a volume of 2,000 mL at a temperature of −73°C.

78. What is the pressure of 4.00 moles of oxygen gas that occupy a volume of 5.00 L at a temperature of 0°C?

79. A 128-g sample of SO_2 has a pressure of 380 torr and a volume of 500 mL. What is the temperature of this gas?

80. A 92.0-g sample of NO_2 gas has a pressure of 770 torr and a volume of 950 mL. What is the temperature of this gas?

LEARNING GOAL 10: *Ideal Gas Law and Molecular Mass*

*81. A gaseous compound is made up of 83% carbon and 17% hydrogen by mass. When 1.2 g of this gas are placed in a 500-mL container at 25°C, the pressure of the gas is found to be 1.0 atm.
(a) What is the empirical formula of the gas?
(b) What is the molecular mass of the gas?
(c) What is the molecular formula of the gas?

82. Given that 4.00 g of a certain gas have a volume of 500.0 mL at a pressure of 2,005 torr and a temperature of 27.0°C, calculate the molecular mass of the gas.

83. Calculate the molecular mass of a gas, given that 2.0 g of the gas have a volume of 400 mL at a pressure of 3,040 torr and a temperature of 127°C.

*84. Using the Ideal Gas Law equation and remembering that density equals mass divided by volume, calculate the density of hydrogen gas at STP. (Report your answer in grams per liter.)

*85. Using the Ideal Gas Law equation and remembering that density equals mass divided by volume, calculate the density of oxygen gas at STP. (Report the answer in grams per liter.)

86. A 20.0-g sample of a gas has a pressure of 4.00 atm and a volume of 1,500.0 mL at a temperature of 30.0°C. What is the molecular mass of the gas?

◄ 87. A 14.63-g sample of a gas has a pressure of 6.00 atm and a volume of 2,500 mL at a temperature of 27°C. What is the molecular mass of this gas?

◄ 88. Given that 150.0 g of a certain gas have a volume of 5.00 L at a pressure of 2,000 torr and a temperature of 150.0°C, calculate the molecular mass of the gas.

*89. A gaseous compound is 69.57% oxygen and 30.43% nitrogen by mass. When 1.44 g of this gas are placed in a $25\overline{0}$-mL flask at 20.0°C, the pressure of the gas is found to be 1.50 atm.
  (a) What is the empirical formula of the gas?
  (b) What is the molecular mass of the gas?
  (c) What is the molecular formula of the gas? (Use 16.00 as the molecular mass of oxygen and 14.01 as the molecular mass of nitrogen.)

90. Given that 75.0 g of a certain gas have a volume of 4.50 L at a pressure of $1,56\overline{0}$ torr and a temperature of 127°C, calculate the molecular mass of the gas.

EXTRA EXERCISES

◀ 91. Explain Boyle's Law and Charles's Law using the kinetic theory of gases.

92. One-half of a 10.0-L tank of hydrogen gas at 25°C and 10.0 atm is released into a balloon at constant temperature. The atmospheric pressure is $76\overline{0}$ torr. Determine the volume of the balloon.

93. The balloon in Problem 92 is released. It rises 10.0 km into the atmosphere, where it bursts. What was its final volume, given that the temperature was −50.0°C and the pressure was $23\overline{0}$ torr?

94. An industrial process for the production of ammonia involves reacting hydrogen gas with nitrogen gas. How many liters of ammonia gas could be produced from 1.00 kg of nitrogen gas, given that the process operates at a temperature of 30.0°C and a pressure of $75\overline{0}$ torr?

95. Change the following temperatures to kelvins:
  (a) $4\overline{0}$°F   (b) $18\overline{0}$°F   (c) 212°F

96. Calculate the volume occupied by 1.00 mole of oxygen gas at STP (density = 1.429 g/L).

97. Calculate the density of CH_4 gas at STP.

◀ 98. Here is a very interesting experiment (see if your instructor will do it): Boil a small amount of water in a tin can. While the water is boiling, seal the can and remove the heat. As the can cools, it collapses. Explain this in terms of the gas laws you are now familiar with.

99. A volume of 272 mL of oxygen is collected at 27.0°C. What volume would the oxygen occupy at $20\overline{0}$°C, assuming that the pressure is held constant?

100. The following data are collected on a gas sample: P = 1.00 atm, V = 2.00 L, t = 27.0°C, mass = 2.56 g. What is the molecular mass of the gas? Suggest an element or compound that the gas might be if it is known to be diatomic.

THE LIQUID AND SOLID STATES

14

It was a hot August day in Phoenix, Arizona. At 1:00 P.M. the temperature reached 38.9 degrees Celsius (102 degrees Fahrenheit). Ryta Ross was at a company picnic with her husband, John, and their two children. Ryta volunteered to barbecue the hamburgers and hot dogs even though she had felt overheated and sweaty during the morning. Over the next few hours she stopped sweating. Suddenly she felt weak, dizzy, and nauseous; her muscles began to cramp; and her heart was pounding in her chest. As she began to feel faint, Ryta left the barbecue pit to tell her husband of her symptoms.

John, a paramedic with several years' experience, suspected that Ryta's symptoms were caused by heatstroke. He examined her pupils and found them to be dilated, and Ryta also was now becoming confused. John immediately laid her down under an opened umbrella to block the sun and sent their 13-year-old son to summon the rescue squad; before their son returned, Ryta had already passed out. Within minutes the paramedics arrived and immediately transported Ryta to the hospital, where an official diagnosis of heatstroke was made. The hospital staff recorded her temperature to be 40.6 degrees Celsius (105 degrees Fahrenheit). She was covered with a blanket soaked in water, and her skin was massaged vigorously as her temperature fell. Every 10 minutes Ryta's temperature was taken so that doctors could be assured that it did not fall below 38.3 degrees Celsius (101 degrees Fahrenheit). Massage was continued because it aids in the acceleration of heat loss and stimulates the return of cooled peripheral blood to the brain. Ryta was later placed in a cool, well-ventilated room with a fan blowing cool air toward her. Blood electrolyte determinations had been made when she arrived so that replacement therapy could take place. Ryta responded well.

When exposed to heat, the body calls upon a number of mechanisms to maintain normal temperature. Sweating takes place so that the molecules of water with the highest energy content can be lost through evaporation. With the high-energy water molecules gone, those with lower energy content remain behind, and the body is cooled. In Ryta's case, her heat maintenance system broke down. When she stopped sweating, her body lost the ability to cool itself. Instead of maintaining normal body temperature, she experienced a rise in temperature, causing a retention of heat. Other heat-maintenance mechanisms include the dilation of peripheral blood vessels and the decrease of total circulating blood volume.

Ryta remained in the hospital for several days before being discharged and was warned against subsequent exposure to heat.

LEARNING GOALS

After you've studied this chapter, you should be able to:

1. Discuss the properties that distinguish gases, liquids, and solids from one another.
2. Apply the kinetic theory to describe the behavior of liquids and solids.
3. Define and illustrate the terms *evaporation, equilibrium vapor pressure*, and *boiling point*.
4. Explain what happens as a solid changes to liquid and the liquid changes to gas.
5. Explain the processes of distillation and fractional distillation, and define the relevant terms.
6. Describe what is meant by crystalline solids and amorphous solids.
7. Discuss the three kinds of crystalline solids and the forces of attraction in each.
8. Define and give an example of a crystal lattice and a unit cell.
9. List some chemical and physical properties of water.
10. Write equations for the reaction of water with oxides of metals and nonmetals.
11. Explain what a hydrate is, give examples of hydrates, and calculate the percentage of water in a hydrated salt.
12. Define the terms *anhydrous, hygroscopic, deliquescence*, and *efflorescence*.

■■■ **INTRODUCTION**

Matter can exist in any of three physical states: gas, liquid, and solid. The early chemists—the alchemists—had trouble dealing with gases, and it was only in the seventeenth century that research into gases really began. The alchemists were, however, able to study liquids and solids extensively, and most of what we know of their work is related to these two states of matter.

In Chapter 13, we discussed the special properties of gases; in this chapter, we do the same for liquids and solids. We begin by using the kinetic theory (which was applied to gases in Section 13.1) to explain some of the differences among the three states.

14.1   The Kinetic Theory Extended to Liquids and Solids

■ LEARNING GOAL 1

Properties of gases, liquids, and solids

■ LEARNING GOAL 2

Kinetic theory of liquids and solids

Table 14.1 lists three physical properties of gases, liquids, and solids. In Chapter 13, we explained these properties of gases according to the kinetic theory of gases. We noted that gas molecules are relatively far apart, are always in motion, and thus show little attraction for each other. It is for these reasons that a gas has no shape of its own, will move through and fill any volume, and is easily compressed.

In a liquid, however, the molecules are very close to each other. Therefore, according to the kinetic theory, the forces of attraction between liquid molecules are much greater than those between gas molecules. The liquid molecules tend to "stick together" and have a definite volume, but they do not maintain a specific shape. And, because the molecules in a liquid are so close together, liquids tend to be incompressible. When an ocean liner goes through the locks of a canal, it does not crush the water down. Water molecules can't be squeezed any closer together than they already are.

The particles composing a solid are not only close together but are also held together in a fixed position in relation to one another. Later in this chapter, we will discuss the forces that hold the particles together. For now,

TABLE 14.1

Properties of Gases, Liquids, and Solids

PROPERTY	GASES	LIQUIDS	SOLIDS
Volume	Have no definite volume (gases expand to fill their containers)	Have a definite volume	Have a definite volume
Shape	Have no definite shape	Have no definite shape (liquids assume the shape of their containers)	Have a definite shape (most solids have a well-defined crystalline arrangement)
Compressibility	Are easily compressed	Tend to be incompressible	Cannot be compressed (they are practically incompressible)

we'll just say that the fact that *the particles of a solid are held in a fixed position* explains why a solid has a definite shape, has a definite volume, and is practically incompressible.

14.2   Evaporation of Liquids

■ LEARNING GOAL 3
Evaporation, equilibrium vapor pressure, and boiling point

If a liquid is placed in an open container, it slowly disappears. That is, the liquid **evaporates** as its molecules change from the liquid state to the gaseous (or vapor) state. For any liquid, the speed of evaporation depends on the temperature of the liquid and its exposed surface area.

Before we discuss the process of evaporation, let us perform an experiment. We place a cube of sugar in a container of water and leave the container undisturbed for a few hours. Later, when we examine the container, we see that the cube of sugar has disappeared. It has dissolved in the water. If we taste the solution, we discover that the molecules of sugar have been distributed evenly throughout the water (Figure 14.1). All parts of the solution have the same sweetness. This could happen only if the molecules had moved around. From experiments like this, we conclude that *the molecules of a liquid are constantly in motion.* We can use this knowledge to explain evaporation.

Because liquid molecules are constantly in motion, each must have a specific amount of energy. And when the molecules collide, energy can be transferred from one molecule to another. If a molecule on the surface of a liquid collides with a molecule that is on its way up toward the surface, the surface molecule may obtain enough energy from the collision to escape from the liquid entirely (Figure 14.2). That molecule has then changed state, from liquid to vapor. This is similar to what happens in a game of pool. When the cue ball strikes another ball, it passes energy to that other ball. The struck ball then starts to move. And if it is struck hard enough, it can leave the table entirely.

■ FIGURE 14.1

A solution of sugar and water

A Dynamic Equilibrium

If we place a cover on a container that is partly filled with water, the evaporation process begins, but it seems to stop after a time. Here is what happens.

As Figure 14.3 shows, water molecules begin to escape from the surface of the liquid. These molecules occupy the space above the water in the

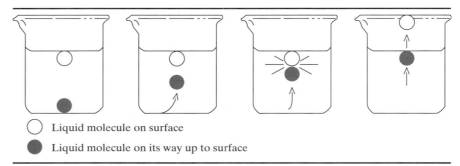

■ FIGURE 14.2

How liquids evaporate

○ Liquid molecule on surface

● Liquid molecule on its way up to surface

○ Escaping molecules

■ **F I G U R E   14.3**

Liquid molecules escaping
from the surface of the liquid

○ Escaping molecules
● Returning molecules

■ **F I G U R E   14.4**

A dynamic equilibrium

partly filled container. As evaporation continues, more and more escaping
molecules occupy this same space. The molecules are now closer together,
they collide with one another more often, and some of them are bumped
back into the water. At first, many molecules escape and few return to the
water. But after a while, as the concentration of molecules in the vapor state
increases, the number of molecules returning to the liquid also increases.
Eventually, the number of molecules escaping from the water equals the
number of molecules returning to it (Figure 14.4). The net result is that no
additional liquid seems to be evaporating from the container. But in reality,
there is a *dynamic equilibrium* between evaporation and return to the liquid
state. *These two opposing processes are occurring at the same rate.*

Equilibrium Vapor Pressure

The molecules that have escaped the liquid in Figure 14.4 exert a pressure
on the container, the cover, and the surface of the liquid. This pressure is
called the *equilibrium vapor pressure*. We can define **equilibrium vapor pressure** as *the pressure exerted by a vapor when it is in equilibrium with its liquid
at a given temperature.*

Every liquid has its own characteristic equilibrium vapor pressure, and
this equilibrium pressure depends on the temperature. To see this, consider
what happens when we raise the temperature of the water in Figure 14.4.
Because we are adding heat to raise the temperature, we are also supplying
additional energy to the liquid molecules. As a result, a greater number of
molecules obtain sufficient energy to escape from the surface of the liquid
and become vapor (Figure 14.5). The equilibrium that had existed is upset,
but a new equilibrium is soon established at the new temperature. And,
because there are now more molecules in the space above the liquid, the
new equilibrium vapor pressure is higher than the original one. That is,
the higher the temperature, the greater the equilibrium vapor pressure for
any substance.

The more easily a liquid evaporates, the more **volatile** it is said to be.
Liquids that have high equilibrium vapor pressures tend to evaporate easily
and so are considered volatile. Some examples are alcohol, gasoline, and
ether. Liquids that have low equilibrium vapor pressures tend to evaporate
less easily. Examples include motor oil, water, and glycerin.

14.3   Boiling of Liquids

If a beaker of water is heated over a flame, bubbles soon form on the bottom
of the beaker. These first bubbles are merely dissolved air (oxygen, nitrogen,
and so forth) being driven out of the water by the heat.

As the temperature rises, however, the water molecules become more
active. Soon other bubbles form at the bottom of the beaker and begin to
rise. These bubbles are water vapor (steam). They form on the *bottom* of
the beaker because the water is hottest there. At first, these steam bubbles
disappear when they reach the cooler water toward the top of the beaker
(Figure 14.6). After a while, however, they begin to rise all the way to the

Glass
cover

Heat source
○ Escaping molecules
● Returning molecules

■ **FIGURE 14.5**

The higher the temperature, the more molecules escape.

surface. At this point, we say that the water is *boiling*. What we mean is that the vapor pressure exerted by the bubbles of steam has become equal to the atmospheric pressure (Figure 14.7). So the bubbles of steam are able to rise up and out of the water.

The **boiling point** of a liquid is *the temperature at which the equilibrium vapor pressure of the liquid equals the atmospheric pressure.* This means that the boiling point depends on the atmospheric pressure: increasing the atmospheric pressure increases the boiling point, and decreasing the atmospheric pressure decreases the boiling point. So, when we discuss the boiling points of liquids, we mean the *normal* boiling points, at the *normal* pressure of 1 atm, unless we state otherwise.

The temperature of a liquid remains constant as it boils at constant pressure. For example, if you heat a container of water, the temperature of the water increases to 100°C (at 1 atm pressure), and the water then starts to boil. But the temperature of the liquid does not increase beyond 100°C *no matter how much heat you add*. What happens is this: The heat you add goes first to raising the temperature of the water to 100°C. After that, all the heat that is added is used to vaporize the water molecules. This heat is called the **heat of vaporization.** That is, energy is required for the water molecules to move fast enough to escape the liquid, and this energy is supplied by the heat. The more heat you add, the faster the water is vaporized. But the temperature of the water remains at the boiling point until the last molecule escapes.

14.4 Freezing of Liquids

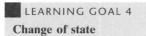

LEARNING GOAL 4
Change of state

■ **FIGURE 14.6**

The first bubbles of water vapor

When heat is removed from a liquid, its temperature decreases. At a certain temperature, the liquid begins to **freeze,** or change to a solid. This temperature is called the **freezing point** of the liquid. As more heat (or energy) is removed, more and more of the liquid freezes *at this same temperature.* During the change of state from liquid water to solid ice, all of the heat removed from the system is used to turn the liquid into solid. This is called the **heat of crystallization (solidification).** Once all the liquid has been transformed to the solid state, its temperature will decrease if more heat is removed. A term known as the *heat of fusion* is used for the reverse process. It is the heat required to melt a solid. It is equal to the value of the heat of crystallization, but has an opposite sign.

For example, the freezing point of pure water is 0°C. As liquid water is cooled, its temperature decreases. At 0°C, the liquid begins to turn to ice—or solid water. As more heat is removed by cooling, more of the liquid turns to ice, but the temperature stays at 0°C. Finally, when all water has been frozen, the temperature of the ice will drop below 0°C if it cooled further.

When a solid is heated, its temperature increases until its freezing point is reached. At that temperature, the solid begins to **melt,** or change to a liquid. (For this reason, the freezing point of a substance is also called its

melting point.) As more heat is added, more of the solid turns to liquid *at this same temperature.* Once all the solid has become liquid, its temperature will increase if more heat is added. Figure 14.8 shows the complete process—solid to liquid to gas—for ice at 1 atm pressure.

Some substances can change directly from the solid state to the gaseous state without first becoming a liquid. This process is called **sublimation.** An example is dry ice, which is solid carbon dioxide. It goes directly from the solid state to the gaseous state at 1 atm pressure.

14.5   The Process of Distillation

LEARNING GOAL 5

Distillation and fractional distillation

The fact that liquids boil gives rise to an interesting method for separating one liquid from another, or even separating a solid from a liquid in which it is dissolved. This process is called *distillation.*

Figure 14.9 shows a simple distillation apparatus. It works this way: Suppose you have a solution of sodium chloride and water. You want to separate the sodium chloride from the water and at the same time collect the purified water. You place the sodium chloride–water solution in the distillation flask and heat it to boiling. As the solution boils, water in the form of vapor fills the flask and travels toward the distilling head. From there, the hot water vapor travels down the double-jacketed, water-cooled condenser. As the vapor travels down within the inner walls of the condenser, it is cooled by tap water running between the outer walls of the condenser. The water vapor **condenses,** or changes back to its liquid form. The condensed liquid, known as the *distillate,* is collected in a receiving flask. Because only the water, and not the sodium chloride, vaporizes and passes through the condenser, the distillate contains only pure water. The sodium chloride remains behind in the distillation flask.

Miscible liquids are those that form solutions when mixed. Two or more miscible liquids that have different boiling points may also be separated by the process of distillation. For example, when a mixture of ethyl alcohol (boiling point, 78.5°C) and water (boiling point, 100°C) is heated, the component with the lower boiling point (ethyl alcohol) vaporizes more readily than the component with the higher boiling point (water). Therefore the distillate is richer in ethyl alcohol than in water. And it can be distilled again to yield even purer ethyl alcohol. If this process is repeated enough times, a distillate of essentially pure ethyl alcohol is obtained.

This technique, called *fractional distillation,* is extremely useful in petroleum refining. The different components of crude petroleum all have different boiling points, so they can be separated by distillation. Petroleum companies use huge distilling columns, known as *fractionating columns,* to separate the various petroleum products (Figure 14.10). When the crude oil is heated, the components (or fractions) with lower boiling points rise to the top of the column, those with higher boiling points rise to the middle of the column, and solids remain at the bottom of the column. As the components rise

■ **FIGURE 14.7**

Boiling water

■ FIGURE 14.8

How water at 1 atm pressure changes from a solid to a liquid to a gas. Below 0°C, water is in the solid state—ice. In this diagram, we begin with ice at −10°C. Energy (heat) is supplied so that the temperature of the ice rises. When it reaches 0°C, the ice starts to melt into water. As more energy is supplied, more ice melts into water. This water remains at 0°C until all the ice melts. After all the ice has melted, the energy supplied starts to warm the water. This continues until the water temperature reaches 100°C. At this point, the water starts to vaporize into steam. The temperature of the water remains constant at 100°C as the liquid is turned into gas.

■ FIGURE 14.9

A distillation apparatus

The following labels appear on the figure:

Gasoline vapors

Condenser

Gases

Crude petroleum vapors

Crude petroleum

Furnace

Gasoline and petroleum ether

Kerosene

Heating oil and diesel fuel

Lubricating oils and greases

Detail of interior

Bubble caps

Trays

Vapors

Overflow pipe

Steam in

Residue (asphalt, tar)

	Molecular size range	Boiling point range
←	C_1–C_4	less than 30°C
←	C_5–C_{12}	30–200°C
←	C_{12}–C_{16}	175–275°C
←	C_{15}–C_{18}	175–375°C
←	C_{16}–up	above 350°C
←	C_{35}–up	

■ **FIGURE 14.10**

A fractionating column for processing crude petroleum

up the column and condense, they are drawn off, to be purified further if necessary.

14.6   Solids

■ LEARNING GOAL 6

Crystalline and amorphous solids

Solids have two main characteristics that distinguish them from gases and liquids: *fixed shape* and *fixed volume*. Most solids owe these unique properties to their well-defined crystal structure. (We'll explain the term *crystal* shortly.) However, a few solids, such as glass and paraffin, do not have a well-defined crystal structure, although they do have definite shape and volume. *Such solids are called* **amorphous,** *which means that they have no definite internal structure or form.* This makes them interesting to research scientists. However, we shall concentrate on crystalline solids.

TABLE 14.2

Some Properties of Crystalline Solids

TYPE	MELTING POINT	HARDNESS	CONDUCTIVITY	EXAMPLE
Ionic	High	Hard and brittle	Nonconductors of electricity	Sodium chloride, NaCl
Molecular	Low	Soft	Nonconductors of electricity	Dry ice, CO_2
Atomic				
Metallic	High*	Varying hardness	Good conductors of electricity	Copper, Cu
Nonmetallic	High	Hard and brittle	Nonconductors of electricity	Diamond, C

*With some exceptions, like mercury

Particles That Make Up Solids

Solids can be composed of three types of particles: atoms, molecules, and ions. Solids such as carbon, silicon, copper, and gold are composed of atoms. Solids such as dry ice (CO_2), iodine, and water (ice) are composed of molecules. Solids such as sodium chloride, calcium chloride, and sodium bromide are composed of ions. The three types of solids have very different properties. Table 14.2 lists some of the differences.

LEARNING GOAL 7

The kinds of crystalline solids

What Is a Crystal?

A **crystal** is *a solid that has a fixed, regularly repeating, symmetrical internal structure. This symmetrical structure, formed by the particles in a crystal, is called the* **crystal lattice.** A crystal lattice is not an actual object, but a *structure* or pattern. The particles that make up the pattern are different in different types of crystals, but in each case they occupy specific *lattice positions*. In an atomic solid, *atoms* occupy the lattice positions; in a molecular solid, *molecules* occupy them; and in an ionic solid, *ions* occupy them (Figure 14.11). It is this internal structure that results in the plane (flat) surfaces exhibited by crystals (see Figure 14.12).

LEARNING GOAL 8

Crystal lattice and unit cell

Each repeated unit in a crystal lattice is called a **unit cell** (Figure 14.13), and the unit cell is repeated over and over again in the crystal. Because the whole crystal is made up of repetitions of the lattice structure, we know the three-dimensional pattern of the crystal when we know the lattice. Figure 14.14 shows six different types of crystals—each precise, orderly, and capable of existing side by side with exact copies of itself. Figure 14.15 is a photograph taken with a microscope, showing the cubic structure of sodium chloride crystals.

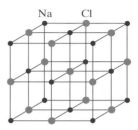

Diamond
Atomic solid
(atoms occupy the
lattice positions)

Carbon dioxide
Molecular solid
(molecules occupy
the lattice positions)

Sodium chloride
Ionic solid
(ions occupy the
lattice positions)

■ **FIGURE 14.11**

Examples of different types of crystalline solids

■ **FIGURE 14.12**

A quartz crystal
(Sean Brady)

■ **FIGURE 14.13**

The unit cell of diamond, a form of carbon

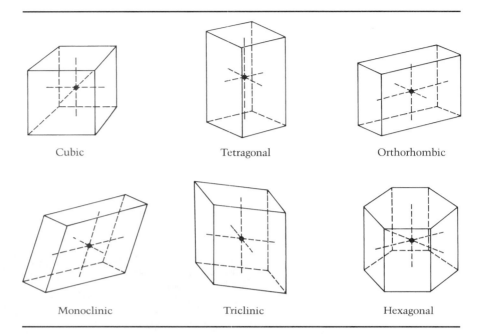

Cubic   Tetragonal   Orthorhombic

Monoclinic   Triclinic   Hexagonal

■ FIGURE 14.14

Six types of crystals

What Holds a Crystal Together?

The properties of a crystalline solid are partly the result of the forces that bind together the particles that make up the crystal. Let's look briefly into the bonding forces within the three types of crystals (ionic, molecular, and atomic).

In an *ionic* crystal, the attractive forces between oppositely charged ions hold the crystal together. This kind of bonding force is very strong, so ionic solids have very high melting points. For example, table salt (NaCl) has a melting point of 800°C. A great deal of heat energy is required to force those ions apart.

Molecular crystals, on the other hand, are usually held together by weak electrical forces. Only molecular crystals whose molecules are polar have binding forces that have some degree of strength. The forces binding most molecular crystals together are so weak that these crystals frequently have much lower melting points than ionic crystals. For example, ice (a molecular crystal) melts at 0°C.

Atomic crystals have strong binding forces—Table 14.3 shows just *how* strong. When the atoms are nonmetallic, they are held together by covalent bonds. An example is diamond, made up of carbon atoms that bond covalently. It takes a temperature of 3,500°C to break the bonds that unite carbon atoms in diamond.

In atomic crystals that are composed of metallic atoms, it is the attraction of opposite charges that makes the bonds so strong. Scientists think of the

■ FIGURE 14.15

Sodium chloride crystals
(Courtesy, Morton Salt)

TABLE 14.3

Melting Points of Some Atomic Crystals

SUBSTANCE	MELTING POINT
Copper	1,083°C
Silicon	1,410°C
Iron	1,535°C
Diamond	3,500°C

metallic lattice as composed of positive ions surrounded by a cloud of electrons. The electrons are given off by the metallic atoms, but they are considered part of the entire crystal—electrons at large, you might say. Because the electrons are free to wander throughout the crystal, they are able to conduct electricity. This is why metals are good electrical conductors.

14.7   Water

Water is the world's most important liquid. You might be able to live a month without food, but you could survive only two or three days without water. All vegetation disappears without water: In West Africa, an area the size of the United States is becoming a desert because there has been almost no rain for some years. The human body is composed of about two-thirds water, and many fruits and vegetables are as much as 80% or 90% water. Milk, one of our basic foods, is more than 90% water.

Yet the supply of water is *not* unlimited. Today the entire population of the United States has only about the same quantity of drinkable water that was available when Columbus landed in 1492. Americans use an estimated 2,000 gallons per person per day, and the trend is to use still more. (This figure of 2,000 gallons includes water used by industry, by power plants, and so forth.)

Physical Properties of Water

LEARNING GOAL 9

Chemical and physical properties of water

Pure water is a tasteless, odorless, colorless substance that has a melting point of 0°C and a boiling point of 100°C at 1 atm pressure. These two temperatures serve as reference points on the Celsius temperature scale.

The density of water is another reference standard. It relates metric units of volume (milliliters) to metric units of mass (grams). At 4°C, water has a density of 1 gram per milliliter (1 g/mL). This means that 1 gram of water at 4°C occupies a volume of 1 milliliter. If we measured out exactly 1 mL of 4°C water, we would find that it has a mass of 1 g (Figure 14.16).

Water has intrigued humanity since the days of the ancient Greeks. One of its most interesting properties is the way in which its volume changes

Establishing the density of water (at 4°C). The experiment shows that 1 mL of water has a mass of 1 g (in other words, the density of water is 1 g/mL).

when its temperature is changed. Suppose we start with a certain amount of water at room temperature. If we warm it, the volume of the water increases. If we cool the water below room temperature, its volume at first decreases. This is exactly the behavior we would expect from a liquid. However, as we cool the water below 4°C, the volume of the water starts to *increase*. And its volume continues to increase until it freezes at 0°C. As water freezes to ice, the volume becomes even larger. This behavior is very unusual, because the volumes of most liquids decrease continuously as the liquids are cooled.

One result of this strange behavior of water is that the density of ice is less than the density of water. Remember that density D = mass/volume. At 4°C, the volume of, say, 10 g of water is 10 mL. But at 0°C, the volume of the ice (solidified water) has increased to 10.9 mL. Therefore

$$D \text{ (at 4°C)} = 10 \text{ g/10 mL} = 1 \text{ g/mL}$$

$$D \text{ (at 0°C)} = 10 \text{ g/10.9 mL} = 0.92 \text{ g/mL}$$

Because the density of ice is less than the density of water, ice floats on water. Most other substances reach their minimum liquid volume (and maximum density) at the freezing point, not above it. As they freeze to solid, the volume becomes even smaller. This causes the solid form to sink in the liquid form.

It is fortunate for us that ice floats on water. If it didn't, when lakes, rivers, and seas freeze and water turns to ice in the winter, the ice would sink to the bottom. The water above the ice would protect it from the sun in the summer, so the ice would never melt. Eventually, the entire body of water would become frozen, and all life forms in it would be killed. If life on earth originated in the sea, then human beings would never have evolved. However, ice *does* float on water, and the layer of floating ice insulates the rest of the water and prevents it from freezing. This explains why an entire body of water does not freeze, even though the air temperature above it stays below freezing for many months (Figure 14.17).

Water also has a much higher freezing point and boiling point than

−20°C air temperature

ICE

Ice

Water temperature 10°C

Sand

Lake bottom

■ FIGURE 14.17

Floating ice insulates the lake water below.

■ FIGURE 14.18

The structure of a water molecule

would normally be expected. The unusual behavior of water results from its bonding and the three-dimensional shape of the water molecule. If we examine a single water molecule, we find that:

1. The bonds between the hydrogen and oxygen atoms are polar. Each hydrogen atom seems to have a slight positive charge, and the oxygen atom seems to have a slight negative charge.
2. The bond angle between the hydrogen atoms is about 105° (Figure 14.18).

The slightly positively charged hydrogen atom of one water molecule can attract the slightly negatively charged oxygen atom of another water molecule. The two molecules then are held together by a force called a **hydrogen bond** (Figure 14.19). Thousands of water molecules may bond together in this way to form long chains (Figure 14.20). It is the hydrogen bonds, which hold the water molecules together, that give water its unusually high boiling point.

But how do we explain the fact that water expands as it freezes? If we looked at the three-dimensional structure of a crystal of ice, it would seem to be hexagonal, or six-sided (Figure 14.21). To attain this type of geometry while maintaining their 105° bond angle, the molecules have to move farther apart in ice than they are in liquid water. Therefore, the water that is in the form of ice has a greater volume than water that is in the form of a liquid.

Chemical Properties of Water

LEARNING GOAL 10
Reactions of water

Water is one of the most stable compounds—so stable that for many years it was considered one of the basic *elements* of matter. Even when heated as high as 2,000°C, water shows little decomposition. However, if an electric current is passed through it, water breaks down.

$$2H_2O \xrightarrow{\text{Electricity}} 2H_2 + O_2$$

Water reacts with oxides of *non*metals to produce *acids*. For example, in the case of sulfur trioxide,

$$SO_3 + H_2O \rightarrow H_2SO_4$$

Sulfuric
acid

and in the case of sulfur dioxide,

$$SO_2 + H_2O \rightarrow H_2SO_3$$

Sulfurous
acid

Water reacts with oxides of metals to produce *bases*. For example,

$$CaO + H_2O \rightarrow Ca(OH)_2$$

$$Na_2O + H_2O \rightarrow 2NaOH$$

LEARNING GOAL 11
Hydrates and water of hydration

Many salts combine with water to form compounds called *hydrates*. A **hydrate** is *a crystalline compound that contains a definite proportion of water molecules loosely held.* Some common hydrates are

$Na_2B_4O_7 \cdot 10H_2O$	Borax	$MgSO_4 \cdot 7H_2O$	Epsom salts
$CaSO_4 \cdot 2H_2O$	Gypsum	$Na_2CO_3 \cdot 10H_2O$	Washing soda
$KAl(SO_4)_2 \cdot 12H_2O$	Alum	$CuSO_4 \cdot 5H_2O$	Blue vitriol

The centered dot between the formula for the salt and the formula for the water is not a decimal point or a multiplication sign. It is a way of showing that the water molecules maintain their integrity in the crystal structure. When a hydrated salt is heated, the water is driven off. For example,

$$MgSO_4 \cdot 7H_2O \xrightarrow{\text{Heat}} MgSO_4 + 7H_2O$$

$$CuSO_4 \quad \cdot \quad 5H_2O \xrightarrow{\text{Heat}} \quad CuSO_4 \quad + \quad 5H_2O$$

A blue
compound
(hydrated salt)

A white
compound
(anhydrous salt)

■ **FIGURE 14.20**

Hydrogen bonding among many water molecules. (*Note:* True bond angles are not shown.)

The salt is left in its **anhydrous** form—the form in which the water of hydration is no longer attached to the salt.

EXAMPLE 14.1

Anhydrous nickel chloride, a golden-yellow salt, is used as an absorbent for ammonia in gas masks. When the anhydrous salt picks up water, it changes color to form nickel chloride hexahydrate, which is green. The reaction is as follows:

$$NiCl_2 \ + \ 6H_2O \rightarrow NiCl_2 \cdot 6H_2O$$
Golden-yellow   Green

Calculate the percentage of water (by mass) in nickel chloride hexahydrate.

■ **FIGURE 14.21**

The crystal structure of ice. (*Note:* Not all the hydrogen atoms are shown, because some of them would be behind the plane of the paper.)

Solution

UNDERSTAND THE PROBLEM

We are asked to determine the percentage of water (by mass) in a hydrated salt.

DEVISE A PLAN

Our strategy will be to determine the molecular mass of the hydrated salt, then divide the mass of the water molecules in the hydrated salt by the molecular mass of the hydrated salt itself. Multiplying the result by 100 will give the percentage of water (by mass).

CARRY OUT THE PLAN

Calculate the molecular mass of the hydrated salt as follows:

$$NiCl_2 \cdot 6H_2O$$

$$58.7 + 71.0 + 6(18.00)$$

$$129.7 + 108.00 = 237.7$$

Now determine the percentage of water (by mass):

$$\text{Percent } H_2O \text{ (by mass)} = \frac{\text{mass of water's contribution}}{\text{mass of hydrated salt}} \times 100$$

$$= \frac{108.0}{237.7} \times 100$$

$$= 45.44\%$$

LOOK BACK

We check to see whether the solution makes sense. Since 108 is slightly less than half of 237, our answer of 45.44% is reasonable.

Practice Exercise 14.1   Pale-green anhydrous chromium(III) nitrate also exists in a deep-violet nonahydrate form, as shown in the following reaction:

$$Cr(NO_3)_3 + 9H_2O \rightarrow Cr(NO_3)_3 \cdot 9H_2O$$

Pale green   Deep violet

Calculate the percentage of water (by mass) in chromium(III) nitrate nonahydrate.

LEARNING GOAL 12
Definitions of anhydrous, hygroscopic, deliquescence, and efflorescence

Chemical samples being dried or kept dry

Anhydrous calcium chloride

■ FIGURE 14.22

A desiccator is used to dry chemical samples and to keep them in a moisture-free atmosphere.

On exposure to air, some anhydrous salts (as well as other substances) can absorb water from the atmosphere. Such a substance is said to be **hygroscopic.** Some substances are so hygroscopic that they absorb enough water from the atmosphere to form a solution. These substances are said to be **deliquescent.** One such substance is sodium hydroxide (NaOH). A small piece of solid sodium hydroxide, if left in an open beaker, soon absorbs enough water from the atmosphere to form a solution.

Some hygroscopic substances, such as calcium chloride ($CaCl_2$), are used in the laboratory as drying agents. Such substances are called **desiccants.** If you want to keep some chemicals in a dry atmosphere, you can place them in a container called a **desiccator** (Figure 14.22). Anhydrous calcium chloride in the desiccator ensures that the atmosphere in the desiccator remains moisture-free. (Of course, in time, the calcium chloride absorbs as much water as it can, and it must be changed.)

Desiccant pellets are sometimes placed in vitamin containers. The pellets absorb any moisture that might accumulate in the containers possibly causing the vitamins to decompose.

We have described hygroscopic substances as those that can spontaneously absorb water from the atmosphere. There are also substances that can do the opposite. A few hydrated salts can spontaneously lose their water of hydration when exposed to the atmosphere. This process is known as **efflorescence.** For example,

$$Na_2SO_4 \cdot 10H_2O \xrightarrow[\text{temperature}]{\text{Room}} Na_2SO_4 + 10H_2O$$

Efflorescence can occur when the vapor pressure of water in the atmosphere is less than the vapor pressure of the hydrate. In the case of the sodium sulfate decahydrate, shown above, it is possible to watch the crystals lose their water and form a noncrystalline-looking powder.

■■■ SUMMARY

Gases have no definite shape or volume and are easily compressed. Liquids have a definite volume but no definite shape, and they tend to be incompressible. Solids have a definite volume and shape and are practically incompressible.

The kinetic theory helps explain the properties of liquids and solids. The forces of attraction between liquid molecules and between solid molecules are much greater than those between gas molecules. As a result, the molecules of a liquid are very close to each other. This gives the liquid a definite volume but no specific shape. The particles composing a solid are not only close together but are also held in a fixed position. This gives the solid both a definite volume and a definite shape.

The molecules of a liquid are constantly in motion. When molecules collide, energy is transferred from one molecule to another. A molecule near the surface of the liquid can obtain enough energy in a collision to evaporate, or escape from the liquid entirely. Molecules that evaporate (or vaporize) have changed from the liquid state to the gaseous state.

When the liquid is in a covered container, the molecules that have escaped the liquid remain above its surface. Some of these molecules return to the liquid and others escape it. Eventually, a dynamic equilibrium is attained in which the rate of escape equals the rate of return to the liquid. The pressure exerted by the escaped molecules at this point is called the equilibrium vapor pressure of the liquid.

The boiling point of a liquid is the temperature at which its equilibrium vapor pressure equals the atmospheric pressure. A heated liquid will continue to boil at its boiling point until it has evaporated completely. The boiling phenomenon is used in the process of distillation, in which liquids are separated by being evaporated in turn.

The melting point (or freezing point) of a substance is the temperature at which it begins to change from the solid state to the liquid state (or from the liquid state to the solid state). The temperature of a substance remains at its melting point (or freezing point) as long as it is melting (or freezing).

Most solids owe their fixed shape and volume to their crystal structure, although a few solids, such as glass and paraffin, are amorphous (do not have a definite internal structure) rather than crystalline. A crystal is a solid that has a definite, repeating, symmetrical internal structure. The form of this internal structure is called the crystal lattice.

In ionic crystals, ions occupy the fixed lattice positions, and the attractive forces between oppositely charged ions hold the crystal together. Ionic solids have high melting points because these attractive forces are so strong. Molecular crystals, in which molecules occupy the lattice positions, are usually held together by weak electrical forces. Most molecular crystals have much lower melting points than those of ionic crystals. Only molecular crystals whose molecules are polar have binding forces that are fairly strong. Atomic crystals, in which atoms occupy the lattice positions, have strong binding forces. These crystals can be composed of either nonmetallic or metallic ions.

Water is the world's most important liquid. Pure water is a colorless, odorless, tasteless substance that has a melting point of 0°C and a boiling point of 100°C at 1 atm pressure. At 4°C, water has a density of 1 g/mL. Because of the way in which hydrogen and oxygen atoms bond to form water, and because of the three-dimensional shape of the water molecule, the volume of water *increases* as water is cooled below 4°C. For this reason, ice is less dense than liquid water and floats on water. Water is a polar molecule: the positively charged hydrogen atom of one water molecule can attract the negatively charged oxygen atom of another water molecule to form a *hydrogen bond*. Such bonds can hold together long strings of water molecules, which accounts for the high boiling point of water. Water reacts with oxides of nonmetals to produce acids and with oxides of metals to produce bases. Water combines chemically with many salts to form hydrates.

KEY TERMS

amorphous **(14.6)**

anhydrous **(14.7)**

boiling point **(14.3)**

condense **(14.5)**

crystal **(14.6)**

crystal lattice **(14.6)**

deliquescent **(14.7)**

desiccant **(14.7)**

desiccator **(14.7)**

efflorescence **(14.7)**

equilibrium vapor pressure **(14.2)**

evaporate **(14.2)**

freeze **(14.4)**

freezing point **(14.4)**

heat of crystallization **(14.4)**

heat of fusion **(14.4)**

heat of vaporization **(14.3)**

hydrate **(14.7)**

hydrogen bond **(14.7)**

hygroscopic **(14.7)**

melt **(14.4)**

melting point **(14.4)**

miscible **(14.5)**

sublimation **(14.4)**

unit cell **(14.6)**

volatile **(14.2)**

CAREER SKETCH

Dental Hygienist

If you want to help people achieve and maintain good oral health, you may be interested in becoming a *dental hygienist.* As a member of the dental health team, you will perform preventive and therapeutic dental services under the supervision of a dentist. Your responsibilities will vary depending on the laws of the state in which you work. However, your duties will typically include removing deposits and stains from your patients' teeth, instructing patients in how to care for their teeth, giving them dietetic and nutritional counseling, and applying various dental agents to prevent tooth decay. You will also take your patients' medical and dental histories, make impressions of teeth for the dentist to study, and prepare other diagnostic aids (such as x-ray films) for the dentist. In some states, you may also assist the dentist in pain-controlling and restorative procedures.

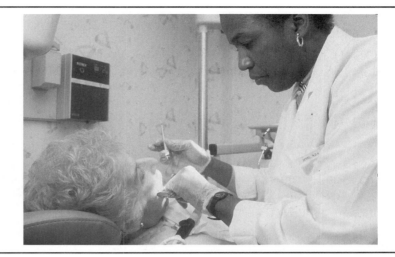

As members of the dental health team, dental hygienists perform preventive and dental therapeutic services under the supervision of a dentist.
(Yoav Levy/Phototake)

As a dental hygienist, you may also want to work in a school system. There, you will examine children's teeth, help the dentist determine the treatment needed, and report findings to parents. You may also scale and polish teeth and instruct children in proper mouth care. You may also present programs on oral health to students.

Another option is to work with a team of dentists performing advanced dental research. With an advanced degree, you may teach in a college that offers a dental hygiene program.

CHEMICAL FRONTIERS

*Fill It Up with the Microalgae, Please**

When you pull into the gas station in the near future, you might be filling your car with a single-celled algae! Researchers at the National Renewable Energy Laboratory in Golden, Colorado, have found a way to convert single-celled algae, or microalgae, into something they call Biodiesel, a substitute for conventional diesel fuel.

Biodiesel is currently being produced in Europe from the lipids of soybeans, rapeseeds, and other oil seeds. Through a simple chemical reaction, parts of these seeds can be changed into a substance that has the viscosity and combustion properties of conventional diesel. In fact, the substances are so similar that the Biodiesel can be burned in regular diesel engines. The new fuel is nontoxic and biodegradable—both great for the environment.

There are a few problems that have kept Biodiesel from being mass-produced. This new form of fuel costs about five times more than conventional petroleum diesel. Also, most of the vegetable oil produced in the world is used for food, which means that there isn't enough extra vegetable oil to produce large quantities of Biodiesel. So scientists have turned to microalgae to solve these problems.

Microalgae are cheap to grow and can be grown practically anywhere in the world. All that is needed is saltwater, such as that found in the ocean. The microalgae produce up to 30 times more oil than land plants of the same size, and one acre of the microalgae can produce up to 400 barrels of oil in a year. Considering how simple and inexpensive microalgae production is, you might soon be filling your car with Biodiesel.

*Adapted from *Chemical & Engineering News*, April 4, 1994, pp. 28–29

SELF-TEST EXERCISES

LEARNING GOAL 1: *Properties of Gases, Liquids, and Solids*

◀ *1.* Explain from a molecular basis why the boiling point of a liquid decreases with decreasing atmospheric pressure and increases with increasing atmospheric pressure.

◀ 2. A sample of a gas has no definite shape and no definite volume. Explain.

3. How is the boiling point of a liquid affected by a decrease in atmospheric pressure?

4. Gases are easily compressed, but solids and liquids are not. True or false?

5. State whether each of the following statements refers to a solid, liquid, or gas:
   (a) It has definite volume but no definite shape.
   (b) It has no definite volume or shape.
   (c) It has both definite volume and definite shape.

6. Solids, liquids, and gases all have a definite volume. True or false?

LEARNING GOAL 2: *Kinetic Theory of Liquids and Solids*

◀ 7. Explain each of the following properties of liquids in terms of the kinetic theory: (a) No definite shape (b) Definite volume (c) Incompressibility

8. Explain each of the following properties of solids in terms of the kinetic theory: (a) Definite shape (b) Definite volume (c) Incompressibility

9. A man heats a pot of potatoes. When the water starts to boil, he turns up the heat "to cook the potatoes faster." Is he wasting gas, or will the potatoes actually cook faster? Explain.

*10. The directions on a box of fudge brownie mix say, "High-Altitude Directions (3,500 to 6,500 feet): Stir 2 tablespoons of all-purpose flour into brownie mix (dry). Increase baking time to 35 to 40 minutes." Why are these high-altitude directions necessary? If you wanted to bake cake-like brownies, no adjustment would be necessary. Why?

11. It takes 15 minutes to cook a hard-boiled egg at sea level. Explain why it takes more time to cook a hard-boiled egg on a mountaintop.

◀ 12. A kettle full of water is boiling on a stove. If you turn up the heat, will the water get hotter? Explain.

*13. Adding salt to water raises the boiling point of water. Give a possible explanation for this phenomenon.

14. You have a bottle of rubbing alcohol on the shelf and you forget to put the cap on the bottle. What will happen to the alcohol in the bottle?

15. As liquid evaporates from an open container, the temperature of the remaining liquid drops. Give a possible explanation for this occurrence.

16. A house heated by a forced-hot-air system has air

that is very dry. How could the situation be remedied?

17. Will an equilibrium vapor pressure be reached in the space above a liquid in an open container (for example, a half-filled glass of water)? Explain.
18. Name a liquid that evaporates slowly.
◀19. Describe an experiment that you could perform to show that the molecules of a liquid are constantly in motion.
◀20. A thermometer containing mercury, a toxic liquid metal, breaks. Why is it dangerous to collect the mercury and let it sit in an open container?
21. If you place equal amounts of gasoline and water in separate glasses, the gasoline evaporates first. Explain why.
22. Which system would have a higher equilibrium vapor pressure: a covered jar half-filled with water at 25°C or a covered jar half-filled with water at 75°C?
23. Adding salt to water lowers its freezing point. Give an explanation for this occurrence.
24. Have you ever been inside a vehicle whose fuel tank is being filled with gasoline and smelled the fumes from the gasoline? If a quart of oil were being added to this same vehicle, you would not smell any fumes from the oil. Why is this so?

LEARNING GOAL 3: *Evaporation, Equilibrium Vapor Pressure, and Boiling Point*

◀**25.** Match the term on the left with its definition on the right.

(a) Normal boiling point
1. Molecules of liquid escaping from a container

(b) Equilibrium vapor pressure
2. Two opposing processes occurring at the same rate

(c) Evaporation
3. The pressure exerted by a vapor when it is in equilibrium with its liquid

(d) Dynamic equilibrium
4. The temperature at which the equilibrium vapor pressure of a liquid equals the atmospheric pressure

26. Which liquid do you think would boil first, ethyl alcohol or salad oil? Explain your answer.
◀27. Define the term *dynamic equilibrium* as it applies to vapor pressure.
28. What happens to the boiling point of a liquid if the weather changes from sunny and dry to rainy? (*Hint:* Keep in mind that sunny and dry weather usually means a high atmospheric pressure system is in the area, while rainy weather usually means a low atmospheric pressure system is in the area.)
29. Explain the difference between the terms *boiling point* and *normal boiling point.*
30. A beaker of water is boiling at 1 atm pressure. The temperature of the water is 100°C. What is the temperature of the water vapor at the boiling point?

LEARNING GOAL 4: *Change of State*

◀31. Explain what happens to molecules of water as they go from solid (ice) to liquid (water) to gas (steam).
*32. When dry ice (solid carbon dioxide) is placed in a beaker of water, vigorous bubbling is seen and a white vapor leaves the beaker. Explain what is happening.

LEARNING GOAL 5: *Distillation and Fractional Distillation*

33. Explain the following processes:
(a) Distillation   (b) Fractional distillation
◀34. Water and isopropyl alcohol are mixed and distilled. Which liquid is collected first?
35. Define each of the following:
(a) Condensation   (b) Distillate   (c) Distilling head   (d) Condenser
36. What is the difference between distillation and fractional distillation?

LEARNING GOAL 6: *Crystalline and Amorphous Solids*

37. Solids that have no well-defined crystalline structure are said to be _____.
38. Name two solids with a well-defined crystalline structure.
39. Crystalline solids are said to exist in a _____ structure.
40. Name two amorphous solids.

LEARNING GOAL 7: **The Kinds of Crystalline Solids**

41. List the properties associated with each of the following:
    (a) Solids composed of oppositely charged ions
    (b) Solids composed of atoms
    (c) Solids composed of molecules

42. Compare the melting points of ionic solids, molecular solids, and atomic solids.

43. The compound naphthalene is a soft solid substance with a low melting point. What type of solid (ionic, molecular, or atomic) is naphthalene? Explain.

44. Ionic solids are hard and brittle with low melting points. True or false?

45. The compound cesium chloride has a melting point of 646°C. It is brittle substance. What type of crystalline solid is it?

46. Molecular solids are nonconductors of electricity. True or false?

◄ **47.** For each of the following statements, classify the solid as ionic, molecular, or atomic (metallic or nonmetallic):
    (a) The solid has a high melting point and is a good conductor of electricity.
    (b) The solid has a high melting point, does not conduct electricity, and can be decomposed into simpler substances.

48. Which types of solids can be good conductors of electricity? Which types can be good nonconductors? What determines whether a solid is a conductor or a nonconductor?

49. Fill in the blanks.
    (a) In atomic solids, _____ occupy the lattice positions.
    (b) In molecular solids, _____ occupy the lattice positions.
    (c) In ionic solids, _____ occupy the lattice positions.

50. Which types of solids are hard and brittle?

51. Describe the forces that hold the crystals together in (a) Atomic solids, (b) Molecular solids, (c) Ionic solids.

52. A certain compound is a soft solid with a low melting point and does not conduct electricity. Describe the particles that occupy its lattice positions.

LEARNING GOAL 8: **Crystal Lattice and Unit Cell**

53. Trace the unit cell of sodium chloride from the drawing in Figure 14.11 and show how several unit cells can build up into a crystalline structure.

◄ 54. Name six types of crystals.

55. Define each of the following: (a) Crystal lattice (b) Unit cell

56. Which type of crystal structure is exhibited by sodium chloride?

LEARNING GOAL 9: **Chemical and Physical Properties of Water**

57. Why is it incorrect to say that rainwater is "pure" water?

58. The density of ice at 0°C is 0.92 g/cm^3. What is the volume of a cube of ice that has a mass of 18.4 g?

◄ 59. A glass is filled to capacity with ice cubes and water. When the ice cubes melt, will the water overflow? (Assume that there is no evaporation.) Explain.

60. The density of water at 4°C is 1.00 g/cm^3. What is the mass of 535 cm^3 of water?

61. The density of ice at 0°C is 0.92 g/cm^3. What is the mass of an ice cube that measures 4.0 cm on each side?

62. What would happen to our rivers, lakes, and seas if ice did not float on water?

63. Draw a picture of a hydrogen bond between two water molecules.

64. How large is the bond angle between the hydrogen atoms in a water molecule?

65. The density of water at 4°C is 1.00 g/cm^3. What is the volume of 1.00 kg of water?

66. What is a hydrogen bond?

LEARNING GOAL 10: **Reactions of Water**

67. Predict the products formed when water combines with (a) MgO, (b) SO_3, (c) Na.

◄ 68. Predict the products formed when water combines with (a) CaO, (b) SO_2, (c) K.

69. Complete the following equations and then balance them:
    (a) $SrO + H_2O \rightarrow$?
    (b) $SO_2 + H_2O \rightarrow$?
    (c) $Li + H_2O \rightarrow$?

70. Complete the following equations and then balance them:
    (a) $Na_2O + H_2O \rightarrow$?
    (b) $CO_2 + H_2O \rightarrow$?
    (c) $Rb + H_2O \rightarrow$?

LEARNING GOAL 11: *Hydrates and Water of Hydration*

◀ **71.** Determine the percentage of water (by mass) in $CuSO_4 \cdot 5H_2O$.

72. Determine the percentage of water (by mass) in $CaSO_4 \cdot 2H_2O$.

73. A commercial for a well-known cleaning powder says it "shakes out white, then turns blue." What type of reaction is occurring? What compound could be used to produce this effect?

74. What is the difference between a hydrated salt and an anhydrous salt?

75. Determine the percent composition by mass of water in each of the following hydrated salts:
    (a) $MgSO_4 \cdot 7H_2O$
    (b) $Na_2CO_3 \cdot 10H_2O$
    (c) $KAl(SO_4)_2 \cdot 12H_2O$

◀ 76. Determine the percentage by mass of water in each of the following hydrated salts:
    (a) $Ca(ClO_3)_2 \cdot 2H_2O$
    (b) $CaCO_3 \cdot 6H_2O$
    (c) $CaCl_2 \cdot 6H_2O$

77. Complete the following equations and then balance them:

(a) $CaSO_4 \cdot 2H_2O \xrightarrow{\text{Heat}}$?

(b) $KAl(SO_4)_2 \cdot 12H_2O \xrightarrow{\text{Heat}}$?

78. Complete the following equations and then balance them:

(a) $Mg(C_2H_3O_2)_2 \cdot 4H_2O \xrightarrow{\text{Heat}}$? + ?

(b) $MgBr_2 \cdot 6H_2O \xrightarrow{\text{Heat}}$? + ?

*79. There are three hydrated forms of the salt $CoSO_4$. They are $CoSO_4 \cdot 1H_2O$, $CoSO_4 \cdot 6H_2O$, and $CoSO_4 \cdot 7H_2O$. From the following data, determine which form of $CoSO_4 \cdot nH_2O$ is being analyzed: When a chemist heats 52.58 g of hydrated salt, 21.60 g of water are produced.

*80. The two hydrated forms of the salt $Mg(NO_3)_2$ are $Mg(NO_3)_2 \cdot 2H_2O$ and $Mg(NO_3)_2 \cdot 6H_2O$. From the following data, determine which of the hydrated salts is being analyzed: When 25.64 g of hydrated salt are heated, 10.80 g of water are produced.

LEARNING GOAL 12: *Definitions of Anhydrous, Hygroscopic, Deliquescence, and Efflorescence*

81. Define and give an example of each of the following: (a) Anhydrous salt (b) Hygroscopic substance (c) Deliquescent substance (d) Substance that undergoes efflorescence

◀ 82. Complete the following sentences:
    (a) A salt that has lost its water of hydration is called a(n) _____ salt.
    (b) A substance that can absorb water from the atmosphere is _____.
    (c) A substance that absorbs enough water from the atmosphere to form a solution is _____.
    (d) A salt that can spontaneously lose its water of hydration to the atmosphere is a(n) _____ salt.

EXTRA EXERCISES

83. Explain, in terms of the kinetic theory, why the temperature of a pure substance remains constant during a change of state.

84. In what way is a unit cell like a molecule?

85. Is it possible for a substance *not* to have a normal boiling point? Explain.

86. Compare the freedom of motion of a molecule of a gas with that of a molecule of a liquid and with that of a molecule of a solid.

87. Using Figure 14.8, explain why steam at 100°C produces more severe burns than liquid water at 100°C.

88. Explain what is happening when a glass rod is heated and begins to sag.

89. Calcium chloride, $CaCl_2$, is often used as a desiccant (drying agent) for wet chemicals. Explain.

90. Certain weather indicators operate by color changes. For example, one indicator turns red if the air is moist and turns blue in dry weather. Explain how this happens.

91. Compare and contrast the characteristics of the three types of solids.

92. State whether each of the following characteristics refers to a solid, a liquid, or a gas. (*Note:* Each characteristic may refer to more than one state of matter.)

(a) Has definite volume

(b) Has definite shape

(c) Has no definite volume

93. Describe an experiment that you could perform to show that the molecules of a gas are constantly in motion.

94. Do liquids that evaporate easily at a particular temperature have *higher* or *lower* equilibrium vapor pressures than liquids that do not evaporate easily at the same temperature? Explain.

95. Even though water is a relatively small molecule, it has a very high boiling point compared to molecules of similar size. Can you explain this in terms of the structure of water?

1. Distinguish between endothermic and exothermic reactions.

◀ 2. How much heat is needed to raise the temperature of 500.0 g of water from 15.0°C to 35.0°C?

3. How much heat is required to raise the temperature of 25.0 g of copper from 30.0°C to 90.0°C?

4. How much heat is required to raise the temperature of 25.0 mL of ethyl alcohol from 25.0°C to 50.0°C? (The density of ethyl alcohol is 0.800 g/mL.)

5. Why must we use standard-state conditions when calculating changes in heat?

6. How much heat is needed to raise the temperature of 1.000 kg of water from 50.0°C to 75.0°C?

◀ 7. Write the thermochemical equation for the formation of 1 mole of water from hydrogen gas and oxygen gas.

8. Use Table 12.2 to determine ΔH for the reaction

$$H_2O(l) + SO_3(g) \rightarrow H_2SO_4(aq)$$

9. Calculate the heat released when 254 g of $SO_2(g)$ are produced from its elements. Use Table 12.2.

◀ 10. Calculate the amount of heat released when 10.2 g of $H_2O_2(l)$ decompose to form $H_2O(l)$ and $O_2(g)$. Use Table 12.2.

11. A gas has an initial volume of 55 mL at a pressure of 2.0 atm. What will be its final volume if the pressure of the gas is increased to 4.5 atm and the temperature remains constant?

◀ 12. Suppose a gas occupies a space of 15.05 mL under a pressure of 2.50 atm at 110.0°C. What will be the volume of the gas when the pressure is decreased to 2.00 atm? (Assume the temperature remains constant.)

13. A gas occupies a volume of 0.200 L under a pressure of 770 torr at 50.0°C. What pressure will cause the gas to occupy a volume of 250.0 mL at the same temperature?

◀ 14. A gas occupies a volume of 1,500.0 mL at 27°C. What will be its volume at 227°C, assuming its pressure remains constant?

15. When a gas is cooled at constant pressure, it contracts from 100.0 mL to 55.0 mL. Its final temperature is 180.0°C. What was its initial temperature?

16. A gas at 50.5°C and 1.50 atm pressure fills a 2.50-L container. What volume will the gas occupy at 250.0°C and 775 torr?

17. A gas collected at 127°C has a volume of 10.5 L and a pressure of 775 torr. What will be its volume at STP?

18. Suppose a sample of oxygen is collected over water at 20.0°C and a pressure of 775 torr. The gas fills a 400-mL container but has water vapor mixed in with it. What would be the volume of the *dry* gas at STP?

19. Suppose 5.00 L of hydrogen gas are collected over water at 28.0°C and a pressure of 725 torr. What would be the volume of the *dry* hydrogen gas at STP?

◀ 20. A gas occupies a volume of 358.4 L at STP. How many moles of gas are in this sample?

21. Determine the volume of 15.0 moles of a gas at STP.

22. Twenty (20.0) moles of a gas exert a pressure of 2,280 torr at a temperature of 227°C. What volume does the gas occupy under these conditions?

23. A gas exerts a pressure of 550.0 torr at a temperature of 155°C. If the volume of the gas is 2.00 L, how many moles of gas are present?

24. Determine the molecular mass of a gas, given the following experimental data: mass = 3.55 g, volume = 255 mL, pressure = 1,520 torr, and $T = 227°C$.

◀ 25. When ethane (C_2H_6) reacts with oxygen, carbon dioxide and water vapor are formed. How many liters of carbon dioxide are produced from 5.00 L of ethane?

26. Using the Ideal Gas Law and the formula for density ($D = m/V$), calculate the density of chlorine gas at STP. Report your answer in grams per liter.

27. Once released, a weather balloon travels to the upper atmosphere. Why does its volume increase there?

◀ 28. An old tube tire has a volume of 2.55 L at 25°C and 1.39 atm pressure. During a trip, the temperature of the tire increases by 12°C and the pressure increases to 1.41 atm. What is the new volume of the tire?

29. A 25.0-g sample of a gas has a pressure of 4.50 atm and a volume of 1.45 L at a temperature of 25.0°C. What is the molecular mass of the gas?

30. Examining the physical states of the halogens at room temperature, we find that fluorine is a gas,

bromine a liquid, and iodine a solid. Use the kinetic theory of gases to explain these properties.

◀ **31.** A balloon is filled with 4.0 L of air at standard pressure. What will happen to the volume of air in the balloon if it is placed in a hyperbaric chamber at 3.5 atm pressure? (Assume the temperature doesn't change.)

32. A sample of intestinal gas is analyzed and found to have the following approximate percentage composition by volume: 42.0% CO_2, 40.0% H_2, 16.9% N_2, and 0.00300% CH_4. If the total pressure in the intestine is 815 torr, what is the partial pressure of each gas?

33. Match each of the following reactions with its description:

(a) $SO_2 + H_2O \rightarrow H_2SO_3$
    1. Electrolysis of water

(b) $CaO + H_2O \rightarrow Ca(OH)_2$
    2. Metal oxide + water

(c) $2H_2O \rightarrow 2H_2 + O_2$
    3. Nonmetal oxide + water

34. A 20.8-g sample of calcium carbonate is heated until all the water is driven off. The remaining salt weighs 10.00 g. Determine the formula of the hydrate.

◀ **35.** Calculate the percentage of water in calcium chloride hexahydrate.

36. Calculate the percentage of water in calcium sulfate dihydrate.

◀ **37.** A hydrate of barium chloride is analyzed and found to contain 14.7% water by mass. Determine the formula of this hydrate.

38. Calculate the formula for the hydrate of sodium carbonate that is 63% water by mass.

39. How many grams of water are needed to produce 25.00 g of oxygen by electrolysis?

40. Match each of the following formulas with the name of the proper hydrate:

(a) $MgSO_4 \cdot 7H_2O$
    1. Alum

(b) $KAl(SO_4)_2 \cdot 12H_2O$
    2. Gypsum

(c) $CaSO_4 \cdot 2H_2O$
    3. Epsom salts

41. Match each of the following properties with the correct term:

(a) Has no definite shape but has a definite volume
    1. Solid

(b) Is easily compressed
    2. Liquid

(c) Has definite shape and volume
    3. Gas

42. Eggs are being cooked in boiling water. The heat is increased. Which of the following statements is true in this situation?

(a) The eggs cook faster.

(b) The water evaporates more quickly.

◀ **43.** On a sunny day following a winter snowstorm, the volume of snow decreases, yet the temperature has not allowed melting to take place. What is the most probable explanation for this occurrence?

44. A crystalline solid has a low melting point and is soft. Is it more likely to be a molecular or an atomic solid?

45. A crystalline solid has a high melting point and is hard and brittle. Is it more likely to be an ionic or a molecular solid?

46. Is a solid with no definite shape more likely to be a crystalline or an amorphous solid?

◀ **47.** Using the kinetic theory, explain what happens to water as it changes from ice to vapor.

◀ **48.** Benzene evaporates more easily than water. Which liquid has the higher equilibrium vapor pressure?

49. Why are some solids capable of conducting electricity and others are not?

50. Why are atomic crystals that are composed of metallic atoms so strong?

THE CHEMISTRY OF SOLUTIONS

15

This is the story of Jim Lind, a 30-year-old accountant who suffered a kidney-stone attack and decided to ensure that he wouldn't suffer another one in the future. Kidney stones are formed in the kidneys or urinary tract when certain salts precipitate out of the urine. Two types of stones may form: uric acid stones and calcium oxalate stones. A combination of hereditary factors and diet make some people stone formers. A kidney-stone attack can be very painful if the stone is large enough to get trapped in the kidneys or urinary tract and impede the flow of urine.

Some kidney stones pass through the urinary tract on their own. However, if the stones are too large, there are several methods that physicians may use to facilitate their removal. If the stone is very large and the kidney is damaged, surgical removal of the kidney may be necessary. If the stone is lodged in the urinary tract, a method called cystoscopy may be used to remove the stone. This method involves the insertion of a device into the urethra to catch the stone and pull it out. A new technique that employs sound waves to pulverize the stone in the kidney or urinary tract can also be used. Once the stone is pulverized, the smaller pieces flow out with the urine.

Jim's stone was removed using cystoscopy, a procedure requiring an overnight hospital stay. The stone was analyzed and found to be composed of calcium oxalate. Jim asked his doctor how he could help prevent future stones from forming. The doctor instructed him to drink several liters of water a day to help keep his urine dilute. The doctor also told Jim that there was no guarantee that he would not suffer another attack.

Unhappy with his doctor's prognosis, Jim decided to do some research of his own. He had some background in chemistry and remembered that certain salts, like calcium oxalate, are alkaline. He wondered if there was a way to keep his urine acidic, so that the alkaline calcium oxalate stones would stay dissolved. Jim spoke to several faculty members in the food-science department of a nearby university, who told him that there were ways to keep his urine acidic by following an appropriate diet. They also told him that there were diets that could lower his calcium intake and the amount of oxalate in his body. Jim shared this information with his doctor and they decided to try this nutritional approach. Ten years have passed without Jim suffering another kidney-stone attack.

LEARNING GOALS

After you have studied this chapter, you should be able to:

1. Define the terms *solution, solute,* and *solvent.*
2. Predict whether substances will form solutions using the idea that "like dissolves like."
3. Define the terms *saturated solution, unsaturated solution,* and *supersaturated solution.*
4. Calculate concentrations of solutions in percent by mass, percent by volume, and percent by mass-volume.
5. Calculate the molarity, number of moles, or volume of a solution when you are given the other two quantities.
6. Explain the concept of equivalents for an acid or base and calculate the number of equivalents when you are given the number of grams or moles.
7. Calculate the normality, number of equivalents, or volume of a solution when you are given the other two quantities.
8. Calculate concentrations of normal and molar solutions after they have been diluted.
9. Explain what electrolytic and nonelectrolytic solutions are and give examples of each.
10. Define the terms *ion* and *ionization.*
11. Explain what is meant by the molality of a solution.
12. Calculate the molality of a solution, given mass of solute and mass of solvent.
13. Define the term *colligative properties* and explain how a solute affects the boiling point and freezing point of a solution.
14. Define the terms *freezing-point depression constant* (K_f) and *boiling-point elevation constant* (K_b).

15. Calculate the boiling and freezing points of a solution when you are given the molality of the solution, its K_f, and its K_b.

16. Define the following terms: *diffusion, osmosis, osmotic pressure*, and *osmometry*.

INTRODUCTION

Solutions are all around us. We drink them, we breathe them, we swim in them, we are even composed of them. Every time you drink a cup of tea or a soda, you swallow a solution. Each time you breathe, you inhale a solution—air. When you swim in the ocean, you're swimming in a solution of salt in water. Even your blood is a solution.

There is a lot more to solutions than just two things put together. In this chapter, we look at the types of solutions, some of their properties, and some special units that apply to solutions.

15.1   Like Dissolves Like

LEARNING GOAL 1

Definitions of solution, solute, and solvent

In Chapter 3, we defined a *solution* as a homogeneous mixture. A solution can contain two substances, three substances, or more. The most common types of solutions are made by dissolving a solid in a liquid—for example, a salt in water. However, solutions can be made up from combinations of all three states of matter. Table 15.1 shows eight possible types of solutions and gives an example of each.

Because solutions are always composed of at least two substances, we need to be able to identify the role that each substance plays. The **solute** is

TABLE 15.1

Examples of Different Types of Solutions

| SOLUTE | SOLVENT | | |
	Solid	*Liquid*	*Gas*
Solid	Copper metal dissolved in silver metal (for example, coins)	Salt dissolved in water	—
Liquid	Mercury in silver (dental fillings)	Ethyl alcohol dissolved in water	Water vapor in air
Gas	Hydrogen dissolved in platinum metal	Carbon dioxide dissolved in water (soda water)	Oxygen gas dissolved in nitrogen gas

the substance that is being dissolved, whereas the **solvent** is *the substance that is doing the dissolving.* For example, in a solution of salt water, salt is the solute and water is the solvent.

When a solution is composed of two substances in the same state (as is a liquid–liquid solution), it is difficult to establish which substance is the solute and which is the solvent. One rule of thumb is that the substance present in the larger amount is the solvent. In a solution of 10 mL of ethyl alcohol and 90 mL of water, the ethyl alcohol is the solute and the water is the solvent.

EXAMPLE 15.1

Determine which is the solute and which is the solvent in each of the following solutions:

(a) sugar and water

(b) hydrogen chloride gas and water

(c) 75.0 mL of ethyl alcohol and 25.5 mL of water

(d) 80 mL of nitrogen and 20 mL of hydrogen

(e) soda water

Solution

(a) Sugar is the solute; water is the solvent.

(b) Hydrogen chloride is the solute; water is the solvent.

(c) Water is the solute; ethyl alcohol is the solvent.

(d) Hydrogen is the solute; nitrogen is the solvent.

(e) Carbon dioxide is the solute; water is the solvent.

Practice Exercise 15.1   Which is the solute and which is the solvent in each of the following solutions?

(a) salt and water

(b) sulfur trioxide gas and water

(c) $3\overline{0}$ mL of isopropyl alcohol and 25 mL of water

(d) 40 mL of hydrogen gas and 50 mL of oxygen gas

(e) carbonated mineral water

■ FIGURE 15.1

Two immiscible liquids

— Oil

— Vinegar

LEARNING GOAL 2

Substances that form solutions

Before we go on, we should note that some pairs of liquids do not form solutions when mixed; such liquids are said to be **immiscible.** For example, gasoline and water do not readily form a solution. The same holds true for oil and vinegar, as in salad dressing (Figure 15.1). There are also substances that do mix, but only to a slight degree. (The opposite of immiscible is miscible. Pairs of liquids that form solutions when mixed are said to be *miscible.*)

How can we determine whether two substances will form a solution? The answer is found in the statement "like dissolves like." To understand its significance, recall that a molecule may be either polar or nonpolar. "Like dissolves like" means that polar substances dissolve other polar substances and that nonpolar substances dissolve other nonpolar substances. But a polar and a nonpolar substance generally do not form a solution. Alcohol and water mix because both are composed of polar molecules. Gasoline and water do not mix, because gasoline is composed of nonpolar molecules and water is composed of polar molecules.

EXAMPLE 15.2

State whether the following pairs of substances form solutions. (*Hint:* Refer to Chapter 7.)
(a) methane (CH_4) and water
(b) HBr and water
(c) *trans*-dichloroethene and carbon tetrachloride

Solution

(a) Methane is nonpolar; water is polar. They do not form a solution.

(b) Both HBr and water are polar. They do form a solution.

(c) *trans*-dichloroethene is nonpolar, and so is carbon tetrachloride. They do form a solution.

Practice Exercise 15.2 State whether the following pairs of substances form solutions. (*Hint:* You may want to refer to Chapter 7.)
(a) oil and water
(b) hydrogen chloride gas and water
(c) benzene and carbon tetrachloride

15.2 Saturated, Unsaturated, and Supersaturated Solutions

LEARNING GOAL 3
Saturated, unsaturated, and supersaturated solutions

Suppose we take a glass of water and decide to add salt. We begin with a teaspoonful of salt. As the salt and water mix, the salt begins to dissociate. Here's how we think the process happens. We know that water molecules are polar, and the negative end of the water molecule (the oxygen portion) can align itself toward the positive sodium ions. The positive end of the water molecule (the hydrogen portion) can align itself toward the negative chloride ions. Thus, the polar water molecules surround ions on the surface of the sodium chloride crystal, forcing them to break away from the actual surface of the crystal. The first ions to break away become hydrated sodium ions and hydrated chloride ions, meaning that each ion is surrounded by a group of water molecules. A newly exposed surface is formed and is ready to be acted on by other water molecules. The process continues until all of the sodium chloride crystals become hydrated sodium ions and hydrated chloride ions. The salt is then said to have dissolved in the water. We can stir the salt water to increase the rate at which the salt dissolves. Stirring helps distribute the hydrated ions uniformly throughout the solution.

We can add another teaspoonful of salt, stir, and watch the salt crystals disappear into solution. As the number of hydrated ions in solution increases, the chance of ions colliding with each other increases. Besides colliding with each other, ions can collide with solvent molecules, and they can collide with any remaining undissolved solid. When they collide with the undissolved solute, they leave the solution. If we continue to repeat the process we will eventually reach the point at which, no matter how much we stir, no more salt will dissolve. For any given temperature there is a point at which no more solute can dissolve in a particular quantity of solvent. Once this point is reached, the solution is said to be **saturated.** When a solution is saturated, the rate at which solid dissolves into the solution is the same as the rate at which dissolved solid crystallizes out of the solution. Ions leave the solution to return to the undissolved solute at the same rate at which ions enter the solution.

An **unsaturated** solution contains less solute than can be dissolved at a particular temperature. In an unsaturated solution, more solute can be dissolved into the solvent without changing the temperature. This happens when the number of ions in the solution is low, making the chance of collision between hydrated ions and the undissolved solute low. The result is that the undissolved solute won't have much opportunity to recapture ions, so they stay in solution.

Although it appears as if nothing is happening once a solution is saturated, we know that a dynamic process is occurring at the molecular level. The undissolved solute is in equilibrium with the dissolved solute. This dynamic equilibrium can be represented as follows:

$$\text{Undissolved solute} \rightleftharpoons \text{dissolved solute}$$

The double arrow means the reaction is reversible.

Table 15.2 gives the solubility of several substances in water. **Solubility** is the amount of a solute that dissolves in a particular amount of solvent to give a saturated solution. Note that solubility varies according to the temperature, that is, a particular solution that is saturated at one temperature is not necessarily saturated at another temperature. In general, an increase in temperature causes more solute to be dissolved.

It is possible to prepare solutions that contain a greater concentration of solute than is needed for saturation. Suppose we start with a teaspoonful of sugar and a glass of water at room temperature. We add the sugar to the water and stir. The sugar dissolves in the water. We repeat the process until no more sugar will dissolve. At this point we find sugar resting on the bottom of the glass, and we know that the solution is saturated. We decide to heat the solution. The sugar that was resting on the bottom of the glass now dissolves. We add more sugar, and it also dissolves when stirred. We allow the solution to cool slowly to room temperature. The solid does not precipi-

TABLE 15.2

Solubility of Various Substances in Water at 0°C and 50°C

SOLUTE	SOLUBILITY (g SOLUTE/100 g SOLVENT)	
	$0°C$	$50°C$
$CuSO_4$	14.3	33.3
NaCl	35.7	37.0
$AgNO_3$	122.0	455.0
CsCl	161.4	218.5
$C_{12}H_{22}O_{11}$	203.9 (20°C)	260.4

tate out of the solution, and what we have is a **supersaturated** solution. Such a solution is unstable and will revert to a saturated solution if disturbed. Such action will cause the excess solute to crystallize rapidly from the solution. If we were to tie a string around a tiny sugar crystal and suspend the string in the supersaturated sugar–water solution, we could grow rock candy, which is made of large sugar crystals.

EXAMPLE 15.3

Determine whether each of the following solutions is unsaturated, saturated, or supersaturated (see Table 15.2).
(a) 200.0 g of $AgNO_3$ dissolved in 200.0 g of water at 0°C
(b) 50.0 g of $CuSO_4$ dissolved in 150.0 g of water at 50°C
(c) 220.0 g sucrose ($C_{12}H_{22}O_{11}$) dissolved in 75.00 g of water at 50°C

Solution

UNDERSTAND THE PROBLEM

In each case, we are given an amount of substance dissolved in a given amount of water at a certain temperature. We are asked whether the resulting solution is saturated (that is, there is just enough solute to achieve dynamic equilibrium), unsaturated (more of the solute could dissolve in the solution), or supersaturated (the solution contains more dissolved substance than a saturated solution).

DEVISE A PLAN

For each example use the factor-unit method and calculate the grams of solute per 100 g of water. Compare this value to the value in Table 15.2.

CARRY OUT THE PLAN

(a) $? \text{ g AgNO}_3 = (100.0 \text{ g } \cancel{\text{H}_2\text{O}}) \left(\dfrac{200.0 \text{ g AgNO}_3}{200.0 \text{ g } \cancel{\text{H}_2\text{O}}} \right)$

$= 100.0 \text{ g AgNO}_3 \text{ (at 0°C)}$

This solution is *unsaturated;* Table 15.2 states that 122.0 g $AgNO_3$ can be dissolved in 100.0 g H_2O at 0°C.

(b) $? \text{ g CuSO}_4 = (100.0 \text{ g } \cancel{\text{H}_2\text{O}}) \left(\dfrac{50.0 \text{ g CuSO}_4}{150.0 \text{ g } \cancel{\text{H}_2\text{O}}} \right)$

$= 33.3 \text{ g CuSO}_4 \text{ (at 50°C)}$

This solution is *saturated;* Table 15.2 states that 33.3 g of $CuSO_4$ can be dissolved in 100.0 g H_2O at 50°C.

(c) $? \text{ g sucrose} = (100.0 \text{ g } \cancel{\text{H}_2\text{O}}) \left(\dfrac{220.0 \text{ g sucrose}}{75.00 \text{ g } \cancel{\text{H}_2\text{O}}} \right)$

$= 293.3 \text{ g sucrose (at 50°C)}$

This solution is *supersaturated;* Table 15.2 states that 260.4 g of sucrose can be dissolved in 100.0 g H_2O at 50°C.

LOOK BACK

Recheck your calculations and see if your answers are reasonable.

Practice Exercise 15.3   Determine whether each of the following solutions is unsaturated, saturated, or supersaturated (see Table 15.2).
(a) 60.0 g of CsCl dissolved in 25.0 g of water at 50°C
(b) 75.0 g of CsCl dissolved in 50.0 g of water at 0°C
(c) 185.0 g of NaCl dissolved in 500.0 g of water at 50°C

15.3 Concentrations of Solutions by Percent

LEARNING GOAL 4

Percent concentrations

Because solutions are mixtures, their components can be present in different ratios. For example, we can make many different salt-and-water solutions, each with a different **concentration,** or ratio of solute to solvent. The concentrations or relative quantities of solutions can vary widely, so we must have a way of describing them. (It is important to remember, however, that at any given temperature there is a limit to the amount of solute that can be dissolved in a solution. This concentration, called the **saturation point,** is the point at which no more solute dissolves to form a stable solution.)

One method of defining the concentrations of solutions is based on the percent of solute in the solution. This method can cause confusion, because there can be three types of percent concentrations:

1. Percent by mass
2. Percent by volume
3. Percent by mass–volume

A 30%-by-mass solution of alcohol and water is not the same as a 30%-by-volume solution of alcohol and water. Therefore a label that reads "30% alcohol in water" tells us nothing. The solution could be a 30%-by-mass solution, a 30%-by-volume solution, or a 30%-by-mass–volume solution. Chemists seldom use the percent method of defining concentrations; however, biologists and medical workers do use percent by mass–volume, so we will discuss all three types.

Percent-by-Mass Solutions

To find percent by mass, we divide the *mass of the solute* by the *total mass of the solution* and multiply the result by 100. (The mass of the solution equals the mass of the solute plus the mass of the solvent.) For example, suppose we made a solution of alcohol in water by mixing 30 g of alcohol with enough water to make 100 g of solution (Figure 15.2). The concentration of our solution would be 30% by mass.

FIGURE 15.2

Preparing a 30%-by-mass alcohol–water solution

$$\text{Percent alcohol by mass} = \frac{\text{mass of alcohol}}{\text{total mass of solution}} \times 100$$

$$= \frac{30 \text{ g}}{100 \text{ g}} \times 100 = 30\%$$

EXAMPLE 15.4

Find the concentrations of the following solutions in percent by mass.
(a) $2\overline{0}$ g of salt with enough water to make $6\overline{0}$ g of solution
(b) $5\overline{0}$ g of sugar with enough water to make $35\overline{0}$ g of solution

Solution

(a) Percent salt by mass $= \dfrac{\text{mass of salt}}{\text{total mass of solution}} \times 100$

$$= \frac{2\overline{0} \text{ g}}{6\overline{0} \text{ g}} \times 100 = 33\%$$

(b) Percent sugar by mass $= \dfrac{\text{mass of sugar}}{\text{total mass of solution}} \times 100$

$$= \frac{5\overline{0} \text{ g}}{35\overline{0} \text{ g}} \times 100 = 14\%$$

Practice Exercise 15.4   Find the concentrations of the following solutions in percent by mass:
(a) 40.0 g of salt with enough water to make 90.0 g of solution
(b) 75.0 g of sugar with enough water to make 250.0 g of solution

Percent-by-Volume Solutions

To find percent by volume, we divide the *volume of the solute* by the *total volume of the solution* and multiply the result by 100. For example, suppose we mixed $3\overline{0}$ mL of alcohol with enough water to make $10\overline{0}$ mL of solution (Figure 15.3). The concentration of the solution would be $3\overline{0}\%$ alcohol by volume.

$$\text{Percent alcohol by volume} = \frac{\text{volume of alcohol}}{\text{total volume of solution}} \times 100$$

$$= \frac{3\overline{0} \text{ mL}}{10\overline{0} \text{ mL}} \times 100 = 3\overline{0}\%$$

EXAMPLE 15.5

Find the concentrations of the following solutions in percent by volume:

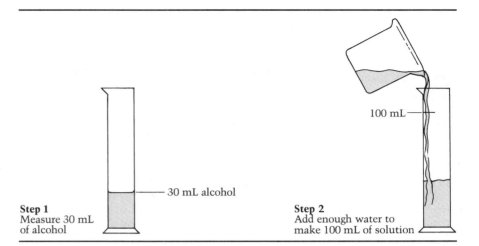

■ FIGURE 15.3

Preparing a 30%-by-volume alcohol–water solution

— 30 mL alcohol

Step 1
Measure 30 mL
of alcohol

100 mL —

Step 2
Add enough water to
make 100 mL of solution

(a) 50.0 mL of alcohol with enough water to make $4\overline{00}$ mL of solution
(b) $1\overline{0}$ mL of benzene is added to $4\overline{0}$ mL of carbon tetrachloride to make $5\overline{0}$ mL of solution

Solution

(a) Percent alcohol by volume $= \dfrac{\text{volume of alcohol}}{\text{total volume of solution}} \times 100$

$$= \dfrac{50.0 \text{ mL}}{4\overline{00} \text{ mL}} \times 100 = 12.5\%$$

(b) Percent benzene by volume $= \dfrac{\text{volume of benzene}}{\text{total volume of solution}} \times 100$

$$= \dfrac{1\overline{0} \text{ mL}}{(1\overline{0} \text{ mL} + 4\overline{0} \text{ mL})} \times 100$$

$$= \dfrac{1\overline{0} \text{ mL}}{5\overline{0} \text{ mL}} \times 100 = 2\overline{0}\%$$

Practice Exercise 15.5   Find the concentrations of the following solutions in percentage by volume:
(a) 60.0 mL of alcohol with enough water to make 300.0 mL of solution
(b) 40.0 mL of carbon tetrachloride in 50.0 mL of benzene

Percent-by-Mass–Volume Solutions

To find percent by mass–volume, we divide the *mass of the solute* in grams by the *volume of the solution* in milliliters and multiply the result by 100.

For example, suppose that we mix $3\overline{0}$ g of alcohol with enough water to make $10\overline{0}$ mL of solution (Figure 15.4). The concentration of the solution would be $3\overline{0}$% alcohol by mass–volume.

$$\text{Percent alcohol by mass–volume} = \frac{\text{mass of alcohol (g)}}{\text{volume of solution (mL)}} \times 100$$

$$= \frac{3\overline{0} \text{ g}}{10\overline{0} \text{ mL}} \times 100 = 3\overline{0}\%$$

EXAMPLE 15.6

Find the concentrations of the following solutions in percent by mass–volume:
(a) 2.0 g of iodine in enough carbon tetrachloride to make $8\overline{0}$ mL of solution
(b) $13\overline{0}$ g of sugar in enough water to make $65\overline{0}$ mL of solution

Solution

(a) Percent iodine by mass–volume $= \dfrac{\text{mass of iodine}}{\text{volume of solution}} \times 100$

$$= \frac{2.0 \text{ g}}{8\overline{0} \text{ mL}} \times 100 = 2.5\%$$

(b) Percent sugar by mass–volume $= \dfrac{\text{mass of sugar}}{\text{volume of solution}} \times 100$

$$= \frac{13\overline{0} \text{ g}}{65\overline{0} \text{ mL}} \times 100 = 20.0\%$$

Practice Exercise 15.6   Find the concentrations of the following solutions in percent by mass–volume:
(a) 4.5 g of *trans*-dichloroethene in enough carbon tetrachloride to make 100.0 mL of solution
(b) 150.0 g of sucrose in enough water to make 750.0 mL of solution

15.4   Molarity

Chemists run many reactions in solutions. Because most of these reactions take place between the solute *particles*, we need a concentration unit that expresses the *number of particles of solute* present in a given amount of solution. The concentration unit chemists use is called *molarity*. The **molarity** of a solution is *the number of moles of solute per liter of solution*. Molarity (abbreviated *M*) is equal to

$$\frac{\text{number of moles of solute}}{\text{number of liters of solution}} \qquad \text{so that} \qquad M = \frac{\text{moles}}{\text{liter}} \qquad \text{or} \qquad M = \frac{n}{V}$$

There can be no confusion when the concentrations of solutions are expressed in these terms. A solution made by dissolving 117 g (2.00 moles) of NaCl in enough water to obtain 1 L of solution (Figure 15.5) would be a 2-molar NaCl solution (usually written as 2 *M* NaCl). (For a review of moles, see Chapter 9.) To calculate the molarity of this solution, we would write

$$M = \frac{\text{moles}}{\text{liter}} \qquad ?\,M = \frac{2\ \text{moles}}{1\ \text{liter}} = 2\,M$$

Chemists often need to prepare solutions that have particular concentrations. *Volumetric flasks* are helpful in preparing such solutions. These flasks are calibrated to contain a specific volume of a solution at a particular temperature (usually 20°C), as shown in Figure 15.5.

Step 1
Weigh out 2 moles of NaCl

117 g of NaCl — 117-g mass

■ **FIGURE 15.5**

Preparing 1 L of a 2 *M* NaCl solution

Step 2
Transfer NaCl to a 1-liter volumetric flask

Step 3
Add water and fill to the 1-liter mark

H_2O

1 liter

1 liter

EXAMPLE 15.7

Calculate the molarity of each of the following solutions:
(a) 3.65 g of HCl in enough water to make $50\overline{0}$ mL of solution
(b) 160 g of NaOH in enough water to make 6.0 L of solution
(c) 36.8 g of ethyl alcohol (C_2H_6O) in enough water to make $1,60\overline{0}$ mL of solution

Solution

(a) First determine the number of moles of HCl.

$$3.65 \, \cancel{g} \text{ of HCl} \times \frac{1 \text{ mole of HCl}}{36.5 \, \cancel{g} \text{ of HCl}} = 0.100 \text{ mole of HCl}$$

Then convert the volume of solutions to liters.

$$50\overline{0} \text{ mL} \times \frac{1 \text{ L}}{1000 \text{ mL}} = 0.500 \text{ L}$$

Then,

$$M = \frac{n}{V}$$

$$= \frac{0.100 \text{ mole}}{0.500 \text{ L}}$$

$$= 0.200 \, M$$

(b) We have 6.0 L of solution, and we first calculate that

$$160 \text{ g of NaOH} = 4.0 \text{ moles}$$

Then,

$$M = \frac{\text{moles}}{\text{liter}} = \frac{n}{V} \qquad ? \, M = \frac{4.0 \text{ moles}}{6.0 \text{ L}} = 0.67 \, M$$

(c) We first calculate that

$$36.8 \text{ g of } C_2H_6O = 0.800 \text{ mole}$$

$$1,60\overline{0} \text{ mL of solution} = 1.600 \text{ L}$$

Then,

$$M = \frac{\text{moles}}{\text{liter}} = \frac{n}{V} \qquad ? \, M = \frac{0.800 \text{ mole}}{1.600 \text{ L}} = 0.500 \, M$$

Practice Exercise 15.7 Calculate the molarity of each of the following solutions:
(a) 9.80 g of H_2SO_4 in enough water to make 0.500 L of solution
(b) 4.60 g of KOH in enough water to make 5.00 L of solution
(c) 73.6 g of ethyl alcohol (C_2H_6O) in enough water to make 1.500 L of solution

EXAMPLE 15.8

Suppose that you have a supply of 3.0 M NaOH solution in your laboratory. You take $\overline{200}$ mL of this solution and evaporate it to dryness. How many grams of *solid* NaOH remain?

Solution

First solve the molarity formula,

$$M = \frac{n}{V}$$

for n, the number of moles:

$$n = MV$$

Then substitute the values.

$$n = \frac{3.0 \text{ mole}}{\text{liter}} \times 0.200 \text{ liter}$$

$$= 0.60 \text{ mole of NaOH}$$

Convert the moles of NaOH to grams.

$$0.60 \text{ mole NaOH} \times \frac{40.0 \text{ g NaOH}}{1 \text{ mole NaOH}} = 24 \text{ g of NaOH}$$

Practice Exercise 15.8   Suppose that you have a supply of 2.00×10^{-3} M Ca(OH)$_2$ solution in your laboratory. You evaporate 250.0 mL of this solution to dryness. How many grams of *solid* Ca(OH)$_2$ do you obtain?

EXAMPLE 15.9

Suppose that you have a supply of 0.500 M Ca(NO$_3$)$_2$ solution in your laboratory. How many milliliters of this solution must be evaporated to obtain 32.8 g of solid Ca(NO$_3$)$_2$?

Solution

Change the 32.8 g of Ca(NO$_3$)$_2$ to moles and then solve for the number of liters of 0.50 M Ca(NO$_3$)$_2$ solution needed to obtain this number of moles.

$$32.8 \text{ g of Ca(NO}_3)_2 \times \frac{1 \text{ mole of Ca(NO}_3)_2}{164.1 \text{ g of Ca(NO}_3)_2} = 0.200 \text{ mole of Ca(NO}_3)_2$$

Solve the molarity formula,

$$M = \frac{n}{V}$$

for *V*, the number of liters:

$$V = \frac{n}{M}$$

Then substitute the values.

$$V = \frac{0.200 \text{ mole}}{0.500 \text{ mole}/\text{L}}$$

$$= 0.400 \text{ L} \qquad \text{or} \qquad 4\overline{00} \text{ mL}$$

You would need to evaporate 0.400 L ($4\overline{00}$ mL) of a 0.500 *M* $Ca(NO_3)_2$ solution to obtain 32.8 g of solid $Ca(NO_3)_2$.

Practice Exercise 15.9 Suppose that you have a supply of 0.200 *M* $Mg(NO_3)_2$ solution in your laboratory. How many milliliters of this solution must be evaporated to obtain 1.48 g of solid $Mg(NO_3)_2$?

15.5 Normality

Chemists have developed another important concentration concept: *normality*. The definition of normality is similar to that of molarity, but normality is easier to use in certain chemical calculations. It is most often applied to solutions of acids and bases. The **normality** of a solution *is the number of equivalents of solute per liter of solution.* Normality (abbreviated *N*) is equal to

$$\frac{\text{number of equivalents of solute}}{\text{number of liters of solution}} \qquad \text{so that} \qquad N = \frac{\text{equivalents}}{\text{liter}}$$

$$\text{or } N = \frac{\text{Eq}}{\text{L}}$$

LEARNING GOAL 6
Equivalents of acids and bases

Of course, the question "What is an equivalent?" immediately comes to mind. Because we'll use normality for calculations involving acids and bases, we will define an equivalent in terms of an acid or a base. An **equivalent of an acid** is *the mass of the acid that contains Avogadro's number of hydrogen ions.* An **equivalent of a base** is *the mass of the base that contains Avogadro's number of hydroxide ions.*

LEARNING GOAL 7
Calculations involving normality

To find the number of grams in 1 equivalent of an acid or a base, first obtain the molecular mass of the substance. Then divide the molecular mass in grams by the number of replaceable hydrogen ions (for an acid) or the number of replaceable hydroxide ions (for a base). This is called the equivalent mass, For example, sulfuric acid (H_2SO_4) has a molecular mass of 98.1. Because H_2SO_4 contains two replaceable hydrogen ions, the equivalent mass,

TABLE 15.3

The Mass of 1 Mole and of 1 Equivalent for Some Common Acids and Bases

SUBSTANCE	MASS OF 1 MOLE	MASS OF 1 EQUIVALENT	NUMBER OF REPLACEABLE H^+ OR OH^- IONS
HCl (hydrochloric acid)	$\dfrac{36.5\ g}{1\ mole}$	$\dfrac{36.5\ g}{1\ equivalent}$	1
$HC_2H_3O_2$ (acetic acid)	$\dfrac{60.0\ g}{1\ mole}$	$\dfrac{60.0\ g}{1\ equivalent}$	1
H_2SO_4 (sulfuric acid)	$\dfrac{98.1\ g}{1\ mole}$	$\dfrac{49.1\ g}{1\ equivalent}$	2
H_3PO_4 (phosphoric acid)	$\dfrac{98.0\ g}{1\ mole}$	$\dfrac{32.7\ g}{1\ equivalent}$	3
NaOH (sodium hydroxide)	$\dfrac{40.0\ g}{1\ mole}$	$\dfrac{40.0\ g}{1\ equivalent}$	1
$Ca(OH)_2$ (calcium hydroxide)	$\dfrac{74.1\ g}{1\ mole}$	$\dfrac{37.1\ g}{1\ equivalent}$	2

or the grams per equivalent, is as follows:

$$\text{Equivalent mass of } H_2SO_4 = \frac{\text{molecular mass}}{\text{number of replaceable hydrogen ions}}$$

$$= \frac{98.1\ g}{2\ Eq} = \frac{49.1\ g}{Eq}$$

This means that a 1 N solution of H_2SO_4 contains 49.1 g of H_2SO_4 per liter of solution. Table 15.3 lists the masses of 1 mole and of 1 equivalent for some common acids and bases.

EXAMPLE 15.10

What is the normality of the solution formed by dissolving 196 g of H_2SO_4 in enough water to make $\overline{800}$ mL of solution? What is the molarity of the solution?

Solution

First convert the grams of H_2SO_4 to equivalents and then use the normality formula. H_2SO_4 has

$$\frac{49.1\ g}{1\ equivalent} \quad \text{or} \quad \frac{1\ equivalent}{49.1\ g}$$

Therefore

$$? \text{ equivalents of } H_2SO_4 = (196 \text{ g}) \left(\frac{1 \text{ equivalent}}{49.1 \text{ g}} \right)$$

$$= 3.99 \text{ equivalents of } H_2SO_4$$

$$N = \frac{Eq}{L} = \frac{\text{equivalents}}{\text{liter}} \qquad ? N = \frac{3.99 \text{ equivalents}}{0.800 \text{ L}} = 4.99 \text{ } N$$

To find the molarity of this solution, we first convert grams of H_2SO_4 to moles and then find the number of moles per liter. (The molecular mass of H_2SO_4 is 98.1.)

$$? \text{ moles of } H_2SO_4 = (196 \text{ g}) \left(\frac{1 \text{ mole}}{98.1 \text{ g}} \right) = 2.00 \text{ moles of } H_2SO_4$$

$$M = \frac{\text{moles}}{\text{liter}} \qquad ? M = \frac{2.00 \text{ moles}}{0.800 \text{ L}} = 2.50 \text{ } M \text{ solution of } H_2SO_4$$

Practice Exercise 15.10   What is the normality of the solution formed by dissolving 60.0 g of $HC_2H_3O_2$ in enough water to make 500.0 mL of solution? What is the molarity of the solution?

EXAMPLE 15.11

What are the normality and the molarity of the solution formed by dissolving 19.6 g of H_3PO_4 in enough water to make $3\overline{00}$ mL of solution?

Solution

Change the 19.6 g of H_3PO_4 to equivalents and then use the normality formula. H_3PO_4 has

$$\text{Equivalent mass of } H_3PO_4 = \frac{\text{molecular mass}}{\text{number of replaceable } H^+ \text{ ions}}$$

$$= \frac{98.0 \text{ g}}{3 \text{ Eq}}$$

$$= \frac{32.7 \text{ g}}{Eq}$$

$$\frac{32.7 \text{ g}}{1 \text{ equivalent}} \qquad \text{or} \qquad \frac{1 \text{ equivalent}}{32.7 \text{ g}}$$

Therefore

$$? \text{ equivalents of } H_3PO_4 = (19.6 \text{ g}) \left(\frac{1 \text{ equivalent}}{32.7 \text{ g}} \right)$$

$$= 0.599 \text{ equivalent of } H_3PO_4$$

$$N = \frac{\text{equivalents}}{\text{liter}} \qquad ?\, N = \frac{0.599 \text{ equivalent}}{0.300 \text{ L}} = 2.00 \text{ N}$$

To find the molarity of the solution, change grams of H_3PO_4 to moles and then find the number of moles per liter.

$$?\text{ moles of } H_3PO_4 = (19.6 \text{ g}) \left(\frac{1 \text{ mole}}{98.0 \text{ g}} \right) = 0.200 \text{ mole of } H_3PO_4$$

$$M = \frac{\text{moles}}{\text{liter}} = \frac{n}{V} \qquad ?\, M = \frac{0.200 \text{ mole}}{0.300 \text{ L}} = 0.667 \text{ M solution of } H_3PO_4$$

Practice Exercise 15.11   What are the normality and the molarity of the solution formed by dissolving 7.23 g of HCl in enough water to make 450.0 mL of solution?

EXAMPLE 15.12

How many grams of $Ca(OH)_2$ are there in $5\overline{0}$ mL of a 3.0 N solution?

Solution

Solve the normality formula,

$$N = \frac{Eq}{L}$$

for Eq, the number of equivalents:

$$Eq = N \times L$$

Then substitute the values.

$$Eq = 3.0 \text{ Eq/L} \times 0.050 \text{ L}$$

$$= 0.15 \text{ Eq of } Ca(OH)_2$$

Next, find the equivalent mass of $Ca(OH)_2$.

$$\text{Equivalent mass of } Ca(OH)_2 = \frac{\text{molecular mass}}{\text{number of replaceable } OH^{1-} \text{ ions}}$$

$$= \frac{74.1 \text{ g}}{2 \text{ Eq}}$$

$$= 37.1 \frac{\text{g}}{\text{Eq}}$$

Convert the 0.15 Eq of $Ca(OH)_2$ to grams.

$$0.15 \text{ Eq } Ca(OH)_2 \times \frac{37.1 \text{ g } Ca(OH)_2}{1 \text{ Eq } Ca(OH)_2} = 5.6 \text{ g } Ca(OH)_2$$

Practice Exercise 15.12   How many grams of NaOH are there in 75.0 mL of a 4.00 N NaOH solution?

15.6 Dilution of Solutions

■ LEARNING GOAL 8
Calculations for dilutions of solutions

Solutions of acids and bases are usually purchased from chemical supply houses. For the sake of economy, the suppliers furnish most of these solutions in highly concentrated form. (This is like buying concentrated soups and adding water to them.) Chemists have to know how to dilute concentrated solutions to the strengths they want. If the concentration is expressed as a normality, the formula for doing this is

$$N_c V_c = N_d V_d$$

where N_c is the normality of the concentrated solution, V_c is the volume of the concentrated solution, N_d is the normality of the dilute solution, and V_d is the volume of the dilute solution. If the concentration is expressed in terms of molarity, the formula becomes

$$M_c V_c = M_d V_d$$

The volume V can be expressed in any convenient unit. But the units used, such as liters or milliliters, must be the same on both sides of the equation. This formula can be applied when solving practical laboratory problems like those in the next two examples.

EXAMPLE 15.13

How would you prepare 2 L of 3 *M* HCl from a 12 *M* HCl solution?

Solution

You need to find the amount V_c of concentrated HCl that, when diluted with water, will give 2 L of 3 *M* HCl.

$$M_c V_c = M_d V_d \qquad \text{so} \qquad V_c = \frac{M_d V_d}{M_c}$$

$$= \frac{(3\ \cancel{M})(2\ \text{L})}{12\ \cancel{M}} = 0.5\ \text{L}$$

Therefore, to make the final solution, measure 0.5 L of concentrated HCl into a flask and dilute with water until you have 2 L. (But see the cautionary note in Example 15.14.)

Practice Exercise 15.13   How would you prepare 3.00 L of a 4.00 *M* HCl solution from a stock solution of 6.00 *M* HCl?

EXAMPLE 15.14

How would you prepare 300 mL of a 0.6 *N* H_2SO_4 solution from a 36 *N* solution?

FIGURE 15.6

An electrolytic solution conducts electricity.

Solution

You need the amount of V_c of concentrated H_2SO_4 that, when diluted with water, will give 300 mL of 0.6 N H_2SO_4.

$$N_c V_c = N_d V_d \qquad \text{so} \qquad V_c = \frac{N_d V_d}{N_c}$$

$$= \frac{(0.6 \ \cancel{N})(300 \text{ mL})}{36 \ \cancel{N}} = 5 \text{ mL}$$

Therefore, to make the final solution, measure 5 mL of concentrated H_2SO_4 into a flask and dilute with water to 300 mL. This will be a 0.6 N H_2SO_4 solution. (A cautionary note: *Always add acid to water, never the reverse. In the case of H_2SO_4, adding water to the concentrated acid will make the acid react violently. In the problem we just solved, it would be wise to put most of the water in the flask before adding the H_2SO_4.*)

Practice Exercise 15.14 How would you prepare 500.0 mL of a 0.400 N H_3PO_4 solution from a stock solution of 12.0 N H_3PO_4?

15.7 Ionization in Solutions

LEARNING GOAL 9
Electrolytic and nonelectrolytic solutions

LEARNING GOAL 10
Ions and ionization

● Sodium ion, Na^{1+}

● Chloride ion, Cl^{1-}

⌬ Water molecule

FIGURE 15.7

Dissociation of NaCl in water

For centuries, scientists have tried to find out what happens when one substance dissolves in another. The chemists of the 1800s were puzzled by the fact that some solutions would conduct electric current and others would not. The English physicist Michael Faraday tried to explain this phenomenon. He classified solutions generally in the following way: **Electrolytic solutions** are *solutions that conduct electric current*. A solute that produces ions in solution is an electrolyte. **Nonelectrolytic solutions** are *solutions that do not conduct electric current*. A solute that does not produce ions in solution is a nonelectrolyte.

Faraday said that electrolytic solutions—such as sodium chloride in water—contain charged particles. He called these particles *ions* (from a Greek word that may be translated as "wanderer"). He said that ions wander through the solution carrying electric current (Figure 15.6). But questions arose about the nature of these ions. For example, why are there ions in a salt-and-water solution but no ions in a sugar-and-water solution? Another scientist, the Swedish chemist Svante Arrhenius, came up with the answer.

In 1884, in his Ph.D. thesis, Arrhenius advanced his ideas about ions. He suggested that Faraday's ions were really simple atoms (or groups of atoms) carrying a positive or a negative charge. He said that when some substances dissolve in solution, they break up into ions, for example:

$$NaCl(s) \xrightarrow{\text{H}_2\text{O}} Na^{1+}(aq) + Cl^{1-}(aq)$$

He called this process **ionization**. Actually, the solid sodium chloride **dissociates** when placed in water, because it is composed of ions (Figure 15.7).

■ **FIGURE 15.8**

Sugar dissolves in water without ionizing.

When placed in water, the ions separate from each other and are free to move about in the solution. Solutions that contain ions are electrolytic because they contain charged particles that can carry electric current. Substances in solution that conduct electricity are called *electrolytes.* A solution of sodium chloride actually contains a 1:1 mixture of sodium *ions* and chloride *ions.* A bottle label reading "1 *M* NaCl" refers to how the solution was made and not what is really in the bottle.

On the other hand, some substances dissolve in solution and *do not ionize.* They simply break up into their neutral molecules and become surrounded by the molecules of solvent:

$$C_6H_{12}O_6(s) \xrightarrow{\text{H}_2\text{O}} C_6H_{12}O_6(aq)$$
$$\text{(Glucose)}$$

These solutions are nonelectrolytic because they contain no charged particles to carry electric current (Figure 15.8). Substances that do not ionize in solution are called *nonelectrolytes.*

Electrolytes and nonelectrolytes acting as solutes have some very interesting effects on the boiling points and freezing points of solutions. However, before we can examine these effects, we must discuss another way of expressing the concentration of solutes in solutions.

15.8 Molality

We can express the concentration of a solute in a solution in terms of the *number of moles of solute per kilogram of solvent.* This concentration unit is known as **molality** (*m*).

$$\text{Molality } (m) = \frac{\text{number of moles of solute}}{\text{kilogram of solvent}} = \frac{n}{\text{kg}}$$

We can determine the molality of a solution if we know either the number of moles or the mass of a solute dissolved in a given mass of solvent.

EXAMPLE 15.15

Calculate the molality of each of the following solutions:
(a) 36 g of glucose ($C_6H_{12}O_6$) dissolved in $50\overline{0}$ g of water
(b) 1.03 g of sodium bromide (NaBr) dissolved in $25\overline{0}$ g of water
(c) 3.40 g of ammonia (NH_3) dissolved in $30\overline{0}$ g of water

Solution

(a) First change the 36 g of $C_6H_{12}O_6$ (molecular mass = 180.0) to moles.

$$? \text{ moles} = (36 \text{ g}) \left(\frac{1 \text{ mole}}{180.0 \text{ g}} \right) = 0.20 \text{ mole}$$

Next change $50\overline{0}$ g of water to kilograms.

$$? \text{ kg} = (50\overline{0} \text{ g}) \left(\frac{1 \text{ kg}}{1,000 \text{ g}} \right) = 0.500 \text{ kg}$$

Finally, substitute these numbers into the molality formula.

$$m = \frac{\text{number of moles of solute}}{\text{kilogram of solvent}}$$

$$? \, m = \frac{0.20 \text{ mole of glucose}}{0.500 \text{ kg of water}} = 0.40 \, m$$

(b) First change 1.03 g of NaBr (molecular mass = 102.9) to moles.

$$? \text{ moles} = (1.03 \text{ g}) \left(\frac{1 \text{ mole}}{102.9 \text{ g}} \right) = 0.0100 \text{ mole}$$

Next change $25\overline{0}$ g of water to kilograms.

$$? \text{ kg} = (25\overline{0} \text{ g}) \left(\frac{1 \text{ kg}}{1,000 \text{ g}} \right) = 0.250 \text{ kg}$$

Finally, substitute these numbers into the molality formula.

$$m = \frac{\text{number of moles of solute}}{\text{kilogram of solvent}}$$

$$? \, m = \frac{0.0100 \text{ mole of NaBr}}{0.250 \text{ kg of } H_2O} = 0.0400 \, m$$

(c) First change the 3.40 g of NH_3 (molecular mass = 17.0) to moles.

$$? \text{ moles} = (3.40 \text{ g}) \left(\frac{1 \text{ mole}}{17.0 \text{ g}} \right) = 0.200 \text{ mole}$$

Next change $3\overline{00}$ g of water to kilograms.

$$? \text{ kg} = (3\overline{00} \text{ g}) \left(\frac{1 \text{ kg}}{1,000 \text{ g}} \right) = 0.300 \text{ kg}$$

Finally, substitute these numbers into the molality formula.

$$m = \frac{\text{number of moles of solute}}{\text{kilogram of solvent}}$$

$$? \, m = \frac{0.200 \text{ mole of } NH_3}{0.300 \text{ kg of } H_2O} = 0.667 \, m$$

Practice Exercise 15.15   Calculate the molality of each of the following solutions:
(a) 72 g of glucose ($C_6H_{12}O_6$) dissolved in $4\overline{00}$ g of water
(b) 1.17 g of NaCl dissolved in $5\overline{00}$ g of water
(c) 0.680 g of NH_3 dissolved in $25\overline{0}$ g of water

Chemists find that molality is a useful concept when they must deal with the effect of a solute on the boiling point and freezing point of a solution. The next section explains what this means.

15.9 Colligative (Collective) Properties of Solutions

Certain properties of solutions depend more on the number of solute particles present than on the type of solute particles. In other words, these properties are the same regardless of what substance is the solute. *These properties, called* **colligative properties,** *depend on the number of solute particles present in a given quantity of solvent.* For example, a mole of sugar molecules will have the same effect on the physical properties of a solution as will a mole of soluble starch or glycerol molecules.

One important colligative property of solutions is related to the vapor pressure of the solution. When a solute is dissolved in a solvent, the vapor pressure of the solution *decreases.* In other words, if we compared the vapor pressure of pure water at 50°C with the vapor pressure of a sugar–water solution at 50°C, we would find that the sugar–water solution had a lower vapor pressure than the pure water (Figure 15.9).

Two other colligative properties of solutions result from this vapor-pressure–reduction property.

1. The boiling point of a solution is *higher* than the boiling point of the pure solvent alone. This is called the **boiling-point–elevation** property.
2. The freezing point of a solution is *lower* than the freezing point of the pure solvent alone. This is called the **freezing-point–depression** property.

For example, when salt is added to water, the boiling point of the water is raised, and food cooks in the water at a higher temperature. And, by adding salt to the ice that is used to freeze ice cream, we lower its freezing point. We both raise the boiling point and lower the freezing point of automobile radiator water when we add antifreeze/coolant.

■**FIGURE 15.9**

At a given temperature, the vapor pressure of the pure solvent is always greater than that of the solution of a nonvolatile solute.

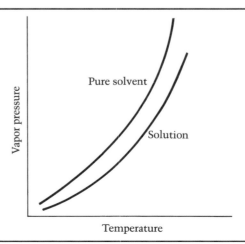

15.10 Boiling-Point Elevation, Freezing-Point Depression, and Molality

LEARNING GOAL 14
Definition of K_f and K_b

The amount by which the boiling point of a solution is elevated or the freezing point lowered depends on (1) the solvent and (2) the concentration of the solute particles. For water, the **boiling-point–elevation constant K_b** is 0.52°C per mole of solute particles per kilogram of water. That is, a 1-molal (1 m) solution of sugar in water raises the boiling point of the water from 100°C to 100.52°C. The **freezing-point–depression constant K_f** for water is 1.86°C per mole of solute particles per kilogram of water. So a 1 m solution of sugar in water would have a freezing point of −1.86°C. Other solvents have different values for the constants K_b and K_f.

The values of K_f and K_b for water are the same for all non-ionic solutes. Thus a 1 m solution of glycerin in water also raises the boiling point of water to 100.52°C and lowers the freezing point to −1.86°C. (However, if we dissolve 1 mole of NaCl in 1,000 g of water, the effect on the boiling point and freezing point would be about double, because the NaCl ionizes into Na^{1+} ions and Cl^{1-} ions. In other words, 1 mole of NaCl yields 2 moles of solute particles. However, here we'll discuss only non-ionizing solutes.) For a solute that does not ionize in water, we can make this statement: *A 1-molal solution of molecular solute lowers the freezing point of 1 kg of water by 1.86°C and raises the boiling point by 0.52°C.*

Each additional mole of solute per kilogram of solvent lowers the freezing point of the solution by an additional 1.86°C and raises the boiling point of the solution by an additional 0.52°C. We can state this mathematically as

$$\Delta t = mK_f \qquad \text{(for freezing-point depression)}$$

$$\Delta t = mK_b \qquad \text{(for boiling-point elevation)}$$

where Δt is the boiling-point elevation or freezing-point depression in Celsius degrees.

EXAMPLE 15.16

LEARNING GOAL 15
Calculations involving boiling and freezing points of solutions

Calculate the boiling point of each of the following solutions:
(a) 36 g of glucose dissolved in $\overline{500}$ g of water
(b) 9.2 g of glycerol ($C_3H_8O_3$) in 250 g of water

Solution

(a) First we need to determine the molality of a solution of 36 g of glucose in $\overline{500}$ g of water. If you look back at Example 15.15(a), you'll see that we calculated this molality to be 0.40 m.

Now use the formula for boiling-point elevation to calculate the change in temperature Δt.

$$\Delta t = mK_b = (0.40 \; m)(0.52°C/m) = 0.21°C$$

If pure water boils at 100°C, then this 0.40 m solution boils at 100.21°C.

(b) First determine the molality of a solution of 9.2 g of $C_3H_8O_3$ (molecular mass = 92.0) in $25\overline{0}$ g of water.

$$? \text{ moles } C_3H_8O_3 = (9.2 \text{ g}) \left(\frac{1 \text{ mole}}{92.0 \text{ g}} \right) = 0.10 \text{ mole}$$

$$? \text{ kg water} = (25\overline{0} \text{ g}) \left(\frac{1 \text{ kg}}{1,000 \text{ g}} \right) = 0.250 \text{ kg}$$

$$? \, m = \frac{0.10 \text{ mole}}{0.250 \text{ kg}} = 0.40 \, m$$

Now use the formula for boiling-point elevation to calculate the change in temperature Δt.

$$\Delta t = mK_b = (0.40 \, m)(0.52°C/m) = 0.21°C$$

If pure water boils at 100°C, then this 0.40 m solution boils at 100.21°C.

Practice Exercise 15.16   Calculate the boiling point of each of the following solutions:

(a) 72.0 g of glucose ($C_6H_{12}O_6$) dissolved in $25\overline{0}$ g of water

(b) 1.84 g of glycerol ($C_3H_8O_3$) dissolved in $50\overline{0}$ g of water

---

EXAMPLE 15.17

Calculate the freezing point of each of the solutions in Example 15.16.

Solution

(a) We already know that the molality of this solution is 0.40 m. We also know that K_f for water is 1.86°C. Therefore

$$\Delta t = mK_f = (0.40 \, m)(1.86°C/m) = 0.74°C$$

Because pure water freezes at 0°C, this 0.40 m solution freezes at −0.74°C.

(b) We already know that the molality of this solution is 0.40 m. Therefore

$$\Delta t = mK_f = (0.40 \, m)(1.86°C/m) = 0.74°C$$

This 0.40 m solution freezes at −0.74°C.

Practice Exercise 15.17   Calculate the freezing point of each of the solutions in Practice Exercise 15.16.

15.11   The Processes of Diffusion and Osmosis

LEARNING GOAL 16

Diffusion and osmosis

Diffusion and osmosis are important colligative properties in living systems and in the chemistry laboratory as well. In this section, we'll take a brief look at each of these processes.

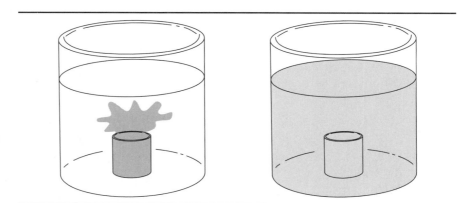

A dye will diffuse through a solution from areas of high dye concentration to areas of low dye concentration. Eventually, the dye will distribute itself evenly throughout the solution.

If you place a few drops of dye in a glass of water, what happens? Very quickly the dye disperses through the water. Given enough time, the dye will distribute itself evenly throughout the water (Figure 15.10). This process is known as **diffusion.** It occurs because the dye moves from an area of high dye concentration to an area of low dye concentration. In other words, there is a *concentration gradient*, or gradually changing difference in concentration, between the place in the water where the dye is dropped and the rest of the water. The diffusion process stops when the dye is evenly distributed throughout the water, so that the concentration gradient no longer exists.

It is this process of diffusion that moves most of the substances in our body. For example, the breathing process is based on the diffusion of gases from areas of high concentration to areas of low concentration. And the process of diffusion is the system by which electrolytes in the body are carried to where they are needed.

A special type of diffusion process called **osmosis** involves the *passage of water through a semipermeable membrane, like cellophane or cell walls.* In this process, *water* moves from an area of low solute concentration to an area of high solute concentration. The semipermeable membrane allows only the water, not the solute, to pass through it.

For example, consider a beaker of water having two compartments separated by a semipermeable membrane (Figure 15.11). Compartment A is filled with a 5% sucrose solution. Compartment B is filled with an equal volume of a 20% sucrose solution. Soon after the solutions are placed in the beaker, we notice a shift in the water levels of the compartments. Water moves from the compartment containing the 5% sucrose solution to the compartment containing the 20% sucrose solution, in an attempt to equalize the sucrose concentrations in both compartments (Figure 15.12). This movement of water is the process of osmosis.

You may be thinking, "Why doesn't the sucrose in the 20% solution move through the semipermeable membrane to equalize the concentrations of the two solutions?" This doesn't happen because an osmotic membrane

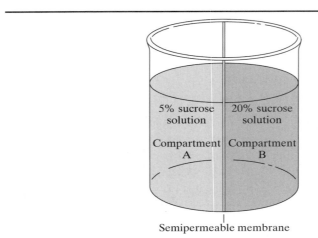

■ FIGURE 15.11

A beaker with two compartments separated by a semipermeable membrane

does not allow solute particles to pass through it but does allow solvent particles to pass freely.

Osmotic Pressure

Consider the two sucrose solutions that we've just discussed. After the process of osmosis has proceeded for a while, it finally stops. Osmosis ends when the rate of water molecules passing from compartment A to compartment B equals the rate of water molecules passing from compartment B to compartment A (Figure 15.13). In other words, a dynamic equilibrium is in progress. But why does this occur?

■ FIGURE 15.12

This is what happens to our two solutions after osmosis has proceeded for a while. Water has left compartment A, increasing the concentration of that sucrose solution. The water has gone into compartment B, diluting that sucrose solution.

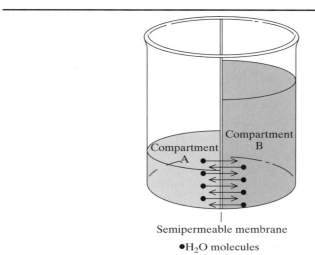

The process of osmosis stops when the number of water molecules moving from compartment A to compartment B equals the number of water molecules moving from compartment B to compartment A.

Semipermeable membrane

•H$_2$O molecules

The answer lies in looking at the height of the water in each compartment. The height of the water in compartment B is greater than that in A (Figure 15.14). At first, water molecules moved from A to B in an attempt to equalize the concentrations of the two sucrose solutions. But eventually the height of the water in compartment B caused a pressure to be exerted on the water molecules in that compartment, pushing them back into compartment A. When these two opposing forces became equal, the process of osmosis stopped. The pressure exerted by the water in compartment B at this point is called the *osmotic pressure* and is related to the concentration of the solute in the solution. **Osmotic pressure** can be defined as *the amount of pressure*

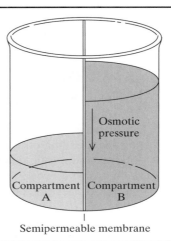

Osmotic pressure

Compartment A   Compartment B

Semipermeable membrane

Osmotic pressure is due to the height of the water column in compartment B.

that must be applied to prevent the flow of water through a membrane. This concept is extremely important in the transport of body fluids.

Like other colligative properties, osmotic pressure can be used to determine the molecular mass of a solute in a solution. A Dutch chemist named van't Hoff developed a mathematical expression that relates the osmotic pressure to the molar concentration of the solution. A technique based on this equation, called **osmometry,** has proved to be far more sensitive for molecular mass determinations than freezing-point depression or boiling-point elevation. In this method, the difference in the height of the water columns in the two compartments is used to obtain the osmotic pressure of the system. From this information the molecular mass of the solute is determined. Although we won't pursue this calculation here, chemists have found this technique quite useful in the laboratory.

EXAMPLE 15.18

Two solutions of equal volume are placed in a beaker similar to the one in Figure 15.11. Compartment A of the beaker contains a 10% NaCl solution. Compartment B of the beaker contains a 5% NaCl solution. Which way does the water move, from A to B or from B to A?

Solution

The water moves from compartment B to compartment A. The solvent moves from the side of low solute concentration to the side of high solute concentration.

Practice Exercise 15.18   Cells in the body have intracellular and extracellular fluid. The intracellular fluid has an osmotic pressure equivalent to 5% glucose. If an individual is given an intravenous solution that has an osmotic pressure equivalent to 10% glucose, would water move into or out of the cell? (*Hint:* The intravenous solution replaces the extracellular fluid.)

SUMMARY

A solution is a homogeneous mixture composed of at least two substances. The solute is the substance that is dissolved, and the solvent is the substance in which it is dissolved. Solutions may be saturated, unsaturated, or supersaturated, depending on the quantity of solute dissolved in a given quantity of solvent. Pairs of liquids that do not form solutions when mixed are said to be immiscible, whereas pairs of liquids that form solutions are said to be miscible. In general, polar substances dissolve polar substances, and nonpolar substances dissolve nonpolar substances.

Because solutions are mixtures, they may have various ratios of solvent to solute. These ratios, or concentrations, may be measured as percentage by mass, percentage by volume, or percentage by mass–volume. In addition, the concentration of a solution may be measured as its molarity: the number of moles of solute per liter of solution. Concentration may also be measured as normality: the number of equivalents of solute per liter of solution.

An equivalent of any acid is the mass of that acid that contains Avogadro's number of hydrogen ions. An equivalent of a base is the mass of that base that contains Avogadro's number of hydroxide ions. The number of grams in one equivalent of an acid or a base is computed by dividing the molecular mass of the substance by the number of replaceable hydrogen ions (for an acid) or replaceable hydroxide ions (for a base) in one molecule of the substance.

Electrolytic solutions are solutions that conduct electricity, and nonelectrolytic solutions are solutions that do not conduct electricity. Electrolytic solutions are capable of conducting electricity because they contain ions, which are charged atoms or groups of atoms. Substances that ionize, or break up into ions, when they dissolve are called electrolytes.

The molality of a solution is the number of moles of solute per kilogram of solvent. Molality is useful in dealing with colligative properties of solutions, which are properties that depend mainly on the number of solute particles present (in a given quantity of solvent) rather than on the type of solute particles. Vapor-pressure reduction, boiling-point elevation, and freezing-point depression are examples of colligative properties. In general, a solution exhibits a lower vapor pressure, a higher boiling point, and a lower freezing point than the pure solvent alone.

Diffusion is the movement of a solute in a solution from an area of high solute concentration to an area of low solute concentration. Osmosis is the passage of water through a semipermeable membrane from the less concentrated to the more concentrated solution.

KEY TERMS

boiling-point elevation **(15.9)**

boiling-point–elevation constant **(15.10)**

colligative properties **(15.9)**

concentration **(15.3)**

diffusion **(15.11)**

dissociation **(15.7)**

electrolytic solution **(15.7)**

equivalent (of an acid or a base) **(15.5)**

freezing-point depression **(15.9)**

freezing-point–depression constant **(15.10)**

immiscible **(15.1)**

ionization **(15.7)**

molality **(15.8)**

molarity **(15.4)**

nonelectrolytic solution **(15.7)**

normality **(15.5)**

osmometry **(15.11)**

osmosis **(15.11)**

osmotic pressure **(15.11)**

saturated solution **(15.2)**

saturation point **(15.3)**

solubility **(15.2)**

solute **(15.1)**

solvent **(15.1)**

supersaturated solution **(15.2)**

unsaturated solution **(15.2)**

Astronomer

As an *astronomer*, sometimes called an *astrophysicist*, you will explore the fundamental nature of the universe. You will search for explanations of how the universe originated and how the solar system evolved. You will use the principles of physics and mathematics to study the behavior of matter and energy in distant galaxies. You will also use this information to support or disprove theories about the nature of matter and energy.

Astronomers explore the fundamental nature of the universe.
(Ed Young/Science Photo Library/ Photo Researchers, Inc.)

To observe the universe, you will use large earth-based telescopes, space-based telescopes, radio telescopes, and other sophisticated instrumentation that can detect electromagnetic radiation from distant celestial sources. For example, you may use a spectroscope to analyze light from a star so that you can determine its chemical composition. You may also use instruments to determine whether a star is emitting radio waves, x-rays, or cosmic rays.

As an astronomer, you will use computers to process astronomical data so that you can calculate the orbits of asteroids and comets or guide spacecraft to specific celestial bodies. Or you may use computers to process and analyze data collected about a star to determine the star's structure or age.

CHEMICAL FRONTIERS

Teardrops and Contact Lenses *

A team of chemists based at the Mayo Clinic in Rochester, Minnesota, has been analyzing peoples' teardrops, hoping to develop a relatively noninvasive method of determining a person's physical condition. In the course of their research, they appear to be on the path to discovering the reason why some people suffer severe irritation from contact lenses.

The researchers actually are analyzing, not only teardrops, but also the aqueous humor, the fluid that fills the eyeball. They developed a method to separate the molecules of the eye fluids into separate components. Here's how it works: First, the tears and aqueous humor are collected from the study subject. (The aqueous humor was extracted with a needle from patients undergoing cataract surgery.) The sample is then injected into an apparatus that separates the fluid into various components. The result is a collection of proteins, salts, peptides, and organic molecules. The process is easy to use, according to the researchers, and can be used with other fluids as well.

The materials are then analyzed for different properties, depending on what the researchers are looking for. Some analyses, for example, have given researchers hints about why contact lenses irritate some people but not others. Further exploration in this area will focus on studying exactly what happens, on a chemical level, when one puts a contact lens in the eye. Other research is focusing on the possibility that tears contain viral proteins. If this is proven, one day soon your tears may be able to determine if you are suffering from a virus or disease. But first, the chemists want to determine what constitutes normal eye fluid. Once this is accomplished, they will have a base line from which they can compare others' eye fluids, hopefully leading to healthier humans with less-irritated eyes!

*Adapted from *Science News*, Sept. 20, 1997, p. 188

SELF-TEST EXERCISES

LEARNING GOAL 1: *Definitions of Solution, Solute, and Solvent*

◄ *1.* Match the word on the left with its definition on the right.

(a) Solute   1. A homogeneous mixture

(b) Solvent   2. The substance in a solution that gets dissolved

(c) Solution   3. Substances that do not form solutions when mixed

(d) Immiscible   4. The substance in a solution that causes another substance to dissolve

2. When a solution is composed of two or more substances in the same state, how do chemists determine which is the solute and which is the solvent?

◄ *3.* Which substance is the solute and which is the solvent in each of the following solutions?
(a) 10%-by-weight hydrochloric acid
(b) Seltzer water (also known as club soda or soda water)
(c) 4%-by-volume benzene in carbon tetrachloride

4. A coin composed of two metals that were melted, mixed together, and solidified is an example of a solution. True or false?

5. Give an example of (a) A solid–solid solution (b) A solid–liquid solution (c) A solid–gas solution (d) A liquid–liquid solution

◄ 6. Which is the solute and which the solvent in each of the following solutions?
(a) 10%-by-mass sodium chloride in water
(b) 5%-by-mass benzene in carbon tetrachloride
(c) 10%-by-mass–volume methyl alcohol in ethyl alcohol

LEARNING GOAL 2: *Substances That Form Solutions*

7. "Like dissolves like." True or false? What does this statement refer to in terms of molecular structure?

◄ 8. You are served a cup of coffee in a restaurant. Assuming that no sugar has been added to the coffee, you add 2 teaspoons of sugar, stir well, and notice a layer of sugar at the bottom of the cup that won't dissolve. What type of solution have you produced?

9. The point at which no more solute dissolves in a solution is called the _____ point.

10. What are the meanings of the terms *miscible* and *immiscible*?

LEARNING GOAL 3: *Saturated, Unsaturated, and Supersaturated Solutions*

11. (a) A solution that contains less solute than can be dissolved in a given quantity of solvent at a particular temperature is a(n) _____ solution.
(b) A solution that contains more solute than is needed for saturation for a given quantity of solvent at a particular temperature is a(n) _____ solution.

12. Determine whether each of the following solutions is unsaturated, saturated, or supersaturated. (See Table 15.2.)
(a) 15.0 g of NaCl in 50.0 g of water at 50°C
(b) 45.5 g of $AgNO_3$ in 50.0 g of water at 50°C

◄ 13. You are given a saturated solution that contains undissolved solute. The solute is constantly dissolving, yet the concentration of the solute in the solution remains the same. Explain.

14. You are given a solution by your instructor who asks you to determine whether or not it is supersat-

urated. Explain the procedure you would use to test for supersaturation.

LEARNING GOAL 4: *Percent Concentrations*

◄ **15.** What is the concentration in percent by mass of the following solutions?
(a) 5.00 g of magnesium nitrate in enough water to make 80.0 g of solution.
(b) 60.0 g of ethanol in enough water to make 300.0 g of solution.
(c) 2.00 g of sodium chloride *plus* 23.0 g of water.

◄ 16. What is the concentration in percent by mass of each of the following solutions?
(a) 80.0 g of sodium chloride with enough water to make $25\overline{0}$ g of solution
(b) 7.50 g of methyl alcohol with enough water to make $40\overline{0}$ g of solution
(c) 50.0 g of potassium nitrate *plus* $15\overline{0}$ g of water

17. How many grams of KOH are there in $20\overline{0}$ g of $1\overline{0}$%-by-mass KOH solution?

18. How many grams of NaOH are there in 500.0 g of a 15.0%-by-mass NaOH solution?

19. What is the percent by volume of the solute in each of the following solutions?
(a) $20\overline{0}$ mL of ether in enough carbon tetrachloride to make 1.0 L of solution
(b) 5 mL of alcohol in enough water to make 80 mL of solution

20. What is the percent by volume of the solute in each of the following solutions?
(a) 300.0 mL of ether in enough carbon tetrachloride to make 2.00 L of solution
(b) 25.0 mL of alcohol in enough water to make 100.0 mL of solution

*21. Find the volume of alcohol present in $10\overline{0}$ g of 10.0%-by-mass alcohol–water solution. (*Hint:* The density of alcohol is 0.800 g/mL.)

*22. Find the volume of alcohol present in 250.0 g of a 10.0%-by-mass alcohol–water solution. (*Hint:* The density of alcohol is 0.800 g/mL.)

23. How many milliliters of chloroform are there in $50\overline{0}$ mL of a 4.0%-by-volume chloroform–ether solution?

24. How many milliliters of carbon tetrachloride are there in 250.0 mL of a 5.00%-by-mass–volume carbon tetrachloride–benzene solution?

◄ **25.** What are the percent by mass–volume of the solutes in the following solutions?
(a) $5\overline{0}$ g of sodium chloride in 1.5 L of solution
(b) $1\overline{0}$ g of ammonia dissolved in $20\overline{0}$ g of solution (Assume that the density of the solution is 1.00 g/cm³.)

◄ 26. What is the percent by mass–volume of the solute in each of the following solutions?
(a) 25.0 g of sodium nitrate in 1.00 L of solution
(b) 15.0 g of sodium hydroxide dissolved in 100.0 g of solution (Assume that the density of the solution is 1.00 g/cm³.)

27. How many grams of sugar are there in $10\overline{0}$ mL of a 15%-by-mass–volume sugar–water solution?

28. How many grams of lactose (milk sugar) are there in 150.0 mL of a 10%-by-mass–volume lactose–water solution?

*29. If a person's blood-sugar level falls below 60 mg per 100 mL, insulin shock can occur. What is the percent by mass of sugar in the blood at this level? (Assume that the density of blood is 1.2 g/mL.)

◄ 30. A person's blood-sugar level is calculated to be 120.0 mg per $10\overline{0}$ mL blood. What is the percentage by mass of sugar in the blood at this level? (The density of blood is 1.20 g/mL.)

31. What are the concentrations in percent by mass of the following solutions?
(a) 25.0 g of NaCl in enough water to make $20\overline{0}$ g of solution
(b) 8.00 g of sugar in enough water to make $16\overline{0}$ g of solution

32. What is the concentration in percent by mass of each of the following solutions?
(a) 55.0 g of Na_2CO_3 in enough water to make 400.0 g of solution
(b) 14.5 g of sugar in enough water to make 290.0 g of solution

33. How many grams of KCl are there in $50\overline{0}$ g of a 6.0%-by-mass KCl solution?

34. How many grams of KNO_3 are there in 750.0 g of a 7.50%-by-mass KNO_3 solution?

35. What are the percents by volume of the solutes in the following solutions?
(a) $15\overline{0}$ mL of ethyl alcohol in enough water to make 3.0 L of solution
(b) $50\overline{0}$ mL of methyl alcohol (dry gas) in enough gasoline to make 60.0 L of solution (This rep-

resents a can of dry gas in a nearly full tank of gasoline in a standard-size automobile.)

36. What is the percent by volume of the solute in each of the following solutions?
    (a) 175 mL of isopropyl alcohol in enough water to make 4.00 L of solution
    (b) 600.0 mL of methyl alcohol (dry gas) in enough gasoline to make 60.0 L of solution

◀ **37.** Determine the volume of ether present in $50\overline{0}$ g of a 4.0%-by-mass ether–carbon tetrachloride solution. (*Hint:* The density of ether is 0.714 g/cm³.)

*38. Determine the volume of ether in 750.0 g of a 10.0%-by-mass ether–carbon tetrachloride solution. (*Hint:* The density of ether is 0.714 g/cm³.)

39. How many milliliters of ethyl alcohol are there in 2.00 L of a 10.0%-by-volume ethyl alcohol–water solution?

40. How many milliliters of isopropyl alcohol are there in 4.50 L of a 20.0%-by-volume isopropyl alcohol–water solution?

41. What are the percents by mass–volume of the solutes in the following solutions?
    (a) 25.0 g of sodium nitrate in $80\overline{0}$ mL of solution
    (b) 0.800 g of potassium chloride in 16.0 mL of solution

42. What is the percent by mass–volume of the solute ◀ in each of the following solutions?
    (a) 50.0 g of sodium carbonate in 500.0 mL of solution
    (b) 0.160 g of potassium nitrate in 80.0 mL of solution

43. How many grams of sodium chloride are there in $25\overline{0}$ mL of a 5.00%-by-mass–volume solution of sodium chloride and water?

44. How many grams of sodium chloride are there in 100.0 mL of a 9.00%-by-mass–volume solution of sodium chloride and water?

LEARNING GOAL 5: *Calculations Involving Molarity*

◀ **45.** What is the molarity of each of the following solutions?
    (a) 5.4 g of HCl in enough water to make $50\overline{0}$ mL of solution
    (b) 117 g of sodium chloride in enough water to make 4 L of solution

46. What is the molarity of each of the following solutions?
    (a) 1.18 g of HCl in enough water to make 2.00 L of solution
    (b) 23.40 g of sodium chloride in enough water to make 5.00 L of solution

*47. What is the molarity of a 10.0%-by-mass HCl solution? (Assume that the density of the solution is 1.00 g/mL.)

*48. What is the molarity of a 5.00%-by-mass HCl solution? (Assume that the density of the solution is 1.00 g/cm³.)

49. How many grams of KOH are obtained by evaporating to dryness $5\overline{0}$ mL of a 3.0 *M* KOH solution?

50. How many grams of NaOH are obtained by evaporating to dryness 25.0 mL of a 2.00 *M* NaOH solution?

*51. The *Merck Manual* reports that a blood-alcohol level of $15\overline{0}$ to $20\overline{0}$ mg per $10\overline{0}$ mL produces intoxication and that a level of $30\overline{0}$ to $40\overline{0}$ mg per $10\overline{0}$ mL produces unconsciousness. At a blood-alcohol level above $50\overline{0}$ mg per $10\overline{0}$ mL, a person may die. What is the molarity of the blood with respect to alcohol at the level of $50\overline{0}$ mg per $10\overline{0}$ mL? (*Hint:* The molecular mass of ethyl alcohol is 46.0.)

*52. A person has been drinking ethyl alcohol. What ◀ is the molarity of the alcohol in this person's blood if the alcohol concentration is $40\overline{0}$ mg per $10\overline{0}$ mL of blood?

◀ **53.** Salt (NaCl) is a necessary ingredient of our diets, and the body has a system to maintain a delicate balance of salt. In certain illnesses, this salt balance can be lost, and a physician or nurse must administer salt intravenously, using a 0.85% solution. What is the molarity of a 0.85%-by-mass–volume sodium chloride solution?

◀ 54. How many milliliters of water are needed to make a 2.00 *M* NaCl solution if 5.85 g of NaCl are present?

55. What is the molarity of each of the following solutions?
    (a) 24.5 g of H_2SO_4 in enough water to make 1.50 L of solution
    (b) 10.1 g of KNO_3 in enough water to make $50\overline{0}$ mL of solution

56. What is the molarity of each of the following solutions?

(a) 49.0 g of H_2SO_4 in enough water to make 2.00 L of solution

(b) 2.02 g of KNO_3 in enough water to make 200.0 mL of solution

*57. What is the molarity of a 5.00%-by-mass magnesium chloride solution? (Assume that the density of the solution is 1.00 g/cm^3.)

*58. What is the molarity of a 7.50%-by-mass sodium nitrate solution? (Assume that the density of the solution is 1.00 g/cm^3.)

59. How many grams of Na_3PO_4 are obtained by evaporating to dryness $25\overline{0}$ mL of a 0.400 M Na_3PO_4 solution?

60. How many grams of $NaC_2H_3O_2$ are obtained by evaporating to dryness 400.0 mL of a 0.500 M $NaC_2H_3O_2$ solution?

LEARNING GOAL 6: *Equivalents of Acids and Bases*

◀ 61. Determine the number of equivalents in each of the following samples: (a) 4.9 g of H_2SO_4 (b) 7.2 g of HCl (c) 32.1 g of $Fe(OH)_3$

62. Determine the number of equivalents in each of the following samples: (a) 9.8 g of H_2SO_4 (b) 3.6 g of HCl (c) 6.42 g of $Fe(OH)_3$

63. For each of the following acids and bases, determine the number of grams per mole and the number of grams per equivalent: (a) H_2SO_3 (b) $Fe(OH)_3$ (c) $Fe(OH)_2$ (d) H_3BO_3

64. For each of the following acids and bases, determine the number of grams per mole and the number of grams per equivalent: (a) $HC_2H_3O_2$ (b) NaOH (c) $Mg(OH)_2$ (d) HNO_3

65. Determine the number of equivalents in each of the following: (a) 1.23 g of H_2SO_3 (b) 26.7 g of $Fe(OH)_3$

66. Determine the number of equivalents in each of the following: (a) 246 g of H_2SO_3 (b) 4.00 g of NaOH

◀ 67. Determine the number of grams in each of the following: (a) 25.0 equivalents of $Fe(OH)_2$ (b) 0.450 equivalent of H_3BO_3

◀ 68. Determine the number of grams in each of the following: (a) 0.250 equivalent of $Mg(OH)_2$ (b) 2.00 equivalents of $HC_2H_3O_2$

LEARNING GOAL 7: *Calculations Involving Normality*

◀ 69. Determine the normality of each of the following solutions:
(a) $97\overline{0}$ g of H_3PO_4 in enough water to make 10.0 L of solution
(b) 3.7 g of $Ca(OH)_2$ in enough water to make $80\overline{0}$ mL of solution
(c) Calculate the molarities of the solutions in (a) and (b).

◀ 70. Determine the normality of each of the following solutions:
(a) 465 g of H_3PO_4 in enough water to make 8.00 L of solution
(b) 740.0 g of $Ca(OH)_2$ in enough water to make 740.0 mL of solution
(c) Calculate the molarities of the solutions in (a) and (b).

71. How many grams of $Ca(OH)_2$ are there in $80\overline{0}$ mL of a 2.0 N $Ca(OH)_2$ solution?

72. How many grams of $Mg(OH)_2$ are there in $75\overline{0}$ mL of a 3.00 N $Mg(OH)_2$ solution?

73. How many milliliters of a 0.60 N H_3PO_4 solution do you need to get 4.9 g of H_3PO_4?

74. How many milliliters of a 0.450 N H_3PO_4 solution do you need to get 98.0 g of H_3PO_4?

75. Determine the normality of each of the following solutions:
(a) 11.23 g of $Fe(OH)_2$ in enough water to make $25\overline{0}$ mL of solution
(b) 4.12 g of H_3BO_3 in enough water to make $40\overline{0}$ mL of solution
(c) Calculate the molarities of the solutions in (a) and (b).

76. Determine the normality of each of the following solutions:
(a) 2.246 g of $Fe(OH)_2$ in enough water to make 500.0 mL of solution
(b) 16.48 g of H_3BO_3 in enough water to make 500.0 mL of solution
(c) Calculate the molarities of the solutions in (a) and (b).

◀ 77. How many grams of $Al(OH)_3$ are there in $25\overline{0}$ mL of a 0.500 N $Al(OH)_3$ solution?

◀ 78. How many grams of $Fe(OH)_3$ are there in 500.0 mL of a 0.250 N $Fe(OH)_3$ solution?

79. How many milliliters of 0.100 N H_2SO_4 do you need to obtain 2.94 g of H_2SO_4?

80. How many milliliters of a 0.200 N H_3PO_4 solution do you need to obtain 29.40 g of H_3PO_4?

LEARNING GOAL 8: *Calculations for Dilutions of Solutions*

◀ **81.** How would you make $50\overline{0}$ mL of 4.0 N H_2SO_4 solution from a 16 N solution?

◀ 82. How would you make 250.0 mL of 8.00 N H_3PO_4 solution from a 10.0 N solution?

83. How would you make 3 L of a 0.1 N NaOH solution from a 6 N solution?

84. How would you make 4.00 L of 0.200 N $Ca(OH)_2$ solution from 2.00 N solution?

85. How would you prepare $25\overline{0}$ mL of a 5.00 M HCl solution from a 12.0 M solution?

86. How would you prepare 450.0 mL of a 7.50 M HNO_3 solution from an 11.0 M solution?

87. How would you prepare 5.00 L of 0.0500 N H_2SO_4 solution from a 3.00 M solution?

88. How would you prepare 4.00 L of a 0.0200 N $HC_2H_3O_2$ solution from a 2.00 N solution?

LEARNING GOALS 9 AND 10: *Electrolytic and Nonelectrolytic Solutions; Ions and Ionization*

◀ 89. Define the terms *ion* and *ionization* and give examples of each.

90. True or false: (a) Ions are groups of charged atoms. (b) Charged elements are not ions.

◀ 91. Explain why some solutions conduct electric current and others do not.

◀ 92. Solution A conducts electricity, but solution B does not. If solutions A and B are mixed and no chemical reaction occurs, will the resulting solution conduct electricity?

◀ 93. Define and give an example of an electrolytic solution and of a nonelectrolytic solution.

94. Which will conduct electricity, a glass of salt water or a cup of tea with sugar? Explain.

LEARNING GOAL 11: *Definition of Molality*

◀ 95. Define *molality*. What information is necessary to calculate the molality of a solution?

◀ 96. How is *molality* different from *molarity*?

LEARNING GOAL 12: *Calculation of Molality*

*97. Calculate the molality of the intravenous salt solution in Self-Test Exercise 53. (Assume that there is 0.85 g of NaCl per $10\overline{0}$ g of water.)

◀ 98. A solution contains 0.373 g of KCl per $10\overline{0}$ g of water. What is its molality?

◀ **99.** Determine the molality of each of the following solutions:
(a) 6.4 g of methanol (CH_3OH) in $25\overline{0}$ g of water
(b) 90.0 g of glucose ($C_6H_{12}O_6$) in $1,50\overline{0}$ g of water

100. Determine the molality of each of the following solutions:
(a) 0.128 g of CH_3OH in $50\overline{0}$ g of water
(b) $45\overline{0}$ g of $C_6H_{12}O_6$ in $2,0\overline{00}$ g of water

LEARNING GOAL 13: *Definition of Colligative Properties and Effect of Solute on Boiling and Freezing Points*

◀ 101. Explain how a solute affects the boiling point and the freezing point of a solution.

◀ 102. What would happen to the engine of an automobile if antifreeze were not added to the water in the radiator during a cold winter?

LEARNING GOAL 14: *Definition of K_f and K_b*

◀ 103. Define the following terms and give an example of each: (a) K_f (b) K_b

◀ 104. What are the values of K_f and K_b for water? What do they mean?

LEARNING GOAL 15: *Calculations Involving Boiling and Freezing Points of Solutions*

◀ **105.** You can winterize a typical car radiator system by adding 1.00 gal of antifreeze to 1.00 gal of water. At what temperature does this water–antifreeze solution begin to freeze? (Assume that the antifreeze is pure ethylene glycol, $C_2H_6O_2$, which has a density of 1.10 g/cm^3.) (*Hint:* You must change the 1.00 gal of ethylene glycol solute to grams and then to moles. Also, the 1.00 gal of water must be changed to kilograms.)

106. Calculate the boiling point of a solution prepared by dissolving 5.40 g of glucose in $5\overline{0}0$ g of water. Glucose is $C_6H_{12}O_6$.

107. Calculate the boiling point of the solution in Self-Test Exercise 105.

◀ 108. Calculate the boiling point of a solution prepared by dissolving 0.184 g of glycerol ($C_3H_8O_3$) in $1,\overline{0}00$ g of water.

109. Calculate the boiling point and the freezing point of a solution prepared by dissolving 27.0 g of glucose, $C_6H_{12}O_6$, in $1\overline{0}0$ g of water.

110. Calculate the freezing point of the solution in Self-Test Exercise 108.

LEARNING GOAL 16: *Diffusion and Osmosis*

111. (a) The passage of water through a semipermeable membrane is the process of _____ .
(b) The movement of a solute in a solvent from an area of high solute concentration to an area of low solute concentration is the process of _____ .

◀ 112. A 50.0-g sample of a freshly peeled raw potato is placed in a beaker containing a saturated NaCl solution. The potato is left in the beaker for two hours. It is then removed, dried, and weighed. Will the mass of the potato be lower, higher, or the same? Explain.

◀ 113. Two solutions of equal volume are placed in a beaker similar to the one in Figure 15.11. Compartment A of the beaker contains a 5% KCl solution. Compartment B of the beaker contains a 20% KCl solution. Which way does the water move, from A to B or from B to A?

114. Cells in the body have both intracellular and extracellular fluid. The intracellular fluid has an osmotic pressure equivalent to 0.9% sodium chloride. If an individual is given an intravenous solution that has an osmotic pressure equivalent to 2.0% sodium chloride, would water move into or out of the cell? (*Hint:* The intravenous solution replaces the extracellular fluid.)

EXTRA EXERCISES

◀ *115. Use the kinetic theory to explain how and why a solute affects the boiling and freezing points of a solution.

116. Calculate the molarity of a solution of HCl that has a hydrogen-ion concentration of 10^{-2} M.

117. Calculate the molarity and the normality of a solution of sulfuric acid that has a $[H^{1+}]$ of 10^{-2} M.

118. How many grams of H_2SO_4 are necessary to make 10.0 L of a 5.00%-by-mass sulfuric acid solution? (Assume that the density of the solution is 1.00 g/mL.)

119. Find the molarity and the normality of the solution in Extra Exercise 118.

120. How does a solution differ from a mixture?

◀ 121. In Chicago, salt is sometimes used on roads to keep the roads free of ice. In northern Minnesota, however, people use sand to avoid slipping on ice. Why don't the Minnesota residents try to use salt to get rid of their ice?

◀ 122. Which of the following would lower the freezing point of $1,\overline{0}00$ g of water by the greatest amount? Explain your choice.
(a) $1\overline{0}0$ g of glucose ($C_6H_{12}O_6$)
(b) $1\overline{0}0$ g of sucrose ($C_{12}H_{22}O_{11}$)
(c) $1\overline{0}0$ g of ethanol (C_2H_6O)

*123. Concentrated hydrochloric acid is delivered as 37.00% HCl by mass. The density of this solution is 1.19 g/mL. What volume of this acid is needed to produce 1.00 L of a 0.500 M HCl solution?

124. You dissolve a 15.0-g sample of an unknown substance (nonelectrolyte) in $1\overline{0}0$ g of water. The solution freezes at $-1.86°C$. What is the molecular mass of the substance?

125. A chemist prepares a solution by taking 1.00 L of water and adding 2.00 moles of solute. Is this solution a 2.00 M solution or a 2.00 m solution? Explain.

*126. You add a 10.0-g sample of a substance to $1,\overline{0}00$ g of water. You find that the freezing point of the solution is $-5.00°C$. What is the molecular mass of the substance?

127. You add 5.85 g of NaCl to $1,\overline{0}00$ g of water. What is the approximate boiling point of the solution?

128. Determine the number of equivalents in each of the following: (a) 73.0 g HCl (b) 98.1 g of H_2SO_4

129. Determine the number of equivalents in each of the following: (a) 1.00 mole of HCl (b) 1.00 mole of H_2SO_4

ACIDS, BASES, AND SALTS

16

This is the story of Mary Johnson—accountant, wife, and mother—and how some knowledge gained in a chemistry class, along with quick thinking, helped save the lives of her sons.

It was spring cleaning time at the Johnson house one Saturday morning in April. Mary, her husband Kevin, and their children Eric, age 8, and Kevin Jr. age 11, had divided the chores for the day. Kevin was to paint the master bedroom on the upper level of the house; Mary was to shampoo the carpet in the family room; and Eric and Kevin Jr. were going to help by doing the laundry.

Above the washing machine were shelves on which were various cleaning products that the Johnsons used around the house. Among these products were detergent, liquid chlorine bleach, baking soda, and household ammonia cleaner. Eric had already taken the chlorine bleach from the shelf, opened it, and placed it on the dryer, next to the washing machine. As he was reaching for the detergent, he accidentally knocked the bottle of ammonia cleaner off the shelf. The bottle came crashing down on the dryer and broke open, spilling its contents on the dryer and onto the tile floor. As Eric jumped back to avoid getting hit by the bottle of ammonia cleaner, he knocked over the open bottle of chlorine bleach, which also spilled all over the top of the dryer and onto the floor.

Hearing the commotion, Mary ran into the laundry room and saw that the bleach and ammonia had mixed, both on top of the dryer and on the floor. Realizing that chlorine gas was being generated by the reaction of the two products, she grabbed her stunned children and pulled them into the family room, closing the door behind her. She then told her children to get out of the house, leaving the door open as they exited, and wait for their parents at the end of the driveway. Mary called to Kevin, who was upstairs, and explained the situation to him. They decided to ventilate the laundry room by opening (from the outside) the two laundry-room doors that led outside the house, and then they called the Department of Health to inspect the house to find the best way to clean up the spill. They also were examined by their family doctor for any effects from the chlorine gas. All of the Johnsons received a clean bill of health, thanks to Mary's knowledge of the situation and her quick action.

LEARNING GOALS

After you have studied this chapter, you should be able to:

1. Define the terms *acid* and *base* in the way Arrhenius did.
2. Define the terms *acid* and *base* according to the Brønsted–Lowry concept.
3. List the important characteristics of acids and bases.
4. Write equations for the preparation of acids and bases.
5. Write equations for important chemical reactions of acids and bases.
6. Define the terms *strong acid* and *strong base* and give examples of each.
7. Define the terms *weak acid* and *weak base* and give examples of each.
8. Explain the concept of dynamic equilibrium.
9. Write an equation for the ionization of water.
10. Define pH and explain the different values on the pH scale.
11. Determine the pH of a solution, given its hydrogen-ion or hydroxide-ion concentration in moles per liter.
12. State the relationship between the hydrogen-ion concentration and the hydroxide-ion concentration in aqueous solutions.
13. Explain what the term *salt* means and give some examples.
14. Cite the most important properties of salts.
15. List methods and equations for the preparation of various salts.
16. Define the term *titration*.

INTRODUCTION

Three important groups of compounds—acids, bases, and salts—are the subject of this chapter. You are probably familiar with many compounds found in each of these groups. Acetic acid is the substance that gives vinegar its characteristic taste and odor, and citric acid is responsible for the tartness of lemons and grapefruits. Many popular household cleansers contain the base ammonia. Bases are the substances that make soaps and cleansers feel slippery. In this chapter, we also discuss titration, an analytical procedure that enables the chemist to determine how much of a substance is present in solution. And finally, we examine salts, which are widely used in both home and industry.

16.1 Acids

Three different definitions of acids and bases have been developed in the last 100 years, and each has its place in explaining observed behavior. Two of these definitions will serve as the basis for our discussion of the properties, reactions, and uses of acids.

The Arrhenius Definition

LEARNING GOAL 1

Arrhenius acids

In 1884 the Swedish scientist Svante Arrhenius described an **acid** as a substance that releases hydrogen ions (H^{1+}) when it is dissolved in water. This description was part of his ionic theory of solutions, discussed in Chapter 15.

Let's look at how an acid dissolves in water. In Chapter 7, we saw that water is a polar molecule: Its oxygen end is negatively charged and its hydrogen end is positively charged. When we bubble hydrogen chloride gas (HCl) into a beaker of water, the gas seems to dissolve in the water, but actually much more is happening. The HCl gas is reacting with the water (Figure 16.1). The covalently bonded hydrogen chloride molecules are being broken apart (ionized) by the water into hydrogen ions (H^{1+}) and chloride ions (Cl^{1-}), and the hydrogen ions are adding on to the negative end of the water molecules, forming an ion. This ion, $(H_3O)^{1+}$, is called a **hydronium ion**. The reaction is

$$HCl + H_2O \rightarrow (H_3O)^{1+} + Cl^{1-}$$

Similar reactions occur when the other acids listed in Table 16.1 are dissolved in water. For example,

$$H_2SO_4 + 2H_2O \rightarrow 2(H_3O)^{1+} + (SO_4)^{2-}$$

According to Arrhenius, HCl is an acid because it undergoes the following reaction when placed in water:

$$HCl + H_2O \rightarrow (H_3O)^{1+} + Cl^{1-}$$

■ **FIGURE 16.1**

The positive end of the HCl molecule is attracted by the negative end of the water molecule. The attraction is strong enough to break the bond between the H and the Cl, forming a hydronium ion, $(H_3O)^{1+}$, and a chloride ion, Cl^{1-}.

TABLE 16.1	
Some Common Arrhenius Acids	
NAME OF ACID	FORMULA
Hydrochloric acid	HCl
Nitric acid	HNO_3
Sulfuric acid	H_2SO_4
Phosphoric acid	H_3PO_4
Acetic acid	$HC_2H_3O_2$

or, in a simpler form,

$$HCl \xrightarrow{\text{Water}} H^{1+} + Cl^{1-}$$

Table 16.1 lists some additional Arrhenius acids. When we confine ourselves to the Arrhenius definition of an acid, then, our discussion is limited to aqueous (water) solutions and to substances that liberate hydrogen ions in solution. Because of these limitations, other definitions for the term *acid* were proposed. However, the Arrhenius definition is adequate for most of our purposes.

The Brønsted–Lowry Definition

LEARNING GOAL 2
Brønsted–Lowry acids

In 1923 two chemists, J. N. Brønsted and T. M. Lowry, working independently, proposed a much broader definition of the term *acid*. They described an acid as any substance that is a proton donor. According to this definition, an **acid** is any substance that can donate a proton (an H^{1+} ion) to another substance. This definition includes all substances that qualify as acids under the Arrhenius definition and many other compounds and ions as well (Table 16.2). For example, the ammonium ion is a Brønsted–Lowry acid because

$$(NH_4)^{1+} + H_2O \rightleftharpoons NH_3 + (H_3O)^{1+}$$

or, more simply,

$$\underset{\text{Acid}}{(NH_4)^{1+}} \underset{\text{}}{\overset{\text{Water}}{\rightleftharpoons}} NH_3 + H^{1+}$$

Properties of Acids

LEARNING GOAL 3
Characteristics of acids

The acids listed in Table 16.1 have many characteristics in common:

1. *Each has a sour taste.* (The word *acid* comes from the Latin *acidus*, which means "sour.")
2. *Each can change the color of certain dyes.* For example, blue litmus dye or blue litmus paper turns red in the presence of an acid. These dyes

TABLE 16.2

Some Common Brønsted–Lowry Acids

NAME OF ACID	FORMULA
Hydrochloric acid	HCl
Nitric acid	HNO_3
Sulfuric acid	H_2SO_4
Phosphoric acid	H_3PO_4
Acetic acid	$HC_2H_3O_2$
Hydronium ion	$(H_3O)^{1+}$
Hydrogen sulfate ion	$(HSO_4)^{1-}$
Ammonium ion	$(NH_4)^{1+}$
Water	H_2O

are called *indicators* because they indicate whether a substance is an acid or a base (Figure 16.2).

3. *Each can react with certain active metals, such as zinc and magnesium, to produce hydrogen gas.* For example,

$$Zn + 2HCl \rightarrow ZnCl_2 + H_2(g)$$

$$Zn + H_2SO_4 \rightarrow ZnSO_4 + H_2(g)$$

4. *Each can neutralize bases.* (We will discuss this property later.)

One of the important characteristics of acids is the one we used in the Arrhenius definition: *They dissolve in water to produce hydrogen ions* (H^{1+}). These ions are responsible for the other properties of acids.

■ **FIGURE 16.2**

The reaction of an acid–base indicator. Hydrochloric acid turns blue litmus red.

Preparation of Acids

LEARNING GOAL 4
Preparation of acids

There are many different ways of preparing acids. One of the most important methods involves the reaction of a nonmetal oxide with water.

$$SO_3 + H_2O \rightarrow H_2SO_4$$

Sulfur   Sulfuric
trioxide   acid

$$N_2O_5 + H_2O \rightarrow 2HNO_3$$

Nitrogen(V)   Nitric
oxide   acid

The acids that result from some of these reactions are called *mineral* acids (because they are prepared from minerals). Other acids, called *organic* acids, are obtained from living things such as citrus fruits or fermented grapes or are found in animal matter.

Reactions of Acids

LEARNING GOAL 5
Chemical reactions of acids

Acids undergo many types of reactions. Two of these are important for our purposes:

1.   The reactions of acids with metals
2.   The reactions of acids with bases

We have already discussed the reaction of an acid with a metal. You learned in Chapter 10 that this type of reaction is a *single-replacement reaction* involving the replacement of hydrogen atoms in an acid by atoms of a metal (which must lie above hydrogen in the activity series, Table 10.1). For example,

$$Mg + H_2SO_4 \rightarrow MgSO_4 + H_2$$

$$Mg + 2HBr \rightarrow MgBr_2 + H_2$$

Hydrobromic
acid

We have also already discussed the reaction of an acid with a base (in Section 10.4). We'll come back to this type of reaction, sometimes called a neutralization reaction, later in this chapter when we cover the reactions of bases.

Uses of Some Common Acids

Acids are important substances with many household and industrial uses. Some acids are very toxic or can burn the skin and must be handled carefully. Let's look at some of the modern uses of acids.

Hydrochloric acid, a strong mineral acid known commercially as muriatic acid, is used industrially to clean metals that are about to be coated or plated.

(This cleaning is known as *pickling*.) Hydrochloric acid is also used to clean brick and cement.

Nitric acid, another mineral acid, is highly reactive. When a fair-skinned person happens to spill some on his or her skin, it turns the skin yellow. This is because of a reaction between the nitric acid and the protein in skin. Nitric acid is used mainly in the production of fertilizers but is also used to make dyes, plastics, and explosives.

Sulfuric acid is another very strong mineral acid that is used in fertilizers. In addition, it is used to make smokeless powder (a type of gunpowder) and nitroglycerin. Automobile batteries contain a diluted form of sulfuric acid. Such batteries work because there is a reaction between their metal plates and the acid, creating a movement of electrons and therefore an electric current. The Russians are manufacturing sulfuric acid from the oxides of sulfur that result when fossil fuels are burned to produce electricity. In other words, sulfuric acid can be made as a byproduct of electric power plants.

Phosphoric acid is a weak mineral acid. Its uses range from flavoring soft drinks to producing superphosphates for fertilizers. And the false teeth of millions of people are held in place by dental cements containing phosphoric acid.

Acetic acid is found in vinegar. According to the pure-food laws of the United States, vinegar must contain no less than 4% acetic acid. Concentrated acetic acid is used in making synthetic fibers called *acetates*. It is also used in manufacturing various plastics. (Remember, though, that acetic acid—unlike the other acids we have just discussed—is an organic acid.)

16.2 Bases

As in the case of acids three different definitions of bases have been developed over the last 100 years. Once again we will look at two useful definitions—those of Arrhenius and of Brønsted and Lowry.

The Arrhenius Definition

■ LEARNING GOAL 1
Arrhenius bases

Arrhenius's definition of an acid applies only to aqueous solutions, and the same is true of his definition of a base. A **base** is a substance that releases hydroxide ions, $(OH)^{1-}$, when that substance is dissolved in water. According to Arrhenius, NaOH is a base because it undergoes the following reaction when placed in water:

$$NaOH \xrightarrow{\text{Water}} Na^{1+} + (OH)^{1-}$$

Many of the common bases found in the laboratory are composed of metallic ions plus hydroxide ions—as is sodium hydroxide (NaOH). Table 16.3 lists some additional Arrhenius bases.

Incidentally, there is another word that people often use for a base: *alkali.* An **alkali** is any substance that dissolves to yield a basic (or alkaline) solution.

The Brønsted–Lowry Definition

■ LEARNING GOAL 2
Brønsted–Lowry bases

Just as Brønsted and Lowry defined an acid as a proton donor, they defined a base as a proton acceptor. In other words, according to Brønsted and Lowry, a **base** is any substance that accepts protons (H^{1+} ions). This important definition includes all substances that qualify as bases under the Arrhenius definition, since the OH^{1-} ion from the base accepts an H^{1+} ion from an acid to form water. Many other compounds and ions act as bases as well (Table 16.4). For example, the compound ammonia is a Brønsted–Lowry base because

$$NH_3 + H^{1+} \rightarrow (NH_4)^{1+}$$

Properties of Bases

■ LEARNING GOAL 3
Characteristics of bases

The bases listed in Table 16.3 have many characteristics in common:

1. *Each has a bitter taste.*
2. *Each can change the color of certain dyes.* For example, red litmus turns blue in the presence of a base (Figure 16.3).
3. *Each feels slippery or soapy on the fingers.*
4. *Each can neutralize acids.* (This property is discussed later.)

One of the important characteristics of bases is the one we used in the Arrhenius definition: *They dissolve in water to produce hydroxide ions,* $(OH)^{1-}$. These hydroxide ions are responsible for the other properties of bases.

TABLE 16.3

Some Common Arrhenius Bases

NAME OF BASE	FORMULA
Lithium hydroxide	LiOH
Sodium hydroxide	NaOH
Potassium hydroxide	KOH
Calcium hydroxide	$Ca(OH)_2$
Magnesium hydroxide	$Mg(OH)_2$
Iron(III) hydroxide	$Fe(OH)_3$

■ **FIGURE 16.4**

The dissociation of NaOH

How do bases dissolve in water to produce hydroxide ions? Consider, for example, solid sodium hydroxide (NaOH). Sodium hydroxide is an ionic compound made up of sodium ions and hydroxide ions, or Na^{1+} and $(OH)^{1-}$. The ions are held together by the attraction of their opposite charges. When we add sodium hydroxide to water, the negative end of each water molecule (the oxygen end) attracts the sodium ion, and the positive end (the hydrogen end) attracts the hydroxide ion. This attraction breaks the bonds between the sodium ions and the hydroxide ions (Figure 16.4). The result is that there are now free sodium ions and free hydroxide ions in solution. The reaction can be represented by the equation

$$NaOH \xrightarrow{\text{Water}} Na^{1+} + (OH)^{1-}$$

TABLE 16.4

Some Common Brønsted–Lowry Bases

NAME OF BASE	FORMULA	NAME OF BASE	FORMULA
Ammonia	NH_3	Chloride ion	Cl^{1-}
Hypochlorite ion	ClO^{1-}	Carbonate ion	$(CO_3)^{2-}$
Hypobromite ion	BrO^{1-}	Sulfate ion	$(SO_4)^{2-}$
Cyanide ion	CN^{1-}	Acetate ion	$(C_2H_3O_2)^{1-}$
Phosphate ion	$PO_4{}^{3-}$	Hydroxide ion	$(OH)^{1-}$
Hydride ion	H^{1-}	Water	H_2O

LEARNING GOAL 4

Preparation of bases

Preparation of Bases

Bases can be prepared in many ways. One major method involves the reaction of a metal oxide with water.

$$Na_2O + H_2O \rightarrow 2NaOH$$

Sodium   Sodium
oxide   hydroxide

$$CaO + H_2O \rightarrow Ca(OH)_2$$

Calcium   Calcium
oxide   hydroxide

Reactions of Bases

Bases undergo many types of reactions. Two that are important for our purposes are:

1. The reactions of bases with nonmetal oxides
2. The reactions of bases with acids

LEARNING GOAL 5
Chemical reactions of bases

When a base reacts with a nonmetal oxide, it produces a salt and water:

$$2NaOH + CO_2 \rightarrow Na_2CO_3 + H_2O$$

$$Ca(OH)_2 + CO_2 \rightarrow CaCO_3 + H_2O$$

When a base reacts with an acid, it again produces a salt and water:

$$NaOH + HCl \rightarrow NaCl + H_2O$$

$$2NaOH + H_2SO_4 \rightarrow Na_2SO_4 + 2H_2O$$

$$3Ca(OH)_2 + 2H_3PO_4 \rightarrow Ca_3(PO_4)_2 + 6H_2O$$

These latter reactions are sometimes called *neutralization reactions*, because when we slowly add an acid to a base, the H^{1+} and $(OH)^{1-}$ ions neutralize each other by forming water.

Uses of Some Common Bases

Like acids, bases play an important role in our daily lives, and they are just as dangerous to our skin and clothes. They are used in our homes, in industrial processes, and in manufacturing drugs and medicines.

Sodium hydroxide (NaOH) is a base used in drain cleaners, or what we call lye. It is also used to make soaps and cellophane. The petroleum industry uses sodium hydroxide to neutralize acids in the refining process.

Calcium hydroxide, $Ca(OH)_2$, sometimes called slaked lime, is important in the construction industry. It is used in mortar, plaster, cement, and many other building and paving materials.

Ammonia (NH_3) dissolved in water is the household cleaner we know as ammonia water, and it has an intense, almost suffocating odor. Ammonia is the chief ingredient of smelling salts. It is excellent for removing stains, and it is also used in manufacturing textiles (such as rayon), plastics, and fertilizers.

Magnesium hydroxide, $Mg(OH)_2$, dispersed in water, commonly called milk of magnesia, has many uses as a medicine. In small doses (a few hundred milligrams), magnesium hydroxide acts as an antacid (bases counteract acid just as reliably in the stomach as in the laboratory). In large doses (2 to 4 grams), it acts as a laxative.

16.3 The Strengths of Acids and Bases

LEARNING GOAL 6
Strong acids and bases

Acids and bases can both be subdivided into two categories: (1) *strong* acids and *strong* bases and (2) *weak* acids and *weak* bases. The **strong acids and bases** are those that ionize completely in dilute solution. For example, hydrochloric acid is a strong acid. One mole of hydrogen chloride gas dissolved in 1 L of water contains essentially 1 mole of hydrogen ions and 1 mole of chloride ions. There are no hydrogen chloride molecules to speak of. Table 16.5 gives some examples of strong acids and bases.

LEARNING GOAL 7
Weak acids and bases

Weak acids and bases are those that do not ionize completely. Acetic acid is a weak acid. If you were to analyze a solution containing 1 mole of acetic acid ($HC_2H_3O_2$) in 1 L of solution, you would find that most of the acetic acid stays in the form of molecules. Only about 0.4% ionizes to form hydrogen ions and acetate ions. We can represent the dissociation of this weak acid in the following manner:

$$HC_2H_3O_2 \underset{\text{Water}}{\rightleftharpoons} H^{1+} + (C_2H_3O_2)^{1-}$$

The *double arrow* indicates that in a given solution of acetic acid, an *equilibrium* (or balance) exists among the acetic acid molecules, the hydrogen ions, and the acetate ions. This equilibrium lies in the direction of (provides more of) the undissociated $HC_2H_3O_2$. That's why the arrow that points to the left is longer than the arrow that points to the right. (If the situation were the other way around, the arrow pointing to the right would be longer.) In weak acids and weak bases, the equilibrium concentration of the undissociated species is favored. Table 16.6 offers some examples of weak acids and bases. (Also see Figure 16.5.)

TABLE 16.5

Strong Acids and Bases

	NAME	FORMULA
Acids	Hydrochloric acid	HCl
	Sulfuric acid	H_2SO_4
	Nitric acid	HNO_3
Bases	Sodium hydroxide	NaOH
	Potassium hydroxide	KOH

TABLE 16.6

Weak Acids and Bases

	NAME	FORMULA
Acids	Acetic acid	$HC_2H_3O_2$
	Carbonic acid	H_2CO_3
	Boric acid	H_3BO_3
Bases	Ammonia (in water)	NH_3
	Aluminum hydroxide	$Al(OH)_3$

Since they produce ions in aqueous solution, acids and bases are electrolytes. Strong acids and strong bases are considered to be strong electrolytes, because they are about 100% dissociated into ions when placed in aqueous solution. Weak acids and weak bases are considered to be weak electrolytes, because they dissociate slightly into ions when placed in aqueous solution. The strong electrolytes contain mostly ions in aqueous solution, whereas the weak electrolytes contain mostly covalently bonded molecules in equilibrium with a small number of ions in aqueous solution.

16.4   Weak Acids and Bases and Dynamic Equilibrium

LEARNING GOAL 8
Dynamic equilibrium

The equilibrium that exists in a solution of a weak acid or weak base is not a static (unchanging) equilibrium but rather a *dynamic* equilibrium. That is, some molecules are always dissociating to form ions, and some ions are always recombining to form molecules. But these dissociations and recombinations take place at the same rate, so the proportion of molecules and ions remains constant—they remain in equilibrium.

As an analogy, consider a popular restaurant during the busy lunch hour. The restaurant can seat 200 people, and it is full at noon. But some people are leaving the restaurant and others are entering. A quarter of an hour

■ **FIGURE 16.5**

One difference between a weak acid and a strong acid

No reaction visible

Vigorous reaction
$Zn + H_2SO_4 \rightarrow ZnSO_4 + H_2(g)$

later, the situation is the same: 200 people in the restaurant, some people entering and others leaving. And again, 15 minutes later, the same situation prevails. During the lunch hour, those who enter the restaurant take the places of those who have eaten and left, maintaining a population equilibrium of 200 diners. But the equilibrium is dynamic (changing), because individual diners are constantly being replaced by other diners.

A chemical equilibrium is dynamic in the same sense. For example, in a solution of acetic acid, there are some hydrogen ions (H^{1+}), some acetate ions ($C_2H_3O_2$), and many acetic acid molecules ($HC_2H_3O_2$). But at any given time, some hydrogen ions are combining with acetate ions to form acetic acid molecules. And some acetic acid molecules are ionizing to form hydrogen ions and acetate ions. As long as these two processes go on at the same rate, the concentrations of the various substances don't change, and a dynamic equilibrium is maintained.

Chemists use various methods to measure the extent of dissociation of acids and bases. There are also quantitative ways of describing the extent of dissociation.

16.5   The Ionization of Water

Like weak acids and bases, water has a tendency to ionize. One way to represent the ionization of water is as follows:

LEARNING GOAL 9
Ionization of water

$$H\text{---}OH + H_2O \rightleftharpoons (H_3O)^{1+} + (OH)^{1-} \qquad \text{Equation 16.1}$$
$$\text{Hydronium} \quad \text{Hydroxide}$$
$$\text{ion} \qquad\quad \text{ion}$$

Here the formula for the first water molecule is written as $H\text{---}OH$ so that we can see how the ionization occurs. Equation 16.1 shows that a hydronium ion and a hydroxide ion are formed.

We can also represent the ionization of water in this simpler way:

$$H_2O \rightleftharpoons H^{1+} + (OH)^{1-} \qquad \text{Equation 16.2}$$

This equation shows that a hydrogen ion and a hydroxide ion are formed. Even though Equation 16.1 is more accurate (technically), Equation 16.2 is easier to use. However, as the double arrow indicates, the ionization of water occurs only to a slight extent. In fact, in exactly 1 L of water, which is 55.6 moles of water, only 0.0000001 mole (which can be written as 10^{-7} mole) actually dissociates into hydrogen and hydroxide ions. In other words, there are 10^{-7} mole of hydrogen ions and 10^{-7} mole of hydroxide ions present in 1 L of water. Because there are equal amounts of hydrogen ions and hydroxide ions in pure water, it is said to be neutral.

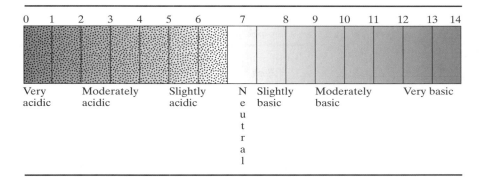

■ **FIGURE 16.6**
The pH scale

16.6   The pH Scale

LEARNING GOAL 10
pH and pH scale

The pH scale is used to measure and express the acidity or basicity (acid or base strength) of a solution. The **pH** of a solution is *equal to the negative of the logarithm of its hydrogen-ion concentration in moles per liter.* This equation is written as

$$pH = -\log [H^{1+}]$$

where the brackets, [ ], mean *concentration in moles of H^{1+} per liter of solution.* The pH scale runs from 0 to 14 (Figure 16.6).

Recall from our discussion of the ionization of water that the concentration of hydrogen ions in pure water is 10^{-7} mole per liter. It happens that this concentration corresponds to a pH of 7. Moreover, we noted that pure water is neutral (neither acidic nor basic). Therefore any solution with a pH of 7 is considered neutral.

A solution that has a pH between 0 and 2 is considered strongly acidic. A solution that has a pH between 2 and 5 is moderately acidic, and one with a pH between 5 and 7 is slightly acidic. A solution with a pH of 7 is, as we saw, neutral. A solution that has a pH between 7 and 9 is slightly basic, one with a pH between 9 and 11 is moderately basic, and one with a pH between 12 and 14 is strongly basic. Table 16.7 lists the pH values of some common substances.

We can determine the pH of a solution by many methods. One simple method is to place a few drops of a chemical dye or a piece of paper tape that contains the dye into the solution. The dye or paper turns a color that depends on the pH of the solution. There is also a device called a pH meter, which directly reads the pH of a solution.

It is easy to find the pH of a solution when its hydrogen-ion concentration (in moles per liter) is given as a power of 10. For example, for pure water, with a hydrogen-ion concentration of 10^{-7} mole/liter,

$$pH = -\log [H^{1+}] = -\log 10^{-7} = -(-7) = 7$$

TABLE 16.7

The pH Values of Some Common Substances

SUBSTANCE	pH	SUBSTANCE	pH
1 M HCl	0.0	Rainwater	6.2
0.5 M HNO$_3$	0.3	Milk	6.5
Lemons	2.3	Pure water	7.0
Vinegar	2.8	Sea water	8.5
Soft drinks	3.0	Milk of magnesia	10.0
Oranges	3.5	0.1 M NH$_4$OH	11.1
Tomatoes	4.2	1 M NaOH	14.0

LEARNING GOAL 11

Determining pH from hydrogen-ion or hydroxide-ion concentration

Note that the logarithm of a power of 10 is simply that power of 10. For example, log $10^{-2} = -2$. (*Note:* pH values don't have to be whole numbers. However, in this text we will deal only with whole-number pH values and with concentrations to the nearest power of 10.)

EXAMPLE 16.1

Determine the pH of each of the following solutions. Then use Figure 16.6 to determine how acidic or basic the solution is.
(a) a solution whose [H^{1+}] is 10^{-3}
(b) a solution whose [H^{1+}] is 10^{-13}
(c) a solution whose [H^{1+}] is 10^{0}

Solution

pH = $-\log$ [H^{1+}]. Therefore,
(a) pH = $-\log 10^{-3} = -(-3) = 3$, which is moderately acidic.
(b) pH = $-\log 10^{-13} = -(-13) = 13$, which is very basic.
(c) pH = $-\log 10^{0} = 0$, which is very acidic.

Practice Exercise 16.1   Determine the pH of each of the following solutions. Then use Figure 16.6 to determine how acidic or basic the solution is.
(a) a solution whose [H^{1+}] is 10^{-5}
(b) a solution whose [H^{1+}] is 10^{-10}
(c) a solution whose [H^{1+}] is 10^{-1}

16.7 Relationship Between the Hydrogen-Ion and Hydroxide-Ion Concentrations

■ LEARNING GOAL 12
Relationship between hydrogen-ion and hydroxide-ion concentrations

There is an interesting relationship between the hydrogen-ion concentration and the hydroxide-ion concentration in aqueous solutions: The product of $[H^{1+}]$ and $[OH^{1-}]$ is always equal to 10^{-14}. In other words,

$$[H^{1+}][OH^{1-}] = 10^{-14}$$

Remember that pure (neutral) water has an H^{1+} concentration of 10^{-7} *M* and an $(OH)^{1-}$ concentration of 10^{-7} *M*. The product of these two numbers is 10^{-14} M^2:

$$[H^{1+}][OH^{1-}] = (10^{-7} \ M)(10^{-7} \ M) = 10^{-14} \ M^2$$

The value 10^{-14} is the *dissociation constant* of water. It is usually given the symbol K_w. Therefore,

$$[H^{1+}][OH^{1-}] = K_w = 10^{-14}$$

(We generally don't bother writing the M^2 after 10^{-14}, but remember that it is understood to be there all the same.) A solution that has a hydrogen-ion concentration of 10^{-5} *M* has to have a hydroxide-ion concentration of 10^{-9} *M*, because

$$(10^{-5} \ M)\,[OH^{1-}] = 10^{-14} \ M^2 \quad \text{so} \quad [OH^{1-}] = \frac{10^{-14} \ \cancel{M^2}}{10^{-5} \ \cancel{M}} = 10^{-9} \ M$$

Table 16.8 summarizes the relationships among hydrogen-ion concentration, hydroxide-ion concentration, and pH.

EXAMPLE 16.2

Determine the pH of each of the following strong electrolytes:
(a) a 0.001 *M* HCl solution
(b) a 0.01 *M* HNO_3 solution
(c) a solution whose $[OH^{1-}]$ is 10^{-3}
(d) a solution whose $[OH^{1-}]$ is 10^{-11}
(e) a 0.0001 *M* NaOH solution

Solution
(a) 0.001 *M* HCl solution releases 0.001 mole of hydrogen ions per liter of solution, so its $[H^{1+}]$ is 10^{-3}. Therefore its pH is 3.

$$pH = -\log [H^{1+}] = -\log 10^{-3} = 3$$

TABLE 16.8

The Relationship Among Hydrogen-Ion Concentration, Hydroxide-Ion Concentration, and pH

$[H^{1+}]$	pH	$[OH^{1-}]$
$10^0 = 1$	0	10^{-14}
$10^{-1} = 0.1$	1	10^{-13}
$10^{-2} = 0.01$	2	10^{-12}
$10^{-3} = 0.001$	3	10^{-11}
$10^{-4} = 0.0001$	4	10^{-10}
$10^{-5} = 0.00001$	5	10^{-9}
$10^{-6} = 0.000001$	6	10^{-8}
$10^{-7} = 0.0000001$	7	10^{-7}
$10^{-8} = 0.00000001$	8	10^{-6}
$10^{-9} = 0.000000001$	9	10^{-5}
$10^{-10} = 0.0000000001$	10	10^{-4}
$10^{-11} = 0.00000000001$	11	10^{-3}
$10^{-12} = 0.000000000001$	12	10^{-2}
$10^{-13} = 0.0000000000001$	13	10^{-1}
$10^{-14} = 0.00000000000001$	14	10^0

(b) A 0.01 *M* HNO_3 solution releases 0.01 mole of hydrogen ions per liter of solution, so its $[H^{1+}]$ is 10^{-2}. Therefore its pH is 2.

$$pH = -\log [H^{1+}] = -\log 10^{-2} = 2$$

(c) A solution whose $[OH^{1-}]$ is 10^{-3} has $[H^{1+}] = 10^{-11}$ (see Table 16.8). Therefore its pH is 11.

$$pH = -\log [H^{1+}] = -\log 10^{-11} = 11$$

(d) A solution whose $[OH^{1-}]$ is 10^{-11} has $[H^{1+}] = 10^{-3}$ (see Table 16.8). Therefore its pH is 3.

$$pH = -\log [H^{1+}] = -\log 10^{-3} = 3$$

(e) A 0.0001 *M* NaOH solution releases 0.0001 mole of hydroxide ions per liter of solution, so its $[OH^{1-}]$ is 10^{-4}. A solution whose $[OH^{1-}]$ is 10^{-4} has $[H^{1+}] = 10^{-10}$ (see Table 16.8). Therefore its pH is 10:

$$pH = -\log [H^{1+}] = -\log 10^{-10} = 10$$

Practice Exercise 16.2 Determine the pH of each of the following solutions:

(a) a 0.0001 M HCl solution

(b) a 0.001 M HCl solution

(c) a solution whose $[OH^{1-}]$ is 10^{-5}

(d) a solution whose $[OH^{1-}]$ is 10^{-11}

(e) a 0.10 M KOH solution

16.8 Salts

Definition of Salt

We have seen that acids and bases react with each other in what are called neutralization reactions. The product of these reactions is a salt. Water is also produced if a hydroxide ion is in the formula of the base.

$$HCl + NaOH \rightarrow \underset{\substack{\text{Sodium} \\ \text{chloride}}}{NaCl} + H_2O$$

$$HC_2H_3O_2 + NaOH \rightarrow \underset{\substack{\text{Sodium} \\ \text{acetate}}}{NaC_2H_3O_2} + H_2O$$

$$HCl + NH_3 \rightarrow \underset{\substack{\text{Ammonium} \\ \text{chloride}}}{NH_4Cl}$$

LEARNING GOAL 13
Definition of a salt

In these reactions, the hydrogen ions from the acid combine with the hydroxde ions from the base to produce water. At the same time, the positive ions of the base combine with the negative ions of the acid to produce a compound that we call a *salt* (Figure 16.7). A **salt** is a compound composed of the positive ion of a base and the negative ion of an acid.

FIGURE 16.7

A neutralization reaction

$$HCl + NaOH \longrightarrow NaCl + H_2O$$

Most salts are composed of a metal ion combined with a nonmetal (or polyatomic) ion that has a negative charge. Examples of such salts are sodium chloride ($NaCl$), which is table salt, silver bromide ($AgBr$), potassium sulfate (K_2SO_4), and iron(III) phosphate ($FePO_4$). Some salts are composed of a polyatomic ion that has a positive charge combined with a nonmetal (or polyatomic) ion that has a negative charge. Examples of such salts are ammonium chloride (NH_4Cl) and ammonium nitrate (NH_4NO_3).

Properties, Preparations, and Some Uses of Salts

LEARNING GOAL 14
Important properties of salts

Different salts have different properties, but *with few exceptions salts are ionic substances* (which means that they are bonded ionically). Their solubilities in water may vary, but their aqueous solutions always contain ions. For example, sodium chloride dissociates into sodium ions and chloride ions when it is dissolved in water. Most salts are strong electrolytes.

$$NaCl \xrightarrow{\text{Water}} Na^{1+} + Cl^{1-}$$

There are many ways to prepare salts. However, we'll limit our discussion to three methods: (1) direct combination of two elements, (2) reaction of an active metal with an acid, and (3) neutralization of an acid by a base.

1. When we heat sodium metal in the presence of an excess of chlorine gas, a vigorous reaction takes place. The product is pure sodium chloride. Many salts can be prepared by heating elements together.

$$2Na + Cl_2 \rightarrow 2NaCl$$

2. In Chapter 10, we noted that hydrogen could be produced by the reaction of an active metal with an acid. The other product formed in such a reaction is a salt. We can use this type of reaction to produce the salt we want by choosing the appropriate metal and acid. If we want to prepare magnesium chloride, for example, we use magnesium metal and hydrochloric acid.

$$Mg + 2HCl \rightarrow MgCl_2 + H_2$$

An important fact to keep in mind is that the metal we choose must be able to replace the hydrogen in the acid. Table 10.1, the activity series, reveals which metals can replace hydrogen.

3. To form a salt by a neutralization reaction, we choose a base that contains the positive ion of the salt and an acid that contains the negative ion of the salt. For example, if we want to prepare potassium bromide, we can use the reaction between potassium hydroxide and hydrobromic acid:

$$KOH + HBr \rightarrow KBr + H_2O$$

Salts have many different uses. Many salts, such as sodium chloride, are necessary for life itself. Calcium phosphate is the main ingredient of our bones. Other salts, such as calcium phosphate dihydrate $[Ca_3(PO_4)_2 \cdot 2H_2O]$, are used for building materials. The variety of uses for salts is matched only by the number of salts known.

16.9   Titration

When explaining double-replacement reactions in Section 10.4, we discussed neutralization as one kind of double-replacement reaction. *Neutralization* is the reaction between an acid and a base to form water and a salt. A neutralization reaction is often represented by the general equation

$$HX + MOH \rightarrow MX + H_2O$$

where HX (an acid) and MOH (a base) form a salt (MX) and water.

The most common use that chemists make of neutralization reactions is called titration. **Titration** is *a volumetric analytical procedure that enables*

■ FIGURE 16.8

A titration setup. A measured volume of one solution is placed in the flask. The buret above it contains another solution. The second solution is added gradually until the end point is reached. The stopcock controls the rate of flow of the solutions into the flask.

the chemist to find out how much of a solute is present in solution. In the laboratory, titration is sometimes used to find the concentrations of chemicals in urine, blood, plasma, and drinking water. When a neutralization titration is performed, the amount of acid or base can be determined.

In an acid–base titration, a measured volume of a solution of acid or base of known concentration (called the titrant) is added to a measured volume of a solution of acid or base whose concentration is to be determined. A device called a buret is used for dispensing the measured quantities of acid or base (Figure 16.8). The buret enables the chemist to stop adding titrant by means of a stopcock at the bottom of the buret. The stopcock can be closed to stop the flow of solution at any time. Calibration marks etched onto the surface of the buret reveal exactly how much titrant has been added. The titrant used is generally a *standard solution* of acid or base, which is a solution whose concentration is accurately known.

How does the chemist know when enough titrant has been added? In other words, when is it evident that the neutralization is complete? *When equal numbers of equivalents of acid and base have been added to the reaction vessel (which is usually an Erlenmeyer flask), the* **equivalence point** *has been reached.* At this point, the volume and concentration of the titrant that has been added can be used to calculate the concentration of the acid or base being titrated.

In order to determine whether the equivalence point has been reached, the chemist must know (1) the pH of a solution of the salt formed by complete neutralization of the particular acid and base involved in the titration and (2) its concentration. In general, the pH at the equivalence point varies with the salt formed by the particular acid and base used in the neutralization reaction and with the concentration of the salt formed.

For example, if we are neutralizing a solution of a strong base, such as sodium hydroxide, with a standard solution of a strong acid, such as nitric acid, we begin with a solution of sodium hydroxide of unknown concentration and quite high pH. As the nitric acid is added, the pH decreases, because $(OH)^{1-}$ is removed by the reaction $H^{1+} + (OH)^{1-} \rightarrow H_2O$. When the equivalence point is reached, the numbers of equivalents of acid and base are the same and there are equal amounts of H^{1+} and $(OH)^{1-}$. The pH at the equivalence point should be 7, because when $[H^{1+}] = [OH^{1-}]$, each value is 10^{-7} M. When a strong acid and a strong base take part in a titration reaction, the equivalence point is reached when the pH reaches 7. At that point, neutralization is complete and no more titrant should be added. The salt sodium nitrate is neutral.

What happens when a strong acid and a weak base take part in a titration? At the equivalence point, when the numbers of equivalents of acid and base are the same, is the pH equal to 7? No, it is not, because when the salt of a strong acid and a weak base forms, it further reacts with water to produce a solution with a pH less than 7. For example, a solution of NH_4Cl is slightly acidic.

Similarly, when a strong base and a weak acid take part in a titration, at the equivalence point the numbers of equivalents of acid and base are the same, but the solution is slightly basic. When the salt of a strong base and a weak acid forms, it further reacts with water to produce a solution with a pH greater than 7. Sodium acetate is an example of a salt that gives a slightly basic solution.

Chemicals called indicators can be carefully selected to show the point at which all the available hydrogen ions have reacted with all the available hydroxide ions in a solution. One must, however, take into account the pH of the solution of salt formed.

16.10   **Indicators**

An **indicator** is *a dye whose color depends on its pH.* Paper saturated with an indicator is called indicator paper. Litmus is the most common indicator. It turns blue when the solution in which it is placed is basic; it turns red when the solution is acidic. *In general, indicators change color within a certain pH range, and the range of pH values within which the color change takes*

TABLE 16.9

Common Indicators

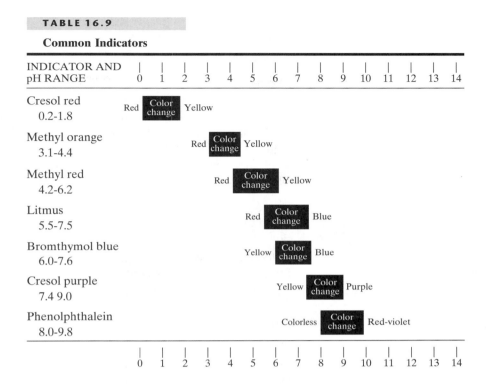

INDICATOR AND pH RANGE															
	0	1	2	3	4	5	6	7	8	9	10	11	12	13	14

Cresol red 0.2-1.8 — Red / Color change / Yellow

Methyl orange 3.1-4.4 — Red / Color change / Yellow

Methyl red 4.2-6.2 — Red / Color change / Yellow

Litmus 5.5-7.5 — Red / Color change / Blue

Bromthymol blue 6.0-7.6 — Yellow / Color change / Blue

Cresol purple 7.4 9.0 — Yellow / Color change / Purple

Phenolphthalein 8.0-9.8 — Colorless / Color change / Red-violet

place depends on the particular indicator. For example, litmus changes from red to blue within the pH range of 5.5 to 7.5. It is blue above 7.5 and red below 5.5. Another common indicator, phenolphthalein, is colorless in acid solution and red-violet in alkaline solution. It changes color within the pH range of 8.0 to 9.8. Table 16.9 lists common indicators and the pH ranges within which their color changes take place.

When an indicator is added to solution that is being titrated and the indicator just changes color, what we call the **end point** has been reached. To do an acid–base titration, chemists estimate the pH at the equivalence point, select an indicator that changes color in the appropriate pH range, and add a few drops of that indicator to the solution being titrated.

16.11   The Mathematics of Acid–Base Titrations

In the laboratory, units of normality are used to simplify calculations. Equal numbers of equivalents of acid and base react with each other. That is,

number of equivalents of acid = number of equivalents of base

Remember, normality (N) is the number of equivalents of acid or base present in 1 liter of solution, so

$$\text{number of equivalents} = \text{normality } (N) \times \text{volume (liters)}$$

At the equivalence point,

$$N_a \times V_a = N_b \times V_b$$

where N_a is the concentration of the acid, V_a is the volume of the acid, N_b is the concentration of the base, and V_b is the volume of the base. Liters do not have to be used as the unit of volume. Any unit may be used as long as it is the same for both V_a and V_b. This equation is often used in calculations involving titrations.

EXAMPLE 16.3

Calculate the normality (or molarity; see the alternative solution that follows) of an HNO_3 solution if 25.0 mL of the solution are neutralized by 100.0 mL of 0.200 N (0.100 M) $Ca(OH)_2$ solution.

Solution

UNDERSTAND THE PROBLEM

We are asked to calculate the normality of an acidic solution, a given volume (25.0 mL) of which is neutralized by 100.0 mL of 0.200 N basic solution. Organizing the data will help you see what is given and what must be found. Let N_a refer to the normality of the acid and N_b to the normality of the base.

$$N_a = ? \qquad N_b = 0.200 \ N$$

$$V_a = 25.0 \text{ mL} \qquad V_b = 100.0 \text{ mL}$$

DEVISE A PLAN

We can use the equation given in the text to solve for the unknown, N_a.

CARRY OUT THE PLAN

Rewrite the formula

$$N_a \times V_a = N_b \times V_b$$

in terms of N_a:

$$N_a = \frac{N_b V_b}{V_a}$$

Then solve the equation.

$$N_a = \frac{(0.200 \ N)(100.0 \ \text{mL})}{25.0 \ \text{mL}} = 0.800 \ N$$

The concentration of the acid is 0.800 N.

LOOK BACK

We can reason that since 25.0 mL of acidic solution is neutralized by 100.0 mL of 0.200 N basic solution, we can expect a higher concentration for the acid. In fact, we see that its concentration is four times greater than that of the base.

Alternative Solution Using Molarity

You may also perform titration calculations by using only molarity. Here's how it's done. First calculate how many moles of $Ca(OH)_2$ solution it took to neutralize the HNO_3 solution.

$$M_{Ca(OH)_2} = 0.100 \frac{\text{mole}}{\text{L}} \qquad V_{Ca(OH)_2} = 100.0 \text{ mL, or } 0.1000 \text{ L}$$

$$M = \frac{\text{moles}}{\text{liter}} \qquad \text{or} \qquad M = \frac{n}{V}$$

$$\text{moles} = (M)(\text{liter}) \qquad \text{or} \qquad n = MV$$

$$? \text{ moles} = \left(0.100 \frac{\text{mole}}{\cancel{\text{L}}}\right)(0.1000 \cancel{\text{L}})$$

$$= 0.0100 \text{ mole } Ca(OH)_2$$

Next write the balanced equation for the reaction between the HNO_3 and the $Ca(OH)_2$.

$$2HNO_3 + Ca(OH)_2 \rightarrow Ca(NO_3)_2 + 2H_2O$$

The balanced equation tells us that 2 moles of HNO_3 are neutralized by 1 mole of $Ca(OH)_2$. Use this information and the factor-unit method to calculate the moles of HNO_3 neutralized by the 0.0100 mole of $Ca(OH)_2$.

$$? \text{ moles } HNO_3 = (0.0100 \text{ mole } \cancel{Ca(OH)_2})\left(\frac{2 \text{ moles } HNO_3}{1 \text{ mole } \cancel{Ca(OH)_2}}\right)$$

$$= 0.0200 \text{ mole } HNO_3$$

Finally, calculate the molarity of the HNO_3 solution from this information and the volume of the HNO_3 (25.0 mL, or 0.0250 L).

$$M = \frac{\text{moles}}{\text{liter}} \qquad \text{or} \qquad M = \frac{n}{V}$$

$$? M = \frac{0.0200 \text{ mole}}{0.0250 \text{ L}} = 0.800 \ M$$

Practice Exercise 16.3   Calculate the normality (or molarity) of an HCl solution if 150 mL of the solution are neutralized by $5\bar{0}$ mL of 1.0 N (1.0 M) NaOH solution.

EXAMPLE 16.4

A quality-control laboratory is analyzing a sample of vinegar to determine its acetic acid content. The technicians find that it requires 14.0 mL of 1.25 N (1.25 M) NaOH solution to neutralize a 25.0-mL sample of vinegar. What is the normality (or molarity) of the vinegar solution?

Solution

Organize the data.

$$N_a = ? \qquad\qquad N_b = 1.25 \ N$$

$$V_a = 25.0 \text{ mL} \qquad V_b = 14.0 \text{ mL}$$

Substitute the values into the equation and solve it for N_a.

$$N_a \times V_a = N_b \times V_b$$

$$N_a = \frac{N_b V_b}{V_a} = \frac{(1.25 \ N)(14.0 \ \cancel{\text{mL}})}{25.0 \ \cancel{\text{mL}}} = 0.700 \ N$$

Alternative Solution Using Molarity

First calculate how many moles of NaOH it took to neutralize the vinegar.

$$M_{\text{NaOH}} = 1.25 \ \frac{\text{moles}}{\text{L}} \qquad V_{\text{NaOH}} = 14.0 \text{ mL, or } 0.0140 \text{ L}$$

$$M = \frac{\text{moles}}{\text{liter}} \qquad \text{or} \qquad M = \frac{n}{V}$$

$$\text{moles} = (M)(\text{liter}) \qquad \text{or} \qquad n = MV$$

$$? \text{ moles} = \left(1.25 \ \frac{\text{moles}}{\cancel{\text{L}}}\right)(0.0140 \ \cancel{\text{L}})$$

$$= 0.0175 \text{ mole NaOH}$$

Next write the balanced equation for the reaction between $HC_2H_3O_2$ and the NaOH.

$$HC_2H_3O_2 + NaOH \rightarrow NaC_2H_3O_2 + H_2O$$

The balanced equation reveals that 1 mole of $HC_2H_3O_2$ is neutralized by 1 mole of NaOH. Use this information and the factor-unit method to calculate the moles of $HC_2H_3O_2$ neutralized by the 0.0175 mole of NaOH.

$$? \text{ moles } HC_2H_3O_2 = (0.0175 \text{ mole } \cancel{\text{NaOH}}) \left(\frac{1 \text{ mole } HC_2H_3O_2}{1 \text{ mole } \cancel{\text{NaOH}}}\right)$$

$$= 0.0175 \text{ mole } HC_2H_3O_2$$

Finally, calculate the molarity of the vinegar from this information and the volume of the vinegar solution (25.0 mL, or 0.0250 L).

$$M = \frac{\text{moles}}{\text{liter}}$$

$$? M = \frac{0.0175 \text{ mole}}{0.0250 \text{ L}} = 0.700 \text{ } M$$

Practice Exercise 16.4   Calculate the normality (or molarity) of a solution of acetic acid if 43.50 mL of vinegar are neutralized by 25.00 mL of 0.15 N (0.15 M) NaOH.

EXAMPLE 16.5

A 50.00-mL sample of $Mg(OH)_2$ is titrated with 65.00 mL of 0.500 N (0.500 M) HCl. What is the normality (or molarity) of the $Mg(OH)_2$ solution?

Solution

Organize the data.

$$N_a = 0.500 \text{ } N \qquad N_b = ?$$
$$V_a = 65.00 \text{ mL} \qquad V_b = 50.00 \text{ mL}$$

Substitute the values into the equation and solve it for N_b.

$$N_a \times V_a = N_b \times V_b$$

$$N_b = \frac{N_a V_a}{V_b} = \frac{(0.500 \text{ } N)(65.00 \text{ mL})}{50.00 \text{ mL}} = 0.650 \text{ } N$$

Alternative Solution Using Molarity

First calculate how many moles of HCl it took to neutralize the $Mg(OH)_2$ solution.

$$M_{\text{HCl}} = 0.500 \frac{\text{mole}}{\text{L}} \qquad V_{\text{HCl}} = 65.00 \text{ mL, or } 0.06500 \text{ L}$$

$$M = \frac{\text{moles}}{\text{liter}}$$

$$\text{moles} = (M)(\text{liter})$$

$$? \text{ moles} = \left(0.500 \frac{\text{mole}}{\text{L}}\right)(0.06500 \text{ L})$$

$$= 0.0325 \text{ mole HCl}$$

Next write the balanced equation for the reaction between the HCl and the $Mg(OH)_2$.

$$2HCl + Mg(OH)_2 \rightarrow MgCl_2 + 2H_2O$$

The balanced equation reveals that 2 moles of HCl are neutralized by 1 mole of $Mg(OH)_2$. Use this information and the factor-unit method to calculate the number of moles of $Mg(OH)_2$ neutralized by the 0.0325 mole of HCl.

$$? \text{ moles Mg(OH)}_2 = (0.0325 \cancel{\text{ mole HCl}})\left(\frac{1 \text{ mole Mg(OH)}_2}{2 \cancel{\text{ moles HCL}}}\right)$$

$$= 0.0163 \text{ mole Mg(OH)}_2$$

Finally, calculate the molarity of the $Mg(OH)_2$ solution from this information and the volume of the $Mg(OH)_2$ solution (50.00 mL, or 0.05000 L).

$$M = \frac{\text{moles}}{\text{liter}}$$

$$? M = \frac{0.0163 \text{ mole}}{0.05000 \text{ L}} = 0.326 \text{ } M$$

Practice Exercise 16.5   A 25.65-mL sample of NaOH solution is titrated with 100.0 mL of 0.105 N (0.105 M) hydrochloric acid solution. What is the normality (or molarity) of the NaOH solution?

EXAMPLE 16.6

A chemistry student was determining the normality of an $Fe(OH)_3$ solution. It required 30.00 mL of a 0.200 N (0.200 M) HCl solution to neutralize 10.00 mL of the $Fe(OH)_3$ solution. Calculate the normality (or molarity) of the $Fe(OH)_3$ solution.

Solution
Organize the data.

$$N_a = 0.200 \text{ } N \qquad N_b = ?$$
$$V_a = 30.00 \text{ mL} \qquad V_b = 10.00 \text{ mL}$$

Substitute the values into the equation and solve it for N_b.

$$N_a \times V_a = N_b \times V_b$$

$$N_b = \frac{N_a V_a}{V_b} = \frac{(0.200 \text{ } N)(30.00 \cancel{\text{ mL}})}{10.00 \cancel{\text{ mL}}} = 0.600 \text{ } N$$

Alternative Solution Using Molarity
First calculate how many moles of HCl it took to neutralize the $Fe(OH)_3$ solution.

$$M_{HCl} = 0.200 \frac{\text{mole}}{\text{L}} \qquad V_{HCl} = 30.00 \text{ mL, or } 0.03000 \text{ L}$$

$$M = \frac{\text{moles}}{\text{liter}}$$

$$moles = (M)(liter)$$

$$? \, moles = \left(0.200 \, \frac{mole}{\cancel{L}}\right)(0.03000 \, \cancel{L})$$

$$= 0.00600 \, mole \, HCl$$

Next write the balanced equation for the reaction between HCl and $Fe(OH)_3$.

$$3HCl + Fe(OH)_3 \rightarrow FeCl_3 + 3H_2O$$

The balanced equation reveals that 3 moles of HCl are neutralized by 1 mole of $Fe(OH)_3$. Use this information and the factor-unit method to calculate the number of moles of $Fe(OH)_3$ neutralized by the 0.00600 mole of HCl.

$$? \, moles \, Fe(OH)_3 = (0.00600 \, \cancel{mole \, HCl})\left(\frac{1 \, mole \, Fe(OH)_3}{3 \, \cancel{moles \, HCL}}\right)$$

$$= 0.00200 \, mole \, Fe(OH)_3$$

Finally, calculate the molarity of the $Fe(OH)_3$ solution from this information and the volume of the $Fe(OH)_3$ solution (10.00 mL, or 0.01000 L).

$$M = \frac{moles}{liter}$$

$$? \, M = \frac{0.00200 \, mole}{0.01000 \, L} = 0.200 \, M$$

Practice Exercise 16.6 In a titration, it took 20.4 mL of 0.1330 N (0.0665 M) sulfuric acid solution to neutralize 26.5 mL of potassium hydroxide solution of unknown concentration. Calculate the normality (or molarity) of the potassium hydroxide solution.

SUMMARY

Arrhenius defined an acid as a substance that releases hydrogen ions, H^{1+}, when dissolved in water. Brønsted and Lowry defined an acid as a proton or H^{1+} donor. Some common Arrhenius acids are HCl, HNO_3, H_2SO_4, and $HC_2H_3O_2$. These are also Brønsted–Lowry acids, as are such substances as $(H_3O)^{1+}$, $(HSO_4)^{1-}$, $(NH_4)^{1+}$, and H_2O. Generally, acids taste sour, change the color of certain dyes, react with certain metals to produce hydrogen gas, and neutralize bases. Most important, they dissolve in water to increase the hydrogen-ion concentration. Acids are commonly prepared by the reaction of a nonmetal oxide with water. Acids react with metals as well as with bases. They are used in the manufacturing of a variety of products.

Bases, like acids, can be defined in more than one way. Arrhenius defined a base as a substance that releases hydroxide ions, $(OH)^{1-}$, when dissolved in water. Brønsted and Lowry defined a base as a proton or H^{1+} acceptor. Some common Arrhenius bases are LiOH, KOH, NaOH, $Ca(OH)_2$, and $Mg(OH)_2$. These are also Brønsted–Lowry bases, as are such substances as fluoride ions, carbonate ions, sulfate ions, acetate ions, and water. Generally, bases taste bitter, change the color of certain dyes, feel soapy or slippery, and neutralize acids. Most important, they dissolve in water to produce hydroxide ions. Bases can be produced by the reaction of a metal oxide with water. They react with nonmetal oxides and with acids. Like acids, they have a number of uses in industry and in our daily lives.

Strong acids and bases are those that essentially ionize completely in a solution. Weak acids and bases are those that don't ionize completely. In solution, the ions and molecules of a weak acid or a weak base are in dynamic equilibrium.

Water ionizes partially, just like a very weak acid or a very weak base. When a water molecule ionizes, a hydronium ion $(H_3O)^{1+}$ and a hydroxide ion are formed. In 1 L of water, only 10^{-7} mole of water actually dissociates, forming 10^{-7} mole of hydronium ions and 10^{-7} mole of hydroxide ions.

The pH scale provides a means of expressing the very dilute concentrations of acids and bases. The pH of any solution is equal to the negative of the logarithm of its hydrogen-ion concentration in moles per liter. The pH of pure water and that of any other neutral solution is 7. An acidic solution has a pH below 7, and a basic solution has a pH above 7. The product of the hydrogen-ion concentration and the hydroxide-ion concentration in an aqueous solution is always 10^{-14}.

A salt is a compound composed of the positive ion of a base and the negative ion of an acid. In the solid state the positive and negative ions are held together by an ionic bond. In aqueous solution, they are not held together at all. Salts can be prepared by direct combination of two elements, by reaction of an active metal with an acid, or by neutralization of an acid by a base.

Titration is an analytical procedure that enables the chemist to determine the concentration of a substance present in solution. A measured volume of a solution of unknown concentration is found. Acid–base titrations involve a neutralization reaction.

KEY TERMS

alkali **(16.2)**	indicator **(16.10)**
Arrhenius acid **(16.1)**	pH **(16.6)**
Arrhenius base **(16.2)**	salt **(16.8)**
Brønsted–Lowry acid **(16.1)**	strong acid **(16.3)**
Brønsted–Lowry base **(16.2)**	strong base **(16.3)**
end point **(16.10)**	titration **(16.9)**
equivalence point **(16.9)**	weak acid **(16.3)**
hydronium ion **(16.1)**	weak base **(16.3)**

Wastewater Treatment Chemist

As a *wastewater treatment chemist*, you will analyze water samples from streams, examine raw and treated wastewater, and look at sludge and other byproducts of wastewater treatment. You will also work to determine the efficiency of the processes used in wastewater treatment plants to ensure that local, state, and federal standards are met.

Wastewater treatment chemists can work in municipal or privately owned wastewater treatment plants. You can also work in a laboratory, where you will use sophisticated analytical equipment. You will test samples of water and wastewater for chemical, bacteriological, and physical components. You will also monitor and regulate the discharge of wastewater into sewer treatment systems and receiving streams.

*Creating a Better Bandage**

Victims of severe burns are often susceptible to life-threatening infections. If these people lost 20% or more of their skin, they must immediately be protected from infection and covered with a bandage that will quickly heal the charred skin. Chemists in Washington have been working with physicians at the University of Washington in Seattle to design a bandage that will perform these two tasks.

When the team of researchers first met, they decided that the best type of bandage would be one that would cover the wound, support healing, prevent infection, and then dissolve away. The substances they chose as a basis for their bandage were naturally occurring chitosan, which is derived from crustacean shells, and calcium alginate, which is derived from seaweed.

Once the bandage is applied to the burn victim, it will release an epidermal growth factor. The growth factor will be derived from mouse salivary glands and will cause the skin to begin regenerating. It will be microencapsulated in the calcium alginate (from the seaweed), which will allow it to be released slowly over a period of a few days. The growth factor will also protect the body from infection-causing fungi and bacteria. After a few days on the wound, the bandage, which will be similar to a soft contact lens, will dissolve.

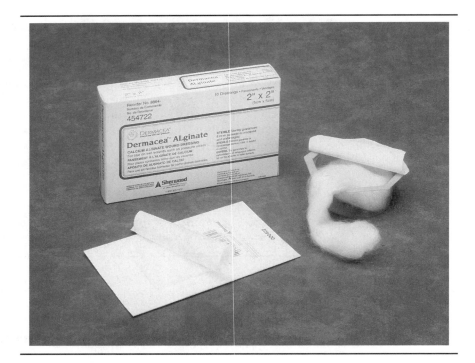

A new bandage helps burn victims generate new skin. (Courtesy Sherwood Davis & Geck).

Testing of the new bandage has already begun. It is hoped that results will show a much greater chance of survival for victims of serious burns.

*Adapted from *Science News*, March 26, 1994

SELF-TEST EXERCISES

LEARNING GOAL 1: *Arrhenius Acids and Bases*

◀ 1. Describe the various attributes of
(a) an Arrhenius acid, (b) an Arrhenius base.

<u>**2.**</u> Match the acid or base on the left with its description on the right.

(a) Acetic acid
(b) Hydro-
chloric acid
(c) Sulfuric acid
(d) Sodium hy-
droxide
(e) Magnesium
hydroxide

1. Commonly known as "lye"
2. Found in automobile batteries
3. Found in "milk of magnesia"
4. Found in vinegar
5. Commercially known as muriatic acid

<u>**3.**</u> Name an acid found in soda pop.
◀ 4. What are the limitations of the Arrhenius definition of acids and bases?

LEARNING GOAL 2: *Brønsted–Lowry Acids and Bases*

◀ 5. Using the Brønsted–Lowry definition, define
(a) An acid, (b) A base.
6. Why is water both an acid and a base according to the Brønsted–Lowry definitions?

*<u>**7.**</u> Complete the following equation, which shows why ammonia (NH_3) is a Brønsted–Lowry base:

$$NH_3 + H_2O \rightleftharpoons$$

◀ *8. Complete the following equations, which show why each substance is a Brønsted–Lowry acid or base:

(a) $(HSO_4)^{1-} \rightarrow ? + ?$
Acid

(b) $Cl^{1-} + ? \rightarrow ?$
Base

LEARNING GOAL 3: *Characteristics of Acids and Bases*

◀ 9. What are the characteristic properties of acids, bases, and salts?

10. What do we call a substance that is used to reveal whether a chemical is an acid or a base?

◀ 11. What is the difference between hydrogen chloride gas and hydrochloric acid?

◀ 12. What is a hydronium ion?

13. Why do people store acids in containers made of nonmetal materials?

14. A substance that dissolves in water to produce hydroxide ions is a(n) _____.

15. The two types of acids are mineral acids and _____ acids.

16. A substance that dissolves in water to produce hydrogen ions is a(n) _____.

17. Place the words *red* and *blue* in the proper spaces.
(a) _____ litmus paper turns _____ in an acid.
(b) _____ litmus paper turns _____ in a base.

18. An acid feels soapy or slippery. True or false?

19. Write a reaction showing the formation of a hydronium ion from HBr.

20. Write an equation for the formation of hydronium ions from HNO_3.

21. Write a reaction showing the formation of hydroxide ions when calcium hydroxide dissolves in water.

22. Write an equation for the formation of hydroxide ions when magnesium hydroxide dissolves in water.

LEARNING GOAL 4: *Preparation of Acids and Bases*

◀ **23.** Write a reaction for the preparation of each of the following bases from a metal oxide and water:
(a) KOH (b) $Mg(OH)_2$

24. Write a reaction for the preparation of each of the following bases from a metal oxide and water:
(a) LiOH (b) $Ca(OH)_2$

25. Write a reaction for the preparation of each of the following acids from a nonmetal oxide and water: (a) H_2SO_3 (sulfurous acid) (b) H_2CO_3 (carbonic acid)

◀ 26. Write a reaction for the preparation of each of the following acids from a nonmetal oxide and water: (a) HNO_3 (b) H_3PO_4

LEARNING GOAL 5: *Chemical Reactions of Acids and Bases*

27. Complete and balance the following equations:
(a) $H_3PO_4 + Mg(OH)_2 \rightarrow$
(b) $P_2O_5 + H_2O \rightarrow$
(c) $K_2O + H_2O \rightarrow$
(d) $Mg(OH)_2 + CO_2 \rightarrow$
(e) $Zn + H_3PO_4 \rightarrow$

28. Complete and balance the following equations:
(a) $HC_2H_3O_2 + Ca(OH)_2 \rightarrow$
(b) $Na_2O + H_2O \rightarrow$
(c) $Mg + HCl \rightarrow$
(d) $Sr(OH)_2 + CO_2 \rightarrow$
(e) $N_2O_5 + H_2O \rightarrow$

◀ **29.** Suppose you spill some sodium hydroxide on your clothes. What could you do to keep this chemical from ruining them?

◀ *30. Suppose that you accidentally drank a liquid containing sodium hydroxide. What should you do to neutralize its effect?

31. Do vinegar and lye react? If a reaction does take place, write a chemical equation for it.

*32. Power plants that burn coal with a high sulfur content send huge quantities of SO_2 into the atmosphere. When the SO_2 reacts with the moisture in the clouds, acid rain results. Write an equation for the formation of acid rain.

◀ 33. Paper mills emit huge quantities of sulfur dioxide (SO_2) into the atmosphere at the mill site. What can happen to exposed metal machinery when rain falls near an operating paper mill?

34. Write a balanced chemical equation for each of the following word equations:
(a) Magnesium + hydrochloric acid $\rightarrow$? + ?
(b) Sulfuric acid + sodium hydroxide $\rightarrow$? + ?
(c) ? + ? $\rightarrow$ magnesium sulfate + hydrogen gas
(d) ? + ? $\rightarrow$ zinc phosphate + water

35. Write balanced chemical equations for the following word equations:
(a) Zinc + sulfuric acid $\rightarrow$ zinc sulfate + hydrogen gas

(b) Hydrochloric acid + lithium hydroxide → ? + ?

(c) ? + ? → calcium sulfate + H_2

(d) ? + ? → calcium phosphate + water

36. Complete and balance the following equations:
    (a) $Zn + H_3PO_4 \rightarrow$
    (b) $Zn + HCl \rightarrow$
    (c) $Zn + H_2SO_4 \rightarrow$

37. Complete and balance the following equations:
    (a) $Zn + HCl \rightarrow$
    (b) $HCl + Ca(OH)_2 \rightarrow$
    (c) $Mg(OH)_2 + CO_2 \rightarrow$

38. All of the reactions in Exercise 36 can be classified as _____ reactions.

LEARNING GOAL 6: *Strong Acids and Bases*

◄ 39. Define and give an example of (a) A strong acid, (b) A strong base.

◄ 40. What is the difference between a strong acid and a strong base?

41. Write formulas for the following acids: (a) Nitric acid (b) Hydrofluoric acid (c) Hydroiodic acid

42. Write formulas for the following acids:
    (a) Phosphoric acid (b) Sulfuric acid
    (c) Boric acid

◄ **43.** Give names for the following bases: (a) NaOH
    (b) $Ca(OH)_2$ (c) $Fe(OH)_2$ (d) $Al(OH)_3$

◄ 44. Give names for the following bases:
    (a) $Mg(OH)_2$ (b) $Zn(OH)_2$ (c) KOH (d) LiOH

45. Give names for the following acids: (a) HNO_3
    (b) H_3PO_4 (c) HI (d) H_2SO_4

46. Name the following acids: (a) HCl (b) H_2SO_3
    (c) H_2CO_3 (d) $HC_2H_3O_2$

47. Write formulas for the following bases:
    (a) Lithium hydroxide (b) Barium hydroxide
    (c) Iron(II) hydroxide (d) Iron(III) hydroxide

48. Write formulas for the following bases:
    (a) Potassium hydroxide (b) Sodium hydroxide
    (c) Zinc hydroxide (d) Calcium hydroxide

LEARNING GOAL 7: *Weak Acids and Bases*

◄ 49. Define and give an example of (a) A weak acid, (b) A weak base.

◄ 50. A weak acid or a weak base ionizes almost completely. True or false?

LEARNING GOAL 8: *Dynamic Equilibrium*

*51. Explain what is meant by dynamic equilibrium. Show how it is related to weak acids and bases.

◄ *52. Explain how a dynamic equilibrium comes to exist when carbon dioxide is dissolved in water to produce carbonic acid, a weak acid.

LEARNING GOAL 9: *Ionization of Water*

53. Write an equation for the ionization of water.

54. How many moles of hydrogen ions and how many moles of hydroxide ions are present in 2 L of water?

LEARNING GOAL 10: *pH and pH Scale*

55. State how acidic or basic each of the following substances is (very acidic, moderately acidic, very basic, and so on):
    (a) Vinegar, pH = 2.8
    (b) Seawater, pH = 8.5
    (c) Blood, pH = 7.4
    (d) Saliva, pH = 7.0

56. State how acidic or basic each of the following substances is (very acidic, moderately acid, very basic, and so on):
    (a) Oranges, pH = 3.5
    (b) Milk, pH = 6.5
    (c) Lemons, pH = 2.3
    (d) Rainwater, pH = 6.2

◄ 57. Why is it important to measure the pH of the water in a swimming pool?

◄ 58. Why must the water in a fish tank be the proper pH?

*59. The pH scale is a logarithmic scale. This means that a difference of one pH unit indicates a multiple of 10 between the actual acidities or basicities of two substances.
    (a) The pH of grapefruit juice is 3; the pH of beer is 5. How many times more acidic is grapefruit juice than beer?
    (b) The pH of lemon juice is 2; the pH of pure water is 7. How many times more acidic is lemon juice than water?

*60. The pH of soft drinks is 3.0; the pH of pure water is 7.0. How many times more acidic is the soft drink than the water?

61. In your own words, give a working definition of pH.

62. What is "pH-balanced shampoo"?

LEARNING GOAL 11: *Determining pH from Hydrogen-Ion or Hydroxide-Ion Concentration*

◀ 63. Determine the pH of each of the following aqueous solutions:
 (a) $[H^{1+}] = 10^{-3}$
 (b) $[H^{1+}] = 10^{-4}$
 (c) $[H^{1+}] = 10^{-2}$
 (d) $[H^{1+}] = 10^{-11}$
 (e) $[OH^{1-}] = 10^{-4}$
 (f) $[OH^{1-}] = 10^{-9}$

 64. Determine the pH of each of the following aqueous solutions:
 (a) $[H^{1+}] = 10^{-5}$
 (b) $[H^{1+}] = 10^{-10}$
 (c) $[H^{1+}] = 10^{-8}$
 (d) $[OH^{1-}] = 10^{-5}$
 (e) $[OH^{1-}] = 10^{-10}$
 (f) $[OH^{1-}] = 10^{-8}$

◀ 65. Determine the $[H^{1+}]$ and $[OH^{1-}]$ of each of the aqueous solutions with the following pH values:
 (a) pH = 5
 (b) pH = 9
 (c) pH = 2
 (d) pH = 12

 66. Determine the $[H^{1+}]$ and $[OH^{1-}]$ of each of the aqueous solutions with the following pH values:
 (a) pH = 4
 (b) pH = 7
 (c) pH = 1
 (d) pH = 13

LEARNING GOAL 12: *Relationship Between Hydrogen-Ion and Hydroxide-Ion Concentrations*

◀ 67. For each of the following aqueous solutions, determine the hydrogen-ion concentration and the pH:
 (a) $[OH^{1-}] = 10^{-5}$
 (b) $[OH^{1-}] = 10^{-8}$
 (c) $[OH^{1-}] = 10^{-10}$
 (d) $[OH^{1-}] = 10^{-2}$

*68. For each of the following aqueous solutions, determine the hydrogen-ion concentration and the pH:
 (a) $[OH^{1-}] = 10^{-4}$
 (b) $[OH^{1-}] = 10^{-11}$
 (c) $[OH^{1-}] = 10^{-9}$
 (d) $[OH^{1-}] = 10^{-3}$

*69. Determine the $(OH)^{1-}$ concentration for each of the following aqueous solutions:
 (a) $[H^{1+}] = 10^{-3}$
 (b) $[H^{1+}] = 10^{-4}$
 (c) $[H^{1+}] = 10^{-8}$
 (d) pH = 9
 (e) pH = 6
 (f) pH = 2

*70. Determine $[OH^{1-}]$ for each of the following aqueous solutions:
 (a) $[H^{1+}] = 10^{-2}$
 (b) $[H^{1+}] = 10^{-9}$
 (c) $[H^{1+}] = 10^{-5}$
 (d) pH = 3
 (e) pH = 4
 (f) pH = 7

LEARNING GOAL 13: *Definition of Salt*

◀ 71. Name each of the following salts: (a) Na_2S (b) KBr (c) Na_2SO_4 (d) KNO_3 (e) $CaCO_3$ (f) BaI_2

 72. Name each of the following salts: (a) Na_2CO_3 (b) KCl (c) $BaCl_2$ (d) $NaNO_3$ (e) $Ca(NO_3)_2$ (f) CaI_2

◀ 73. Write the formula for each of the following salts:
 (a) Silver nitrate (b) Cobalt(II) chloride (c) Copper(I) sulfite (d) Mercury(II) phosphate

◀ 74. Write the formula for each of the following salts:
 (a) Silver chloride (b) Iron(III) nitrate (c) Iron(II) nitrate (d) Copper(I) chloride

LEARNING GOAL 14: *Important Properties of Salts*

◀ 75. List at least two important properties of salts.
◀ 76. What kind of bonding is found in most salts?

LEARNING GOAL 15: *Preparation of Salts*

◀ 77. List at least two ways in which salts can be prepared by chemical reactions.
◀ 78. How would you prepare the salt silver nitrate?

 79. Write an acid–base reaction showing the preparation of the salt lithium sulfate (Li_2SO_4).
 80. Write an acid–base reaction showing the preparation of the salt lithium nitrate ($LiNO_3$).
 81. Write a single-replacement reaction showing the preparation of the salt zinc bromide ($ZnBr_2$).

82. Write a single-replacement reaction showing the preparation of the salt zinc chloride ($ZnCl_2$).

83. Write a single-replacement reaction showing the preparation of each of the following salts: (a) Magnesium chloride (b) Zinc sulfate

84. Write a single-replacement reaction showing the preparation of each of the following salts: (a) Magnesium bromide (b) Copper(II) sulfate

85. Write an acid–base reaction showing the preparation of each of the following salts: (a) Potassium phosphate (b) Calcium nitrate

86. Write and acid–base reaction showing the preparation of each of the following salts: (a) Lithium phosphate (b) Calcium chloride

87. Write a reaction for the preparation of each of the following salts by the direct combination of their elements: (a) KCl (b) $CaBr_2$

88. Write a reaction for the preparation of each of the following salts by the direct combination of their elements: (a) KI (b) $CaCl_2$

LEARNING GOAL 16: *Titration*

◀ 89. Calculate the normality of an HNO_3 solution if 30.0 mL of the solution are neutralized by 75.0 mL of 0.25 *N* NaOH.

◀ 90. Calculate the normality and the molarity of an HCl solution if 125 mL of the solution are neutralized by 25.0 mL of 1.5 *N* NaOH.

91. Calculate the normality and the molarity of a solution of acetic acid if 45.00 mL of the solution of acetic acid are neutralized by 50.00 mL of 0.2500 *N* (0.2500 *M*) NaOH.

92. A 35.0-mL sample of NaOH solution is titrated with 75.0 mL of 0.150 *N* lactic acid. Calculate the normality of the NaOH solution.

93. A 25.0-mL sample of vinegar is titrated with 30.00 mL of 0.150 *N* (0.150 *M*) NaOH solution. Calculate the normality and the molarity of the vinegar.

94. A 35.00-mL sample of KOH solution required 25.00 mL of 1.000 *N* H_2SO_4 to neutralize it. Calculate the concentration of the KOH solution.

95. If 50.0 mL of 1.00 *M* HCl solution neutralize 75.0 mL of $Ca(OH)_2$ solution, what is the molarity of the $Ca(OH)_2$ solution?

96. Calculate the molarity of an NaOH solution if 25.0 mL of 0.150 *M* HCl are needed to titrate 30.0 mL of the NaOH solution.

EXTRA EXERCISES

97. Consider the reaction $PH_3 + NaH \rightarrow NaPH_2 + H_2$.
(a) Identify PH_3 as a Brønsted–Lowry acid or base.
(b) Identify NaH as a Brønsted–Lowry acid or base.

98. Consider the reaction $NH_3 + HCl \rightarrow (NH_4)^{1+} + Cl^{1-}$.
(a) Identify NH_3 as a Brønsted–Lowry acid or base.
(b) Identify HCl as a Brønsted–Lowry acid or base.

99. Consider the reaction $(HSO_4)^{1-} + (H_3O)^{1+} \rightarrow H_2SO_4 + H_2O$.
(a) Identify $(HSO_4)^{1-}$ as a Brønsted–Lowry acid or base.
(b) Identify $(H_3O)^{1+}$ as a Brønsted–Lowry acid or base.

100. Calculate the pH and $[OH^{1-}]$ of a 0.001 *M* solution of NaOH.

101. Complete and balance the following equations:
(a) HCl + NaOH $\rightarrow$
(b) BaO + HCl $\rightarrow$
(c) Zn + H_2SO_4 $\rightarrow$

◀ 102. Show by a reaction with water that each of the following is a base: (a) $(OH)^{1-}$ (b) NH_3 (c) $(HSO_4)^{1-}$

◀ 103. Show by a reaction with water that each of the following is an acid: (a) H_2O (b) $(HCO_3)^{1-}$ (c) HNO_3

◀ 104. Many midwestern industries give off large amounts of sulfur trioxide, SO_3, through their smokestacks. Explain why the pH of lakes in New England is decreasing.

105. Explain why phosphorus trichloride, PCl_3, might be a Brønsted–Lowry base.

◀ 106. Suggest a way to prepare each of the following via a correctly balanced equation: (a) Na_2SO_4 (b) $ZnCl_2$ (c) KCl (d) H_2SO_3 (*Hint:* Choose your reactants and then use them in a combination reaction, single-replacement reaction, or double-replacement reaction.)

◀ _1._ Which is the solute and which is the solvent in each of the following solutions?
   (a) 75 mL of nitrogen and 35 mL of oxygen
   (b) Sulfur dioxide gas and water
   (c) Lemon juice, sugar, and water in lemonade

◀ _2._ State whether each of the following pairs of substances forms a solution:
   (a) Hydrogen fluoride gas and water
   (b) Benzene and _trans_-dichloroethane

_3._ Find the concentration of each of the following solutions in percent by mass:
   (a) 50.0 g of salt in enough water to make 75.0 g of solution
   (b) 175 g of sugar in enough water to make 1,500 g of solution

_4._ Find the concentration of each of the following solutions in percent by volume:
   (a) 25.0 mL of alcohol in enough water to make 455 mL of solution
   (b) 35.0 mL of benzene dissolved in enough carbon tetrachloride to make 95.0 mL of solution

◀ _5._ If 5.0 g of benzene are dissolved in enough carbon tetrachloride to make 75 mL of solution, what is the concentration of the solution, what is the concentration of the solution in percent by mass–volume?

_6._ What is the molarity of the solution formed by dissolving 1.84 g of ethyl alcohol (C_2H_5OH) in enough water to make 1.5 L of solution?

◀ _7._ When 250 mL of 1.33 M NaOH solution are evaporated to dryness, how many grams of NaOH remain?

_8._ How many milliliters of 1.00 M $CaCl_2$ solution must one evaporate to obtain 26.0 g of $CaCl_2$?

◀ _9._ Distinguish between normality and molarity.

_10._ How many grams of H_3PO_4 are present in 1.00 L of a 2.00 N solution of H_3PO_4?

_11._ Calculate the normality and the molarity of a solution of calcium hydroxide formed by dissolving 18.5 g of calcium hydroxide in enough water to make 500 mL of solution.

_12._ How would 5.50 L of 1.33 N NaOH solution be prepared from a 36.0 N solution of NaOH?

◀ _13._ A photographer is preparing 2.00 L of 0.150 M silver nitrate solution. How many grams of silver nitrate are needed?

_14._ A solution is a heterogeneous mixture. True or false?

_15._ Calculate the grams of solute in 100 mL of 18.0 M H_2SO_4 solution.

_16._ A solution of sugar in water contains 25.0% sugar, by mass, and has a density of 1.09 g/mL. The molecular formula for sugar is $C_{12}H_{22}O_{11}$.
   (a) Calculate the number of grams of sugar present in 2.00 L of solution.
   (b) Calculate the molarity of the solution.

_17._ Calculate the grams of solute present in 100.0 g of 0.90% by mass NaCl solution.

_18._ To what volume must a solution of 75.0 g of H_2SO_4 in 400 mL of water be diluted to give a 0.200 M solution?

◀ _19._ Calculate the boiling point of a solution prepared by dissolving 100.0 g of ethylene glycol, $C_2H_6O_2$, in 500.0 g of water. K_b for water is 0.520°C/m.

_20._ A solution is prepared by dissolving 12.8 g of naphthalene, $C_{10}H_8$, in 1.00 L of water. Is this solution 0.100 M or 0.100 m?

_21._ State the difference between the Arrhenius and the Brønsted–Lowry definitions of an acid.

◀ _22._ A solution of pH 4 is 100 times more acidic than a solution of pH 7. True or false?

_23._ Complete and balance each of the following equations:
   (a) $HC_2H_3O_2 + Ca(OH)_2 \rightarrow$?
   (b) $N_2O_5 + H_2O \rightarrow$?

_24._ Write an equation for the formation of acid rain in the atmosphere.

_25._ Determine the pH of a solution that has a hydrogen-ion concentration of 1.0×10^{-2} M.

_26._ Calculate the pH of black coffee if its hydrogen ion-concentration is 1.0×10^{-6} M.

_27._ How many milliliters of 0.500 M HCl can one prepare by diluting 750 mL of 12.0 M HCl?

_28._ Calculate the molarity of each of the ions present in a solution prepared by combining 50.0 mL of 2.00 M HCl with 50.0 mL of 2.00 M NaOH.

_29._ If 40.0 mL of 1.50 M NaOH solution are needed to titrate 50.0 mL of HCl, what is the molarity of the HCl?

30. Calculate the molarity of NaOH if 15.0 mL of 0.150 M HCl are needed to titrate 50.00 mL of NaOH.

31. Differentiate between the terms *ionization* and *dissociation*.

32. Which solution is more acidic, 1 M H_2SO_4 or 1 N H_2SO_4?

33. Which solution is more acidic, 1 M HCl or 1 M $HC_2H_3O_2$?

34. Calculate the molarity of the sum of all the ions present in 0.02 M NaCl if the solution is 100% ionized.

35. Calculate the concentration of Sr^{2+} ions in a solution of $SrCl_2$ that has a Cl^{1-} ion concentration of 0.500 M. (Assume 100% ionization.)

36. If 50.00 mL of 0.500 M HCl neutralize 100.0 mL of $Ca(OH)_2$ solution, what is the molarity of the $Ca(OH)_2$ solution? The reaction is

$$Ca(OH)_2(aq) + 2HCl(aq) \rightarrow CaCl_2(aq) + 2H_2O$$

◀ **37.** Do the terms *strong acid* and *strong base* refer to concentration or ionization?

◀ **38.** When a mole of $SrCl_2$ is dissolved in water, are there more Sr^{2+} ions or more Cl^{1-} ions present?

39. Calculate the pH of tomato juice whose H^{1+} concentration is 1.00×10^{-4}.

40. What is the $(OH)^{1-}$ concentration of a solution whose H^{1+} concentration is 10^{-8}?

APPENDIX A
Basic Mathematics for Chemistry

LEARNING GOALS

After you have studied this appendix, you should be able to:
1. Add and subtract algebraically.
2. Multiply and divide fractions.
3. Multiply and divide exponential numbers.
4. Use the factor-unit method in solving problems.
5. Add and subtract decimals.
6. Multiply and divide decimals.
7. Solve simple algebraic equations.
8. Set up and solve proportions.
9. Solve simple density problems.
10. Solve problems involving percentages.
11. Perform mathematical operations using a calculator.

INTRODUCTION

Full appreciation of many ideas in chemistry requires some basic mathematics. (There are also concepts in chemistry that require absolutely *no* mathematics.) This appendix is here so that you can review and master the mathematical skills you need to be successful in chemistry. Whenever you are in doubt about a mathematical problem as you work your way through the book, turn to this appendix for guidance.

You'll find that chemical mathematics is not very difficult, as long as you can perform some basic operations. Work your way through this appendix carefully and try your hand at all the sample problems. Then do the exercises at the end. Remember that the knowledge you gain here will greatly increase your success in chemistry.

A.1   Adding and Subtracting Algebraically

In chemistry we deal with both positive and negative numbers, so it is important to know how to add and subtract both. The process is called algebraic addition and subtraction. Here are the rules to follow.

Addition

1. *When the signs are alike:* When the signs of two numbers are the same, add the numbers and keep the same sign.

$$+4 + 2 = +6$$
$$-4 - 2 = -6$$

2. *When the signs are different:* When the signs of two numbers are not the same, subtract the numbers and keep the sign of the larger number.

$$-4 + 2 = -2$$
$$+4 - 2 = +2$$

Subtraction

Change the sign of the *number being subtracted* and follow the rules for addition. (*Note:* Plus signs before positive numbers are usually left out, but minus signs before negative numbers are always written.)

$4 - (-2)$	becomes	$4 + 2 = 6$
$4 - (+2)$	becomes	$4 - 2 = 2$
$-4 - (+2)$	becomes	$-4 - 2 = -6$
$-4 - (-2)$	becomes	$-4 + 2 = -2$

A.2   Fractions

A ratio of two numbers is called a **fraction.** Some examples of fractions are $\frac{1}{4}$, $\frac{1}{2}$, $\frac{2}{5}$, $\frac{1}{11}$, and $\frac{2}{1}$. A fraction has two parts: the **numerator,** which is the top number, and the **denominator,** which is the bottom number.

Here are some basic rules to follow in dealing with fractions.

Multiplication

Multiply numerators; then multiply denominators.

$$\frac{1}{2} \times \frac{3}{4} = \frac{3}{8} \qquad \frac{5}{9} \times \frac{9}{5} = \frac{45}{45} = 1$$

Division

Invert (turn upside down) the fraction to the right of the division sign, then follow the rules for multiplication.

$$\frac{1}{2} \div \frac{1}{4} \qquad \text{becomes} \qquad \frac{1}{2} \times \frac{4}{1} = \frac{4}{2} = 2$$

$$\frac{3}{4} \div \frac{5}{9} \qquad \text{becomes} \qquad \frac{3}{4} \times \frac{9}{5} = \frac{27}{20}$$

A.3 Exponents

Exponents are numbers written to the right of and above another number (called the **base number**). A positive exponent tells you to multiply the base number by itself. The exponent tells you how many times the base number is taken. For example, $(4)^2$ means take the number 4 twice: 4×4. Here 4 is the base number and 2 is the exponent.

$$(4)^2 \longleftarrow \text{Exponent}$$
$$\uparrow$$
$$\textsf{Base number}$$

Sometimes exponents are negative. For example, we may have

$$(4)^{-2} \qquad \text{which means} \qquad \frac{1}{(4)^2}$$

$$\frac{1}{(4)^{-2}} \qquad \text{which means} \qquad \frac{(4)^2}{1}$$

Example A.1 shows how exponents are used.

EXAMPLE A.1

Perform the indicated operation: (a) $(3)^2$ (b) $(4)^3$ (c) $(3)^1$

(d) $(3)^{-2}$ (e) $\dfrac{1}{(2)^{-3}}$

Solution

(a) $(3)^2 = 3 \times 3 = 9$

(b) $(4)^3 = 4 \times 4 \times 4 = 64$

(c) $(3)^1 = 3$

(d) $(3)^{-2} = \dfrac{1}{(3)^2} = \dfrac{1}{3 \times 3} = \dfrac{1}{9}$

(e) $\dfrac{1}{(2)^{-3}} = \dfrac{(2)^3}{1} = 2 \times 2 \times 2 = 8$

LEARNING GOAL 3

Multiplying and dividing exponential numbers

Sometimes it is necessary to *multiply* exponential numbers that have the *same base*. This is done as shown in Example A.2.

EXAMPLE A.2

Perform the indicated operation:
(a) $(4)^8(4)^2$ (b) $(3)^2(3)^{-1}$ (c) $(8)^2(8)^1$ (d) $(2)^{-2}(2)^{-3}$ (e) $(3)^{-2}(3)^2$

Solution

Keep the same base and *add the exponents algebraically*.
(a) $(4)^8(4)^2 = (4)^{10}$
(b) $(3)^2(3)^{-1} = (3)^1 = 3$
(c) $(8)^2(8)^1 = (8)^3$
(d) $(2)^{-2}(2)^{-3} = (2)^{-5}$
(e) $(3)^{-2}(3)^2 = (3)^0 = 1$ (Any number to the zero power is equal to 1.)

Sometimes it is necessary to *divide* exponential numbers that have the *same base*. This is done as shown in Example A.3.

EXAMPLE A.3

Perform the indicated operation: (a) $\dfrac{(2)^5}{(2)^2}$ (b) $\dfrac{(2)^5}{(2)^{-2}}$ (c) $\dfrac{(2)^{-5}}{(2)^2}$

(d) $\dfrac{(2)^{-5}}{(2)^{-2}}$

Solution

Keep the same base and subtract the exponent in the denominator from the exponent in the numerator *algebraically*.

(a) $\dfrac{(2)^5}{(2)^2} = (2)^{5-2} = (2)^3$

(b) $\dfrac{(2)^5}{(2)^{-2}} = (2)^{5-(-2)} = (2)^{5+2} = (2)^7$

(c) $\dfrac{(2)^{-5}}{(2)^2} = (2)^{-5-(+2)} = (2)^{-5-2} = (2)^{-7}$

(d) $\dfrac{(2)^{-5}}{(2)^{-2}} = (2)^{-5-(-2)} = (2)^{-5+2} = (2)^{-3}$

In Examples A.2 and A.3, only two numbers have been multiplied or divided. However, sometimes there are more than two numbers in the calculation, as in Example A.4.

EXAMPLE A.4

Perform the indicated operation: (a) $(5)^3(5)^3(5)^2$ (b) $(a)^5(a)^3(a)^{-4}$
(c) $\dfrac{(b)^3(b)^6}{(b)^2}$ (d) $\dfrac{(2)^5(2)^{-6}}{(2)^{-9}}$ (e) $\dfrac{(a)^{-3}(a)^4(a)^6}{(a)^{-2}(a)^7}$

Solution

(a) $(5)^3(5)^3(5)^2 = (5)^8$ The exponents were added algebraically.

(b) $(a)^5(a)^3(a)^{-4} = (a)^4$ The exponents were added algebraically.

(c) $\dfrac{(b)^3(b)^6}{(b)^2} = \dfrac{(b)^9}{(b)^2} = (b)^{9-2} = (b)^7$ When you work out a problem one step at a time, it becomes easy.

(d) $\dfrac{(2)^5(2)^{-6}}{(2)^{-9}} = \dfrac{(2)^{-1}}{(2)^{-9}} = (2)^8$ (e) $\dfrac{(a)^{-3}(a)^4(a)^6}{(a)^{-2}(a)^7} = \dfrac{(a)^7}{(a)^5} = (a)^2$

A.4   Working with Units

■ LEARNING GOAL 4

Solving problems using the factor-unit method

Almost all numbers, in chemistry or anywhere else, are accompanied by units. It is important to be able to work with these units as well as with the numbers. Study the following examples carefully.

EXAMPLE A.5

How many feet are there in 36 inches?

Solution

We are going to use what is known as the **factor-unit method** of analysis. Here's how it works. Because there are 12 inches in 1 foot, we can express this relationship mathematically as

$$\frac{12 \text{ inches}}{1 \text{ foot}} \quad \text{(reads ``12 inches per 1 foot'')}$$

or as

$$\frac{1 \text{ foot}}{12 \text{ inches}} \quad \text{(reads ``1 foot per 12 inches'')}$$

These are called factor units. We set the problem up so that when we multiply the quantity in question (36 inches) by the appropriate factor unit, we end up with what we want (feet). Here, we have

$$? \text{ feet} = (36 \text{ inches})\left(\frac{1 \text{ foot}}{12 \text{ inches}}\right) = 3 \text{ feet}$$

Notice how the unit *inches* cancels out in Example A.5 when the numbers are multiplied. You may wonder how you'd know that you must use the factor unit

$$\left(\frac{1 \text{ foot}}{12 \text{ inches}}\right) \quad \text{and not the factor unit} \quad \left(\frac{12 \text{ inches}}{1 \text{ foot}}\right)$$

That's easy. If you used the wrong factor unit, the term *inches* wouldn't cancel out. You would arrive at

$$? \text{ feet} = (36 \text{ inches})\left(\frac{12 \text{ inches}}{1 \text{ foot}}\right) = \frac{432 \text{ (inches)}^2}{1 \text{ foot}}$$

which has no meaning to you. Try the problems in the following examples for practice in using the factor-unit method.

EXAMPLE A.6

How many inches are there in 5 feet?

Solution
Set the problem up this way:

$$? \text{ inches} = (5 \text{ feet})\left(\frac{12 \text{ inches}}{1 \text{ foot}}\right) = 60 \text{ inches}$$

Note how the unit *feet* cancels out when the numbers are multiplied with their units.

EXAMPLE A.7

How many minutes are there in 4 hours? How many seconds are there in 4 hours?

Solution
There are 60 minutes in 1 hour. So the factor unit is

$$\frac{60 \text{ minutes}}{1 \text{ hour}}$$

The problem is solved this way:

$$? \text{ minutes} = (4 \text{ hours})\left(\frac{60 \text{ minutes}}{1 \text{ hour}}\right) = 240 \text{ minutes}$$

There are 60 seconds in 1 minute. Our factor unit for this part of the problem is

$$\frac{60 \text{ seconds}}{1 \text{ minute}}$$

So the problem is solved as follows:

$$? \text{ seconds} = (240 \text{ minutes}) \left(\frac{60 \text{ seconds}}{1 \text{ minute}} \right) = 14{,}400 \text{ seconds}$$

EXAMPLE A.8

How many gallons are there in 60 quarts?

Solution

There are 4 quarts in 1 gallon. Therefore

$$? \text{ gallons} = (60 \text{ quarts}) \left(\frac{1 \text{ gallon}}{4 \text{ quarts}} \right) = 15 \text{ gallons}$$

EXAMPLE A.9

Change 60 miles per hour to feet per second.

Solution

Here you can really see the value of the factor-unit method. However, we have to know that 1 mile equals 5,280 feet and that 1 hour equals 3,600 seconds. Now, 60 miles per hour means

$$60 \frac{\text{miles}}{\text{hour}} \qquad \text{which is the same as} \qquad \frac{60 \text{ miles}}{1 \text{ hour}}$$

So the complete calculation reads

$$? \frac{\text{feet}}{\text{seconds}} = \left(\frac{60 \text{ miles}}{1 \text{ hour}} \right) \left(\frac{5{,}280 \text{ feet}}{1 \text{ mile}} \right) \left(\frac{1 \text{ hour}}{3{,}600 \text{ seconds}} \right)$$

$$= 88 \frac{\text{feet}}{\text{second}}$$

A.5   Decimals

Chemistry is such an exact science that the measurements we take usually consist of a string of numbers with a decimal point—for example, 2.7135 grams. When we are working with decimal numbers, we have to keep certain rules in mind.

LEARNING GOAL 5

Adding and subtracting decimals

Addition and Subtraction

In adding or subtracting numbers, it is important to line up the columns correctly. Keep the decimal points stacked one over the other.

EXAMPLE A.10

Perform the indicated operation: (a) $25.8 + 107.09 + 88.004 + 0.011$ (b) $342.78 - 14.99$

Solution

Line up the numbers properly.

(a)   25.8
   107.09
    88.004
     0.011
  220.905

(b)   342.78
  − 14.99
   327.79

Multiplication

When we multiply 2.4×1.6, where does the decimal point go? First we write the numbers and perform the multiplication, ignoring the decimal point.

$$
\begin{array}{r}
2.4 \\
\times\ 1.6 \\
\hline
144 \\
24\ \ \\
\hline
384
\end{array}
$$

To find out where the decimal point goes, we count the number of digits to the right of the decimal point in each term. (There is one digit to the right of the decimal point in each term.) Then we add these numbers of digits (one plus one equals two) and count off this new number from the right-hand side of the answer. In our example, we count off two digits from the right, and we place the decimal point between the 3 and the 8. This gives us an answer of 3.84.

EXAMPLE A.11

Perform the indicated operation: (a) $(4.25)(5)$ (b) $(7.12)(3.64)$
(c) $(5.222)(4.11)$

Solution

(a)   4.25
  ×   5
  21.25

(b)   7.12
  × 3.64
   2848
  4272
  2136
  25.9168

(c)   5.222
  ×   4.11
   5222
  5222
 20888
 21.46242

Division

Where does the decimal point go when we divide 11 by 0.56? First we write the problem in a more familiar way.

$$0.56\overline{)11}$$

(Here 0.56 is called the **divisor,** and 11 is called the **dividend.**) Our next step is to move the decimal point in the divisor so that all digits are to the left of the decimal point: 0.56 becomes 56. We must also move the decimal point in the dividend the same number of places: 11 becomes 1,100.

$$0.56\overline{)11.00} \qquad \text{becomes} \qquad 56\overline{)1,100.00}$$

We now perform the division, locating the decimal point in the answer directly above the decimal point in the dividend.

```
        19.64
56)1,100.00
     56x xx
     540
     504
      360
      336
       240
       224
        16
```

EXAMPLE A.12

Perform the indicated operation: (a) $\dfrac{9.25}{2.5}$ (b) $\dfrac{24.138}{7.45}$

Solution

(a) $2.5\overline{)9.2\,5}$ (b) $7.45\overline{)24.13\,8}$

```
        3.7                      3.24
   25)92.5              745)2,413.80
      75 x                   2235 xx
      175                      1788
      175                      1490
        0                      2980
                              2980
                                 0
```

A.6 Solving Algebraic Equations

Very often in chemistry, we need to solve algebraic equations ("find the unknown"). So study the following examples carefully.

EXAMPLE A.13

LEARNING GOAL 7

Solving simple algebraic equations

Solve each of the following equations for the unknowns a, y, and z:
(a) $3a = 9$ (b) $4y + 5 = 37$ (c) $9z - 3 = 22$

Solution

To solve an algebraic equation, we try to isolate the unknown.

(a) In $3a = 9$, we can isolate the a by dividing both sides of the equation by 3. This operation removes the 3 from the left-hand side of the equation. Remember, what we do to one side of the equation we *must* do to the other side, in order to maintain the equality.

$$\frac{\cancel{3}a}{\cancel{3}} = \frac{9}{3}$$

After simplifying both sides of the equation, we get the equality

$$a = 3$$

(b) In $4y + 5 = 37$, we want to isolate the y. Our first step is to remove the 5 from the left-hand side of the equation. We do this by subtracting 5 from both sides of the equation.

$$4y + 5 - 5 = 37 - 5 \qquad \text{so} \qquad 4y = 32$$

We now divide both sides of the equation by 4.

$$\frac{\cancel{4}y}{\cancel{4}} = \frac{32}{4}$$

After simplifying both sides of the equation, we get the equality

$$y = 8$$

(c) In $9z - 3 = 22$, the first step is to remove the 3 from the left-hand side of the equation. We do this by adding 3 to both sides of the equation.

$$9z - 3 + 3 = 22 + 3 \qquad \text{so} \qquad 9z = 25$$

Now divide both sides of the equation by 9.

$$\frac{\cancel{9}z}{\cancel{9}} = \frac{25}{9}$$

After simplifying both sides of the equation, we have the equality

$$z = \frac{25}{9}$$

EXAMPLE A.14

Solve each of the following equations for the unknown: (a) $\dfrac{2a}{3} = \dfrac{12}{9}$ (b) $5y + 3 = 2y - 42$

Solution

(a) In the equation

$$\frac{2a}{3} = \frac{12}{9}$$

We can remove the 3 on the left-hand side by multiplying both sides of the equation by 3.

$$(3)\left(\frac{2a}{3}\right) = \left(\frac{12}{9}\right)(3) \qquad 2a = \frac{36}{9} \qquad 2a = 4$$

Now divide both sides of the equation by 2.

$$\frac{\cancel{2}a}{\cancel{2}} = \frac{4}{2}$$

After simplifying both sides of the equation, we have the equality

$$a = 2$$

(b) In the equation

$$5y + 3 = 2y - 42$$

we first gather all the terms containing the unknown on one side. Place all y terms on the left-hand side of the equation by subtracting $2y$ from each side of the equation.

$$5y + 3 - 2y = 2y - 42 - 2y \qquad \text{so} \qquad 3y + 3 = -42$$

Then subtract 3 from each side of the equation.

$$3y + 3 - 3 = -42 - 3 \qquad \text{so} \qquad 3y = -45$$

Now divide each side of the equation by 3.

$$\frac{\cancel{3}y}{\cancel{3}} = \frac{-45}{3}$$

After simplifying both sides of the equation, we have the equality

$$y = -15$$

A.7 Ratios and Proportions

LEARNING GOAL 8

Solving proportions

There are four wheels on a car. This relationship can be expressed mathematically in the following ways:

$$4\ \text{wheels} : 1\ \text{car} \qquad \text{or} \qquad \frac{4\ \text{wheels}}{1\ \text{car}}$$

Both statements express the fact that there are four wheels on one car. Such statements are known as **ratios** because they reveal the numerical relationship between different things. A ratio is just like a factor unit, and it can be used as an alternative to the factor-unit method to solve many types of problems. But to use ratios in problem solving, you must understand the concept of proportions. A **proportion** is *an equality between two ratios*, as in

$$\frac{1}{2} = \frac{5}{10} \qquad \text{or} \qquad 1 : 2 = 5 : 10$$

(When the second expression is used, the symbol : is often read "is to" and the symbol = is read "as.") In a problem involving a proportion, you are usually given three of the four numbers and are asked to determine the missing number. Example A.15 shows how this works.

EXAMPLE A.15

If there are 4 wheels on 1 car, how many wheels are there on 5 cars?

Solution

The proportion is set up in the following way:

$$\frac{4 \text{ wheels}}{1 \text{ car}} = \frac{y \text{ wheels}}{5 \text{ cars}}$$

To solve for *y*, multiply each side of the equation by 5 cars.

$$(5 \text{ cars}) \left(\frac{4 \text{ wheels}}{1 \text{ car}} \right) = \left(\frac{y \text{ wheels}}{5 \text{ cars}} \right) (5 \text{ cars})$$

$$20 \text{ wheels} = y \text{ wheels}$$

In other words, *y* = 20 wheels, so there are 20 wheels on 5 cars.

A.8   Solving Word Equations

LEARNING GOAL 9
Solving simple density problems

Many of the mathematical equations we use in chemistry are word equations, such as

$$\text{Density} = \frac{\text{mass}}{\text{volume}} \qquad \text{sometimes abbreviated as} \qquad D = \frac{m}{V}$$

As written, the equation is set up so that we can solve for the density of a substance if we are given its mass and volume. In some problems, however, we are given the density and volume of a substance and are asked to solve for its mass. We do so as shown in Example A.16.

EXAMPLE A.16

Given the following information, solve for the mass *m*:

$$D = 0.8 \frac{\text{g}}{\text{mL}} \qquad V = 30 \text{ mL}$$

Solution

First write the formula.

$$D = \frac{m}{V}$$

Then isolate the m by multiplying both sides of the equation by V.

$$(D)(V) = \left(\frac{m}{\cancel{V}}\right)(\cancel{V}) \qquad \text{so} \qquad DV = m$$

Now substitute the numbers given in the problem and solve for the mass in grams.

$$\left(0.8\,\frac{g}{\cancel{mL}}\right)(30\,\cancel{mL}) = m \qquad \text{so} \qquad 24\,g = m$$

In some problems, we are given the density and mass of a substance and are asked to solve for its volume.

EXAMPLE A.17

Given the following information, solve for the volume V:

$$D = 2\,\frac{g}{mL} \qquad m = 20\,g$$

Solution

Write the density formula in its original form.

$$D = \frac{m}{V}$$

We must get the V into the numerator by itself. This can be done in two steps:

1. Multiply both sides of the equation by V.

$$(D)(V) = \left(\frac{m}{\cancel{V}}\right)(\cancel{V}) \qquad \text{so} \qquad DV = m$$

2. Divide both sides of the equation by D.

$$\frac{\cancel{D}V}{\cancel{D}} = \frac{m}{D} \qquad \text{so} \qquad V = \frac{m}{D}$$

Now substitute the numbers given in the problem and solve for the volume in milliliters.

$$V = \frac{m}{D} \qquad \text{so} \qquad V = \frac{20\,\cancel{g}}{2\,\cancel{g}/mL} = 10\,mL$$

Do you understand how the unit *gram* cancels out to give milliliters in the answer? We actually have

$$V = 20 \text{ g} \div \frac{2 \text{ g}}{1 \text{ mL}}$$

When we apply the *rules for division of fractions* discussed in Section A.2, this expression becomes

$$20 \cancel{\text{ g}} \times \frac{1 \text{ mL}}{2 \cancel{\text{ g}}} = 10 \text{ mL}$$

and the unit *gram* cancels out.

A.9   Calculating and Using Percentages

LEARNING GOAL 10

Solving problems involving percentages

A **percentage** is *the number of parts of something out of 100 parts.* For example, if a chemistry class has 100 students and 40 of the students are women, the percentage of women is 40 percent (also written as 40%).

$$40 \text{ women per 100 students} = 40\%$$

We can state this mathematically (using the factor-unit method) as follows:

$$\text{Percentage women} = \left(\frac{40 \text{ women}}{100 \cancel{\text{ students}}} \right) (100 \cancel{\text{ students}})$$

$$= 40 \text{ percent women}$$

However, we usually set the formula up as follows:

$$\textbf{Percentage} = \frac{\text{number of items of interest}}{\text{total number of items}} \times 100$$

Suppose that we have a chemistry class with 60 students and that 48 of them are women. What is the percentage of women in the class? By the formula,

$$\text{Percentage} = \frac{\text{number of items of interest}}{\text{total number of items}} \times 100$$

$$= \frac{48 \text{ women}}{60 \text{ students}} \times 100 = 80\% \text{ women}$$

EXAMPLE A.18

A box of mixed vegetables has 25 carrots, 30 tomatoes, and 95 heads of lettuce. What is the percentage of each vegetable in the box?

Solution

First obtain the total number of vegetables in the box.

25 carrots + 30 tomatoes + 95 lettuces = 150 vegetables

Now calculate the percentage of each vegetable, using our formula.

$$\% \text{ carrots} = \frac{25 \text{ carrots}}{150 \text{ vegetables}} \times 100 = 16.7\% \text{ carrots}$$

$$\% \text{ tomatoes} = \frac{30 \text{ tomatoes}}{150 \text{ vegetables}} \times 100 = 20\% \text{ tomatoes}$$

$$\% \text{ lettuces} = \frac{95 \text{ lettuces}}{150 \text{ vegetables}} \times 100 = 63.3\% \text{ lettuces}$$

Note that the total of the percentages equals 100%. In other words, the whole is the sum of the parts.

Sometimes we know the percentage of something and we want to solve for the particular number of items. For example, say that 30% of the people in a particular community own a car. If this community has 800 people, how many people own cars? We can solve this problem using the factor-unit method if we remember that 30% car ownership means

$$\frac{30 \text{ people own cars}}{100 \text{ people in town}}$$

Therefore

$$? \text{ people owning cars} = (800 \text{ people}) \left(\frac{30 \text{ people owning cars}}{100 \text{ people}} \right)$$

$$= 240 \text{ people own cars}$$

However, a simpler method of solving this problem is to move the decimal point on the percent number two places to the left and multiply it by the number of people in town. In this method, 30% becomes 0.30. Therefore

$$800 \text{ people} \times 0.30 = 240 \text{ people own cars}$$

EXAMPLE A.19

A box of fruit contains apples, oranges, and bananas. There are 500 pieces of fruit in the box; 20% are apples, 30% are oranges, and 50% are bananas. How many pieces of each kind of fruit are in the box?

Solution

Turn each percent into its decimal equivalent and multiply it by the total number of fruit in the box.

20% becomes 0.20 (for apples)

30% becomes 0.30 (for oranges)

50% becomes 0.50 (for bananas)

Therefore

$$500 \text{ fruit} \times 0.20 = 100 \text{ apples}$$
$$500 \text{ fruit} \times 0.30 = 150 \text{ oranges}$$
$$500 \text{ fruit} \times 0.50 = 250 \text{ bananas}$$

Note that the total number of pieces of fruit is 500.

A.10 Using the Calculator

■ LEARNING GOAL 11

Performing mathematical operations using a calculator

If they are used correctly, electronic calculators speed up the process of making mathematical computations. One common error is the simple mistake of pushing the buttons incorrectly (in the wrong order, for instance). Many such errors can be detected by first estimating the answer so you can tell whether the calculated answer makes sense.

You should follow the specific directions given for your particular calculator, of course, but the following rules apply to most calculators:

1. Enter the operations in the order in which they are written.

$$\boxed{4}\;\boxed{\times}\;\boxed{5}\;\boxed{=}\;\boxed{20}$$

First press the "4" key, then the "×" key, then the "5" key, followed by the "=" key. The answer "20" will appear in the electronic display position.

EXAMPLE A.20

Perform the following operations:
(a) $\boxed{10}\;\boxed{\times}\;\boxed{5}\;\boxed{=}$
(b) $\boxed{10}\;\boxed{+}\;\boxed{5}\;\boxed{=}$
(c) $\boxed{10}\;\boxed{\div}\;\boxed{5}\;\boxed{=}$
(d) $\boxed{10}\;\boxed{-}\;\boxed{5}\;\boxed{=}$

Solution
(a) 50
(b) 15
(c) 2
(d) 5

---

2. Once the calculator is directed to add, subtract, multiply, or divide two numbers, the results of the mathematical operation are stored in its memory. Another operation key ($+$, $-$, $\times$, or $\div$) can then be pressed and another calculation carried out.

EXAMPLE A.21

Perform the following operations:

(a) $\boxed{10}\;\boxed{\times}\;\boxed{5}\;\boxed{=}\;\boxed{+}\;\boxed{4}\;\boxed{=}$

(b) $\boxed{20}\;\boxed{\div}\;\boxed{5}\;\boxed{-}\;\boxed{4}\;\boxed{=}$

(c) $\boxed{6}\;\boxed{-}\;\boxed{4}\;\boxed{\times}\;\boxed{20}\;\boxed{=}$

(d) $\boxed{8}\;\boxed{+}\;\boxed{6}\;\boxed{\times}\;\boxed{2}\;\boxed{=}$

Solution

(a) 54

(b) 0

(c) 40

(d) 28

3. Your calculator probably has a floating decimal point. Once you reach the proper place for the decimal point, press the $\boxed{\cdot}$ key to position the decimal point.

4. Final zeros to the right of the decimal point are not shown in the electronic display after a calculation is complete.

SELF-TEST EXERCISES

LEARNING GOAL 1: *Adding and Subtracting Algebraically*

1. *Add* the following numbers algebraically:
(a) $5 + 12$ (b) $-5 - 10$ (c) $5 + 10$
(d) $-5 + 10$

2. *Subtract* the following numbers algebraically:
(a) $3 - (+2)$ (b) $3 - (-2)$ (c) $-3 - (+2)$
(d) $-3 - (-2)$

3. *Add* the following numbers algebraically:
(a) $8 + 25$ (b) $8 - 25$ (c) $-8 + 25$
(d) $-8 - 25$

4. *Subtract* the following numbers algebraically:
(a) $55 - (+44)$ (b) $55 - (-44)$
(c) $-55 - (+44)$ (d) $-55 - (-44)$

LEARNING GOAL 2: *Multiplying and Dividing Fractions*

5. Perform the following operations: (a) $\frac{1}{8} \times \frac{1}{2}$

(b) $\frac{1}{8} \div \frac{1}{2}$ (c) 8^3 (d) $(2)^{-5}$

6. Perform the following operations: (a) $\frac{3}{5} \times \frac{2}{7}$

(b) $\frac{20}{50} \times \frac{9}{5}$ (c) $\frac{5}{6} \div \frac{10}{3}$ (d) $\frac{25}{35} \div \frac{7}{5}$

LEARNING GOAL 3: *Multiplying and Dividing Exponential Numbers*

7. Perform the following operations: (a) $(4)^{-2}(4)^{-5}$
(b) $(4)^2(4)^3$ (c) $(6)^4(6)^{-2}$ (d) $(5)^{-8}(5)^9$

8. Perform the following operations: (a) $\frac{(4)^5}{(4)^2}$

(b) $\frac{(4)^5}{(4)^{-3}}$ (c) $\frac{(a)^3(a)^2}{(a)^4}$ (d) $\frac{(b)^6(b)^4}{(b)^3(b)^5}$

9. Perform the following operations: (a) $(25)^2$

(b) $(10)^3$ (c) $(99)^1$ (d) $(8)^{-3}$ (e) $\frac{1}{(4)^{-2}}$

10. Perform the following operations: (a) $(5)^2(5)^3$
(b) $(12)^3(12)^9$ (c) $(10)^2(10)^{-2}$ (d) $(4)^{-8}(4)^{-2}$

11. Perform the following operations: (a) $\frac{(4)^5}{(4)^3}$

(b) $\frac{(12)^3}{(12)^{-2}}$ (c) $\frac{(20)^{-4}}{(20)^3}$ (d) $\frac{(20)^{-4}}{(20)^{-3}}$

12. Perform the following operations: (a) $(4)^4(4)^3(4)^2$
(b) $(x)^2(x)^3(x)^{-4}$ (c) $\dfrac{(y)^2(y)^8}{(y)^3}$ (d) $\dfrac{(p)^2(p)^{-9}}{(p)^{-8}}$
(e) $\dfrac{(10)^{-3}(10)^5(10)^7}{(10)^{-8}(10)^9}$

13. Perform the following operations: (a) $(2)^6$
(b) $(3)^{-4}$ (c) $\dfrac{1}{(4)^3}$ (d) $\dfrac{1}{(4)^{-3}}$

14. Perform the following operations: (a) $(10)^5(10)^{-2}$
(b) $(2)^{-3}(2)^{-2}$ (c) $(3)^{-5}(3)^2$ (d) $(6)^{-8}(6)^{10}$

15. Perform the following operations: (a) $\dfrac{(3)^7}{(3)^5}$
(b) $\dfrac{(2)^4(2)^{-6}}{(2)^7(2)^{-12}}$ (c) $\dfrac{(a)^{12}(a)^{-10}}{(a)^{-7}(a)^{-6}}$ (d) $\dfrac{(b)^{-20}(b)^{-10}}{(b)^{-9}(b)^{-8}}$

LEARNING GOAL 4: ***Solving Problems Using the Factor-Unit Method***

16. $(4 \text{ feet})\left(\dfrac{12 \text{ inches}}{1 \text{ foot}}\right) = ?$

17. $A = l \times w$; therefore $w = ?$ Solve the equation for w in terms of A and l.

18. How many yards are there in 18 feet?

19. How many eggs are there in 4 dozen eggs?

20. (a) How many hours are there in 360 minutes?
(b) How many hours are there in 7,200 seconds?

21. How many pints are there in 8 quarts?

22. Change 1 mile per hour to feet per second.

23. $(72 \text{ inches})\left(\dfrac{1 \text{ foot}}{12 \text{ inches}}\right) = ?$

24. $(15 \text{ feet})\left(\dfrac{12 \text{ inches}}{1 \text{ foot}}\right) = ?$

25. How many feet are there in 100 yards?

26. How many inches are there in 5 yards?

27. How many dozen eggs do you have if you have 288 eggs?

28. How many minutes are there in 24 hours? How many seconds are there in 24 hours?

29. How many gallons are there in 40 quarts?

30. A rocket is traveling at 22,000 feet per second. How many miles per hour is it traveling?

LEARNING GOAL 5: ***Adding and Subtracting Decimals***

31. Perform the following operations:
(a) $15.4 + 117.33 + 16.909 + 0.044$
(b) $171.82 - 30.41$

32. Perform the following operations:
(a) $25.431 + 0.761 + 0.325 + 0.008$
(b) $123.25 - 19.54$ (c) $98.77 - 38.25 + 45.62$
(d) $254.37 - 68.26 - 38.33$

LEARNING GOAL 6: ***Multiplying and Dividing Decimals***

33. Perform the following operations: (a) $(7.33)(4)$
(b) $(9.01)(4.28)$ (c) $(3.111)(8.7)$

34. Perform the following operations: (a) $\dfrac{6.324}{3.1}$
(b) $\dfrac{25.30}{5.06}$

35. Perform the following operations:
(a) $(9.54)(3.27)$ (b) $(6.5)(7.1)$ (c) $\dfrac{80.4}{2.02}$
(d) $\dfrac{640.75}{8.5}$

LEARNING GOAL 7: ***Solving Simple Algebraic Equations***

36. Solve each of the following equations for the unknown: (a) $25a = 100$ (b) $40y + 13 = 26$
(c) $12z - 12 = 24$

37. Solve each of the following equations for the unknown: (a) $\dfrac{6x}{3} = \dfrac{36}{12}$ (b) $12y + 6 = 6y + 12$

38. Solve the equation $PV = nRT$ for (a) P,
(b) n,   (c) T,   (d) n/V.

39. Solve each of the following equations for the unknown: (a) $4t = 64$ (b) $6y + 3 = 45$
(c) $2x - 7 = 33 - 3x$ (d) $40k + 8 = 96 + 51k$

40. Solve the equation $A = bcd$ for   (a) b,   (b) c,
(c) A/b.

LEARNING GOAL 8: ***Solving Proportions***

41. Solve each of the following equations for the unknown: (a) $\dfrac{x}{25} = \dfrac{4}{5}$ (b) $\dfrac{3}{y} = \dfrac{7}{63}$ (c) $\dfrac{4t}{12} = \dfrac{2}{3}$
(d) $\dfrac{2}{5k} = \dfrac{5}{50}$

42. Solve each of the following proportions:

(a) $\dfrac{k}{5} = \dfrac{9}{45}$ (b) $\dfrac{2}{7} = \dfrac{h}{28}$ (c) $\dfrac{72}{f} = \dfrac{9}{2}$ (d) $\dfrac{18}{6} = \dfrac{12}{y}$

43. Solve each equation for the missing value or values: (a) $2:5:9 = x:20:y$ (b) $7:9 = h:54$ (c) $2:3:5:7 = 1:a:b:c$ (d) $2:5 = 5:d$

LEARNING GOAL 9: *Solving Simple Density Problems*

44. Given that $D = \dfrac{m}{V}$, solve this formula for (a) m and (b) V.

45. A substance has a mass of 25 g and a volume of 5 mL. What is its density?

LEARNING GOAL 10: *Solving Problems Involving Percentages*

46. A university has the following enrollments: (a) liberal arts, 200 students; (b) laboratory technology, 100 students; and (c) health science technology, 300 students. This accounts for all the students at the university. What percentage of the students are enrolled in each curriculum?

47. A survey of 2,000 people shows that 40% want to travel to Europe, 30% to Asia, 10% to Alaska, and 20% to Miami. How many people want to travel to each place?

LEARNING GOAL 11: *Performing Mathematical Operations Using a Calculator*

48. Perform the following operations using a calculator:
(a) $11 \times 6 = ?$
(b) $11 + 6 = ?$
(c) $30 \div 5 = ?$
(d) $30 - 5 = ?$

49. Perform the following operations using a calculator:
(a) $(12 \times 5) + 5 = ?$
(b) $(20 \div 4) - 2 = ?$
(c) $(10 + 20) \div (5 \times 6) = ?$
(d) $(100 \times 50) \div (10 \times 40) = ?$

50. Perform the following operations using a calculator:
(a) $(50 \times 6) \div 12 = ?$
(b) $(100 + 50) \div (15 \times 0.01) = ?$
(c) $(0.05 \times 0.30) \div 4 = ?$
(d) $(10,000 \times 5) \div 4,000 = ?$

APPENDIX B
Important Chemical Tables

TABLE B.1

Prefixes and Abbreviations

nano- $= 0.000000001$	$= 10^{-9}$	nanometer $=$ nm	centigram $=$ cg
micro- $= 0.000001$	$= 10^{-6}$	micrometer $= \mu$m (Greek letter mu)	decimeter $=$ dm
milli- $= 0.001$	$= 10^{-3}$	millimeter $=$ mm	decigram $=$ dg
centi- $= 0.01$	$= 10^{-2}$	milliliter $=$ mL	kilometer $=$ km
deci- $= 0.1$	$= 10^{-1}$	milligram $=$ mg	kilogram $=$ kg
deka- $= 10$	$= 10^{1}$	centimeter $=$ cm	
kilo- $= 1,000$	$= 10^{3}$		

TABLE B.2

The Metric System

Length

1 millimeter = 0.001 meter = $\frac{1}{1000}$ meter	or	1 meter = 1,000 millimeters
1 centimeter = 0.01 meter = $\frac{1}{100}$ meter	or	1 meter = 100 centimeters
1 decimeter = 0.1 meter = $\frac{1}{10}$ meter	or	1 meter = 10 decimeters
1 kilometer = 1,000 meters	or	1 meter = 0.001 kilometer

Mass

1 microgram = 0.000001 gram	1 gram = 1,000,000 micrograms
1 milligram = 0.001 gram	1 gram = 1,000 milligrams
1 centigram = 0.01 gram	1 gram = 100 centigrams
1 decigram = 0.1 gram	1 gram = 10 decigrams
1 kilogram = 1,000 grams	1 gram = 0.001 kilogram

Volume

1 milliliter = 0.001 liter

1 milliliter = 1 cubic centimeter*

*Cubic centimeter is abbreviated cm^3 or cc.

TABLE B.3

Conversion of Units (English—Metric)

TO CONVERT	INTO	MULTIPLY BY
Length		
inches	centimeters	2.540 cm/in
centimeters	inches	0.3937 in/cm
feet	meters	0.30 m/ft
meters	feet	3.28 ft/m
Weight (mass)		
ounces	grams	28.35 g/oz
grams	ounces	0.035 oz/g
pounds	grams	454 g/lb
grams	pounds	0.0022 lb/g
Volume		
liters	quarts	1.057 qt/L
quarts	liters	0.9463 L/qt

A-21

TABLE B.4

Solubilities

	ACETATE	ARSENATE	BROMIDE	CARBONATE	CHLORATE	CHLORIDE	CHROMATE	HYDROXIDE	IODIDE	NITRATE	OXIDE	PHOSPHATE	SULFATE	SULFIDE
Aluminum	W	a	W	—	W	W	—	A	W	W	a	A	W	d
Ammonium	W	W	W	W	W	W	W	W	W	W	—	W	W	W
Barium	W	w	W	w	W	W	A	W	W	W	W	A	a	d
Cadmium	W	A	W	A	W	W	A	A	W	W	A	A	W	A
Calcium	W	w	W	w	W	W	W	W	W	W	w	w	w	w
Chromium	W	—	W*	W	—	I	—	A	W	W	a	w	W†	d
Cobalt	W	A	W	A	W	W	A	A	W	W	A	A	W	A
Copper(II)	W	A	W	—	W	W	—	A	a	W	A	A	W	A
Hydrogen	W	W	W	—	W	W	—	—	W	W	W	W	W	W
Iron(II)	W	A	W	w	W	W	—	A	W	W	A	A	W	A
Iron(III)	W	A	W	—	W	W	A	A	W	W	A	w	w	d
Lead(II)	W	A	W	A	W	W	A	w	w	W	w	A	w	A
Magnesium	W	A	W	w	W	W	W	A	W	W	A	w	W	d
Mercury(I)	w	A	A	A	W	a	w	—	A	W	A	A	w	I
Mercury(II)	W	w	W	—	W	W	w	A	w	W	w	A	d	I
Nickel	W	A	W	w	W	W	A	w	W	W	A	A	W	A
Potassium	W	W	W	W	W	W	W	W	W	W	W	W	W	W
Silver	w	A	a	A	W	a	w	—	I	W	w	A	w	A
Sodium	W	W	W	W	W	W	W	W	W	W	d	W	W	W
Strontium	W	w	W	w	W	W	w	W	W	W	A	w	w	W
Tin(II)	d	—	W	—	W	W	A	A	W	d	A	A	W	A
Tin(IV)	W	—	W	—	—	W	W	w	d	—	A	—	W	A
Zinc	W	A	W	w	W	W	w	A	W	W	w	A	W	A

Reprinted with permission from *CRC Handbook of Chemistry and Physics,* 78th ed. Copyright © The Chemical Rubber Co., Boca Raton: CRC Press, Inc.

Abbreviations: W = soluble in water; *A* = insoluble in water, but soluble in acids; *w* = only slightly soluble in water, but soluble in acids; *a* = insoluble in water, and only slightly soluble in acids; *I* = insoluble in both water and acids; *d* = decomposes in water.

*CrBr$_3$

†Cr$_2$(SO$_4$)$_3$

APPENDIX C
Glossary

In each entry, the number in parentheses is the number and section of the chapter in which the term is first discussed.

absolute zero (13.4)   The coldest possible temperature that matter can reach. It is equal to 0 K ($-273°C$).

accuracy (2.4)   The closeness of a measurement to its true value

acid (10.4)   A substance that releases hydrogen ions in solution, counteracts bases, and donates protons

activity series (10.5)   The elements listed in decreasing order of their reactivity, or ability to react chemically

alkali (16.2)   Any substance that dissolves to give a basic solution

alpha (α) particle (4.4)   Particle with a mass of 4 atomic mass units and a charge of $+2$ emanating from an unstable nucleus. This particle is like a helium atom with its electrons removed.

amorphous solid (3.5)   A solid that does not have a well-defined crystalline structure

anhydrous salt (14.7)   A salt molecule that has no water of hydration bonded to it

anion (6.7)   A negatively charged ion

anode (4.3)   A positive electrode in an electrolytic cell

area (2.6)   A measure of the extent of a surface, equal to an object's length times its width

Arrhenius acid (16.1)   A substance that releases hydrogen ions when it dissolves in water

Arrhenius base (16.2)   A substance that releases hydroxide ions when it dissolves in water

atmosphere (atm) (13.2)   A unit of pressure equal to 760 torr or 14.7 pounds per square inch

atom (3.10)   The smallest part of an element that can enter into chemical combinations

atomic mass (3.15)   The mass of an element in relation to the mass of an atom of carbon-12

atomic mass unit (amu) (4.5)   The unit used to compare the relative masses of atoms. One atomic mass unit is one-twelfth the mass of a carbon-12 atom.

atomic number (4.6)   The number of electrons or protons in a neutral atom

atomic radius (6.6)   The distance from the center of the nucleus to the outermost electron

Avogadro's number (9.1)   The number of atoms whose mass is the gram-atomic mass of any element. It is equal to 6.02×10^{23}.

Avogadro's principle (13.9)   Equal volumes of all gases, under the same conditions of temperature and pressure, contain the same number of molecules

balanced equation (10.2)   A chemical equation that has the same number of atoms of each element on the reactant side as on the product side

barometer (13.2)   A device used to measure air pressure

base (10.4)   A substance that releases hydroxide ions in solution, counteracts acids, and accepts protons

base number (2.5)   A number that is raised to a power, for example, 10 in 10^3

binary compound (8.1)   A compound composed of two elements

boiling point (14.3)   The temperature at which the equilibrium vapor pressure of a liquid equals the atmospheric pressure

boiling-point elevation (15.9)   The temperature at which a solution boils, which is higher than the boiling point of the pure solvent alone

boiling-point-elevation constant (15.10)   The boiling point of a solution minus the boiling point of the pure solvent alone divided by the molality of the solution. This constant is unique for each solvent.

Boyle's Law (13.3)   At constant temperature, the volume of a given mass of gas varies inversely with the pressure.

Brønsted–Lowry acid (16.1)   A substance that donates protons to another substance

Brønsted–Lowry base (16.2)   A substance that accepts protons from another substance

calorie (cal) (12.2)   A unit of heat energy representing the amount of heat needed to raise the temperature of 1 gram of water by 1 Celsius

cathode (4.3)   A negative electrode in an electrolytic cell

cation (6.7)   A positively charged ion

Celsius (centigrade) temperature scale (2.11)   A temperature scale on which the freezing point of water is 0 degrees and the normal boiling point of water is 100 degrees

Charles's Law (13.4)   At constant pressure, the volume of a given amount of gas varies directly with the temperature in kelvins.

chemical bond (Chapter 7, Introduction)   The binding force due to electron loss or gain that holds atoms or ions together in molecules or formula units

chemical change (3.6)   A change in the chemical composition of a substance

chemical formula (3.14)   The combination of the symbols of the particular elements that form a chemical compound, showing the number of atoms per element

chemical nomenclature (Chapter 8, Introduction)   A system for naming chemical compounds

chemical property (3.6)   Those properties that show how one substance reacts with another substance

coefficient (10.3)   A number placed before an element or a compound in a chemical equation to balance the equation

colligative property (15.9)   Any of the collective properties of solutions, such as boiling-point elevation and freezing-point depression, that depend mainly on the number of solute particles present in a given quantity of solvent

combination reaction (10.4)   A reaction in which two or more substances combine to form a more complex substance

Combined Gas Law (13.5)   The gas law formula that relates the initial pressure, volume, and temperature of a gas to its final pressure, volume, and temperature. For a given quantity of gas, $P_iV_i/T_i = P_fV_f/T_f$.

common name (Chapter 8, Introduction)   A name for a chemical compound that is derived from common usage or has been handed down through chemical history

compound (3.11)   A chemical combination of two or more elements

concentration (15.3)   The amount of solute in a solution, which can be expressed in terms of percentage, molarity, normality, or molality

condense (14.5)   To change from the gaseous state to the liquid state

continuous spectrum (5.1)   A series of colors in which one color merges into the next, such as a rainbow

coordinate covalent bond (7.3)   A chemical bond in which one atom donates all the electrons used to form the bond

covalent bond (7.2)   A chemical bond formed by the sharing of electrons between two atoms

crystal (14.6)   A solid that has geometrically arranged planar surfaces and a symmetrical structure on the inside

crystal lattice (14.6)   The symmetrical structure formed by the particles in a crystal

crystalline solid (3.5)   A solid that has a fixed, regularly repeating, symmetrical internal structure

Dalton's Law of Partial Pressures (13.7)   The total pressure of a mixture of gases is equal to the sum of the partial pressures of all the gases in the mixture.

decomposition reaction (10.4)   A reaction in which a complex substance is broken down into simpler substances

deliquescence (14.7)   The ability to absorb enough water from the atmosphere to form a solution

density (2.10)   The mass per unit volume of a substance

desiccant (14.7)   A hygroscopic substance that is used as a drying agent

desiccator (14.7)   A container used to hold chemicals and keep them dry

diatomic element (7.2)   An element that is found in molecules made up of two like atoms. For example, hydrogen, oxygen, and nitrogen exist in the diatomic forms H_2, O_2, and N_2, respectively.

diffusion (15.11)   The process by which a solute moves from an area of high solute concentration to an area of low solute concentration

dissociate (10.4)   To separate an ionic substance into ions by the action of a solvent

double covalent bond (7.2)   A chemical bond in which two pairs of electrons are shared between two atoms

double-replacement reaction (10.4) A reaction in which two compounds exchange their ions with each other

efflorescence (14.7) The spontaneous loss of water of hydration by a hydrated salt when exposed to the atmosphere

electrolytic solution (15.7) A solution that conducts electric current

electromagnetic spectrum (5.1) The full range of electromagnetic radiation, including radio waves and x rays

electron (4.3) A particle with a relative negative charge of one unit and a mass of 0.0005486 atomic mass units

electron affinity (6.8) The energy released when an additional electron is added to a neutral atom

electron configuration (5.8) The positions of the electrons in the various energy levels of an atom

electron-dot structure (Lewis electron-dot structure) (7.1) A notation that shows the symbol for an element and, by the use of dots, the number of outer electrons in an atom of the element

electronegativity (7.6) The attraction that an atom has for the electrons it is sharing with another atom

element (3.9) Any of the basic building blocks of matter that cannot be broken down physically or by chemical means into simpler substances

empirical formula (9.2) A chemical formula showing the simplest whole-number ratio of the atoms that make up a molecule of a compound

endothermic reaction (12.1) A reaction that absorbs energy from the surroundings

end point (16.10) The point in a titration at which the indicator changes color

energy (3.2) The ability to do work. Energy appears in many forms—for example, as heat and as chemical, electrical, mechanical, and radiant (light) energy

energy level (5.3) Any of the various regions outside the nucleus of an atom in which electrons move

energy sublevel (subshell) (5.6) Any of the more specific regions within an energy level in which electrons move

enthalpy (12.2) The heat content of a chemical substance, represented by the symbol H

equilibrium vapor pressure (14.2) The pressure exerted by a vapor when it is in equilibrium with its liquid at any given temperature

equivalence point (16.9) The point in an acid–base titration at which the equivalents of acid are equal to the equivalents of base

equivalent (of an acid or a base) (15.5) The mass of the acid or base that contains Avogadro's number of hydrogen ions or hydroxide ions

evaporate (14.2) To change from the liquid state to the gaseous (or vapor) state

excited state (5.3) The state of an electron when it is at an energy level higher than its ground-state level. An electron is in an excited state after it has absorbed energy.

exothermic reaction (12.1) A reaction that releases energy to the surroundings

exponent (2.5) A number written to the right of and above another number (the base number) and indicating the operation of raising to a power, for example, 3 in 10^3

Fahrenheit temperature scale (2.11) A temperature scale on which the freezing point of water is 32 degrees and the normal boiling point of water is 212 degrees

family (6.2) A vertical column of elements in the periodic table

formula mass (3.16) The sum of the atomic masses of all the atoms that make up a formula unit of a compound

formula unit (3.13) For an ionic compound, the smallest part of the compound that retains the properties of that compound

freeze (14.4) To change from the liquid state to the solid state

freezing point (14.4) The temperature at which the solid and liquid states of a substance exist together

freezing-point depression (15.9) The temperature at which a solution freezes, which is lower than the freezing point of the pure solvent alone

freezing-point depression constant (15.10) The freezing point of a solution minus the freezing point of the pure solvent alone, divided by the molality. This constant is unique for each solvent.

gas (3.5) The least compact of the three physical states of matter. Gases have no definite shape or volume; they are easily compressible and will spread to fill the container in which they are placed.

Gay-Lussac's Law of Combining Volumes (13.8) When gases at the same temperature and pressure combine, the volumes of the reactants and products are in the same ratio as the coefficients of the balanced equation for the reaction.

gram-atomic mass (9.1) The atomic mass of a mole of atoms of an element expressed in grams

gram-formula mass (9.3)   The formula mass of a substance expressed in grams

ground state (5.3)   The most stable state of an atom. In this state, the electrons are in their lowest possible energy levels.

group (of elements) (6.2)   A vertical column of elements in the periodic table

Haber process (Chapter 11, Introduction)   A method used to synthesize ammonia from its elements, nitrogen and hydrogen, for commercial purposes

heat of crystallization (solidification) (14.4)   The quantity of heat released to the environment when a quantity of liquid changes to a solid at the freezing point. It is usually expressed in calories per gram.

heat of formation (12.7)   The heat released or absorbed when 1 mole of a compound is formed from its elements

heat of reaction (12.2)   The change in heat content of a substance during a chemical reaction

heat of vaporization (14.3)   The quantity of heat required to change a quantity of liquid to a vapor at the normal boiling point. It is usually expressed in calories per gram.

heterogeneous matter (3.7)   Matter made up of parts with different properties; nonuniform matter

heterogeneous mixture (3.7)   A mixture that consists of two or more substances that retain their own characteristic properties

homogeneous matter (3.7)   Matter made up of parts with similar properties; uniform matter

homogeneous mixture (3.7)   A mixture that consists of two or more substances but is uniform in composition—that is, every part of the mixture is exactly like every other part.

hydrate (14.7)   A compound that contains chemically combined water in definite proportions

hydrogen bond (as applied to water molecules) (14.7)   An unusually strong attractive force formed between the hydrogen atom of one water molecule and the oxygen atom of another water molecule

hydronium ion (16.1)   The ion formed by the addition of a hydrogen ion to a water molecule; written as H_3O^+

hygroscopic substance (14.7)   A substance that can absorb water from the atmosphere

ideal gas (13.1)   A gas that obeys the Ideal Gas Law. For a gas to be "ideal," the molecules of the gas should have no attraction for each other under all conditions of temperature and pressure.

Ideal Gas Law (13.10)   The relationship among the pressure, volume, temperature, and number of moles of a gas expressed by the mathematical formula $PV = nRT$

immiscible (15.1)   Incapable of forming a solution when mixed

indicator (16.10)   A chemical substance that has the ability to change color depending on the pH of the solution in which it is placed

ion (7.4)   An atom or group of atoms that has gained or lost one or more electrons and therefore has a positive or negative charge

ionic bond (7.4)   A bond formed by the transfer of electrons from one atom to another. The atoms involved are always of different elements.

ionization (6.7)   The process by which ions are formed from atoms or molecules by the transfer of electrons

ionization potential (ionization energy) (6.7)   The energy needed to remove an electron—generally an outershell electron—from an isolated atom

isotope (4.7)   One of two or more atoms that have the same number of electrons and protons but different numbers of neutrons

Kelvin scale (2.11)   An absolute temperature scale in which kelvins equal degrees Celsius + 273

kinetic energy (3.4)   Energy that an object possesses by virtue of its motion

kinetic theory of gases (13.1)   A theory that attempts to explain the properties of gases in terms of the forces between the gas molecules and the energy of these molecules

Law of Conservation of Energy (3.3)   Energy can be neither created nor destroyed.

Law of Conservation of Mass (3.3)   Matter can be neither created nor destroyed.

Law of Conservation of Mass and Energy (3.3)   Matter and energy can be neither created nor destroyed, but they can change from one form to another, and the sum of all the matter and energy in the universe always remains the same.

Law of Definite Composition (3.11)   A law that states every compound is composed of elements in a certain fixed proportion

line spectrum (5.1)   A series of colors that shows bright lines separated by dark bands

liquid (3.5)   The physical state of matter in which particles are held together but are free to move about. Liquids have a definite volume but take the shape of the container in which they are placed.

mass (2.9)   A measure of the quantity of matter in an object

mass number (4.7)   The mass of a particular atom in atomic mass units. It is essentially the total number of protons and neutrons in the nucleus of the atom.

matter (3.2)   Anything that occupies space and has mass

melt (14.4)   To change from the solid state to the liquid state

melting point (14.4)   The temperature at which the solid and liquid states of a substance exist together

metal (3.9)   An element that conducts electricity and heat, has luster, and takes on a positive oxidation number when it bonds with another element

metalloid (3.9)   An element that has some of the properties of metals and some of the properties of nonmetals

metric system (2.8)   A system of measurement based on multiples of 10

miscible (14.5)   Capable of forming a solution when mixed

mixture (3.7)   A combination of two or more substances that can be separated by physical means

molality (m) (15.8)   A concentration unit for solutions: moles of solute per kilogram of solvent

molarity (15.4)   A concentration unit for solutions: moles of solute per liter of solution

molar mass (9.3)   A general term used to describe the gram-formula mass or gram-atomic mass of a substance

molar volume of a gas (13.9)   The volume of 1 mole of a gas at 0°C and 1 atm, which is 22.4 liters for an ideal gas

mole (9.1) 6.02×10^{23} items

molecular formula (9.4)   A chemical formula showing the number of atoms of each element in a molecule of a compound

molecular mass (3.16)   The sum of the gram-atomic masses of all of the elements that make up a molecule

molecule (3.12)   The smallest particle of a compound that can enter into chemical reactions and retain the properties of the compound

monatomic ion (7.4)   An ion consisting of a single atom that has taken on a positive or negative charge

negative ion (anion) (6.7)   An atom that has gained one or more electrons and thereby taken on a negative charge

neutralization reaction (10.4)   A reaction in which an acid and a base react to form a salt and water

neutron (4.5)   A particle with no electric charge and a mass of 1.0086650 atomic mass units

nonelectrolytic solution (15.7)   A solution that does not conduct electric current

nonmetal (3.9)   An element that is not a good conductor of heat or electricity and usually takes on a negative oxidation number when it bonds with a metal

nonpolar covalent bond (7.8)   A covalent bond in which the electrons are shared equally by the atoms forming the bond

normality (N) (15.5)   A concentration unit for solutions: equivalents of solute per liter of solution

nucleus (4.4)   The center of an atom, containing most of the mass and one or more units of positive charge

octet rule (7.2)   A general rule stating that atoms with eight valence electrons tend to be nonreactive

orbital (5.4)   A region of space, near the atomic nucleus, in which there is a 95% probability of finding an electron

osmometry (15.11)   A method for determining molecular masses that relates osmotic pressure to the molar concentration of a solute in a solution

osmosis (15.11)   The passage of a solvent through a semipermeable membrane

osmotic pressure (15.11)   The amount of pressure that must be applied to prevent the flow of a solvent through a semipermeable membrane

outermost shell (5.9)   The energy level farthest away from the nucleus of an atom

oxidation (10.6)   The loss of electrons by a substance undergoing a chemical reaction

oxidation number (8.3)   A number that expresses the combining capacity of an element or a polyatomic ion in a compound

oxidation–reduction (redox) reaction (10.6)   A reaction in which one chemical substance is oxidized and another chemical substance is reduced

oxidizing agent (10.6)   A substance that causes something else to be oxidized, or to lose electrons

partial pressure (13.7)   The pressure of a particular gas in a mixture of gases

percentage composition (9.5)   The percentage by mass of each element in a compound

period (6.2)   A horizontal row of elements in the periodic table

pH scale (16.6)   A scale that expresses the acidity of a solution

physical property (3.6)   A characteristic property such as color, odor, taste, boiling point, or melting point that can be measured by nonchemical means

plasma (3.5)   A form of matter composed of electrically charged atomic particles

polar covalent bond (7.8)   A covalent bond in which there is an unequal sharing of electrons by the atoms forming the bond

polar molecule (dipole) (7.8)   A molecule that is partially positive at one end and partially negative at another end

polyatomic ion (8.4)   A charged group of covalently bonded atoms

positive ion (cation) (6.7)   An atom that has lost one or more electrons and thereby taken on a positive charge

potential energy (3.4)   Energy that is stored in an object by virtue of its position or its inherent chemical energy

precipitate (10.4)   A solid substance that separates out of a solution in the course of a chemical reaction

precision (2.4)   The closeness of repeated measurements to each other

product (10.1)   A substance produced in a chemical reaction

proton (4.4)   A particle with a relative positive charge of one unit and a mass of 1.0072766 atomic mass units

pure substance (3.7)   Matter that has a definite and fixed composition. Elements and compounds are pure substances.

quantum (plural, *quanta***)** (5.2)   A specific bundle of energy emitted by an electron as it moves from one energy level to another

quantum mechanics (5.4)   A mathematical model of the atom based on the probability of finding electrons in a particular region of space surrounding the nucleus of an atom

reactant (10.1)   Any of the starting materials in a chemical reaction

real gas (13.1)   Any gas that actually exists, such as oxygen, nitrogen, hydrogen, or helium gas

redox (10.6)   An abbreviation for the term *oxidation–reduction*

reducing agent (10.6)   A substance that causes something else to be reduced, or to gain electrons, while it is oxidized

reduction (10.6)   The gain of electrons by a substance undergoing a chemical reaction

representative element (6.2)   An A-group element. For the A-group elements, the group number indicates how many outer electrons there are.

rounding off (2.4)   A process in which one or more digits at the right end of a number are dropped in order to attain the correct number of significant figures

salt (10.4)   A compound composed of the positive ion of a base and the negative ion of an acid

saturated solution (15.2)   A solution in which no more solute can be dissolved

saturation point (15.3)   The level of concentration at which no more solute can dissolve in a given amount of solvent at a particular temperature

scientific method (3.1)   A series of logical steps used by scientists to approach a problem and solve it effectively

scientific notation (2.5)   A number expressed in exponential notation. The number 0.00625 can be 6.25×10^{-3} in scientific notation.

shells (5.3)   See energy levels.

significant figures (2.4)   Digits that express information that is reasonably reliable

single covalent bond (7.2)   A chemical bond in which a single pair of electrons is shared by two atoms

single-replacement reaction (10.4)   A reaction in which an uncombined element replaces another element that is in a compound is a single replacement reaction.

solid (3.5)   The physical state of matter in which particles are held in a definite arrangement. Solids have a definite shape and definite volume.

solute (15.1)   In a solution, the substance that is being dissolved

solution (3.8)   A homogeneous mixture

solvent (15.1)   In a solution, the substance that is doing the dissolving

specific heat (12.3)   The number of calories needed to raise the temperature of 1 gram of a substance by 1 Celsius degree

standard temperature and pressure (STP) (13.6)   The conditions 273K (0°C) and 1 atm pressure

stoichiometry (11.1)   The calculation of the quantities of substances involved in chemical reactions

strong acid (16.3)   An acid that completely dissociates into ions

strong base (16.3)   A base that completely dissociates into ions

sublimation (14.4)   The process in which a substance changes directly from the solid state to the gaseous state

supersaturated (15.2)   A solution that contains more solute than is needed for saturation for a given quantity of solvent at a particular temperature

synthesize (Chapter 11, Introduction)   To combine reactants to make a product

systematic chemical name (Chapter 8, Introduction)   Any of the names for chemical compounds derived from the naming system developed by the International Union of Pure and Applied Chemistry

temperature (2.11)   A measure of the intensity of heat

ternary compound (8.1)   A compound composed of three elements

thermochemical equation (12.4)   A chemical equation that includes information about the physical state of the reactants and products and the amount of heat released or absorbed during the chemical reaction

titration (16.9)   The process of determining the amount of a substance present in a solution by measuring the volume of a different solution of known strength that must be added to complete a chemical change

torr (13.2)   A unit of pressure equal to 1/760 of an atmosphere

transition metal (6.2)   A B-group element

triple covalent bond (7.2)   A chemical bond in which three pairs of electrons are shared by two atoms

unit cell (14.6)   A repeated unit in a crystal lattice

unsaturated solution (15.2)   A solution that contains less solute than can be dissolved in it at a particular temperature

vacuum (4.3)   An enclosed space from which all matter has been removed

volatile (14.2)   Easily vaporize

volume (2.6)   A measure of the capacity of a three-dimensional object

volumetric flask (15.4)   A type of flask used by chemists to prepare a solution with a particular concentration

weak acid (16.3)   An acid that partially ionizes in aqueous solution

weak base (16.3)   A base that partially ionizes in aqueous solution

weight (2.9)   The gravitational attraction of an object to the earth or any other body

This part of the text contains answers to self-test exercises and extra exercises whose numbers are in color at the end of each chapter. In addition, answers to all practice exercises, cumulative review exercises, and exercises in Appendix A are included.

Answers to Practice Exercises

2.1 27   **2.2** 53.44   **2.3** 6.3 cm^2 **2.4** 10   **2.5** 82
2.6 (a) 2 (b) 4 (c) 6 (d) 4 (e) 2 (f) 5
2.7 (a) 4.2×10^3 (b) 5.60×10^4 (c) 6.023×10^6
(d) 1.23×10^{-3} **2.8** 301 in^3 **2.9** 3,500 cm
2.10 0.350 g   **2.11** 1,800 cm   **2.12** 66 cm
2.13 (a) 5.5 cm (b) 0.055 m (c) 0.000055 km or
5.5×10^{-5} km   **2.14** (a) 4,400 g (b) 44,000 dg
(c) 440,000 cg (d) 4,400,000 mg or 4.4×10^6 mg
2.15 13.8 ft by 4.10 ft   **2.16** 140 L   **2.17** 86,400 s
2.18 2.7 g/cm^3 **2.19** 8.0 g/cm^3 **2.20** 5.50 g/cm^3 liquid A, 2.00 g/cm^3 liquid B. Liquid A is more dense.
2.21 $V = 1,300$ cm^3 **2.22** 118°C   **2.23** 302°F
3.1 (a) 2 carbon atoms, 7 hydrogen atoms, 1 nitrogen atom (b) 2 nitrogen atoms, 8 hydrogen atoms, 1 sulfur atom, 4 oxygen atoms   **3.2** (a) 138.9 (b) 55.8
(c) 39.9 (d) 118.7   **3.3** (a) 106.0 (b) 129.9 (c) 87.0
(d) 92.0   **4.1** (a) 14 p, 14 e, 14 n (b) 15 p, 15 e, 16 n
(c) 3 p, 3 e, 3 n   **4.2** Average mass = 6.941
4.3 ^{100}Y = 20.0%, ^{110}Y = 80.0%   **4.4** 909,200

5.1 (a) 32 electrons (b) 98 electrons   **5.2** (a) 32 electrons (b) 4 sublevels: s, p, d, and f (c) s can hold 2, p can hold 6, d can hold 10, and f can hold 14. (d) s sublevel has 1 orbital, p sublevel has 3 orbitals, d sublevel has 5 orbitals, and f sublevel has 7 orbitals.
5.3 (a) $1s^2\ 2s^2\ 2p^6\ 3s^2$ (b) $1s^2\ 2s^2\ 2p^6\ 3s^2\ 3p^6\ 4s^2\ 3d^{10}$
$4p^4$ **6.1** The order is S, B, N, and O from longest to shortest radius.   **6.2** The order is O, N, B, and S from highest to lowest ionization potential.
7.1 (a) H:N̈:H   (b) H:C̈l: (c) :Ö:S̈
       H   :Ö

7.2 H—O—S—O—H
            ↓
            O

7.3 (a) Covalent (b) Ionic (c) Covalent (d) Covalent
7.4 (a) 84% ionic, 16% covalent (b) 63% ionic, 37% covalent (c) 55% ionic, 45% covalent (d) 59% ionic, 41% covalent   **7.5** CS_2 is most covalent, H_2S is next, then $AsCl_3$ and CO_2 tie for least covalent.
7.6 (a) Nonpolar (b) Polar (c) Polar   **7.7** (a) BF_3 bonds are polar; molecule is nonpolar. (b) Cl_2O bonds are polar; molecule is polar. (c) CO_2 bonds are polar; molecule is nonpolar.
8.1 (a) SO_2 (b) N_2O_5 (c) N_2O_4 (d) PBr_5 **8.2** KF
8.3 MgO   **8.4** SrS   **8.5** K_2O **8.6** $AlBr_3$
8.7 (a) HgO (b) $FeBr_3$ (c) CoI_2 (d) MnO_2 (e) Cu_2S
(f) Fe_2S_3 **8.8** (a) AgCl (b) ZnO   **8.9** (a) $Fe(NO_2)_3$

(b) $Ba_3(PO_4)_2$ (c) $CuSO_4$ (d) K_2CrO_4 **8.10** (a) Diphosphorus pentoxide (b) Oxygen difluoride (c) Sulfur dioxide (d) Carbon monoxide **8.11** 2+
8.12 (a) 4+ (b) 5+ **8.13** (a) Rubidium chloride (b) Gallium sulfide (c) Strontium oxide (d) Zinc iodide (e) Nickel(II) chloride (f) Iron(III) iodide (g) Mercury(II) sulfide (h) Copper(I) oxide
8.14 (a) Ammonium oxide (b) Magnesium cyanide (c) Aluminum nitrite (d) Zinc phosphate (e) Calcium oxalate (f) Nickel(II) borate **8.15** (a) Hydrochloric acid (b) Hydroselenic acid (c) Hydroiodic acid
8.16 (a) Phosphoric acid (b) Chloric acid
8.17 (a) Chlorous acid (b) Nitrous acid **8.18** (a) Perchloric acid (b) Chloric acid (c) Chlorous acid (d) Hypochlorous acid **9.1** (a) 6.00 ft (b) 120 in
9.2 (a) 10.0 moles (b) 32.0 g **9.3** (a) 0.500 mole (b) 0.100 mole (c) 20.0 moles (d) 1.00×10^{-4} mole
9.4 (a) 981 g (b) 16.0 g (c) 120 g **9.5** (a) 9.03×10^{24} atoms (b) 1.20×10^{23} atoms (c) 1.81×10^{24} atoms
9.6 H_2O **9.7** Cr_2S_3 **9.8** ZnN_2O_4
9.9 (a) 5.0 moles (b) 0.0010 mole **9.10** (a) 184 g (b) 5.0 g (c) 395 g (d) 7.9 g **9.11** (a) 2.41×10^{24} (b) 3.0×10^{22} (c) 1.51×10^{24} (d) 3.6×10^{22}
9.12 $C_9H_8O_4$ **9.13** 32.9% K, 67.1% Br
9.14 (a) 85.7% C, 14.3% H (b) 39.3% Na, 60.7% Cl (c) 92.3% C, 7.7% H **9.15** 75.0 g
10.1 $2CuO \rightarrow 2\ Cu + O_2$
10.2 $Zn + H_2SO_4 \rightarrow ZnSO_4 + H_2$
10.3 (a) $2Al + 3Cl_2 \rightarrow 2AlCl_3$
(b) $2Na + 2H_2O \rightarrow 2NaOH + H_2$
(c) $2KNO_3 \rightarrow 2KNO_2 + O_2$
(d) $2HNO_3 + Ba(OH)_2 \rightarrow Ba(NO_3)_2 + 2H_2O$
10.4 (a) $4K + O_2 \rightarrow 2K_2O$
(b) $2Ca + O_2 \rightarrow 2CaO$
(c) $H_2 + Br_2 \rightarrow 2HBr$
10.5 (a) $Sr(OH)_2 \rightarrow SrO + H_2O$
(b) $SrCO_3 \rightarrow SrO + CO_2$
(c) $2KClO_3 \rightarrow 2KCl + 3O_2$
(d) $2KCl \rightarrow 2K + Cl_2$
10.6 (a) $2Li + 2H_2O \rightarrow 2LiOH + H_2$
(b) $2Al + 3H_2SO_4 \rightarrow Al_2(SO_4)_3 + 3H_2$
(c) $3Mg + 2Al(NO_3)_3 \rightarrow 3Mg(NO_3)_2 + 2Al$
10.7 (a) $2BiCl_3 + 3H_2S \rightarrow Bi_2S_3 + 6HCl$
(b) $2Fe(OH)_3 + 3H_2SO_4 \rightarrow Fe_2(SO_4)_3 + 6H_2O$
10.8 (a) No reaction occurs. (b) Reaction does occur.
10.9 (a) Copper is oxidized; silver ion is reduced. (b) Sodium is oxidized; chlorine is reduced. (c) Bromide ion is oxidized; chlorine is reduced. (d) Zinc is

oxidized; sulfur is reduced. **10.10** (a) Copper is the reducing agent; silver nitrate is the oxidizing agent. (b) Sodium is the reducing agent; chlorine is the oxidizing agent. (c) Sodium bromide is the reducing agent; chlorine is the oxidizing agent. (d) Zinc is the reducing agent; sulfur is the oxidizing agent.
11.1 2.43 g Mg, 1.60 g O **11.2** 0.900 g Al, 7.10 g $Al(NO_3)_3$, 3.27 g Zn **11.3** 49.0 g **12.1** 40,000 cal or 40 kcal **12.2** 12,500 cal or 12.5 kcal **12.3** 5,380 cal or 5.38 kcal **12.4** 2,680 cal or 2.68 kcal
12.5 $C_2H_6(g) + 3\frac{1}{2} O_2(g) \rightarrow$
$$2CO_2(g) + 3H_2O(l) + 372\ \text{kcal}$$
12.6 $2C_6H_{14}(l) + 19O_2(g) \rightarrow$
$$12CO_2(g) + 14H_2O(l) + 1,980\ \text{kcal}$$
12.7 204.9 kcal heat are produced.
12.8 $PCl_5(g) \rightarrow PCl_3(g) + Cl_2(g) - 22.2\ \text{kcal}$
12.9 $\Delta H = -75$ kcal **12.10** $\Delta H = -135.4$ kcal
12.11 $\Delta H = -216.6$ kcal **12.12** $\Delta H = -6$ kcal
12.13 $\Delta H = -74.4$ kcal **13.1** (a) 57 torr (b) 7,370 torr **13.2** (a) 0.010 atm (b) 12.0 atm **13.3** 250 mL
13.4 500 mL **13.5** 6,080 torr **13.6** 1,875 mL or 1,880 **13.7** $t_i = 17.0°C$ **13.8** $V_f = 5.15$ L
13.9 $t_f = 12.0°C$ **13.10** $V_f = 2.50$ L **13.11** $V_f = 235$ mL **13.12** $V_f = 187$ mL **13.13** $V_f = 4.70$ L
13.14 40.0 L of HF gas **13.15** 0.0200 mole
13.16 4.48 L **13.17** 36 L **13.18** 0.0217 mole
13.19 M.M. = 410 **14.1** 40.50% **15.1** (a) Salt is the solute and water is the solvent. (b) Sulfur trioxide is the solute and water is the solvent. (c) Water is the solute and isopropyl alcohol is the solvent. (d) Hydrogen gas is the solute and oxygen gas is the solvent. (e) Carbon dioxide is the solute and water is the solvent. **15.2** (a) No (b) Yes (c) Yes **15.3** (a) Supersaturated (b) Unsaturated (c) Saturated
15.4 (a) 44.4% (b) 30.0% **15.5** (a) 20.0% (b) 55.6% **15.6** (a) 4.5% (b) 20.00% **15.7** (a) 0.200 M (b) 0.0164 M (c) 1.07 M **15.8** 3.71×10^{-2} g
15.9 49.9 mL **15.10** 2.00 N and 2.00 M **15.11** 0.440 N and 0.440 M **15.12** 12.0 g **15.13** 2.00 L of the 6.00 N HCl solution are needed. **15.14** 16.7 mL of the 12.0 N H_3PO_4 solution are needed.
15.15 (a) 1.0 (b) 0.0400 (c) 0.160 **15.16** (a) 100.83°C (b) 100.02°C **15.17** (a) −2.98°C (b) −0.0744°C
15.18 Water will move out of the cell.
16.1 (a) pH = 5 (b) pH = 10 (c) pH = 1
16.2 (a) pH = 4 (b) pH = 3 (c) pH = 9 (d) pH = 3 (e) pH = 13 **16.3** 0.33 N **16.4** 0.086 N
16.5 0.409 N **16.6** 0.102 N

Chapter 2

1. (a) Five (b) One (c) Three (d) Four (e) Seven (f) Two (g) One (h) Four (i) Three   **3.** (a) Seven (b) Two (c) One (d) Three (e) Seven (f) Three (g) Six (h) Four (i) Six (j) Four   **5.** 17.7 cm; three significant figures   **7.** 15.0 cm; three significant figures **9.** 20.60 cm   **11.** 214 cm^2 **13.** (a) Two (b) Three (c) Four (d) Four (e) One (f) Two (g) Three (h) Four (i) Four (j) Three   **15.** (a) 6×10^2 (b) 6.00×10^2 (c) 6.000×10^2 (d) 3.2×10^{11} (e) 2×10^{-3} (f) 3.007×10^{-4} (g) 1.5×10^{-7} **17.** (a) 1.0581×10^4 (b) 2.05×10^{-3} (c) 1×10^6 (d) 8.02×10^2 **19.** (a) 8.5×10^8 (b) 6.07×10^{-6} (c) 6.308×10^6 (d) 6.005×10^{-2} (e) 5×10^2 (f) 5.0×10^2 (g) 5.00×10^2 (h) 5.000×10^2 (i) 2.300×10^7 (j) 9.30×10^{-8} **21.** 24.0 m^2, 96.0 m^3 **23.** 426 cm^2 or 66.1 in^2 (depending on how you measure the page)   **25.** $6\overline{0}0$ m^2 **27.** $45\overline{0}$ cm^3 **29.** 42,400 cm^3 **31.** 2,560 ft^3, 7.27×10^7 cm^3 **33.** 7.27×10^4L, 19,300 gal.   **35.** 1,728 in^3 **37.** (a) 250g (b) 2,500 dg (c) 250,000 mg (d) 250,000,000 μg **39.** (a) 31 dm (b) 310 cm (c) 3,100 mm **41.** (a) 14.9 cm (b) 0.149 m (c) 0.000149 km **43.** (a) $7,85\overline{0}$ mm (b) 785.0 cm (c) 78.50 dm (d) 0.007850 km   **45.** (a) 340 m (b) 3,400 dm (c) 34,000 cm (d) 340,000 mm **47.** (a) 218.5 cg (b) 21.85 dg (c) 2.185 g (d) 0.002185 kg   **49.** (a) 3.500 L (b) 35.00 dL   **51.** 1.00×10^6 cm^3 **53.** (a) 82.0 ft (b) 985 in.   **55.** (a) 0.32 gal (b) 1.3 qt   **57.** (a) 0.0110 lb (b) 0.176 oz   **59.** 35.40 ft^3/m^3 **63.** 10.8 ft^2/m^2 **69.** (a) 3.0 m (b) 3.0×10^2 cm (c) 3.0×10^3 mm **71.** (a) 1.83 m (b) 183 cm   **73.** (a) 91.5 m (b) 9,150 cm   **75.** (a) 18.9 L (b) 18,900 mL **79.** 4.80 g/cm^3 **81.** 203 cm^3 or $2\overline{0}0$ cm^3 (for the proper number of significant figures)   **83.** 53 g **85.** 0.019 g   **87.** 6.83 g/cm^3 **89.** 1.25 g/cm^3 **91.** $11\overline{0}$ g   **93.** 162 g   **95.** 9.32 g/cm^3 **97.** 28,400 g or $29,\overline{0}00$ (for the proper number of significant figures)   **99.** 22.1 cm^3 **101.** 3.67 g/cm^3 **103.** 19,500 g **105.** 50.0 cm^3 **107.** 37.8°C, 310.8 K   **109.** −40°C. At this temperature, degrees F equal degrees C. **111.** 5.4°F, 104°F   **113.** (a) 10°C (b) −70°C (c) 215°C (d) −90.0°C   **115.** (a) 203°F (b) −112°F (c) 176°F (d) $41\overline{0}$°F   **118.** No; the person would have to be about 13.1 feet tall and weigh $44\overline{0}$ pounds. **119.** 48 cm^3 **120.** 218°C   **123.** 820   **124.** −8.89°C **125.** (a) 61°F (b) 392°F   **126.** (a) 149°C (b) −101°C **127.** 22.6 miles/hour

Chapter 3

5. (a) 2 (b) 5 (c) 3 (d) 4 (e) 1   **7.** (a) Physical (b) Chemical (c) Chemical (d) Physical (e) Chemical (f) Chemical   **11.** (a) Metal (b) Metalloid (c) Nonmetal (d) Metal (e) Metalloid (f) Metal (g) Metal   **13.** Co is the element cobalt, and CO is the compound carbon monoxide.   **15.** The percent H and O are the same for each experiment.   **17.** (a) 12 carbon atoms, 22 hydrogen atoms, 11 oxygen atoms (b) 2 potassium atoms, 1 chromium atom, 4 oxygen atoms (c) 8 hydrogen atoms, 2 nitrogen atoms, 3 oxygen atoms, 2 sulfur atoms (d) 1 zinc atom, 2 nitrogen atoms, 6 oxygen atoms   **21.** The atomic mass of sulfur would be 2 amu.   **23.** (a) 85.5 (b) 52.0 (c) 238.0 (d) 79.0 (e) 74.9   **25.** (a) 71.8 (b) 159.6 (c) 317.3 (d) 164.0 (e) 58.3 (f) 218.5 (g) 791.8 (h) 96.0 **27.** (a) 60.1 (b) 82.1 (c) 121.6 (d) 104.5 (e) 187.5 (f) 218.7 (g) 149.0 (h) 60.0   **35.** (a) Magnesium and chlorine (b) Nitrogen and oxygen (c) Nitrogen, hydrogen, sulfur, and oxygen (d) Hydrogen, phosphorus, and oxygen   **37.** (a) Element (b) Mixture (c) Compound (d) Mixture (e) Compound   **39.** (a) 254.2 (b) 63.0 (c) 89.8 (d) 601.9

Cumulative Review/Chapters 1–3

1. True   **2.** False   **3.** False   **4.** True   **5.** False **6.** True   **7.** False   **8.** True   **9.** True   **10.** True **11.** 128 m^2 **12.** 4,548 cm^3 **13.** 16,540 cm^3 **14.** (a) 280 g (b) 2,800 dg (c) 280,000 mg **15.** (a) 68 dm (b) 680 cm (c) 6,800 mm **16.** (a) 12.5 cm (b) 0.125 m (c) 0.000125 km **17.** (a) 25.595 L (b) 255.95 dL **18.** (a) 164 ft (b) 1,970 in   **19.** (a) 0.0562 lb (b) 0.899 oz   **20.** 16.4 cm^3/in^3 **21.** 0.363 lb **22.** (a) 45.4 L (b) 1,540 fl oz   **23.** 0.289 g/cm^3 **24.** 0.177 g/mL. The density of air is 1.18×10^{-3} g/mL, so this gas is heavier than air and will not float in air.   **25.** 6.00 g/mL   **26.** 4.00 cm   **27.** (a) 31.8 (b) 9.29 (c) 79.3 (d) 98.3   **28.** (a) 5×10^3 (b) 5×10^{-4} (c) 6.023×10^8 (d) 3.5000×10^7 **29.** (a) 5 (b) 4 (c) 5 (d) 4   **30.** (a) 7.22°C (b) −23.3°C (c) $23\overline{0}$°C (d) −73.33°C   **31.** (a) $19\overline{0}$°F (b) 9.5°F (c) $15\overline{0}$°F (d) 752°F   **32.** Amorphous solids have no definite internal structure or form. Crystalline solids have a fixed, regularly repeating, symmetrical inner structure.   **33.** (a) Chemical (b) Physical (c) Chemical (d) Physical   **34.** Heterogeneous matter is made up of different parts with different properties.

Homogeneous matter is made up of parts with the same properties throughout. **35.** (a) 1 zinc atom, 4 carbon atoms, 6 hydrogen atoms, 4 oxygen atoms (b) 2 nitrogen atoms, 8 hydrogen atoms, 1 chromium atom, 4 oxygen atoms **36.** Molecular mass is used for compounds composed of molecules. Formula mass is used for compounds composed of ions. **37.** Mercury would have a mass of 5 amu. **38.** (a) 173.0 (b) 210.0 (c) 31.0 (d) 107.9 **39.** (a) 232.8 (b) 132.1 (c) 180.0 (d) 251.1 **40.** The formula of $Al_2(SO_4)_3$ means that this formula unit contains 2 aluminum atoms, 3 sulfur atoms, and 12 oxygen atoms. **41.** (a) Mixture (b) Element (c) Compound (d) Compound (e) Element (f) Mixture (g) Mixture **42.** (a) Heterogeneous (b) Homogeneous (c) Heterogeneous (d) Homogeneous **43.** Actinium (Ac), aluminum (Al), americium (Am), argon (Ar), arsenic (As), astatine (At), gold (Au), and silver (Ag) **44.** Hydrogen (H), Helium (He), carbon (C), nitrogen (N), oxygen (O), fluorine (F), neon (Ne), phosphorus (P), sulfur (S), chlorine (Cl), argon (Ar), selenium (Se), bromine (Br), krypton (Kr), iodine (I), xenon (Xe), and radon (Rn) **45.** (a) Potassium and sulfur (b) Silver, chromium, and oxygen (c) Potassium, manganese, and oxygen (d) Mercury, phosphorus, and oxygen **46.** (a) N_2O (b) K_2CrO_4 (c) NH_3 (d) $SrSO_4$ **47.** (a) Chemical (b) Physical (c) Chemical (d) Physical **48.** No is the element nobelium, and NO is the compound nitrogen monoxide. **49.** No! During the burning process, some of the atoms that compose paper react with oxygen in the air and form gaseous compounds.

50. Add water to the mixture and shake. The salt dissolves and the sand settles to the bottom of the container. Filter the sand and collect the salt water (filtrate). To retrieve the pure salt, evaporate the water.

Chapter 4

9. 1,836 electrons **13.** (a) 94 p, 94 e, 150 n (b) 22 p, 22 e, 26 n (c) 103 p, 103 e, 159 n **15.** (a) 8 p, 8 e, and 8 n (b) 8 p, 8 e, and 9 n (c) 8 p, 8 e, and 10 n (d) 10 p, 10 e, and 10 n (e) 10 p, 10 e, and 11 n (f) 10 p, 10 e, and 12 n **17.** (b) is not an isotope of the others **19.** (a) 25 p, 25 e, and 32 n (b) 27 p, 27 e, and 33 n (c) 36 p, 36 e, and 44 n (d) 52 p, 52 e, and 76 n **21.** 3 p, 3 e, and 8 n

23.

SYMBOL	PROTONS	ELECTRONS	NEUTRONS	MASS NUMBER	ATOMIC NUMBER
$^{174}_{70}Yb$	70	70	104	174	70
$^{141}_{59}Pr$	59	59	82	141	59
$^{104}_{44}Ru$	44	44	60	104	44
$^{45}_{21}Sc$	21	21	24	45	21
$^{50}_{22}Ti$	22	22	28	50	22
$^{25}_{12}Mg$	12	12	13	25	12

27. Protons, electrons, neutrons **29.** $^{200}_{80}Hg$ (b) $^{15}_{7}N$ (c) $^{27}_{13}Al$ (d) $^{262}_{107}Bh$ **31.** Atomic mass of Ga = 69.72. **33.** 35.46 amu **35.** 52.00 amu **37.** The percentage abundance of ^{79}Br is 50.87%, and the percentage abundance of ^{81}Br is 49.13%. **39.** $^{121}Sb = 57.696\%$, $^{123}Sb = 42.304\%$ **41.** 8.82×10^7 atoms ^{22}Ne **43.** 2.0×10^7 oxygen atoms **45.** 7.42×10^6 atoms ^6Li **47.** 5.59×10^5 atoms ^{26}Mg **49.** 222 atoms ^{13}C **51.** 3.7×10^{12} atoms ^{15}N **53.** 1.0×10^9 hydrogen atoms **56.** (a) C (b) D (c) 222 (d) D (e) Po

57. They are the same. **59.** The Lenard experiment using ultraviolet light **60.** (a) 6 p, 6 e, 7 n (b) 5 p, 5 e, 5 n (c) 7 p, 7 e, 7 n **61.** Isotopes exist. **62.** 2,570 Ne-21 atoms **63.** 1,836 electrons

Chapter 5

3. By examining the line spectrum of an extraterrestrial body with a spectroscope, scientists can determine which elements are present. **9.** Energy levels in atoms

11. Electrons   **13.** Seven energy levels: *K, L, M, N, O, P, Q* or 1, 2, 3, 4, 5, 6, 7, respectively   **23.** (a) 1 (b) 2 (c) 3 (d) 4   **25.** The number of electrons in an energy level is the sum of the electrons in the sublevels.
27. (a) 6 (b) 72   **31.** Total of four orbitals   **35.** c and d   **37.** (a) $1s^22s^22p^3$ (b) $1s^22s^22p^63s^23p^64s^23d^3$ (c) $1s^22s^22p^63s^23p^64s^23d^{10}4p^65s^24d^{10}5p^6$ (d) $1s^22s^22p^63s^23p^64s^23d^{10}4p^65s^24d^{10}5p^3$
39. (a) $_{56}$Ba (b) $_{20}$Ca (c) $_{38}$Sr (d) $_4$Be   **41.** (a) 31 (b) 8 (c) 13 (d) 10 (e) 31   **43.** Elements 6 and 14, elements 3 and 19, elements 17 and 35   **45.** 98 electrons
47. (a) $1s^22s^22p^63s^23p^64s^23d^{10}4p^1$ (b) $1s^22s^22p^63s^23p^64s^23d^{10}4p^65s^24d^{10}5p^3$ (c) $1s^22s^22p^63s^23p^64s^23d^{10}4p^65s^24d^{10}5p^66s^24f^{14}5d^{10}6p^2$ (d) $1s^22s^22p^63s^23p^64s^23d^{10}4p^65s^24d^{10}5p^66s^24f^{14}5d^{10}6p^67s^2$
49. (a) $_{12}$Mg (b) $_{38}$Sr (c) $_{70}$Yb (d) $_{81}$Tl   **51.** (a) 60 (b) 12 (c) 24 (d) 20 (e) 4 (f) 60   **55.** The Group VIIIA elements tend to be chemically unreactive.
65. The atomic number would be 280.

Chapter 6

7. (a) Ga (b) S (c) Hg (d) Bh   **8.** Period, group or family   **9.** The same number of electrons in the outermost energy level   **11.** The number of electrons in the outermost energy level   **12.** (a) 8 (b) 8 (c) 18 (d) 18 (e) 32 (f) 23   **13.** A-group elements, B-group elements   **15.** Ba, Sr, Te, I, Br,   **17.** P, As, I, Sb, In   **25.** (a) decreases (b) increases (c) decreases   **27.** Cl, S, Ca, Sr, Rb   **29.** Sb, I, As, P   **31.** (d)   **33.** (c)   **35.** (d)   **39.** (a) $_{11}$Na, $1s^22s^22p^63s^1$ (b) $_{13}$Al, $1s^22s^22p^63s^23p^1$ (c) $_8$O, $1s^22s^22p^4$ (d) $_{16}$S, $1s^22s^22p^63s^23p^4$ (e) $_5$B, $1s^22s^22p^1$ (f) $_3$Li, $1s^22s^1$. Therefore, (a) and (f) are similar, (b) and (e) are similar, and (c) and (d) are similar.   **41.** Atomic number 118   **42.** (a) Group IA (b) Group VIIA   **44.** (a) K (b) K (c) K   **48.** (a) K (b) Se   **49.** (a) K (b) Se   **50.** Group IVA

Cumulative Review/Chapters 4–6

1. Protons and neutrons in nucleus, electrons surrounding nucleus   **2.** Cathode rays came from the cathode.   **3.** A magnet would deflect particles, not waves. An electric field would easily deflect particles.   **4.** UV light is directed onto a metal and makes it emit electrons.   **5.** An atom is a sphere of positive electric charge in which electrons are embedded. The positive particles balance the negative electrons.   **6.** Rutherford found that one alpha particle in 20,000 ricocheted. This led him to believe that an atom has a small but dense center of positive charge.   **7.** New evidence disproved Thomson's theory.   **8.** Neutrons   **9.** Isotopes are atoms of the same element with different atomic masses.   **10.** (a) 27 protons (b) 40 protons (c) 34 protons   **11.** 20.17   **12.** 370 ^{15}N atoms   **13.** (a) $^{35}_{17}$Cl (b) $^{28}_{14}$Si   **14.** 75% ^{50}Z and 25% ^{52}Z   **15.** 32 p, 32 e, 41 n   **16.** 26 electrons   **17.** ^{16}O is most abundant.   **18.** 1900 ^{7}Li atoms   **19.** 9.258×10^5 atoms   **20.** 1.00×10^7 atoms   **21.** Electrons   **22.** (a)   **23.** Six electrons   **24.** (a) *K* has 1. (b) *K* has 2 and *L* has 1. (c) *K* has 2, *L* has 8, and *M* has 1. (d) *K* has 2, *L* has 8, *M* has 8, and *N* has 1.   **25.** Two electrons   **26.** Sulfur   **27.** (b)   **28.** It has five electrons in its outermost energy level.   **29.** (a) *K* has 1. (b) *K* has 2 and *L* has 1. (c) *K* has 2, *L* has 8, and *M* has 1. (d) *K* has 2, *L* has 8, *M* has 8, *N* has 1.   **30.** $_{70}$Yb or any element with a higher atomic number.   **31.** Raindrops act as prisms.   **32.** Line spectra are used to analyze the sun and other extraterrestrial bodies.   **33.** Line spectra of his hair indicated the presence of the element arsenic.   **34.** Our major source of radiant energy is the sun.   **35.** Electro-magnetic waves travel through the vacuum of space.   **36.** (c)   **37.** By producing electromagnetic waves that were longer than visible light waves   **38.** Luminous watch dials absorb light energy and reemit it when they glow.   **39.** A sample of an element is heated and it begins to glow.   **40.** Quanta   **41.** The periodic law states that the chemical properties are periodic functions of their atomic numbers. This means that elements with similar chemical properties recur at regular intervals and are placed accordingly on the periodic table.   **42.** Moseley was able to determine the nuclear charge or the atomic numbers of the atoms of known elements.   **43.** $1s^22s^22p^63s^23p^63d^{10}4s^24p^64d^{10}4f^{14}5s^25p^65d^{10}5f^{14}6s^26p^66d^57s^2$. Element 107 is a Group VIIB element.   **44.** $1s^22s^22p^63s^23p^63d^{10}4s^24p^64d^{10}4f^{14}5s^25p^65d^{10}5f^{14}6s^26p^66d^47s^2$. Element 106 is a Group VIB element.   **45.** Elements 18–19, 27–28, 52–53, and 92–93   **46.** False   **47.** True   **48.** False   **49.** Fe, Ru, Os   **50.** $_{11}$Na   **51.** True   **52.** False   **53.** Fluorine   **54.** $1s^22s^22p^63s^23p^2$ **55.** True   **56.** (a) Group IIA (b) Group VIA   **57.** (a) Na (b) Ca   **58.** Mg   **59.** Cs   **60.** Li

Chapter 7

1. (a) ·A̤s· (b) ·C̈l: (c) Ra· (d) Ċs (e) :R̈n:

3. (a) ·Ö: (b) :Ö:²⁻ **5.** (a) Mg· (b) Mg²⁺

9. (a) Mg²⁺ (b) Na¹⁺ (c) Al³⁺

11. :C̈l:¹⁻ or K¹⁺ are two of several examples.

15. Double

21. (a) :F̈—C—F̈: (b) H—C=C—H with F below and H, H below
(c) H—As—H with H below (d) H—S̈: with H below

23. :C̈l: C̈l: Ten electrons surround the P in this compound.
:C̈l—P—C̈l: :C̈l:

25. (a) :Ö=N—Ö—H
(b) H—Ö—S—Ö—H with :Ö: above and arrow up

29. (a) Ionic (b) Covalent (c) Covalent (d) Covalent **31.** (a) 4% ionic, 96% covalent (b) 0% ionic, 100% covalent (c) 51% ionic, 49% covalent (d) 63% ionic, 37% covalent

33. HI, HBr, HCl, HF
Most covalent ——— Least covalent

35. (a) 43% ionic, 57% covalent (b) 0.5% ionic, 99.5% covalent (c) 74% ionic, 26% covalent (d) 47% ionic,

53% covalent **37.** H_2O, H_2S, H_2Se, H_2Te
Least covalent ——— Most covalent

39. CsCl is the most ionic. **41.** (a) Covalent (b) Covalent (c) Ionic (d) Covalent **45.** (a) Bonds are polar; molecule is polar. (b) Bonds are polar; molecule is polar. (c) Bonds are polar; molecule is polar. (d) Bond is nonpolar; molecule is nonpolar.
47. (a) Bonds are polar; molecule is nonpolar. (b) Bonds are polar; molecule is polar. (c) Bonds are polar; molecule is nonpolar. (d) Bond is nonpolar; molecule is nonpolar.

50. (a) [H—P—H with H above (arrow) and H below]¹⁺ (b) O=C=O
(c) O with F and F below

52. O=O N—O C—O
Nonpolar bond — Most polar bond of the three
53. C—S bond
54. (a) [H—N—H with H above (arrow) and H below]¹⁺ (b) Cl—Br **55.** (a) MgO
(b) Al_2O_3 (c) Cs_2S **59.** There is no way that the duet rule can be satisfied for a molecule containing three hydrogen atoms.

Chapter 8

3. Hydrogen oxide or dihydrogen oxide **5.** (a) 7 (b) 2 (c) 3 (d) 8 (e) 1 **7.** (a) P_2S_5 (b) ClO_2 (c) N_2O_4 (d) Cl_2O_7 **9.** (a) Al_2S_3 (b) Li_2O (c) Na_3N (d) Sr_3P_2 (e) AgBr (f) ZnO **11.** (a) IX (b) VIII (c) VII (d) VI **13.** (a) Metal ion with the higher charge (b) Metal ion with the lower charge (c) The suffix for the atom or ion that has a negative charge in a chemical compound. **15.** (a) Cu_2S (b) $HgCl_2$ (c) FeO (d) SnI_2 (e) $CoBr_3$ (f) Hg_3N_2 **17.** (a) $(NH_4)^{1+}$ (b) $(OH)^{1-}$ (c) $(BO_3)^{3-}$ (d) $(HCO_3)^{1-}$ (e) $(C_2O_4)^{2-}$ (f) $(SO_3)^{2-}$ **19.** (a) Arsenate (b) Permanganate (c) Sulfite (d) Ammonium (e) Hydrogen sulfate (f) Acetate **21.** (a) $Hg_3(PO_4)_2$ (b) $Sn_3(AsO_4)_2$ (c) $Fe(C_2H_3O_2)_3$ (d) Li_3PO_4 (e) $Al_2(SO_3)_3$

(f) $Zn(NO_2)_2$ **23.** (a) $PbSO_4$ (b) $Co_3(PO_4)_2$
(c) $(NH_4)_2Cr_2O_7$ (d) CaC_2O_4 (e) $Sn(NO_3)_2$ (f)
$Mg(HSO_4)_2$ **25.** (a) Diphosphorus pentasulfide
(b) Carbon monoxide (c) Silicon dioxide (d) Chlorine
dioxide **27.** (a) Aluminum oxide (b) Sodium iodide
(c) Zinc chloride (d) Magnesium nitride (e) Silver sulfide (f) Lithium iodide **29.** (a) 1− (b) 2.67+ (c) 4+
(d) 5+ (e) 1− **31.** (a) 5+ (b) 1+ (c) 1+ (d) 15+
(e) 3− (f) 1− **33.** (a) Osmium(VIII) oxide (b) Mer-
cury(I) phosphide or mercurous phosphide
(c) Iron(II) sulfide or ferrous sulfide (d) Cobalt(II)
chloride or cobaltous chloride (e) Copper(I) nitride
or cuprous nitride (f) Copper(I) oxide or cuprous oxide **35.** (a) Silver carbonate (b) Mercury(II) phosphate or mercuric phosphate (c) Iron(III) sulfate or
ferric sulfate (d) Sodium nitrate (e) Copper(II) chromate or cupric chromate (f) Zinc hydroxide **37.** (a)
Calcium sulfate (b) Potassium cyanide (c) Aluminum
phosphate (d) Copper(I) oxalate or cuprous oxalate
(e) Iron(III) chromate or ferric chromate (f) Cop-
per(II) nitrite or cupric nitrite **39.** (a) HClO (b)
HBr (c) HNO_3 (d) $HClO_2$ (e) $HBrO_4$ (f) H_2SO_4
41. (a) Hydrocyanic acid (b) Hydrosulfuric acid
(c) Bromic acid (d) Sulfuric acid (e) hypochlorous
acid (f) bromous acid **43.** (a) $Mg(OH)_2$ (b) H_2SO_4
(c) $NaNO_3$ (d) N_2O **45.** (a) Quicklime (b) Marble
(c) Gypsum (d) Oil of vitriol **47.** (a) SrO (b) AlN
(c) Rb_2S **49.** (a) Aluminum nitride (b) Vanadi-
um(V) oxide (c) Iron(III) hydroxide or ferric hydroxide (d) Ammonium sulfide **53.** (a) GaF_3 (b)
$Pd(NO_3)_2$ (c) AuP (d) $La(C_2H_3O_2)_3$ (e) PuO_2 (f)
RuO_4

Cumulative Review/Chapters 7–8

1. (a) $\overset{\cdot}{Ca}\cdot$ (b) $[Ca]^{2+}$ **2.** (a) $\cdot\overset{\cdot\cdot}{\underset{\cdot\cdot}{Cl}}\colon$ (b) $[\colon\overset{\cdot\cdot}{\underset{\cdot\cdot}{Cl}}\colon]^{1-}$
3. An ionic bond is formed by the transfer of elec-
trons from one atom to another. The atoms are always of different elements. A covalent bond is
formed by the sharing of electrons between two
atoms. A coordinate covalent bond is a bond in
which one element donates the electrons to form the
bond.

4. $H\overset{x}{\underset{x}{C}}\cdot\cdot\overset{\cdot}{\underset{\cdot}{C}}\cdot\cdot\overset{x}{\underset{x}{C}}H$ (with H above and below each C)

5. HCl = 0.9, $CaCl_2$ = 2.0, $BaCl_2$ = 2.1,
CaF_2 = 3.0 **6.** CsF **7.** (a) Cu_2SO_3 (b) OsO_4
(c) $(NH_4)_2CO_3$ (d) N_2O

8. (a) W^{5+} (b) Sn^{4+} (c) V^{5+} (d) In^{3+}
9. (a) Ammonium iodide (b) Calcium carbide
(c) Copper(I) chromate or cuprous chromate
10. X_2Y_3 **11.** False **12.** True **13.** True
14. False **15.** True **16.** MgO **17.** C−H
18. Na_2SO_4 **19.** 2−
20. (a) $\left[\begin{array}{c} \overset{xx}{\underset{xx}{O}}\colon N\colon\overset{xx}{\underset{xx}{O}} \\ \overset{x}{\underset{x}{O}} \end{array}\right]^{1-}$ (b) $\left[\begin{array}{c} H \\ H\colon N\colon H \\ H \end{array}\right]^{1+}$
21. (a) 6+ (b) 4+ (c) 7+ (d) 3+ **22.** (a) Hydrogens
have partial positive charges, and oxygen has a partial
negative charge. (b) Hydrogen has a partial positive
charge, and iodine has a partial negative charge.
(c) Chlorines have partial positive charges, and oxygen has a partial negative charge. (d) Phosphorus has
a partial positive charge, and chlorines have a partial
negative charge.
23. (a) $\colon\overset{\cdot\cdot}{F}\colon\overset{\cdot\cdot}{F}\colon$ (b) $\colon\overset{\cdot\cdot}{Cl}\colon C\colon\colon C\colon\overset{\cdot\cdot}{Cl}\colon$ (c) $\colon\overset{\cdot\cdot}{Cl}\colon$ ... P ... $\overset{\cdot\cdot}{Cl}\colon$
(d) $\colon\overset{\cdot\cdot}{Cl}\colon C\colon\colon C\colon\overset{\cdot\cdot}{Cl}\colon$ with $\colon\overset{\cdot\cdot}{Cl}\colon\colon\overset{\cdot\cdot}{Cl}\colon$ below
24. (a) 1+ (b) 1− (c) 2+ (d) 3+ **25.** (a) $Fe_3(PO_4)_2$
(b) $Hg(CN)_2$ (c) $Zn(HCO_3)_2$ (d) $Fe_2(Cr_2O_7)_3$
26. (a) In_2S_3 (b) $Mg_3(AsO_4)_2$ (c) Ga_2O_3 (d) Rb_2Se
(e) FeO (f) $CoCl_2$ (g) Cu_3N (h) Cu_2O **27.** (a) HIO_4
(b) $HClO_3$ (c) HBr (d) $HClO_2$ **28.** A coordinate covalent bond is a bond in which one atom donates
both electrons to form the bond. **29.** Yes. A molecule like carbon tetrachloride has four polar C−Cl
bonds, but due to the shape of the molecule, the dipoles cancel out, because the center of positive
charge in the molecule coincides with the center of
negative charge.
30. (a) Group VIA (b) 6 electrons (c) 2 −

Chapter 9

3. 0.100 moles **5.** 2.53×10^8 dollars or 3×10^8 dollars (for one significant figure) **7.** (a) 0.600 mole
(b) 5.00 moles (c) 0.00250 mole (d) 15.00 moles
9. 50.2 g **11.** 44.2 g
13. (a) 270 g (b) 0.479 g (c) 8,020 g (d) 1.28 g
15. 4.46×10^{22} atoms **17.** 9×10^{-20} dollars/atom
19. 1×10^{10} miles **21.** NaCl, sodium chloride (table
salt) **23.** $C_{27}H_{46}O$ **25.** Empirical formula is CH_3;
molecular formula is C_2H_6. **27.** Empirical formula is
C_2H_4O; molecular formula is $C_4H_8O_2$. **29.** 0.7500
moles **31.** 2.74 moles **33.** 0.0500 mole

35. (a) 40.0 g CH_4 (b) 3.20 g SO_3 (c) 2.05 g $Al_2(SO_4)_3$ (d) 3.0 g $(NH_4)_3PO_4$ **37.** 13 g of SO_2
39. 2,270 g of propane or 5.00 pounds of propane
41. 2.71×10^{24} molecules **43.** 1.50×10^{23} formula units
45. 3.25×10^{24} formula units **47.** $C_6H_8O_6$
49. C_6H_6 **51.** $C_3H_6N_6$ **53.** $C_2H_4O_2$, acetic acid; $C_6H_{12}O_6$, dextrose **55.** K = 40.3%, Cr = 26.8%, O = 33.0% **57.** (a) B = 15.9%, F = 84.1% (b) U = 67.61%, F = 32.39% (c) C = 39.1%, H = 8.7%, O = 52.2% **59.** 34 g of sugar present
61. (a) S = 50.1%, O = 49.9% (b) C = 75.0%, H = 25.0% (c) C = 42.11%, H = 6.43%, O = 51.46% (d) Ca = 40.1%, C = 12.0%, O = 48.0% **63.** 1,280 g of copper **65.** 244.7 g of iron **67.** 199 g of oxygen
69. 304 g of Na_2S and 196 g of Fe_2O_3
71. 1.9×10^{16} years **72.** CH_2Cl **73.** $C_2H_4Cl_2$
74. 362 g **75.** 1.76×10^{22} atoms **76.** Hematite, 69.9% Fe, and magnetite, 72.3% Fe **77.** Cu_5FeS_4 and $CuFeS_2$ **78.** S = 50.1%, O = 49.9%
79. 1.79×10^{-21} g **80.** 4.00 moles **81.** 124 g
82. 1.66×10^{-24} g **83.** (a) 0.20 mole (b) 1.2×10^{23} molecules (c) 0.40 mole C atoms (d) 1.2 moles H atoms (e) 0.20 mole O atoms (f) 2.4×10^{23} C atoms (g) 7.2×10^{23} H atoms (h) 1.2×10^{23} O atoms
85. (a) 0.400 mole H_2O (b) 3.00 moles MnO_2 (c) 2.5×10^{-3} mole N_2 (d) 14.99 moles $(NH_4)_2SO_4$
87. (a) 4.00 moles (b) 2.41×10^{24} molecules (c) 8.00 moles C (d) 20.0 moles H (e) 8.00 moles O (f) 4.00 moles N (g) 96.0 g C (h) 20 g H (i) 128 g O (j) 56.0 g N (k) They should be the same.

Chapter 10

1. (a) 2 K + $2H_2O \rightarrow 2KOH + H_2$
(b) $2HC_2H_3O_2 + Ca(OH)_2 \rightarrow Ca(C_2H_3O_2)_2 + 2H_2O$
(c) Mg + $Cu(NO_3)_2 \rightarrow Mg(NO_3)_2 + Cu$
(d) $Na_2O + H_2O \rightarrow 2NaOH$
(e) 2ZnS + $3O_2 \rightarrow 2ZnO + 2SO_2$
(f) 3KOH + $Al(NO_3)_3 \rightarrow Al(OH)_3 + 3KNO_3$
3. (a) 4Fe + $3O_2 \rightarrow 2Fe_2O_3$
(b) $3H_2SO_4 + 2Al(OH)_3 \rightarrow Al_2(SO_4)_3 + 6H_2O$
(c) $2AgNO_3 + BaCl_2 \rightarrow 2AgCl + Ba(NO_3)_2$
(d) $Cu_2S + O_2 \rightarrow 2Cu + SO_2$
5. (a) Combination (b) Decomposition (c) Double-replacement (d) Single-replacement (e) Double-replacement
7. (a) Decomposition (b) Decomposition (c) Decomposition (d) Combination (e) Combination (f) Single-replacement (g) No reaction (h) Double-replace-

ment **9.** (a) Combination (b) Double-replacement (c) Double-replacement (d) Single-replacement
11. (a) $H_2 + Br_2 \rightarrow 2HBr$
(b) BaO + $H_2O \rightarrow Ba(OH)_2$
(c) 2Na + $Cl_2 \rightarrow 2NaCl$
(d) $N_2 + 2O_2 \rightarrow 2NO_2$
(e) $CaCO_3 \rightarrow CaO + CO_2$
(f) 2KOH $\rightarrow K_2O + H_2O$
(g) $Hg(ClO_3)_2 \rightarrow HgCl_2 + 3O_2$
(h) $PbCl_2 \rightarrow Pb + Cl_2$
(i) 2Li + $2H_2O \rightarrow 2LiOH + H_2$
(j) Zn + $H_2SO_4 \rightarrow ZnSO_4 + H_2$
(k) Ni + $Al(NO_3)_3 \rightarrow$ no reaction (Aluminum is above nickel in the activity series.)
(l) 2Al + $3Hg(C_2H_3O_2)_2 \rightarrow 2Al(C_2H_3O_2)_3 + 3Hg$
(m) $H_2SO_4 + 2NH_4OH \rightarrow (NH_4)_2SO_4 + 2H_2O$
(n) $2AgNO_3(aq) + BaCl_2(aq) \rightarrow$
$$Ba(NO_3)_2(aq) + 2AgCl(s)$$
(o) $3H_2SO_3 + 2Al(OH)_3 \rightarrow Al_2(SO_3)_3 + 6H_2O$
(p) $NaNO_3(aq) + KCl(aq) \rightarrow$ no reaction
13. (a) $2N_2O \rightarrow 2N_2 + O_2$
(b) $H_2CO_3 \rightarrow H_2O + CO_2$
(c) $2NaNO_3 \rightarrow 2NaNO_2 + O_2$
(d) $H_2 + F_2 \rightarrow 2HF$
(e) $N_2 + 3H_2 \rightarrow 2NH_3$
(f) Ca + 2HCl $\rightarrow CaCl_2 + H_2$
(g) Cu + $NiCl_2 \rightarrow$ no reaction
(h) $3AgNO_3(aq) + K_3AsO_4(aq) \rightarrow$
$$3KNO_3(aq) + Ag_3AsO_4(s)$$
15. (a) 2K + $Cl_2 \rightarrow 2KCl$
(b) $K_2O + H_2O \rightarrow 2KOH$
(c) MgO + $H_2O \rightarrow Mg(OH)_2$
(d) CaO + $CO_2 \rightarrow CaCO_3$
(e) $2NH_3 \rightarrow N_2 + 3H_2$
(f) $SrCO_3 \rightarrow SrO + CO_2$
(g) $2NaClO_3 \rightarrow 2NaCl + 3O_2$
(h) $Mg(OH)_2 \rightarrow MgO + H_2O$
17. (a) NaOH and HCl (b) KOH and H_2SO_4
(c) $Al(OH)_3$ and HNO_3
19. (a) Zn + $2HNO_2 \rightarrow Zn(NO_2)_2 + H_2$
(b) Ag + $NiCl_2 \rightarrow$ no reaction
(c) Zn + $2AgNO_3 \rightarrow Zn(NO_3)_2 + 2Ag$
(d) 2Cs + $2H_2O \rightarrow 2CsOH + H_2$
(e) 3HCl + $Al(OH)_3 \rightarrow AlCl_3 + 3H_2O$
(f) $KNO_3 + ZnCl_2 \rightarrow$ no reaction
(g) $Al(NO_3)_3 + 3NaOH \rightarrow Al(OH)_3(s) + 3NaNO_3$
(h) $K_2CrO_4 + Pb(NO_3)_2 \rightarrow PbCrO_4(s) + 2KNO_3$
23. (a) 2K + $Cl_2 \rightarrow 2KCl$ (The K is oxidized; the Cl is reduced.)

(b) $2NH_3 \rightarrow N_2 + 3H_2$ (The N is oxidized; the H is reduced.)
(c) $CuO + H_2 \rightarrow Cu + H_2O$ (The H is oxidized; the Cu is reduced.)
(d) $Sn + 2Cl_2 \rightarrow SnCl_4$ (The Sn is oxidized; the Cl is reduced.)
25. (a) K is the reducing agent; Cl_2 is the oxidizing agent. (b) The NH_3 is both the oxidizing and the reducing agent. (c) H_2 is the reducing agent; CuO is the oxidizing agent. (d) Sn is the reducing agent; Cl_2 is the oxidizing agent. **27.** (a) Zn is the reducing agent; HNO_2 is the oxidizing agent. (b) No reaction took place. (c) Zn is the reducing agent; $AgNO_3$ is the oxidizing agent. (d) Cs is the reducing agent; H_2O is the oxidizing agent. **29.** (a) vanadium pentoxide (b) zinc (c) zinc (d) vanadium **30.** $MgCO_3 \rightarrow$ $MgO + CO_2$ $CaCO_3 \rightarrow CaO + CO_2$
31. (a) $N_2 + 3H_2 \rightarrow 2NH_3$
(b) $3Fe + 4H_2O \rightarrow Fe_3O_4 + 4H_2$ (c) balanced
33. (a) $2C + O_2 \rightarrow 2CO$ (b) $H_2 + Cl_2 \rightarrow 2HCl$
(c) $2Na + 2H_2O \rightarrow 2NaOH + H_2$ (d) no reaction
34. (a) no reaction
(b) $Mg + Zn(NO_3)_2 \rightarrow Mg(NO_3)_2 + Zn$
(c) no reaction
(d) $Zn + Cu(NO_3)_2 \rightarrow Zn(NO_3)_2 + Cu$
36. (a) $Ca(OH)_2 + 2HCl \rightarrow CaCl_2 + 2H_2O$ (double replacement) (b) $Zn + H_2SO_4 \rightarrow ZnSO_4 + H_2$ (single replacement) (c) $2Ba + O_2 \rightarrow 2BaO$ (combination) (d) $Cs_2O + H_2O \rightarrow 2CsOH$ (combination)
37. $4NH_3 + 7O_2 \rightarrow 2N_2O_4 + 6H_2O$ **38.** (a) Fe^{2+}
(b) chromium in $Cr_2O_7^{2-}$ (c) $Cr_2O_7^{2-}$ (d) Fe^{2+}
39.
$12Na$	$+$	P_4	$\rightarrow$	$4Na_3P$
Sodium		Phosphorus		Sodium phosphide

Chapter 11

1. (a) $2Na_2S_2O_3 + AgBr \rightarrow Na_3Ag(S_2O_3)_2 +$ NaBr (b) 79.10 g sodium thiosulfate (c) 100.3 g $Na_3Ag(S_2O_3)_2$, 25.73 g NaBr
3. (a) $NH_4NO_3 \xrightarrow{\text{Heat}} N_2O + 2H_2O$ (b) 4.0 g ammonium nitrate (c) 1.8 g water
5. (a) $C_7H_8 + 3HNO_3 \rightarrow C_7H_5N_3O_6 + 3H_2O$
(b) 405 g of toluene, 832 g nitric acid (c) 238 g water
7. (a) $2NH_4Cl + H_2SO_4 \rightarrow (NH_4)_2SO_4 + 2HCl$
(b) 14.6 g sulfuric acid (c) 19.7 g ammonium sulfate, 10.8 g hydrogen chloride **9.** Percent $KClO_3 = 51.2\%$
11. (a) $3BaO_2 + 2H_3PO_4 \rightarrow 3H_2O_2 + Ba_3(PO_4)_2$
(b) 68.0 g hydrogen peroxide (c) $13\overline{0}$ g phosphoric

acid **13.** 11.9 g hydrogen **15.** 1,460 g air
17. (a) $2C_4H_{10} + 13O_2 \rightarrow 8CO_2 + 10H_2O$
(b) 83.2 g oxygen (c) 70.4 g CO_2 and 36.0 g H_2O
19. 64.0 g ozone **21.** (a) $2Al(OH)_3 + 3H_2SO_4 \rightarrow$ $Al_2(SO_4)_3 + 6H_2O$ (b) 228 g $Al(OH)_3$ and 430 g H_2SO_4 (c) 158 g H_2O **23.** (a) $Fe_2O_3 + 3CO \rightarrow 2Fe +$ $3CO_2$ (b) $65\overline{0}$ g Fe_2O_3 (c) 342 g CO and 536 g CO_2
25. (a) $4NH_3 + 5O_2 \rightarrow 4NO + 6H_2O$
(b) $1,\overline{0}00$ g O_2 (c) $75\overline{0}$ g NO and 675 g H_2O
27. 502 g $Zn(NO_3)_2$ **29.** (a) $CaC_2 + 2H_2O \rightarrow$ $Ca(OH)_2 + C_2H_2$ (b) 148 g $Ca(OH)_2$ and 52.0 g C_2H_2 **31.** (a) $Fe_2O_3 + 3CO \rightarrow 2Fe + 3CO_2$
(b) 223 g of iron, 264 g of carbon dioxide
33. (a) $2H_2 + O_2 \rightarrow 2H_2O$ (b) 72 g water
35. (a) $Ca_3P_2 + 6H_2O \rightarrow 2PH_3 + 3Ca(OH)_2$
(b) 136 g PH_3 (c) 445 g $Ca(OH)_2$ **37.** $2\overline{0}$ moles $CaSiO_3$, 33 moles CO and 6.7 moles P_2
39. (a) $3Fe + 4H_2O \rightarrow Fe_3O_4 + 4H_2$
(b) 10.7 g H_2 (c) $31\overline{0}$ g Fe_3O_4 **41.** (a) $2K + Cl_2 \rightarrow$ $2KCl$, 0.128 mole of Cl_2 (b) 19.1 g
42. (a) $2Cu(NO_3)_2 \rightarrow 2CuO + 4NO_2 + O_2$
(b) 8.54 g (c) $11\overline{0}$ g **43.** (a) oxygen (b) 206 g
44. (a) $2H_2 + O_2 \rightarrow 2H_2O$ (b) 0.8 g H_2 left and 72.0 g of H_2O produced **45.** $39\overline{0}$ g
46. (a) $Zn + H_2SO_4 \rightarrow ZnSO_4 + H_2$ (b) 19,600 g
47. (a) The second experiment, $Pb + Cu(NO_3)_2 \rightarrow$ $Pb(NO_3)_2 + Cu$ (b) 79.8 g $Pb(NO_3)_2$ and 15.3 g Cu
48. 6.7 g **49.** 58.5 g NaCl produced and 40.0 g NaOH left **50.** 112 g

Cumulative Review/Chapters 9–11

1. 0.0100 mole **2.** 0.40 mole **3.** (a) 73.6 g H_2SO_4
(b) 90.0 g H_2O (c) 0.434 g H_2CO_3 (d) 3.000×10^3 moles $HC_2H_3O_2$ **4.** 1.2×10^{23} atoms **5.** SnO_2
6. $C_{40}H_{64}O_{12}$ **7.** $C_6H_{12}O_6$ **8.** (a) 0.200 mole
(b) 1.20×10^{23} molecules (c) 1.80 moles C atoms
(d) 1.60 moles H atoms (e) 0.800 moles O atoms
9. N_2O_4 **10.** (a) 310.3 (b) 16.0 (c) 46.0
(d) 255.3 **11.** (a) 5.9% H, 94.1% S (b) 39.3% Na, 60.7% Cl (c) 24.0% Fe, 30.9% C, 3.9% H, 41.2% O
(d) 44.9% K, 18.4% S, 36.7% O **12.** C_6H_{14}
13. 74.1% C, 8.6% H, 17.3% N **14.** 2.41×10^{22} molecules **15.** 2.50 g $CaCO_3$ **16.** $CaBr_2 + H_2SO_4 \rightarrow$ $2HBr + CaSO_4$ **17.** $2Mg + O_2 \rightarrow 2MgO$
18. $NH_3(g) + HCl(g) \rightarrow NH_4Cl(s)$ **19.** (a) Decomposition (b) Combination (c) Single replacement
(d) Double replacement **20.** (a) $H_2 + Cl_2 \rightarrow 2HCl$
(b) $2NaCl + Pb(NO_3)_2 \rightarrow 2NaNO_3 + PbCl_2$

(c) $2Al + 3H_2SO_4 \rightarrow Al_2(SO_4)_3 + 3H_2$
(d) $SrCO_3 \rightarrow SrO + CO_2$
21. $CaCO_3 \rightarrow CaO + CO_2$ **22.** (a) $TiCl_4 + 2H_2O \rightarrow$
$TiO_2 + 4HCl$ (b) $P_4O_{10} + 6H_2O \rightarrow 4H_3PO_4$
23. (a) $Fe(OH)_3 + H_3PO_4 \rightarrow FePO_4 + 3H_2O$
(b) $Pb(OH)_2 + 2HNO_3 \rightarrow Pb(NO_3)_2 + 2H_2O$
24. $SO_3 + H_2O \rightarrow H_2SO_4$ **25.** $2C_8H_{18} + 25O_2 \rightarrow$
$16CO_2 + 18H_2O$ **26.** $2C_3H_7OH + 9O_2 \rightarrow 6CO_2 +$
$8H_2O$ **27.** $CaSO_4 + Na_2CO_3 \rightarrow CaCO_3(s) + Na_2SO_4$
28. $C_{12}H_{22}O_{11} \rightarrow 12C + 11H_2O$ **29.** True
30. False **31.** True **32.** True **33.** True **34.** True
35. $C + 2H_2SO_4 \rightarrow CO_2 + 2SO_2 + 2H_2O$ **36.** 5
moles CuO **37.** 0.20 mole NaOH **38.** 88 grams
39. 728 grams HCl **40.** 12.6 grams CO_2 **41.** 1,860
grams H_3PO_4 **42.** 77.6 grams $Ca_3(PO_4)_2$, 55.6 grams
$Ca(OH)_2$ **43.** 15.0 moles, 1,810 grams **44.** 119
grams H_2O_2 **45.** 13.9 grams $PbCl_2$ **46.** 96.1 grams
LiOH **47.** 72.9 grams Mg **48.** 171.3 grams
$Ba(OH)_2$, 2.0 grams H_2, 36.0 grams excess H_2O

Chapter 12

1. 1(c) **2.** (a) **3.** (d) **4.** (b) **7.** $3,\overline{0}00$ cal
9. 1,110 cal **11.** 2.09 kcal **13.** 30.0°C **15.** $\Delta H_R =$
-68.5 kcal/mole **17.** (a) $S(s) + 1\frac{1}{2} O_2(g) \rightarrow$
$SO_3(g) + 94.45$ kcal (b) $\frac{1}{2}H_2(g) + \frac{1}{2}Br_2(l) \rightarrow HBr(g) +$
8.7 kcal (c) $C(s) + 2 Cl_2(g) \rightarrow CCl_4(l) + 33.4$ kcal
(d) $\frac{1}{2}N_2(g) + \frac{1}{2}O_2(g) \rightarrow NO(g) - 21.5$ kcal **21.** -14
kcal **23.** 3,210 kcal **25.** (a) $\Delta H_R = -621.4$ kcal
(b) $-62,100$ kcal **27.** (a) $\Delta H_R = -55.7$ kcal
(b) 1,520 g **29.** $\Delta H_f = -183.0$ kcal **31.** ΔH_f
(calculated) $= -94.8$ kcal **33.** $\Delta H_f = -107$ kcal
35. -14.2 kcal **37.** 2.5 kcal **39.** -47.0 kcal
41. (a) -47.0 kcal (b) 141 kcal **43.** -21.36 kcal
45. (a) $PCl_5 + 4H_2O \rightarrow H_3PO_4 + 5HCl$ (b) -139.8
kcal (c) 27.96 kcal **47.** $\Delta H_f = -37.2$ kcal/mole
49. 6.30 kcal **50.** -288.5 kcal **51.** (a) and (d),
Energy is absorbed; (b) and (c), Energy is released
52. Acetylene releases more energy; $\Delta H = -300.2$
kcal/mole **53.** Endothermic **54.** 5,500 kcal **55.** H_2
produces 29 kcal/g, CH_4 produces 13.3 kcal/g, C produces 7.8 kcal/g **57.** 10,700 kcal **58.** 10.5 kcal/mole

Chapter 13

3. (a) 2.45 atm (b) 0.620 atm (c) 8.00 atm (d) 0.164
atm **5.** (a) 3,040 torr (b) 304 torr (c) 1.52 torr
(d) 3,990 torr **7.** (a) 0.10 atm (b) 3.50 atm
(c) 0.00500 atm **9.** (a) 7.60 torr (b) 9,120 torr

(c) 4,180 torr **11.** $V_f = 27\overline{0}$ mL **13.** $V_f = 3,\overline{0}00$ mL
or 3.00 L **15.** $P_f = 9,5\overline{0}0$ torr **17.** $T_f = 1,327$°C
19. $T_f = 527$°C **21.** $V_f = 33.3$ L **23.** (a) 27°C
(b) -253°C **25.** (a) 283 K (b) 263 K (c) $2\overline{0}0$ K
27. (a) -273°C (b) -173°C (c) -73°C (d) 27°C
29. (a) 293 K (b) 273 K (c) 173 K (d) 373 K
31. $P_f = 2.22$ atm **33.** $P_f = 1,590$ torr
35. $V_f = 421$ mL **37.** $V_f = 709$ mL **39.** $V_f = 394$ mL
41. It would increase to 2.7 times its original volume. **43.** $P_f = 59.8$ atm **45.** $V_f = 819$ mL
47. $V_f = 276$ mL **49.** (1) increase (2) increase
(3) increase **51.** $P_{total} = 836$ torr **53.** 608 torr or
610 torr for two sig. figs. **55.** 25.0 L H_2O vapor 32.5
L of oxygen. **57.** 89.6 L **59.** 20.5 g Na **61.** O_2 uncombined $= 40$ mL, $H_2O(g)$ formed $= 80$ mL
63. (a) $2SO_2 + O_2 \rightarrow 2SO_3$ (b) 75.0 L (c) $15\overline{0}$ L
65. (a) $2C_4H_{10} + 13O_2 \rightarrow 8CO_2 + 10H_2O$ (b) 97.5 L
(c) 60.0 L CO_2 and 75.0 L H_2O **67.** (a) $N_2 + 2O_2 \rightarrow$
$2NO_2$ (b) 37.5 mL (c) 75.0 mL **69.** 6.83 g N_2
71. $R = 0.0821 \dfrac{\text{liter-atm}}{\text{K-mole}}$ **73.** 11.2 L **75.** 2.24 g
77. $1\overline{0}0$ atm **79.** 1.52 K or -271.5°C **81.** (a) C_2H_5
(b) MM $= 59$ (c) C_4H_{10} **83.** MM $= 41$ **85.** $D_{O_2} =$
1.43 g/L **87.** MM $= 24.0$ **89.** (a) NO_2 (b) MM $=$
92.3 (c) N_2O_4 **92.** 50.0 L **93.** 124 L **94.** 1,797 L or
$1,8\overline{0}0$ L for three sig. figs. **95.** (a) 277 K (b) 355 K
(c) 373 K **96.** 22.4 L **97.** 0.714 g/L **99.** 429 mL
100. 31.5 g/mole; O_2 is a likely candidate.

Chapter 14

1. The boiling point of a liquid increases with increasing atmospheric pressure because a greater vapor
pressure is needed for molecules to escape from the
liquid. This can happen only at a higher temperature. **5.** (a) Liquid (b) Gas (c) Solid
9. Once the water is boiling, it doesn't get any hotter.
Turning up the heat on the stove doesn't change the
temperature of the water and doesn't speed up the
cooking of the potatoes. **11.** If the mountain is high
enough, the water boils at a significantly lower temperature because of the decrease in atmospheric pressure. Because the temperature of the water is lower
when it begins to boil, it takes longer to cook a hard-
boiled egg on a mountain than at sea level. **15.** It is
the high-energy molecules of liquid that escape from
the container. The low-energy ones get left behind.
The liquid therefore has less energy—in other words,
less heat. The temperature of the liquid drops.

25. The matching pairs are 1. (c), 2. (d), 3. (b), 4. (a) **37.** Amorphous **39.** Well-defined (lattice) **43.** A molecular solid **45.** An ionic solid **47.** (a) Metallic (b) Ionic **49.** (a) Atoms (b) Molecules (c) Ions **61.** 59 g **65.** 1.00 L (1,$\overline{0}$00 mL) **67.** (a) $Mg(OH)_2$ (b) H_2SO_4 (c) $2NaOH + H_2$ **69.** (a) $SrO + H_2O \rightarrow Sr(OH)_2$ (b) $SO_2 + H_2O \rightarrow H_2SO_3$ (c) $2Li + 2H_2O \rightarrow 2LiOH + H_2$ **71.** 36.0% **75.** (a) 51.1% (b) 62.9% (c) 45.5% **77.** (a) $CaSO_4 \cdot 2H_2O \rightarrow CaSO_4 + 2H_2O$ (b) $KAl(SO_4)_2 \cdot 12H_2O \rightarrow KAl(SO_4)_2 + 12H_2O$ **79.** $CoSO_4 \cdot 6H_2O$ **84.** It is a repeating unit. **85.** Yes, some substances are already gases at *normal* atmospheric pressure (1.00 atm), for example, CO_2. The *normal boiling point* is defined as the boiling point at one atmosphere pressure. Therefore many substances do not have normal boiling points. **88.** Glass is an amorphous solid. **92.** (a) Solid and liquid (b) Solid (c) Gas **94.** Higher

Cumulative Review/Chapters 12–14

1. Endothermic is heat taken in from the surroundings; exothermic is heat given off to the environment **2.** 10,$\overline{0}$00 calories **3.** 138 calories **4.** 291 calories **5.** We use standard state conditions so we can measure heat given up or absorbed during thermochemical reactions, as opposed to measuring heat involved in temperature or pressure changes. **6.** 25,$\overline{0}$00 calories **7.** $H_2(g) + \frac{1}{2}O_2(g) \rightarrow H_2O(l) + 68.3$ kcal **8.** -54 kcal **9.** 281 kcal **10.** 7.05 kcal **11.** 24 mL **12.** 18.8 mL **13.** 616 torr **14.** 2,$\overline{5}$00 mL **15.** 551°C **16.** 5.94 L **17.** 7.31 L **18.** 0.371 L **19.** 4.16 L **20.** 16.0 moles **21.** 336 L **22.** 274 L **23.** 0.0412 mole **24.** MM = 286 **25.** 10.0 L **26.** 3.12 g/L **27.** Because $P_f < P_i$, therefore $V_f > V_i$ **28.** 2.62 L **29.** MM = 93.7 **30.** Forces of attraction between gas molecules vary. They increase with increasing molecular weight. **31.** It will decrease. **32.** $CO_2 = 342$ torr, $H_2 = 326$ torr, $N_2 = 138$ torr, $CH_4 = 0.0245$ torr **33.** 3. (a), 2. (b), 1. (c) **34.** $CaCO_3 \cdot 6H_2O$ **35.** 49.3% H_2O **36.** 20.9% **37.** $BaCl_2 \cdot 2H_2O$ **38.** $Na_2CO_3 \cdot 10H_2O$ **39.** 28.1 grams **40.** 3. (a), 1. (b), 2. (c) **41.** 2. (a), 3. (b), 1. (c) **42.** (b) **43.** Sublimation occurred. **44.** Molecular **45.** Ionic **46.** Amorphous **47.** Molecules gain energy; kinetic energy increases; attractive forces between molecules decrease.

48. Benzene **49.** In some crystals, electrons are free to wander throughout the crystal. **50.** Attraction of opposite charges makes the bonds so strong.

Chapter 15

1. The matching pairs are 1. (c), 2. (a), 3. (d), 4. (b) **3.** (a) HCl is the solute, water is the solvent. (b) CO_2 is the solute, water is the solvent. (c) Benzene is the solute, carbon tetrachloride is the solvent. **7.** True **9.** Saturation **11.** (a) Unsaturated (b) Supersaturated **15.** (a) 6.25% by mass (b) 20.0% by mass (c) 8.00 % by mass **17.** $2\overline{0}$ g **19.** (a) $2\overline{0}$% by volume (b) 6% by volume **21.** 12.5 mL **23.** $2\overline{0}$ mL **25.** (a) 3.3% by mass–volume (b) 5.0% by mass–volume **27.** 15 g **29.** 0.05% by mass **31.** (a) 12.5% (b) 5.00% **33.** $3\overline{0}$g **35.** (a) 5.0% (b) 0.833% **37.** 28 mL **39.** $2\overline{0}$0 mL **41.** (a) 3.13% (b) 5.00% **43.** 12.5 g **45.** (a) 0.30 M (b) 0.5 M **47.** 2.74 M **49.** 8.4 g **51.** 0.109 M **53.** 0.15 M **55.** (a) 0.167 M (b) 0.200 M **57.** 0.525 M **59.** 16.4 g **61.** (a) 0.10 equivalent (b) 0.20 equivalent (c) 0.0902 equivalent **65.** (a) 0.0299 equivalent (b) 0.750 equivalent **67.** (a) 1,120 g (b) 9.27 g **69.** (a) 2.97 N (b) 0.13 N (c) 0.99 M, 0.063 M **71.** 59 g **73.** 250 mL **75.** (a) 1.00 N (b) 0.500 N (c) 0.500 M $Fe(OH)_2$, 0.167 M H_3BO_3 **77.** 3.25 g **79.** 599 mL **81.** Take 125 mL of 16 N H_2SO_4; dilute to $5\overline{0}$0 mL with water **83.** Take 50 mL of 6 N NaOH; dilute to 3 L with water. **85.** Take 104 mL of 12.0 M HCl; dilute to $25\overline{0}$ mL with water. **87.** Take 42 mL of 3.00 M H_2SO_4; dilute to 5.00 L with water. **97.** 0.15 molal **99.** (a) 0.80 molal (b) 0.333 molal **105.** -33.1°C **107.** 109.26°C **109.** b.p. = 100.78°C, f.p. = -2.79°C **111.** (a) Osmosis (b) Diffusion **116.** 1×10^{-2} M or 0.01 M **117.** 0.005 M and 0.01 N **118.** $5\overline{0}$0 g **119.** 0.510 M and 1.02 N **122.** Ethanol **123.** 41.3 mL. **124.** $15\overline{0}$ g/mole **125.** 2.00 *molal* solution **126.** 3.72 g/mole **127.** A 0.104°C increase in boiling point; therefore, if the boiling point of the pure water is 100.000°C, the boiling point of the solution is 100.104°C **128.** (a) 2.00 equivalents (b) 2.00 equivalents **129.** (a) 1.00 equivalent (b) 2.00 equivalents

Chapter 16

2. The matching pairs are 1. (d), 2. (c), 3. (e), 4. (a), 5. (b) **3.** Carbonic acid and phosphoric acid **7.** $NH_3 + H_2O \rightleftharpoons NH_4^{1+} + OH^{1-}$ **15.** Organic

17. (a) Blue, red, (b) Red, blue
19. $HBr + H_2O \rightarrow H_3O^{1+} + Br^{1-}$
23. (a) $K_2O + H_2O \rightarrow 2KOH$
(b) $MgO + H_2O \rightarrow Mg(OH)_2$
25. (a) $SO_2 + H_2O \rightarrow H_2SO_3$
(b) $CO_2 + H_2O \rightarrow H_2CO_3$
27. (a) $2H_3PO_4 + 3Mg(OH)_2 \rightarrow Mg_3(PO_4)_2 + 6H_2O$
(b) $P_2O_5 + 3H_2O \rightarrow 2H_3PO_4$
(c) $K_2O + H_2O \rightarrow 2KOH$
(d) $Mg(OH)_2 + CO_2 \rightarrow MgCO_3 + H_2O$
(e) $3Zn + 2H_3PO_4 \rightarrow Zn_3(PO_4)_2 + 3H_2$ **29.** Use a *weak* acid; for example, dilute acetic acid. Under no circumstances should you use a strong mineral acid (even if it is diluted).
35. (a) $Zn + H_2SO_4 \rightarrow ZnSO_4 + H_2$
(b) $HCl + LiOH \rightarrow LiCl + H_2O$
(c) $Ca + H_2SO_4 \rightarrow CaSO_4 + H_2$
(d) $3Ca(OH)_2 + 2H_3PO_4 \rightarrow Ca_3(PO_4)_2 + 6H_2O$
37. (a) $Zn + 2HCl \rightarrow ZnCl_2 + H_2$
(b) $2HCl + Ca(OH)_2 \rightarrow CaCl_2 + 2H_2O$
(c) $Mg(OH)_2 + CO_2 \rightarrow MgCO_3 + H_2O$
41. (a) HNO_3 (b) HF (c) HI **43.** (a) Sodium hydroxide (b) Calcium hydroxide (c) Iron(II) hydroxide (d) Aluminum hydroxide   **45.** (a) Nitric acid (b) phosphoric acid (c) hydroiodic acid (d) sulfuric acid
47. (a) $LiOH$ (b) $Ba(OH)_2$ (c) $Fe(OH)_2$ (d) $Fe(OH)_3$
59. (a) 100 times more acidic (b) 100,000 times more acidic   **63.** (a) 3 (b) 4 (c) 2 (d) 11 (e) 10 (f) 5
65. (a) $10^{-5}\ M$ (b) $10^{-9}\ M$ (c) $10^{-2}\ M$ (d) $10^{-12}\ M$ (All answers for $[H^+]$ concentration)
67. (a) $[H^{1+}] = 10^{-9}\ M$ (b) $[H^{1+}] = 10^{-6}M$
(c) $[H^{1+}] = 10^{-4}\ M$ (d) $[H^{1+}] = 10^{-12}M$
69. (a) $10^{-11}\ M$ (b) $10^{-10}\ M$ (c) $10^{-6}\ M$ (d) $10^{-5}\ M$
(e) $10^{-8}\ M$ (f) $10^{-12}\ M$ **71.** (a) Sodium sulfide (b) Potassium bromide (c) Sodium sulfate (d) Potassium nitrate (e) Calcium carbonate (f) Barium iodide   **73.** (a) $AgNO_3$ (b) $CoCl_2$
(c) Cu_2SO_3 (d) $Hg_3(PO_4)_2$
79. $2LiOH + H_2SO_4 \rightarrow Li_2SO_4 + 2H_2O$
81. $Zn + 2HBr \rightarrow ZnBr_2 + H_2$
83. (a) $Mg + 2HCl \rightarrow MgCl_2 + H_2$
(b) $Zn + H_2SO_4 \rightarrow ZnSO_4 + H_2$
85. (a) $3KOH + H_3PO_4 \rightarrow K_3PO_4 + 3H_2O$
(b) $Ca(OH)_2 + 2HNO_3 \rightarrow Ca(NO_3)_2 + 2H_2O$
87. (a) $2K + Cl_2 \rightarrow 2KCl$
(b) $Ca + Br_2 \rightarrow CaBr_2$ **89.** $0.63\ N$ **91.** $0.2778\ N$
93. $0.180\ N$ **95.** $0.667\ N$ or $0.334\ M$
97. (a) Acid (b) Base   **98.** (a) Base (b) Acid
99. (a) Base
(b) Acid   **100.** pH = 11, $[OH^{1-}] = 10^{-3}$

101. (a) $HCl + NaOH \rightarrow NaCl + H_2O$
(b) $BaO + 2HCl \rightarrow BaCl_2 + H_2O$
(c) $Zn + H_2SO_4 \rightarrow ZnSO_4 + H_2$
102. (a) $OH^{1-} + H_2O \rightarrow H-OH + OH^{1-}$
(b) $NH_3 + H_2O \rightarrow NH_4^{1+} + OH^{1-}$
(c) $HSO_4^{1-} + H_2O \rightarrow H_2SO_4 + OH^{1-}$
103. (a) $H_2O + H-OH \rightarrow H_3O^{1+} + OH^{1-}$
(b) $HCO_3^{1-} + H_2O \rightarrow H_3O^{1+} + CO_3^{2-}$
(c) $HNO_3 + H_2O \rightarrow H_3O^{1+} + NO_3^{1-}$
105. If in a reaction the PCl_3 accepted a proton, it would be a Brønsted-Lowry base.

105. $H^{1+} + :PCl_3 \rightarrow \left[\begin{array}{c} Cl \\ | \\ H \leftarrow P - Cl \\ | \\ Cl \end{array} \right]^{1+}$

Cumulative Review/Chapters 15–16

1. (a) Oxygen is the solute and nitrogen is the solvent. (b) Sulfur dioxide is the solute and water is the solvent. (c) Lemon juice and sugar are solutes and water is the solvent.   **2.** (a) H—F is polar; H_2O is polar. Therefore a solution forms. (b) Benzene is nonpolar; *trans*-dichloroethane is nonpolar. Therefore a solution forms.   **3.** (a) 66.7% by mass salt (b) 11.7% by mass sugar   **4.** (a) 5.49% by volume alcohol in water (b) 36.8% by volume benzene in CCl_4 **5.** 6.7% mass–volume benzene in CCl_4 **6.** $0.027\ M$ **7.** 13.3 g NaOH   **8.** 234 mL   **9.** Normality is the number of equivalents of solute per liter of solution, and molarity is the number of moles of solute per liter of solution.   **10.** 65.4 g H_3PO_4 **11.** $1.00\ N$ or $0.500\ M$
12. Dilute 0.203 L of $36.0\ N$ with water to 5.50 L.
13. 51.0 g $AgNO_3$ **14.** False   **15.** 177 g H_2SO_4
16. (a) 545 g of sugar in 2,180 g solution (b) 0.795 M solution   **17.** 0.90 g solute   **18.** Dilute the solution to a total volume of 3.82 L.
19. 101.7°C   **20.** $0.100\ m$ **21.** Arrhenius acid is a substance that releases hydrogen ions when dissolved in water. Brønsted–Lowry acid is a substance that is a proton donor.   **22.** False
23. (a) $2HC_2H_3O_2 + Ca(OH)_2 \rightarrow$
$\qquad\qquad\qquad\qquad Ca(C_2H_3O_2)_2 + 2H_2O$
(b) $N_2O_5 + H_2O \rightarrow 2HNO_3$
24. $H_2O + SO_3 \rightarrow H_2SO_4$ or
$H_2O + SO_2 \rightarrow H_2SO_3$ **25.** pH = 2   **26.** pH = 6
27. $18,\overline{0}00$ mL   **28.** $1.00\ M\ Na^{1+}$ and $1.00\ M\ Cl^{1-}$
29. $1.20\ M$ HCl   **30.** $M_{NaOH} = 0.0450$
31. Ionization is the process by which ions are

formed from atoms or molecules by the transfer of electrons. Dissociation is the process by which an ionic substance separates into ions by the action of a solvent. **32.** 1 *M* sulfuric acid is more acidic than 1 *N* sulfuric acid **33.** 1 *M* hydrochloric acid is more acidic than 1 *M* acetic acid. **34.** 0.04 *M* ions **35.** If the Cl^{1-} concentration is 0.500 *M*, the Sr^{2+} concentration must be half of that (or 0.250 *M*). **36.** 0.125 *M* **37.** The term refers to ionization. **38.** Twice as many Cl^{1-} as Sr^{2+} **39.** pH = 4 **40.** $[OH^{1-}] = 10^{-6}$

Appendix A

1. (a) 17 (b) −15 (c) 15 (d) 5 **2.** (a) 1 (b) 5 (c) −5 (d) −1 **3.** (a) 33 (b) −17 (c) 17 (d) −33 **4.** (a) 11 (b) 99 (c) −99 (d) −11 **5.** (a) $\frac{1}{16}$ (b) $\frac{1}{4}$ (c) 512 (d) $\frac{1}{32}$ **6.** (a) $\frac{6}{35}$ (b) $\frac{18}{25}$ (c) $\frac{1}{4}$ (d) $\frac{25}{49}$ **7.** (a) $(4)^{-7}$ (b) $(4)^5$ (c) $(6)^2$ (d) $(5)^1$ **8.** (a) $(4)^3$ (b) $(4)^8$ (c) $(a)^1$ (d) $(b)^2$ **9.** (a) 625 (b) 1,000 (c) 99 (d) $\frac{1}{512}$ (e) 16 **10.** (a) $(5)^5$ (b) $(12)^{12}$ (c) $(10)^0 = 1$ (d) $(4)^{-10}$ **11.** (a) $(4)^2$ (b) $(12)^5$ (c) $(20)^{-7}$ (d) $(20)^{-1}$ **12.** (a) $(4)^9$ (b) x (c) $(y)^7$ (d) p (e) $(10)^8$ **13.** (a) 64 (b) $\frac{1}{81}$ (c) $\frac{1}{64}$ (d) 64 **14.** (a) $(10)^3$ (b) $(2)^{-5}$ (c) $(3)^{-3}$ (d) $(6)^2$ **15.** (a) $(3)^2$ (b) $(2)^3$ (c) $(a)^{15}$ (d) $(b)^{-13}$ **16.** 48 inches **17.** $w = A/l$ **18.** 6 yards **19.** 48 eggs **20.** (a) 6 hours (b) 2 hours

21. 16 pints **22.** 1.47 ft/sec **23.** 6 feet **24.** 180 inches **25.** 300 feet **26.** 180 inches **27.** 24 dozen **28.** 1,440 minutes, 86,400 seconds **29.** 10 gallons **30.** 15,000 mi/hr **31.** (a) 149.683 (b) 141.41 **32.** (a) 26.525 (b) 103.71 (c) 106.14 (d) 147.78 **33.** (a) 29.32 (b) 38.5628 (c) 27.0657 **34.** (a) 2.04 (b) 5 **35.** (a) 31.20 (b) 46.15 (c) 39.80 (d) 75.38 **36.** (a) $a = 4$ (b) $y = \frac{13}{40} = 0.325$ (c) $z = 3$ **37.** (a) $x = \frac{3}{2}$ (b) $y = 1$ **38.** (a) $P = \dfrac{nRT}{V}$ (b) $n = \dfrac{PV}{RT}$ (c) $T = \dfrac{PV}{nR}$ (d) $\dfrac{n}{V} = \dfrac{P}{RT}$ **39.** (a) $t = 16$ (b) $y = 7$ (c) $x = 8$ (d) $k = -8$ **40.** (a) $b = \dfrac{A}{cd}$ (b) $c = \dfrac{A}{bd}$ (c) $\dfrac{A}{b} = cd$ **41.** (a) $x = 20$ (b) $y = 27$ (c) $t = 2$ (d) $k = 4$ **42.** (a) $k = 1$ (b) $h = 8$ (c) $f = 16$ (d) $y = 4$ **43.** (a) $x = 8$, $y = 36$ (b) $h = 42$ (c) $a = 1.5$, $b = 2.5$, $c = 3.5$ (d) $d = 12.5$ **44.** (a) $m = DV$ (b) $V = \dfrac{m}{D}$

45. 5 g/mL **46.** LA = 33.3%, LT = 16.7%, HT = 50.0% **47.** 800 to Europe, 600 to Asia, 200 to Alaska, 400 to Miami **48.** (a) 66 (b) 17 (c) 6 (d) 25 **49.** (a) 65 (b) 3 (c) 1 (d) 12.5 **50.** (a) 25 (b) 1,000 (c) 0.00375 (d) 12.5

INDEX

Table of Atomic Numbers and Atomic Masses

Name	Symbol	Atomic number	Atomic mass
Actinium	Ac	89	(227)
Aluminum	Al	13	26.98
Americium	Am	95	(243)
Antimony	Sb	51	121.8
Argon	Ar	18	39.95
Arsenic	As	33	74.92
Astatine	At	85	(210)
Barium	Ba	56	137.3
Berkelium	Bk	97	(247)
Beryllium	Be	4	9.012
Bismuth	Bi	83	209.0
Bohrium	Bh	107	(262)
Boron	B	5	10.81
Bromine	Br	35	79.90
Cadmium	Cd	48	112.4
Calcium	Ca	20	40.08
Californium	Cf	98	(251)
Carbon	C	6	12.01
Cerium	Ce	58	140.1
Cesium	Cs	55	132.9
Chlorine	Cl	17	35.45
Chromium	Cr	24	52.00
Cobalt	Co	27	58.93
Copper	Cu	29	63.55
Curium	Cm	96	(247)
Dubnium	Db	105	(262)
Dysprosium	Dy	66	162.5
Einsteinium	Es	99	(252)
Erbium	Er	68	167.3
Europium	Eu	63	152.0
Fermium	Fm	100	(257)
Fluorine	F	9	19.00
Francium	Fr	87	(223)
Gadolinium	Gd	64	157.3
Gallium	Ga	31	69.72
Germanium	Ge	32	72.59
Gold	Au	79	197.0
Hafnium	Hf	72	178.5
Hassium	Hs	108	(265)
Helium	He	2	4.003
Holmium	Ho	67	164.9
Hydrogen	H	1	1.008
Indium	In	49	114.8
Iodine	I	53	126.9
Iridium	Ir	77	192.2
Iron	Fe	26	55.85
Krypton	Kr	36	83.80
Lanthanum	La	57	138.9
Lawrencium	Lr	103	(260)
Lead	Pb	82	207.2
Lithium	Li	3	6.941
Lutetium	Lu	71	175.0
Magnesium	Mg	12	24.31
Manganese	Mn	25	54.94

A value in parentheses is the mass number of the isotope of longest half-life.